ION TRANSPORT IN PLANT CELLS AND TISSUES

# Ion transport
# in plant cells and tissues

*Editors :*

## D.A. Baker

and

## J.L. Hall

*School of Biological Sciences, University of Sussex,
Brighton, Sussex BNI 9QG, United Kingdom*

1975


NORTH-HOLLAND PUBLISHING COMPANY, AMSTERDAM · OXFORD

AMERICAN ELSEVIER PUBLISHING COMPANY, INC., NEW YORK

*Library of Congress Catalog Card number : 75–22025*

*North-Holland ISBN : 0 7204 4519 1*
*American Elsevier ISBN : 0 444 10915 3*

PUBLISHERS :

NORTH-HOLLAND PUBLISHING COMPANY – AMSTERDAM
NORTH-HOLLAND PUBLISHING COMPANY, LTD. – LONDON

SOLE DISTRIBUTORS FOR THE U.S.A. AND CANADA :

AMERICAN ELSEVIER PUBLISHING COMPANY, INC.
52 VANDERBILT AVENUE, NEW YORK, N.Y. 10017

Printed in the Netherlands

# Preface

Ion transport studies have been conducted with a wide range of plant materials and investigators have evolved methods, terminology and concepts which sometimes differ with the material employed. This is not merely the result of differences in approach to the same basic problem but reflects the variation of the ion transporting properties of cell membranes which have selectively evolved in response to both internal and external environmental pressures. Thus it is not surprising that such divergences of thought and methods of investigation now exist that the investigator of ion uptake by, say, isolated mitochondria may find only limited common ground with someone who is studying ion transport within the root. This is in part a result of the vast amount of literature which each investigator must digest in order to keep abreast of his own specialist topic. How much greater the problem is for the advanced undergraduate student, the graduate student during his research or the teacher of advanced students preparing a comprehensive course on ion transport. Although these topics are frequently reviewed, the review articles are usually written specifically for those with a specialist knowledge and do not fully meet the requirements of people in the above categories. It is the aim of the present text to present a bridge, or rather a series of stepping stones, which will enable the advanced student of ion uptake processes to further his knowledge in a comprehensive way.

It would be virtually impossible for one or two individuals to attempt to write such a text for the very reasons outlined above and therefore we have adopted the multi-author approach with its obvious advantages of bringing together authors with specialist knowledge to write individual chapters within their particular areas of study. Each contributor was invited to write about his particular system within certain editorial guidelines which provided a knowledge of what was to be covered in the other chapters. This has enabled ready cross-referencing and reduced unnecessary duplication of common subject material. The individual chapters are not intended to be encyclopaedic in nature but present the current concepts and development of ideas in their particular areas.

The first two chapters are intended to provide a coverage of introductory material (historical, biophysical and biochemical), which is basic to ion transport studies and the fundamental membrane phenomena involved. The remaining chapters each deal with a specific organelle, tissue or whole organism on which transport studies have been undertaken. Mitochondria and chloroplasts, the two major cell organelles which are known to have active transport processes, are discussed in Chapters 3 and 4 with particular emphasis on the energy requirements for this transport. Giant algal cells, which have proved to be excellent experimental material for the characterization and localization of ion fluxes, are the subject of Chapter 5. Storage tissues have provided a very useful homogenous material for ion transport and are discussed in Chapter 6 which, in addition, provides detailed discussions of compartmental analysis and the regulation of transport. Roots, because of their primary function in ion accumulation by the plant, have been the subject of considerable study and are discussed here in relation to the characterization of carrier systems in excised root material (Chapter 7) and to the radial transport and release within the stele of those ions which undergo long distance transport (Chapter 8). The whole plant, and in particular the control mechanisms involved in ion distribution and circulation, is the subject matter of Chapter 9. The role of endogenous growth substances is discussed in relation to these regulatory processes. Salinity is a major agricultural problem and in recent years the mechanism by which halophytes can survive at high salinity levels has been the subject of investigation in relation to ion transport processes (Chapter 10). The salt-secreting glands of a number of species provide a unique experimental material and the mechanism by which these glands secrete relatively large quantities of ions is the subject of Chapter 11. Ions, particularly potassium, have been implicated in stomatal movements and this has led to an intensive study of the ion-transporting properties of the guard cells. The results of these studies are discussed in Chapter 12.

In some instances the treatment of a particular topic may seem more extensive in relation to other chapters than warranted by the amount of investigation and published work in that area. This course has been taken where such subjects have not been previously presented in detail in an advanced text. S.I. units have been used throughout the text (the familiar mM being replaced by Mol m$^{-3}$) although in some instances, where data are reproduced, the original units are given.

We would like to thank our fellow researchers who have contributed their specialist knowledge in this book. Their preparation of the individual chapters has made the editing a stimulating and pleasurable task. We are also grateful to Dr. Jack Franklin, Chief Scientific Editor of A.S.P. Biological and Medical Press for guidance during the production of the text.

Sussex, 1975 D.A.B. and J.L.H.

# List of contributors

W.P. Anderson, Research School of Biological Sciences, Australian National University, P.O. Box 475, Canberra City, ACT 2601, Australia.

D.A. Baker, School of Biological Sciences, University of Sussex, Brighton, BN1 9QG, Sussex, England.

W.J. Cram, School of Biological Sciences, University of Sydney, Sydney 2006, N.S.W., Australia.

T.J. Flowers, School of Biological Sciences, University of Sussex, Brighton, BN1 9QG, Sussex, England.

J.L. Hall, School of Biological Sciences, University of Sussex, Brighton, BN1 9QG, Sussex, England.

J.B. Hanson, Department of Botany, 289 Morrill Hall, University of Illinois, Urbana, Illinois 61801, U.S.A.

R.G.Wyn Jones, Department of Biochemistry and Soil Science, University College of North Wales, Bangor, Gwynedd LL57 2UW, Wales.

D.E. Koeppe, Department of Agronomy, University of Illinois, Urbana, Illinois 61801, U.S.A.

U. Lüttge, Botanisches Institut der Technischen Hochschule Darmstadt, D-6100 Darmstadt, Germany.

P.S. Nobel, Department of Biology, University of California, Los Angeles, California 90024, U.S.A.

M.G. Pitman, School of Biological Sciences, University of Sydney, Sydney 2006, N.S.W., Australia.

J.A. Raven, Department of Biological Sciences, University of Dundee, Dundee, Scotland.

D.A. Thomas, Station Centrale de Bioclimatologie, I.N.R.A., Route de Saint-Cyr, Versailles, France.

# Abbreviations

| | |
|---|---|
| ABA | abscisic acid |
| AdN | adenine nucleotide |
| ADP | adenosine diphosphate |
| ATP | adenosine triphosphate |
| ATPase | adenosine triphosphatase |
| BSA | bovine serum albumin |
| CAM | crassulacean acid metabolism |
| CCCP | carbonyl cyanide-$m$-chlorophenylhydrazone |
| CN | cyanide |
| CNS | thiocyanate |
| DCMU | 3-(3,4-dichlorophenyl)-1,1-dimethylurea |
| DHAP | dihydroxyacetonephosphate |
| DNA | deoxyribonucleic acid |
| DNP | 2,4-dinitrophenol |
| EDTA | ethylenediaminetetraacetic acid |
| EPA | electron probe analyser |
| FCCP | carbonyl cyanide $p$-trifluoromethoxyphenylhydrazone |
| GA$_3$ | gibberellic acid |
| IA | iodoacetic acid |
| IAA | indole acetic acid |
| NAD$^+$ | nicotinamide adenine dinucleotide (oxidised form) |
| NADH | reduced form of NAD |
| NADP$^+$ | nicotinamide adenine dinucleotide phosphate (oxidised form) |
| NADPH | reduced form of NADP |
| PEP | phosphoenol pyruvate |
| 3-PGA | 3-phosphoglyceric acid |
| P$_i$ | inorganic phosphate |
| PMF | proton motive force |

| | |
|---|---|
| PTA | phosphotungstic acid |
| RNA | ribonucleic acid |
| SDS | sodium dodecylsulphate |
| Tris | tris(hydroxymethyl)aminomethane (2-amino-2-hydroxy-methyl-propane-1,3-diol) |
| UTP | uridine triphosphate |

# Contents

For a detailed list of contents the reader is referred to the first page of each chapter.

*Ion transport in plant cells and tissues*
*edited by D.A. Baker and J.L. Hall*
© *North-Holland Publishing Company, 1975*



# CHAPTER 1

# Ion transport – introduction and general principles

## D.A. BAKER and J.L. HALL

## Contents



## *1.1 Historical survey*

The uptake and transport of ions by plant cells and tissues is a subject whose early study was inextricably linked with the problem of plant mineral nutrition and it was not until the middle of the nineteenth century, when a basic understanding of plant nutrition had been established, that investigations into the mechanism of ion transport could be meaningfully undertaken. Thus most of the studies before about 1850 are of historical interest only as far as ion transport studies are concerned and will only be briefly dealt with here. Readers who seek more detailed historical information will find within the books by Russell (1974) and Epstein (1972) useful surveys of the history of research on plant nutrition and an introduction to the literature on this subject.

Until the seventeenth century the view of Aristotle that all matter consisted of earth, air, fire and water prevailed virtually unchallenged. The first recorded quantitative study of plant mineral nutrition was performed by van Helmont (1577–1644) who grew a willow cutting weighing 5 lb in 200 lb of earth watered with rain or, when necessary, distilled water. After five years the plant weighed 169 lb 3 oz and the earth had lost 2 oz dry weight. This experiment disproved the widely held Aristotlian humus theory that plants absorb elaborated food from the soil, but van Helmont wrongly concluded that the 164 lb of plant material was the product of water only and failed to relate the 2 oz net decrease of the soil to mineral depletion. It is of interest that a similar experiment is reported to have been carried out earlier by Leonardo da Vinci (1452–1519) but was not published and that Boyle (1627–1691) describes similar experiments from which he reached the same conclusions as van Helmont. However, Nicholas of Cusa (1401–1464) may well have preceded van Helmont, at least in conception of his experiment, by some 150 years as a 1650 translation of his writings indicates:

'If a man should put an hundred weight of earth into a greath earthen pot, and then should take some Herbs, and Seeds, and weight them, and then plant or sow them in that pot, and then should let them grow there so long, untill hee had successively by little and little, gotten an hundred weight of them, hee would finde the earth but very little diminished, when he came to weigh it againe : by which he might gather, that all the aforesaid Herbs, had their weigh from the water. Therefore the waters being ingrossed (or impregnated) in the earth, attracted a terrestreity, and by the opperation of the Sunne, upon the Herb were condensed (or were condensed into an Herb.) If those Herbs bee then burn't to ashes, mayest not thou guesse by the diversity of the weights of all; How much earth though foundest more than the hundred weight, and then conclude that the water brought all that?...'

The suggestion that the substance of plants is derived from water, and that this can be proved through the use of a balance had originally been made at least a thousand years before the time of Nicholas of Cusa, in a work certainly known to him and probably to van Helmont also. In the eighth book of the pseudo-Clementine Recognitions, which were translated into Latin by Rufinus of Aquileia soon after 400 A.D., is the following passage the origin of which is unknown, although it was probably composed between 50 B.C. and 200 A.D.:

'By manifest fact and example let us prove that nothing is supplied to seeds from the substance of the earth, but that they are entirely derived from the element of water and the spirit (spiritus) that is in it. Suppose for example, that into some barrel of enormous size we put a hundred talents (about three tons) of earth. Now let different sorts of seeds of herbs or bushes be planted in it, and enough water supplied to keep them moist. For several years take good care of it; collect all the seed that develops, the wheat and the barley and other kinds separately, year by year, until the pile of each amounts to a hundred talents. Then uproot the plants and weigh them. When they have all been removed the barrel will still present its hundred talents without loss. But where did all that bulk come from, that mass of different sorts of seeds and vegetation? Is it not obvious that it came from the water?...'

It is unlikely that the experiment was ever performed as described. The amount of earth involved would demand a pot of enormous size, the whole to sit on a balance for several years. The procedure of waiting for the produce to weigh as much as the original earth is obviously unnecessary, though Nicholas kept it. Thus it is probable that both the Recognitions and the writings of Nicholas refer to this experiment only as a thought-provoking suggestion and it was van Helmont who took the quantification seriously and actually performed the experiment. More details of these profound observations can be found in interesting publications by Howe (1965) and Krikorian and Steward (1968).

The suggestions of Nicholas of Cusa were further developed in the experiments of Woodward, published in 1699, in which the importance of mineral matter for the growth of plants was stressed. Woodward observed that plant growth was better in river than in rain water and better still in water to which soil had been added (Table 1.1). From these results he inferred that earth and not water was the matter which constituted plants. Hales, in his book 'Vegetable Staticks' published in 1727, partially recognised that the air contributed something to plant matter through the activity of the leaves, but it was Senebier (1782) who was the first to recognise that part of the increased weight of the plant in van Helmont's experiment came from fixed air.

It was not until the 'phlogiston' theory had been finally dispelled by the new chemistry of Lavoisier (1804) that de Saussure was able to draw attention to the absorption of inorganic nutrients including nitrogen from the soil

Table 1.1.

Growth of Spearmint plants in water obtained from various sources (Woodward, 1699).

| Source of water | Weight of plants | | Gained in 77 days | Expense of water* | Proportion of increase of plant to expense of water |
|---|---|---|---|---|---|
| | When put in | When taken out | | | |
| | grains | grains | grains | grains | |
| Rain water | $28\frac{1}{4}$ | $45\frac{3}{4}$ | $17\frac{1}{2}$ | 3004 | 1 to 171 $\frac{23}{55}$ |
| River Thames | 28 | 54 | 26 | 2493 | 1 to 95 $\frac{23}{26}$ |
| Hyde Park conduit | 110 | 249 | 139 | 13140 | 1 to 94 $\frac{74}{139}$ |
| Hyde Park conduit plus $1\frac{1}{2}$ oz garden mould | 92 | 376 | 284 | 14950 | 1 to 52 $\frac{182}{284}$ |

* transpiration

(Recherches Chimiques sur la Végétation, 1804). These views were much debated during the early decades of the nineteenth century and it was only after von Leibig published his famous address to the British Association for the Advancement of Science (Organic Chemistry in its Applications to Agriculture and Physiology, 1840) that the ideas of de Saussure were given complete credence. With the development of the technique of solution or water culture by Sachs and by Knop in the 1860s the stage was set for investigations to find the essential inorganic elements of plants.

Once it was fully recognised that plants utilise essential inorganic elements speculation commenced as to the mechanism of absorption and transport. The early view that absorption was a purely passive response to water movement into the plant was shown to be inadequate by the demonstration of de Saussure that some selective mechanism operates. Mulder (1851) stressed the importance of diffusion in response to gradients of concentration, suggesting that adsorption of particular molecules resulted in their continued accumulation thus providing a basis for selectivity. The discovery of plasmolysis by Nägeli (1855) promoted a number of permeability studies of plant cell protoplasts and, with the realisation that only water need be exchanged in this process (Hofmeister, 1867), the concept of semipermeability evolved, although the term was not introduced until several years later by van 't Hoff.

During the same period artificial semipermeable membranes were being developed by Traube (1867) and Pfeffer made a detailed study of one of these, the copper ferrocyanide membrane, the results of which were published in 1877. Pfeffer stressed that the high resistance of protoplasts was a feature of the thin plasma membrane and developed a general theory of selective permeability based on not only the respective dimensions of the membrane pores and of the permeating molecules, but also on the affinity of the diffusing substance for the membrane material. Further, he stated that if water-filled pores are within a membrane then the diffusion through these pores will be affected by the interfacial forces at the water–membrane interface.

The extension of these investigations using plant protoplasts indicated that substances as varied as ammonia (de Vries, 1871), basic dyes (Pfeffer, 1886), glycerol (Klebs, 1887) and urea (de Vries, 1888) readily enter the cell. However, it was the detailed studies of Overton from about 1890 onwards which led to an explicit distinction between cell permeability and the permeability of inanimate membranes. Overton recognised that the interchange of substances between a cell and its surroundings is regulated by the passive resistance to diffusion of substances through the cell membrane and to a metabolically dependent transport, which he termed 'adenoid'. The recognition that specific active transport processes were brought about by cellular metabolism was further elaborated by Pfeffer (1900) who asserted that living organisms have the ability to transport substances across membranes and from cell to cell in the absence of concentration gradients and that combination with cell constituents may be involved in these processes:

'... the nature of the plasma is such as to render it possible that a substance may combine chemically with the plasmatic elements, thus being transmitted internally, and then set free again...' (Pfeffer, 1900).

Despite this clear postulation of both active transport and the concept of carrier compounds at the beginning of the twentieth century, it was over two decades before investigators demonstrated more than a superficial awareness of the significance of these phenomena.

During the intervening period researchers concentrated on resolving physical mechanisms and recognised that mineral salt absorption was an ionic rather than a molecular process. The unequal uptake of two ions of a single salt was observed with both isolated plant tissue (Meurer, 1909; Ruhland, 1909) and for whole plants (Pantanelli, 1915), although Hoagland (1919) reported that under some conditions two ions of a salt can be absorbed in equivalent amounts. The description of the Donnan principle in 1911 provided an interpretation for these data which could not be explained purely by consideration of permeability. This approach reached its height of application in the systematic studies of Stiles and Kidd (1919) who investigated absorption from a variety of salt solutions by discs of storage tissue and introduced the concept of the absorption ratio as a measure of accumulation of ions by these tissues. This ratio gives a measure of the accumulation during the experimental period rather than the total level of an ion within a tissue which had been used by other investigators. This relative absorption was observed to be greatest from dilute solutions. These detailed investigations also revealed that one ion could influence the relative absorption of another, $K^+$ and $Na^+$ competing with one another and with divalent cations. The observed absorption sequence from salts with chloride as a common anion was found to be $K^+ > Na^+ > Li^+ > Ca^{2+} > Mg^{2+}$; similarly for anions from salts with a common cation the observed sequence was $NO_3^- > Cl^- > SO_4^{2-}$.

It was not, however, until attention was redirected towards the role of cell metabolism in specific transport processes that progress could be made towards elucidating the mechanism of the accumulation process. This change in the trend of thought was brought about by the studies of Hoagland and his co-workers who had turned their attention to large algal coenocytes from which pure vacuolar sap could be readily obtained. Analyses of this sap showed unequivocally that both cations and anions were accumulated by these organisms against existing concentration gradients (Hoagland and Davis, 1923a, b, 1929), and that this accumulation was enhanced by light and showed a temperature dependence similar to that of the cell metabolism. This observed dependence of ion uptake upon cell metabolism brought about recognition of the role of respiration in this process.

This view was further developed by the studies of Steward, who, while recognising a general relationship between ion uptake and respiration, could find no specific quantitative relationship between respiration and ion absorption by slices of various storage tissues (Berry and Steward, 1934). At the same time as Steward was conducting these investigations Lundegårdh and Burström (1933) demonstrated a quantitative relation between the increased respiration rate of a tissue and the amount of anion absorbed by that tissue. This increased respiration was termed 'anion respiration' and may be defined as the concomitant increase in respiration rate when a tissue is transferred from water to a salt solution, the respiration in water being termed the ground respiration. It was further suggested that the salt respiration might be linked to the transport of anions through the cytochrome systems, the cytochromes transporting anions inwards as electrons are moved outwards (Lundegårdh and Burström, 1935). A diagrammatic representation of Lundegårdh's cytochrome theory is shown in Fig. 1.1. Steward (1937) accepted that salt had a stimulating effect on respiration but doubted whether this effect had any specific quantitative importance for accumulation.

Steward put forward the concept that accumulation of ions by plant cells was directly related to the growth of the cell, particularly with that part of the metabolism relating protein synthesis to respiration. The recognition that 'energy rich' phosphorylated compounds could mediate energy changes within living systems such as muscle led to the suggestion by Steward and

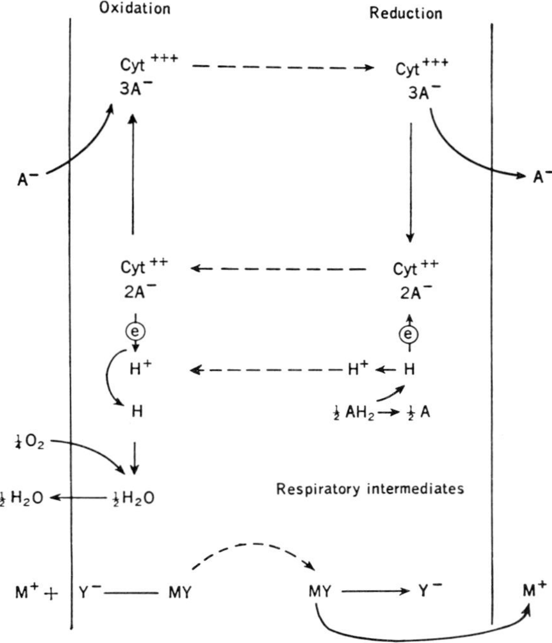



Fig. 1.1. Diagram summarizing Lundegårdh's concept of ion absorption (modified from Lundegårdh, 1954).

Street (1947) that phosphorylated nitrogen compounds might function as carriers. The general relationship between salt accumulation, growth and metabolism as envisaged by Steward is shown in Fig. 1.2.

During the past twenty five years considerable progress has been made towards elucidating the mechanism of ion transport. The approach to the topic has developed a strong biophysical and biochemical emphasis in this period. The biophysical approach has developed from the studies of Mac-Robbie and Dainty (1958a, b) and of Hope and Walker (1960, 1961) who were among the first to make detailed studies of electrochemical potentials and ion fluxes across the membranes of large algal cells using the methods which had been developed with animal cells. This approach was also applied to higher plant cells by Higinbotham and co-workers who established the existence of electropotentials in the cells of higher plants and located the chief electrical barrier at the plasmalemma. These biophysical studies have been well reviewed during this period (see for instance Dainty, 1962; Mac-Robbie, 1970; Higinbotham, 1973) and further details will be given in the relevant chapters here.

　　　　　　　*D.A. Baker and J.L. Hall*

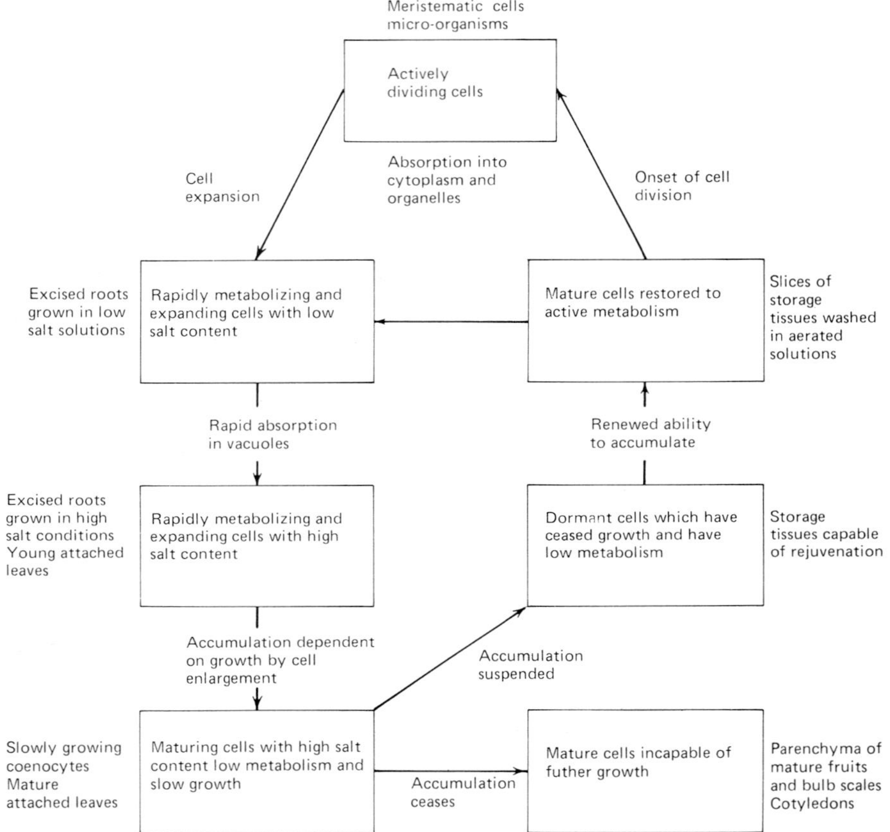



Fig. 1.2. Salt accumulation in relation to growth and metabolism (modified from Steward and Sutcliffe, 1959).

The development of the electron microscope and the great advances in biochemical techniques, particularly those of cell fractionation, have enabled detailed studies to be made upon specific cellular fractions. These studies have revealed the complex nature of the cytoplasm, its boundary membranes and the cell organelles. The finding that isolated mitochondria and chloroplasts can transport ions in vitro has led to a number of exciting findings on the fundamental aspects of the transport process. The impact of these and other developments will be discussed in detail in the following chapters.

## *1.2 Experimental materials and general methods of investigation*

### *1.2.1 Materials*

Ion transport has been studied in a wide variety of plant materials, ranging from isolated membrane fractions to whole flowering plants. This range of experimental materials has been mirrored in the composition of this book and only a brief general summary will be given in this chapter.

*Membranes*   Membrane fractions can be isolated using a number of recently devised techniques and relatively pure plasmalemma fractions can now be obtained. Investigations are currently in progress in a number of laboratories to characterise these membrane fractions with respect to their role in ion transport. Details of these studies are given in Chapter 2.

*Organelles*   Although it is well-established that the bulk of the ions absorbed by plant cells is accumulated in the cell vacuole or vacuoles, there is an appreciable accumulation by other cell organelles, in particular the mito-chondria and chloroplasts. These organelles can be readily isolated and it has been demonstrated that they have not only a different ionic composition to the remainder of the cell but also the ability to accumulate ions in vitro by energy-dependent processes. The behaviour of these organelles with respect to ion transport is described in detail in Chapters 3 and 4. Some attempts have been made to study ion accumulation by isolated protoplasts but to date such studies have not yielded satisfactory results, the isolation procedure probably increasing the permeability of the membranes. However, if prepara-tions similar to erythrocytic 'ghosts' could be obtained it would undoubtedly lead to a greater understanding of the absorption process.

*Tissues*   A number of plant tissues have been utilised for ion transport studies. Slices of storage tissue have been used by numerous investigators, the advantage of this material being the ready availability of large quantities of easily prepared material which accumulates ions under suitable conditions. Among the storage organs that have been widely used are artichokes, red beet root, carrot, potato, swede and turnip. In addition to storage organs some investigators have utilised callus tissue cultures, although to date the tissue culture system has not proved as advantageous in ion transport studies as was at one time hoped. Chapter 6 provides a detailed account of ion accumu-lation by discs of storage material and callus tissue cultures. Excised roots have been used as experimental material by a large number of investigators.

                    *D.A. Baker and J.L. Hall*

Root segments obtained from plants such as barley readily accumulate ions in a reproducible way. In some cases the excised root system has been reduced in complexity by isolating the tissues of the cortex and stele and measuring the ion uptake by these tissues individually. The results of studies with excised roots are discussed in Chapter 7.

*Organs*  Intact plant organs have been employed in a number of studies. In particular, root systems or individual roots have been studied in considerable detail, as described in Chapter 8. Single leaves or shoot segments bearing leaves have also been used, particularly with respect to studies on integration within the whole plant, as discussed in Chapter 9. Specialised structures such as the salt glands of halophytes have been the subject of a number of detailed studies (Ch. 11), while the recent finding that stomatal responses are accompanied by fluxes of monovalent cations has focused attention on these structures in relation to their transport properties (Ch. 12).

*Whole organisms*  Micro-organisms, multicellular algae, gymnosperms and angiosperms have all been investigated extensively. Unicellular algae have the advantage that they can be cultured under carefully controlled conditions. Multicellular algae have provided one of the most suitable materials for ion transport studies. In particular, the large coenocytic units have enabled experimental manipulations which have provided information of fundamental importance to the study of ion transport, described in detail in Chapter 5. The absorption of ions by intact gymnosperms and angiosperms has been the subject of numerous studies, despite the complexities of the whole system and concomitant difficulties of interpretation; this complex situation is dealt with in Chapter 9. Those plants which are able to grow under conditions of high salinity, the halophytes, are of interest because of their unique tolerance of, and in many cases requirement for, high levels of NaCl; the transport properties of these plants is discussed in Chapter 10.

The wide range of experimental materials outlined above is an indication of the great proliferation of ion transport studies conducted upon plant material. Each type of material has advocates as to its greater suitability, but ultimately the aims of all investigators must be to contribute to the resolution of the fundamental mechanisms of ion transport across the membrane, to relate this to the living cell and finally to build an integrated picture of ion uptake and transport within the whole plant. These aims can only be fulfilled as new techniques and methods of investigation are developed. Within recent years a considerable number of new and powerful techniques have become available. Radioactive isotopes, the electron microscope and the wide range of analytical techniques now available have made the last two decades a period of intense activity in the investigation of ion transport phenomena. Details of some general techniques are outlined below, specific experimental designs being given in the appropriate chapters.

### 1.2.2 General methods of investigation

*Electron microscope* The development of the electron microscope and associated techniques has contributed greatly to our knowledge of membranes, organelles and other cellular structures. Cytochemical techniques have been developed for the localization of certain enzymes and ions and for the specific staining of certain membranes under the electron microscope. The use of these techniques in relation to the role of membranes in ion transport is discussed in Chapter 2.

*Electron probe analysis (EPA)* This recently developed technique permits a complete elemental analysis of small regions of specimens. The spatial resolution of EPA is only about 0.1 $\mu$m in biological specimens, but this permits resolution of ions in plant cells as far as major cell organelles. The elements in a specimen are determined by irradiating the sample with an electron beam focused on the sample surface and then measuring the X-rays which emerge from the irradiated area. These X-rays which are characteristic for the elements present may be separated by wavelength dispersion or energy dispersion (Läuchli, 1973).

*Analytical techniques* The ionic composition of plant material and bathing solutions is determined using a variety of analytical techniques. Metallic cations can be estimated by atomic absorption and emission flame spectrophotometry. Anions are more difficult to determine accurately, macro- and micro-chemical methods generally being utilised. Ion-sensitive electrodes have been developed, allowing estimation of $K^+$, $Na^+$, $Ca^{2+}$ and the halide anions.

### 1.2.3 Electrophysiological methods

#### Electropotential measurements

Plant cell electropotentials are usually measured by inserting a microcapillary glass electrode into a cell by micromanipulation under a microscope. The electrode is filled with $3 \times 10^3$ Mol m$^{-3}$ KCl, which provides a low resistance through the tip, the diameter of which may be less than 1 $\mu$m for small cells and up to 10 $\mu$m for the giant-celled algae. A reference electrode filled with

$3 \times 10^3$ Mol m$^{-3}$ KCl in agar is placed in the solution bathing the cells and the two electrodes are coupled to a high input resistance electrometer (resistance $10^{10} - 10^{14}$ $\Omega$) by Ag/AgCl wires or calomel cells (Fig. 1.3a).

*Resistance measurements*
Resistance, or its reciprocal, conductance, is calculated from the change in electropotential of the cell induced by a small current of known magnitude using the Ohm's law relationship. This normally involves the insertion of two microelectrodes into the cell, one to monitor the cell potential and the other to pass the current. This is technically difficult with the small cells of higher plants and a method has been developed using a single electrode which utilises brief pulses of current of about 50 $\mu$s permitting both potentials and resistance to be measured (Fig. 1.3b).

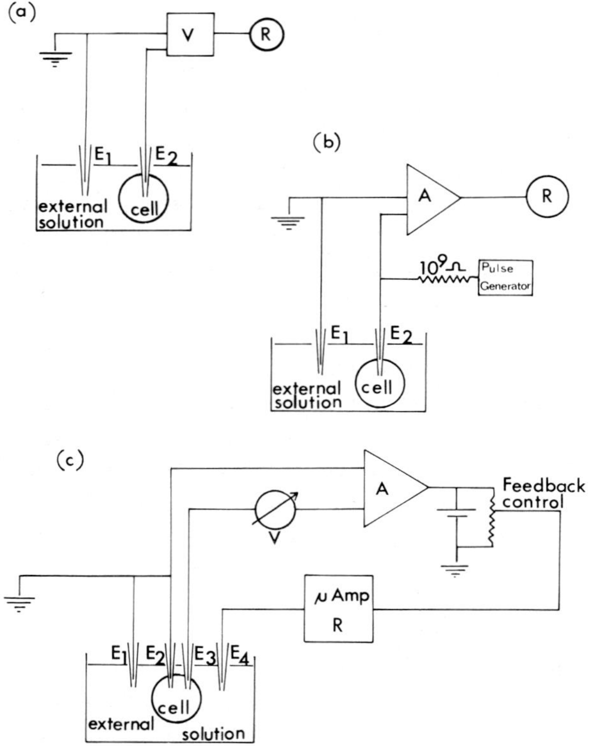


Fig. 1.3. Arrangement for measuring (a) electrical potential difference, (b) resistance and (c) the voltage-clamp technique. V is a high impedance voltmeter, R a recorder or oscilloscope, $E_1 E_4$ external or reference electrodes, $E_2 E_3$ internal microelectrodes and A is a differential amplifier.

*Short-circuit technique*
In this technique an electrical field is applied to the membrane which clamps the potential at zero. Under these conditions any net ion movement must be active and can be measured by electrical current flow (Fig. 1.3c).


## 1.3 The forces moving ions

The following sections are largely devoted to the biophysical parameters which have been developed to describe the transport of ions across membranes. Most of the fundamental relationships are based on models of physical diffusion barriers and passive equilibrium conditions. Only rarely do biological membranes behave in the manner predicted by such models and transport physiologists concentrate on describing the deviation of the living system from the inert physical model. Attempts to characterise the active transport properties of biological membranes using the criteria of irreversible thermodynamics have not so far been particularly successful, the complexities of the living systems giving rise to parameters and interactions which cannot at present be measured. The importance of the biophysical descriptions based on thermodynamics is that they provide a reference point, an equilibrium zero for descriptions of ion transport across plant cell membranes. Thus, they are of major conceptual importance and provide us with a diagnostic tool for describing some of the phenomena of living membranes and their ion transporting properties.

An ion in aqueous solution is acted on by at least two physical forces – one arising from chemical potential gradients and the other from electrical potential differences. Chemical potential is related to the concentration of the ion and electrical potential is the result of the net positive or negative charge carried by the ion. When a salt is added to water it diffuses through the solution from regions of higher concentration to those of lower concentration until a uniform concentration is achieved. In general, one of the ion species will have a higher mobility than the other and will tend to diffuse more quickly than its oppositely charged partner and thus cause a slight separation of charges. This sets up an electrical potential gradient leading to a diffusion potential. As a result the faster moving ion of the pair is slowed down and the slower one speeded up until they both move at the same rate. Thus, a dissociated salt diffusing in a solution behaves as a single substance and has a characteristic diffusion coefficient. According to Fick's law, the rate of diffusion $dv/dt$ is related to the concentration gradient $dc/dx$ thus: $dv/dt = -DA\ dc/dx$ where $D$ is the diffusion coefficient and $A$ the area across which diffusion occurs. The negative sign is a convention to indicate that diffusion occurs from a higher to a lower concentration.

The cell surface membrane or plasmalemma is usually the main barrier for the diffusion of molecules into and out of cells. Diffusion of ions across such a membrane results in the development of a diffusion potential across it which is termed the membrane diffusion potential.

The driving force for ion migration is the gradient of electro-chemical potential. For an ion species j the electrochemical potential $\bar{\mu}_j$ is given by the following component energies:

$$\bar{\mu}_j = \mu_j^* + RT\ln\gamma_j c_j + z_j F\Psi + P\bar{V}_j \tag{1.1}$$

where $\mu_j^*$ is the chemical potential of the ion j in its standard state, $R$ the gas constant, $T$ the temperature in K, $\gamma_j$ the activity coefficient, $c_j$ the chemical concentration, $z_j$ the valency (with sign), $F$ the Faraday constant, $\Psi$ the electric potential, $P$ the pressure in excess of hydrostatic, $\bar{V}_j$ the partial molal volume. $\gamma_j c_j$ is equal to the chemical activity $a_j$, the collective term $RT\ln a_j$ sometimes being used. If the ion is in an ideal solution only the concentration $c_j$ need be employed and the collective term is then $RT\ln c_j$. In practice $P\bar{V}_j$ is rarely used, having a negligible effect except at pressures of hundreds of bars and will be omitted from further consideration here.

The force $X_j$ on an ion species j is the negative of the electrochemical potential gradient, $-\mathrm{d}\bar{\mu}_j/\mathrm{d}x$ obtained by differentiation of Eq. (1.1), i.e.

$$X_j = \frac{-\mathrm{d}\bar{\mu}_j}{\mathrm{d}x} = \frac{-RT}{a_j}\frac{\mathrm{d}a_j}{\mathrm{d}x} - z_j F\frac{\mathrm{d}\Psi}{\mathrm{d}x} \tag{1.2}$$

where $x$ is the distance across which movement is considered.

Thus the driving force consists of two terms, one depending on the concentration gradient and one on the electrical potential gradient.

### 1.3.1 Nernst equation

An ion will be in equilibrium across a membrane when its electrochemical potential is the same on the two sides of the membrane and thus no change in free energy occurs if the ions move from one side to another providing the net flux $\phi_j$ is zero. $\phi_j$ is velocity times concentration where velocity is given by the mobility, $u_j$, multiplied by the force; i.e. $\phi_j = (u_j X_j)c_j$ $\tag{1.3}$

Thus, at equilibrium

$$\phi_j = 0 = -u_j c_j\left(\frac{RT}{a_j}\frac{\mathrm{d}a_j}{\mathrm{d}x} + z_j F\frac{\mathrm{d}\Psi}{\mathrm{d}x}\right) \tag{1.4}$$

and therefore

$$z_j F \frac{\mathrm{d}\Psi}{\mathrm{d}x} = -\frac{RT}{a_j}\frac{\mathrm{d}a_j}{\mathrm{d}x} \tag{1.5}$$

Integration from outside (o) to inside (i) across the membrane and rearranging Eq. (1.5) becomes

$$\Psi^i - \Psi^o = \frac{RT}{z_j F}\ln\frac{a_j^o}{a_j^i} \tag{1.6}$$

Putting $\Psi^i - \Psi^o = E_{N_j}$, inserting actual numerical values for $R$, $F$ and replacing the natural logarithm by 2.303 log where log is the common logarithm to the base 10 Eq. (1.6) becomes

$$E_{N_j} = \frac{59.2}{z_j}\log\left(\frac{a_j^o}{a_j^i}\right) \text{ millivolts (mV) at 25 °C} \tag{1.7}$$

Eq. (1.7) is Nernst's equation and the potential $E_{N_j}$ is the Nernst potential for ion species j. For some calculations the ratio of the activity coefficients $\gamma_j^o/\gamma_j^i$ may be assumed to equal one and the ratio $a_j^o/a_j^i$ becomes $c_j^o/c_j^i$, the ratio of concentrations. This assumption is justified when the ionic strengths on the two sides of the membrane are approximately equal but leads to errors when the outside solution is more dilute and then the internal concentrations must be corrected for activity.

The Nernst potential for an individual ionic species j may be calculated from Eq. (1.7) using the ratio of activities $(a_j^o/a_j^i)$ or concentrations $(c_j^o/c_j^o)$ and can be compared with the measured potential difference across the membrane $E_M$.* The minimum amount of energy needed is proportional to the difference between $E_{N_j}$ and $E_M$ in millivolts (1 millivolt equals 23.06 cal Mol$^{-1}$).

### 1.3.2 Passive ion fluxes across membranes

Eq. (1.4) is the equilibrium condition of the general expression for the net flux of an ion species j moving passively. Neglecting $\gamma$ this can be written:

$$\phi_j = -u_j RT\frac{\mathrm{d}c_j}{\mathrm{d}x} - u_j c_j z_j F\frac{\mathrm{d}\Psi}{\mathrm{d}x} \tag{1.8}$$

In order to calculate passive ion fluxes certain simplifying assumptions must be made. The electrical potential is assumed to be a linear function of

---

* If $E_{N_j}$ differs markedly from $E_M$ then ion species j is not in electrochemical equilibrium and energy must be expended in moving the ions across the membrane.

the distance inside the membrane. This means that $d\Psi/dx$ is a constant equal to $E_M/\delta$, where $E_M$ is the electrical potential difference and $\delta$ is the membrane thickness, and the electrical potential steps at the two boundaries of the membranes with the bathing solutions can be ignored or are equal and opposite. Making these rather arbitrary assumptions Eq. (1.8) can be integrated from one side of the membrane to the other to give:

$$\frac{RT}{z_j FE_M} = \ln \frac{\phi_j + u_j c_j^o z_j FE_M/\delta}{\phi_j + u_j c_j^i z_j FE_M/\delta} \tag{1.9}$$

Taking exponentials of both sides of Eq. (1.9) and rearranging:

$$\phi_j = -\left(\frac{u_j z_j FE_M}{\delta}\right)\left(\frac{c_j^o - c_j^i \exp\left(z_j FE_M/RT\right)}{1 - \exp\left(z_j FE_M/RT\right)}\right) \tag{1.10}$$

Application of Eq. (1.10) is difficult in biological systems. In particular, mobilities, solubilities and $\delta$, the membrane thickness, are usually uncertain and in practice $u_j$ and $\delta$ are combined to give a permeability coefficient $P_j$ with units m s$^{-1}$. Eq. (1.10) then becomes:

$$\phi_j = -P_j \frac{z_j FE_M}{RT}\left(\frac{c_j^o - c_j^i \exp\left(z_j FE_M/RT\right)}{1 - \exp\left(z_j FE_M/RT\right)}\right) \tag{1.11}$$

where $P_j = u_j RT/\delta$.

Although the derivation of Eq. (1.11) involves some cumbersome mathematical manipulations, it is of extreme importance for the description of both passive ion fluxes and membrane diffusion potentials.

### 1.3.3 Ussing–Teorell equation

The net flux of an ion species j is the difference between the influx $\phi_j^{oi}$ and the efflux $\phi_j^{io}$:

$$\phi_j = \phi_j^{oi} - \phi_j^{io} \tag{1.12}$$

It is conventional to identify the unidirectional influx and efflux with the two terms of Eq. (1.11). The rationale behind this is as follows: influx is measured by labelling the outside solution with an appropriate isotope and the internal concentration $c_j^i$ for this isotope equals zero at the beginning of the experiment. The initial net flux ($\phi_j$) of the isotope will thus reflect $\phi_j^{oi}$ only. Hence

$$\phi_j^{oi} = -P_j \frac{z_j FE_M/RT}{1 - \exp\left(z_j FE_M/RT\right)} c_j^o \tag{1.13}$$

By a similar argument, if material is labelled with radioisotope and then transferred to an unlabelled solution of the same composition, the initial net flux ($\phi_j$) of the isotope will reflect $\phi_j^{io}$ only, and therefore

$$\phi_j^{oi} = -P_j \frac{z_j F E_M / RT}{1 - \exp(z_j F E_M / RT)} c_j^i \exp(z_j F E_M / RT) \qquad (1.14)$$

The ratio of the two fluxes is obtained by dividing Eq. (1.13) by Eq. (1.14) which takes on the relatively simple form:

$$\frac{\phi_j^{oi}}{\phi_j^{io}} = \frac{c_j^o}{c_j^i \exp(z_j F E_M / RT)} \qquad (1.15)$$

Eq. (1.15) was derived independently by Ussing (1949) and Teorell (1949) and is known as the Ussing–Teorell equation. It is the basis of an important test for the independent, passive movement of ions across a membrane. By taking logarithms of the two sides Eq. (1.15) can be expressed:

$$RT \frac{\phi_j^{oi}}{\phi_j^{io}} = \bar{\mu}_j^o - \bar{\mu}_j^i \qquad (1.16)$$

when $\bar{\mu}_j^o$ equals $\bar{\mu}_j^i$, $\phi_j = 0$ and Eq. (1.15) reduces to Nernst's equation (1.7). Thus, when $E_M$ equals $E_{N_j}$, $\phi_j^{oi}$ will equal $\phi_j^{io}$ and no net passive flux of ion j is expected and no energy need be expended in moving the ion from one side of the membrane to the other. When Eq. (1.16) is not satisfied, such ions are not moving passively or, in some cases, not moving independently from other fluxes. Interdependent fluxes are describable by irreversible thermodynamics which will be introduced later in this chapter. Active transport, using energy derived from metabolism, is usually the way in which solutes are moved to regions of higher chemical potential, by doing work against the passive gradient.

### 1.3.4 Goldman equation

In many plant cells the total ionic flux is mainly due to movements of $K^+$, $Na^+$ and $Cl^-$ although $H^+$ and $OH^-$ fluxes can also be considerable. The fluxes of these ions across the membrane are caused by gradients in the chemical potentials which create electrical potential differences across the membrane. This electrical potential difference is termed a diffusion potential and arises as a result of the differential mobilities of the ions involved and the condition of electrical neutrality which is maintained. The faster moving ion is slowed up and the slower moving ion of opposite charge is speeded up until they are both moving at the same rate and therefore no electric charge is carried across the membrane.

18                          *D.A. Baker and J.L. Hall*

The potential difference in terms of the concentrations and permeabilities may be expressed when there is no net electric current. For the three ion species $K^+$, $Na^+$ and $Cl^-$ substitution of the net flux of each species into Eq. (1.11) so that $\phi_K + \phi_{Na} - \phi_{Cl} = 0$ is given by

$$P_K\left(\frac{c_K^o - c_K^i \exp(FE_M/RT)}{1 - \exp(FE_M/RT)}\right) + P_{Na}\left(\frac{c_{Na}^o - c_{Na}^i \exp(FE_M/RT)}{1 - \exp(FE_M/RT)}\right)$$
$$+ P_{Cl}\left(\frac{c_{Cl}^o - c_{Cl}^i \exp(-FE_M/RT)}{1 - \exp(-FE_M/RT)}\right) = 0 \tag{1.17}$$

In the above equation valencies have been given actual numerical values and $FE_M/RT$ has been cancelled from each term. Further cancellation of $1/(1 - \exp(FE_M/RT))$ from each term of Eq. (1.17) gives

$$P_K c_K^o - P_K c_K^i \exp(FE_M/RT) + P_{Na} c_{Na}^o - P_{Na} c_{Na}^i \exp(FE_M/RT)$$
$$- P_{Cl} c_{Cl}^o \exp(FE_M/RT) + P_{Cl} c_{Cl}^i = 0 \tag{1.18}$$

Solving Eq. (1.18) for $\exp(FE_M/RT)$ and taking logarithms, the following constant-field equation is obtained for the membrane potential:

$$E_M = \frac{RT}{F}\ln\left(\frac{P_K c_K^o + P_{Na} c_{Na}^o + P_{Cl} c_{Cl}^i}{P_K c_K^i + P_{Na} c_{Na}^i + P_{Cl} c_{Cl}^o}\right)$$

or

$$= \frac{RT}{F}\ln\left(\frac{c_K^o + (P_{Na}/P_K)c_{Na}^o + (P_{Cl}/P_K)c_{Cl}^i}{c_K^i + (P_{Na}/P_K)c_{Na}^i + (P_{Cl}/P_K)c_{Cl}^o}\right) \tag{1.19}$$

Eq. (1.19) is the Goldman or Hodgkin–Katz equation which is widely used in interpreting membrane potential differences. The equation predicts that the membrane potential is determined by the passively moving ions but undoubtedly the active transport of ions will contribute to this potential. If the active transport involves an ion–carrier complex which is charged on one or both of its journeys across the membrane then this process of ion-pumping will contribute to the membrane potential. Such a pump is termed electrogenic and an additional term may be added to the membrane potential $E_M$. Thus in the presence of an electrogenic pump

$$E_M = E_{eq} + E_x \tag{1.20}$$

where $E_{eq}$ is a diffusion equilibrium potential and $E_x$ an additive electrogenic mechanism. $E_x = F(\phi_x/g_m)$ where $\phi_x$ is the electrogenic ion flux and $g_m$ the conductance of the membrane. Thus Eq. (1.19) becomes

$$E_M = \frac{RT}{F}\ln\left(\frac{P_K c_K^o + P_{Na} c_{Na}^o + P_{Cl} c_{Cl}^i}{P_K c_K^i + P_{Na} c_{Na}^i + P_{Cl} c_{Cl}^o}\right) + F\frac{\phi_x}{g_m} \tag{1.21}$$

### 1.3.5 Membrane conductance

When small currents flow through the membrane the membrane potential is not shifted far from its resting value as given by the Goldman Eq. (1.19). The conductance of the membrane in terms of permeabilities and activities is given by

$$g_m = \frac{dJ}{dE_M} = -\frac{F^2}{RT}\frac{\ln(C^o/C^i)\,C^o}{1 - C^o/C^i} \tag{1.22}$$

where $C^{o,i} = P_K c_K^{o,i} + P_{Na} c_{Na}^{o,i} + P_{Cl} c_{Cl}^{o,i}$ if we are concerned only with current carried by these ions. Under these conditions $g_m$ is the slope conductance.

When larger currents are passed, the expression for conductance is more complicated and the chord conductance must be calculated. This conductance should increase when current is passed in one direction and should decrease when passed in the opposite direction, a phenomenon equivalent to rectification. If then the membrane is permeable to monovalent cations only, the chord conductance $g'_m$ is given by the following expression:

$$g'_m = \frac{F^2 E_M C^o_+ \left[1 - \exp(F\Delta E_M/RT)\right]}{RT\Delta E_M \left[1 - \exp(FE_M/RT)\right]} \tag{1.23}$$

When the Nernst condition for electrochemical equilibrium is satisfied the partial conductance expected from an observed unidirectional flux is given by:

$$g_j = (z_j^2 F^2/RT)\phi_j \tag{1.24}$$

This assumes that the influx and efflux of ion species j are independent of each other and of other fluxes.

### 1.3.6 Donnan potential

Another type of electrical potential difference encountered in biological systems is that associated with immobile or fixed charges in a solid phase adjacent to an aqueous phase and this is termed the Donnan potential. Such a phase boundary occurs in the cell walls of plant cells. The region containing the immobile charged particles is generally referred to as the Donnan phase. In cell walls the presence of a large number of immobile carboxyl groups ($RCOO^-$) associated with pectin and other compounds provides a cation-exchange system. The anion-exchange capacity, which is usually less than that for cations, is due to the presence of fixed organic cations, such as amines, within the cell wall matrix. Donnan phases also occur in the cytoplasm where the immobile charges are mainly due to proteins. These proteins are fixed in the sense that they are unable to diffuse across either the plasmalemma or the tonoplast.

When the immobile or fixed ions are arranged in a layer the mobile ions of opposite sign occur in an adjacent layer, the two together being referred to as an electrical double layer, such as is shown in Fig. 1.4. From a know-

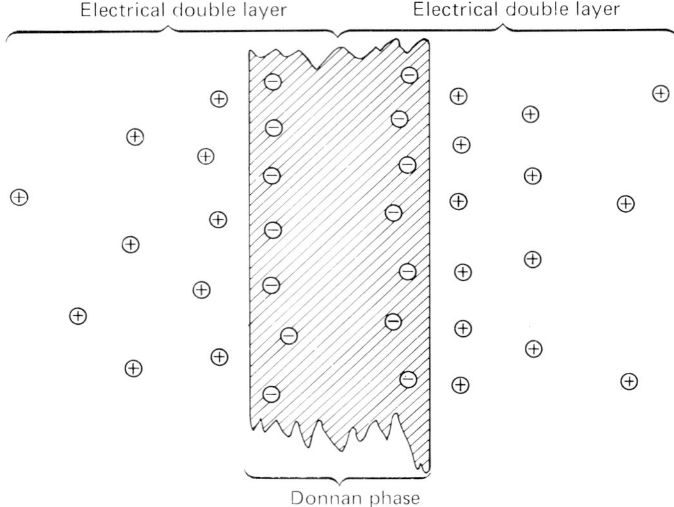


Fig. 1.4. The distribution of positively charged ions, +, occurring on either side of a Donnan phase containing immobile negative charges, −.

ledge of the ratio of the concentrations of any mobile ion across the electrical double layer the electrical potential difference, or Donnan potential can be calculated, again using the Nernst equation. The Donnan potential may be looked upon as a type of diffusion potential as the mobile ions tend to diffuse away from the charges of opposite sign that are fixed in the Donnan phase.

### 1.3.7 Free space uptake

If a plant tissue is first washed in water and then immersed in a solution of a salt, there is a rapid initial uptake which is usually completed in 10–20 minutes followed by a less rapid steady uptake which may continue for several hours or even days (Fig. 1.5). The initial rapid uptake is reversible, non-selective and independent of metabolism. If the tissue is transferred back to water a large proportion of the initial uptake may be washed out and this is referred to as the water-extractable fraction. A further fraction will be extracted if the

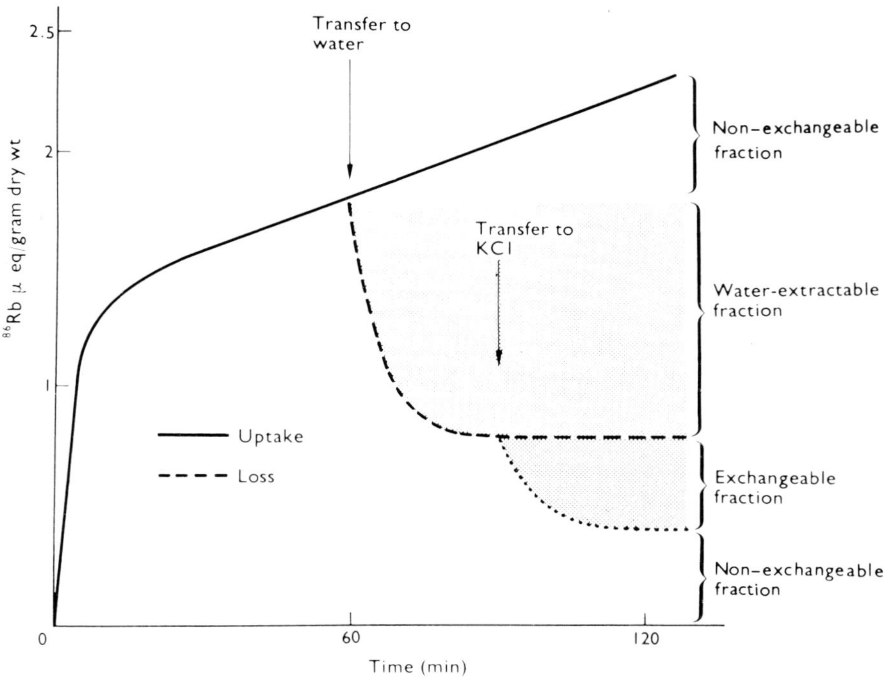

Fig. 1.5. The uptake and loss of potassium by maize roots. The initial rapid uptake may be reversed by washing out the water-extractable and the exchangeable fractions with water and unlabelled KCl respectively. The slow steady uptake process evident after 15 minutes is non-exchangeable. (From Sutcliffe and Baker, 1974.)

tissue is washed in a salt solution with which absorbed ions can exchange. This latter fraction, the exchangeable fraction, may be measured more conveniently if the original solution is labelled with a radioisotope (Fig. 1.5).

It is now generally agreed that the rapid initial uptake represents ion movement into water-filled spaces in the cell walls of the tissue and it is usually referred to as uptake into the free space, that is into the freely accessible part of the cells. The water-extractable fraction consists of mobile ions in the aqueous phase, or water-free space, in the cell wall, while the exchangeable fraction comprises those ions which become adsorbed in the electrical double layer, or Donnan free space, which is also in the cell wall.

There have been attempts in the past to estimate the volume of tissue occupied by the free space. This can be done by assuming that the concentration of solutes in the free space is the same as that in the bathing medium and converting a quantity of ions taken up per unit volume of tissues to a volume of solution it would occupy. As such an assumption is incorrect, due to the high concentration of ions associated with the electrical double layer, the volumes calculated for the free space in this way are greatly in excess of the actual free space volumes. To avoid this error the concept of apparent free space (AFS) was introduced. This may be defined as the calculated volume of a cell or tissue which would be occupied if the ion concentration in that volume were the same as that of the bathing medium. This somewhat cumbersome concept is now generally little used and water-extractable and ion-exchangeable fractions of the free space expressed in quantities, not volumes, are the terms used by modern researchers in this field.

## *1.4 Ion fluxes and compartmentation*

The influx and efflux of ions across plant cell membranes may be measured and compared with the behaviour of model systems. The simplest model is a one compartment system in which a cell of volume $V$, surface area $A$, containing a solution concentration $C_i$ of solute j, is bounded by a membrane. It is assumed that rapid diffusion occurs on either side of the surface boundary and thus the gradient of solute j is across the membrane only. The fluxes of j, or a radioactive version j*, across the boundary membrane are depicted in Fig. 1.6.

Influx is determined by measuring the initial rate of increase of internal radioactivity when an unlabelled cell is placed in a radioactive solution. Thus at time zero, $S_i = 0$ and $S_o = $ const. If the fluxes of solute j across the boundary

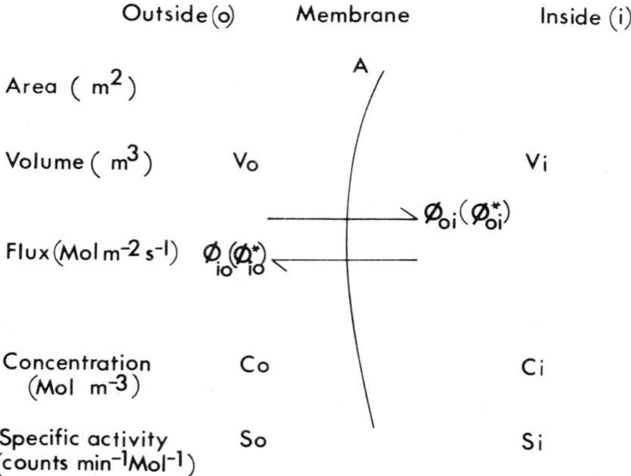


Fig. 1.6. The quantities considered in relating specific activity to time during the course of labelling or eluting a one-compartment system.

are constant and unidirectional then $\phi_{oi} = \phi_{io} = 0$. Putting $Q_i = C_i V_i$, the total quantity of j in the inside compartment. With the above conditions, for uptake

$$\frac{dS_i}{dt} = A(\phi^*_{oi} - \phi^*_{io})/Q_i = A(S_o\phi_{oi} - S_i\phi_{io})/Q_i = A(S_o - S_i)\phi/Q_i \quad (1.25)$$

from which

$$S_i = S_o[1 - e^{-\phi At/Q_i}] \quad (1.26)$$

and thus the internal specific activity will increase exponentially until $S_i = S_o$ at $t = \infty$. The initial uptake is linear and thus

$$\phi = \frac{\Delta(S_iQ_i)}{AS_o\Delta t} \quad (1.27)$$

When $S_i \neq 0$ and tends towards $S_o$

$$\phi_j = \frac{-2.203Q_i \log(1 - S_i/S_o)}{t'A} \quad (1.28)$$

where $t'$ is the duration of the uptake.

Efflux is determined by allowing the cell to accumulate the radioisotope from a bathing solution until the internal specific activity $S_i$ has reached some convenient value $^oS_i$ at $t = 0$. The cell is then transferred to an identical non-radioactive bathing solution which is replaced frequently so that $S_o = 0$.

Under these conditions

$$\frac{dS_i}{dt} = -A\phi^*_{io}/Q_i = -A\phi S_i/Q_i \quad (1.29)$$

Integration gives

$$S_i = {}^oS_i e^{-\phi At/Q_i} \quad (1.30)$$

The initial appearance of radioactivity in the external solution is linear and can be estimated by the following relationship:

$$\phi = \frac{\Delta(S_oQ_o)}{AS_i\Delta t} \quad (1.31)$$

When the log of $S_i$ is plotted against time a straight line is obtained of slope $-\phi A/Q_i$. This is related to the 'half-time' for the exchange by the following:

$$t_{\frac{1}{2}} = 0.693\, Q_i/\phi A \quad (1.32)$$

where $t_{\frac{1}{2}}$ is the time for $S_i$ to reach $S_o/2$ for uptake or for $S_i$ to reach $°S_i$ for efflux. $\phi A/Q_i$ is termed the rate constant $k$, with dimension time$^{-1}$. Substituting $k$ in Eq. (1.32) gives

$$t_{\frac{1}{2}} = 0.693/k \tag{1.33}$$

The simple situation does not pertain, however, even in a single plant cell. There are at least three compartments – cell wall, cytoplasm and vacuole. The non-membrane-bound cell wall equilibrates rapidly with the external solution and can be corrected for leaving two membrane-bound compartments in series, the cytoplasm and the vacuole (see Ch. 6).

The method for compartmental analysis is dependent on the relative permeabilities of the plasmalemma and tonoplast. If the plasmalemma is more permeable than the tonoplast, then during an influx period the cytoplasm will fill up with isotope first and then afterwards the vacuole will fill, enabling the separate influxes to be measured. If the cell is loaded with radioactive tracer and the time course of efflux followed, three phases can be observed in the

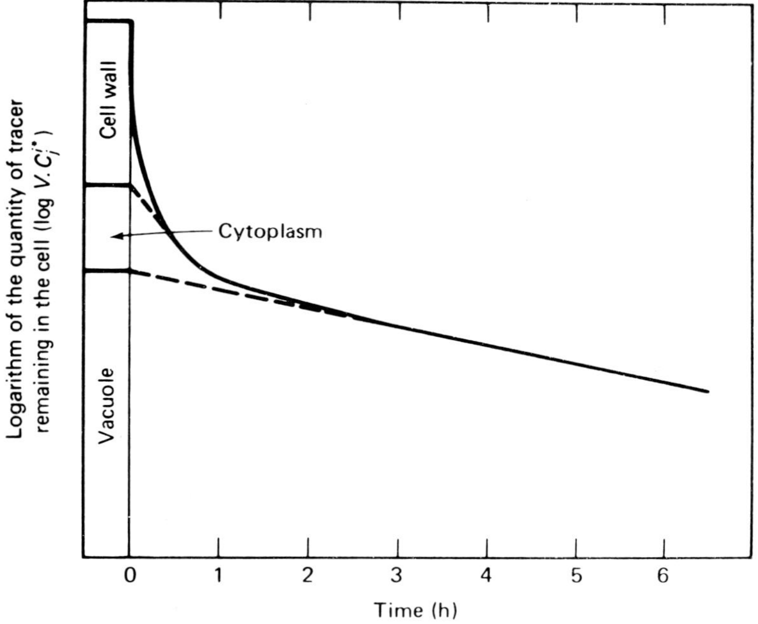


Fig. 1.7. Efflux curves for a plant cell. From the intercepts of the straight lines on the ordinate the amount of radioactivity in each phase at the start of elution can be estimated. (From Clarkson, 1974.)

loss of radioactivity. The logarithm of the amount of isotope remaining in the tissue plotted against time yields a straight line after 2–4 h which is believed to represent the steady rate of efflux from the vacuole (Fig. 1.7). Extrapolation to $t = 0$ gives the initial amount of isotope present in the vacuole and the slope of the line gives $k_v$, the rate constant of efflux from the vacuole, assuming that there is a steady state transfer of isotope through the cytoplasm. By subtraction of the amount of isotope present in the vacuole from the total amount in the cell the time course of the efflux from the cell wall and cytoplasm is obtained, the straight line obtained after about 15 minutes representing the cytoplasm. Extrapolation to $t = 0$ yields the amount of isotope initially present in the cytoplasm and the slope of the line gives $k_c$, the rate constant of the efflux from the cytoplasm.

The situation becomes more complex to analyse if the tonoplast is much more permeable than the plasmalemma. Under these conditions the plasmalemma becomes the rate-limiting membrane for both influx and efflux determinations. To obtain the fluxes across the tonoplast the specific activities of the cytoplasm and the vacuole must be estimated separately. MacRobbie (1966) has shown that at time $t'$ when the specific activities of cytoplasm and vacuole are rising together as a result of bathing the cell in a radioactive solution

$$\phi_t = Q_v S_v / t' (S_c - S_v) \tag{1.34}$$

where subscripts c, t, and v refer to cytoplasm, tonoplast and vacuole respectively.


## 1.5  Kinetics of ion transport

It has been proposed that the kinetics of ion transport are similar to those of enzymic catalysis and that reversible binding to a carrier mediates the accumulation process (Epstein and Hagen, 1952). In both cases, ion transport and enzymic catalysis, the mechanism is believed to involve the attachment of a substrate (ion or enzyme substrate) to an active agent (ion carrier or enzyme) and following transport or catalysis the active agent is released to recombine with the substrate once again.

The rate of carrier-mediated ion transport can be characterised by two factors, one the maximum rate of transport which can be achieved when all available carrier sites are loaded, $V_{max}$ and the other the fraction of the carrier actually loaded at a given substrate concentration, $\theta$. $V_{max}$ can be calculated from the asymptote reached when rate of absorption is measured over a range

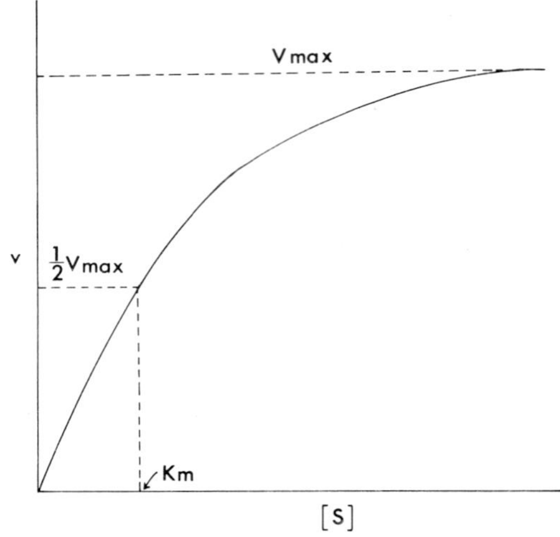


Fig. 1.8. The relationship between the external solute concentration [S] and the rate of absorption, v.

of concentrations (Fig. 1.8) and $\theta$, the fraction of sites occupied at a given ion concentration [S] is obtained from the Langmuir adsorption equation

$$\theta = \frac{[S]}{K_{\mathrm{m}} + [S]} \tag{1.35}$$

where $K_{\mathrm{m}}$ is the dissociation constant of the carrier–ion complex, characteristic of a particular ion crossing a specific membrane and is expressed in units of concentration (Mol m$^{-3}$). The rate of absorption, $v$, is given by the product of the two factors, $V_{\mathrm{max}}$ and $\theta$, to give

$$v = \frac{V_{\mathrm{max}}[S]}{K_{\mathrm{m}} + [S]} \tag{1.36}$$

This relationship is based on the assumption that the model for uptake is as follows

$$S + C \underset{k_{-1}}{\overset{k_1}{\rightleftharpoons}} SC \underset{k_{-2}}{\overset{k_2}{\rightleftharpoons}} C + S \tag{1.37}$$

where S represents the ion, C the carrier and $k_1$, $k_{-1}$, $k_2$ and $k_{-2}$ the rate constants of the reaction. Eq. (1.36) is dependent on there being no counter-flow in the opposite direction, $k_{-2}$ being negligible.

By analogy with enzyme kinetics, $K_{\mathrm{m}}$ is equal to the concentration of the ion $[S]$ at which $v$ reaches half the theoretical maximal rate. Substituting $K_\mathrm{m}$ for $[S]$ in Eq. (1.36) we have:

$$v = \frac{K_\mathrm{m}V_\mathrm{max}}{2K_\mathrm{m}} = \frac{V_\mathrm{max}}{2} \tag{1.38}$$

Eq. (1.36) is the Michaelis–Menten equation which can be rearranged to yield straight-line plots. These are the double reciprocal form due to Lineweaver and Burk (1934):

$$\frac{1}{v} = \frac{K_\mathrm{m}}{V_\mathrm{max}[S]} + \frac{1}{V_\mathrm{max}} \tag{1.39}$$

and the single reciprocal form of Eadie and of Hofstee (1952):

$$\frac{v}{[S]} = \frac{V_\mathrm{max}}{K_\mathrm{m}} - \frac{v}{K_\mathrm{m}} \quad \text{or} \quad v = V_\mathrm{max} - \frac{K_\mathrm{m}}{v[S]}$$

also

$$\frac{[S]}{v} = \frac{K_\mathrm{m}}{V_\mathrm{max}} + \frac{[S]}{V_\mathrm{max}} \tag{1.40}$$

The way in which these equations may be employed for the determination of $V_\mathrm{max}$ and $K_\mathrm{m}$ is indicated in Fig. (1.9). Plotting of data by Eq. (1.39) gives $1/v$ versus $1/[S]$, which appears to have some advantages, but determination

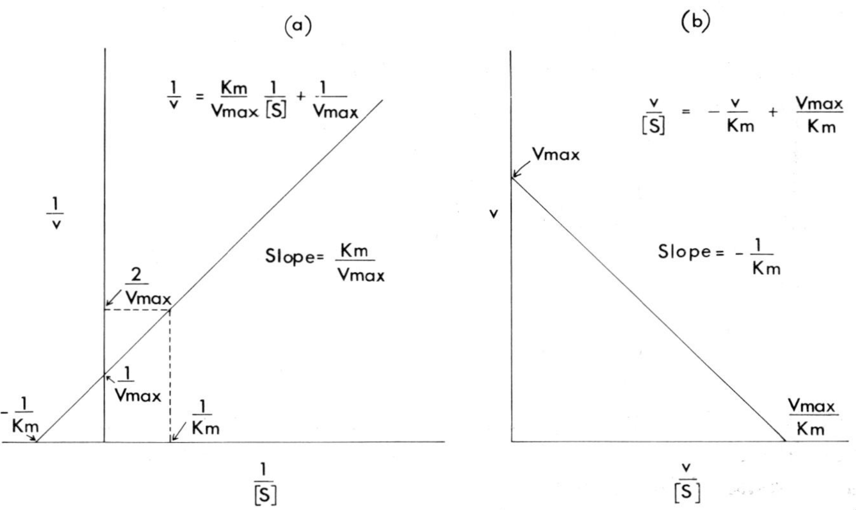


Fig. 1.9. Linear transformations of the Michaelis–Menten equation. (a) Lineweaver and Burk; (b) Eadie, Hofstee.

of values for $V_{max}$ and $K_m$ requires statistical treatment. Data can be fitted to Eq. (1.40), plotting $v/[S]$ versus $v$, by the least squares method, using $v^4$ weighting factors.

Many experiments have yielded results which conform to Michaelis–Menten kinetics, details of which can be found in Epstein (1972). In a number of instances two different $K_m$ and $V_{max}$ values are obtained for a single ion species. This result has led to the conclusion that two separate carriers or carrier sites are involved in the transport of some ion species. One of these sites which is functional at low concentrations ($< 1$ Mol m$^{-3}$) is referred to as system 1; the other site, functional at concentrations in the range 1–50 Mol m$^{-3}$ is called system 2. There has been considerable controversy over the location of these two sites, some investigators subscribing to the view that they are in series – system 1 at the plasmalemma and system 2 at the tonoplast (Laties, 1969) – while others believe that the two sites are in parallel and are both at the plasmalemma (Epstein, 1972). It has also been proposed that single, multiphasic mechanisms mediate ion transport and that the dual mechanisms referred to above reflect an incomplete analysis of the uptake kinetics (Nissen, 1974). See Ch. 7 for a detailed discussion.

There are a number of investigators who have expressed reservations about the above kinetic analysis approach. The fact that ion uptake obeys Michaelis–Menten kinetics is not proof that the carrier model is correct, other processes having been suggested which conform to such kinetics (see Chs. 6 and 7).

An interesting alternative approach to ion-uptake kinetics has been developed by Thellier (1970) who has analysed absorption kinetics, including the dual mechanisms outlined above, without invoking the concept of carrier sites. The overall uptake is envisaged as:

$$S_o \underset{}{\overset{\text{cell}}{\rightleftharpoons}} S_i$$

where $S$ is the substrate (ion) outside and inside the cell. The speed of the process $v$ is formally equivalent to an electric intensity $I$ and the magnitude is given by

$$\Delta E = 2.3 A \log B \frac{[S_o]}{[S_i]} \tag{1.41}$$

where $\Delta E$ is the formal equivalent of the electrical potential difference, $A$ equals $RT/zF$ when electric charges are transferred and $B$ is a constant characteristic of the thermodynamic state of the cell.

With a process which obeys Ohm's law

$$I = \Delta E/r \tag{1.42}$$

where $r$ is the resistance, and then

$$v = 2.3\frac{A}{r}\log\frac{B[S_o]}{[S_i]} \tag{1.43}$$

When the process is non-ohmic as with varistant semi-conductors or with tissue when $[S_o]$ is high

$$I = \Delta E/r + (\lambda\Delta E)^m \text{ where } m > 1 \tag{1.44}$$

and

$$v = 2.3\frac{A}{r}\log B\frac{[S_o]}{[S_i]} + \left(2.3\lambda\ A\log B\frac{[S_o]}{[S_i]}\right)^m \tag{1.45}$$

where $\lambda$ and $m$ are parameters characteristic of cell structures catalyzing the processes.

When the above equations are applied to data on ion absorption showing the dual isotherms, an ohmic process is indicated when $[S_o]$ is low and a non-ohmic process when $[S_o]$ is high. This result implies that systems 1 and 2 are the result of structural changes within the membrane which is seen to behave as a semi-conductor, thus making it unnecessary to invoke the presence of two carrier systems.

## 1.6 Principles of irreversible thermodynamics

Water, ions and other solutes moving through a membrane will exert a frictional drag on each other and their fluxes will therefore be interdependent. That is to say that the flux of a solute will be dependent not only on its own gradient of chemical potential but also on the gradient of the chemical potential of the solvent water. These interdependent fluxes can be quantitatively described by irreversible thermodynamics. Classical thermodynamic treatment is restricted to reversible processes, or systems at equilibrium, and can rarely be applied to biological systems which have few equilibria while living and are therefore open systems. Thus living systems are in a non-equilibrium situation in that they will, in the absence of external influences, spontaneously and irreversibly move towards the equilibrium state.

The general theory of irreversible thermodynamics concerns the relationships between flows or fluxes of matter, or energy, or electrical charge, within a system and the forces responsible for them. Onsager (1931) proposed that the flux of a component is linearly dependent on all the thermodynamic forces operating in the system, although this linearity holds only for processes not going too fast, or not too far from equilibrium. A set of equations can be written relating all the fluxes and forces in the following form:

$$\begin{aligned}
\phi_1 &= L_{11}X_1 + L_{12}X_2 + \cdots\cdots + L_{1n}X_n \\
\phi_2 &= L_{21}X_1 + L_{22}X_2 + \cdots\cdots + L_{2n}X_n \\
\phi_n &= L_{n1}X_1 + L_{n2}X_2 + \cdots\cdots + L_{nn}X_n
\end{aligned} \qquad (1.46)$$

where each flux, $\phi_i$, is dependent on its conjugate force, $X_i$, through a straight coefficient, $L_{ii}$, which is always positive, and may also be linked to non-conjugate forces, $X_j$, through the cross-coefficients, $L_{ij}$ ($i \neq j$), which may be positive or negative as long as the determinant formed by the coefficients is positive. An important reduction in the number of coefficients comes about because of the Onsager reciprocal relation, which states that the 'matrix' of the cross-coefficients is symmetrical so that $L_{ij} = L_{ji}$, corresponding to having equal action and reaction. The reciprocal relation requires that the rate of entropy production times temperature, and per unit volume ($\Phi$), is given by the sum of the products of these fluxes and forces, that is

$$\Phi = \sum_j X_j \phi_j \qquad (1.47)$$

Although Onsager had established a sound and convenient means of coupling linked flows with the forces responsible for them, it was not until the work of Staverman (1951, 1952) that this approach was applied to the transport of materials across membranes. To illustrate the principles involved we will first consider the simple case of a single non-electrolyte solute moving across a membrane barrier and then we will consider the more complex situation which arises when the system contains electrolytes. In order to keep the analysis in a manageable form we will restrict the treatment to isothermal conditions.

If a membrane is permeable to both water (w) and solute (s) the fluxes can be represented by two phenomenological equations.

$$\begin{aligned}
\phi_w &= L_{ww}X_w + L_{ws}X_s \\
\phi_s &= L_{sw}X_w + L_{ss}X_s
\end{aligned} \qquad (1.48)$$

The forces $X_w$ and $X_s$ can be replaced by the 'reduced forces' $\Delta\mu_w$ and $\Delta\mu_s$ (Kedem and Katchalsky, 1963) where $\Delta\mu_w$ and $\Delta\mu_s$ are the differences in the chemical potentials of water and solute respectively.

$$\Delta\mu_w = \bar{V}_w(\Delta P - \Delta\pi_s) \qquad (1.49)$$

where $\bar{V}_w$ is the partial molal volume of water, $\Delta P$ is the difference in hydrostatic pressure and $\Delta\pi_s$ is the osmotic potential corresponding to the difference in concentration $\Delta c_s$

$$\Delta\mu_s = \bar{V}_s\Delta P + \Delta\pi_s/\bar{c}_s \qquad (1.50)$$

where $\bar{c}_s$ is an average concentration within the membrane defined by

$$\bar{c}_s = \Delta\pi_s / RT\Delta\ln a_s \tag{1.51}$$

Thus

$$\begin{aligned}
\phi_w &= L_{ww}\Delta\mu_w + L_{ws}\Delta\mu_s \\
\phi_s &= L_{sw}\Delta\mu_w + L_{ss}\Delta\mu_s
\end{aligned} \tag{1.52}$$

These equations can be transferred in terms of the entropy dissipation function to give

$$\begin{aligned}
\Phi &= \phi_w\Delta\mu_w + \phi_s\Delta\mu_s \\
&= \phi_w\bar{V}_w(\Delta P - \Delta\pi_s) + \phi_s\bar{V}_s\Delta P + \Delta\pi_s/\bar{c}_s
\end{aligned} \tag{1.53}$$

When the membrane is permeable to the solute, $\phi_s$ is finite and volume change on either side of the membrane is due to both $\phi_w$ and $\phi_s$. A volume flow, $\phi_v$, is then observed (Kedem and Katchalsky, 1958).

$$\phi_v = \phi_w\bar{V}_w + \phi_s\bar{V}_s \tag{1.54}$$

Eq. (1.48) can now be rearranged to give

$$\Phi = \phi_v(\Delta P - \Delta\pi_s) + \phi_s(\Delta\pi_s/\bar{c}_s) \tag{1.55}$$

and the new conjugate flows and forces can be written

$$\begin{aligned}
\phi_v &= L_{ww}(\Delta P - \Delta\pi_s) + L_{ws}(\Delta\pi_s/\bar{c}_s) \\
\phi_s &= L_{sw}(\Delta P - \Delta\pi_s) + L_{ss}(\Delta\pi_s/\bar{c}_s)
\end{aligned} \tag{1.56}$$

where $L_{ww}$ and $L_{ss}$ are the 'straight' coefficients linking volume and solute flow to $\Delta P$ and $\Delta\pi_s$ respectively; and $L_{ws}$ and $L_{sw}$ are the 'cross' coefficients which describe the volume flow associated with $\Delta\pi_s$ and the exchange flow associated with $\Delta P$ respectively. It should be remembered that $L_{ws} = L_{sw}$ as before.

When $\Delta\pi_s$ is zero, i.e. when there is no solute concentration across the membrane, $\phi_v = L_{ww}\Delta P$ and the observed flow is the result of $\Delta P$ only. Thus, $L_{ww}$ can be identified as the filtration coefficient (hydraulic conductivity $L_p$, units m s$^{-1}$ bar$^{-1}$).

When $\Delta P - \Delta\pi_s = 0$, i.e. when pressure is applied to stop volume flow $\phi_s = L_{ss}(\Delta\pi_s/\bar{c}_s)$ and $L_{ss}$ can be identified as the solute permeability coefficient $\omega_s$, units Mol s$^{-1}$ m$^{-2}$.

These two coefficients, $L_p$ and $\omega_s$ are only sufficient if the membrane is completely permeable or completely impermeable to the solute molecules. For the range of intermediate situations encountered in biological systems a selectivity or reflection coefficient ($\sigma$) is introduced.

When $\phi_v = 0$,

$$L_{ww}(\Delta P - \Delta\pi_s) = -L_{ws}(\Delta\pi_s/\bar{c}_s)$$

and

$$\sigma = \frac{-L_{ws}}{L_{ww}} = \frac{\Delta P - \Delta\pi_s}{\Delta\pi_s/\bar{c}_s} \tag{1.57}$$

Eq. (1.51) can now be written

$$\phi_v = L_p(\Delta P - \Delta\pi_s) - \sigma L_p\Delta\pi_s/\bar{c}_s \tag{1.58}$$

When $L_{ww} = -L_{ws}$, $\sigma = 1$ and the membrane is impermeable to the solute. When the membrane is non-selective, $\sigma = 0$. For intermediate states $1 \geqslant \sigma \geqslant 0$. The reflection coefficient is dimensionless. Using the coefficients derived above the solute flow $\phi_s$ becomes

$$\phi_s = (1 - \sigma)\bar{c}_s\phi_v + \omega_s\Delta\pi_s/\bar{c}_s \tag{1.59}$$

Thus the three independent coefficients required to provide a correct description of the above system are defined as:

$$L_p = \left(\frac{\phi_v}{\Delta P}\right)_{\Delta\pi_s=0} \qquad \omega = \left(\frac{\phi_s}{\Delta\pi_s/\bar{c}_s}\right)_{\Delta P - \Delta\pi_s=0} \qquad \sigma = \left(\frac{\Delta P - \Delta\pi_s}{\Delta\pi_s/\bar{c}_s}\right)_{\phi_v=0}$$

When the solute under consideration is an electrolyte in addition to the above forces the electrical force must also be included, and the fluxes can be represented by three phenomenological equations and the entropy dissipation function becomes

$$\Phi = \phi_w\Delta\mu_w + \phi_s\Delta\mu_s + J\Delta V \tag{1.60}$$

where $J$ is the flow of electrical current through the membrane (amp m$^{-2}$) and $\Delta V$ is the electromotive force (volts).

The phenomenological equations for an electrolyte are

$$\begin{aligned}
\phi_v &= L_{ww}(\Delta P - \Delta\pi_s) + L_{ws}(\Delta\pi_s/\bar{c}_s) + L_{wv}\Delta V \\
\phi_s &= L_{sw}(\Delta P - \Delta\pi_s) + L_{ss}(\Delta\pi_s/\bar{c}_s) + L_{sv}\Delta V \\
J &= L_{vw}(\Delta P - \Delta\pi_s) + L_{vs}(\Delta\pi_s/\bar{c}_s) + L_{vv}\Delta V
\end{aligned} \tag{1.61}$$

For these three equations six independent Onsager coefficients are required. In addition to $L_p$, $\omega$ and $\sigma$ three electrical coefficients are required. When $\Delta P = 0$ and $\Delta\pi_s = 0$ the coefficients are (Katchalsky and Kedem, 1962)

$$\begin{aligned}
L_{wv} &= \phi_v/\Delta V \qquad \text{the electro-osmotic volume flow} \\
L_{sv} &= \phi_s/\Delta V \qquad \text{the electro-osmotic solute flow} \\
L_{vv} &= J/\Delta V = g \text{ the specific electrical conductance (ohms}^{-1}\text{ m}^{-2}\text{)}
\end{aligned}$$

When $\Delta\pi_s = \Delta V = 0$

$L_{vw} = J/\Delta P$ the streaming current per unit pressure difference and when $\Delta P - \Delta\pi_s = 0$ and $\Delta V = 0$

$L_{vs} = J/(\Delta\pi_s/\bar{c}_s)$ the permeation current.

Other well known parameters are obtained by certain combinations of the above coefficients. When $\phi_v = 0$ and $\Delta\pi_s = 0$

$$\frac{-L_{wv}}{L_{ww}} = \frac{\Delta P}{\Delta V} = P_v \text{ the electro-osmotic pressure at zero flow per unit potential}$$

When $\Delta\pi_s = 0$ and $\Delta P = 0$

$$\frac{L_{wv}}{L_{vv}} = \phi_v/J = \beta \text{ the electro-osmotic permeability}$$

$$\frac{L_{sv}}{L_{vv}} = \phi_s/J = \frac{v}{izF} \text{ where } v \text{ is the transport number, } z \text{ the valency, } i \text{ the number of ions per salt molecule and } F \text{ the Faraday}$$

With the above coefficients and combinations of coefficient, Katchalsky and Kedem (1962) have written the three phenomenological equations for an electrolyte as:

$$\phi_v = L_p(\Delta P - \Delta\pi_s) - \sigma L_p \Delta\pi_s/\bar{c}_s - P_v L_p J/g$$

$$\phi_s = (1 - \sigma)\bar{c}_s \phi_v + \omega_s \Delta\pi_s/\bar{c}_s + \frac{vJ}{izF}$$

$$J = -P_v \phi_v + \frac{gv}{izF}\frac{\Delta\pi_s}{\bar{c}_s} + g\Delta V \tag{1.62}$$

for the flow of volume, solute and electricity, respectively. When there is no flow of electricity $J = 0$ and Eq. (1.62) reduce to Eqs. (1.58) (1.59).


## 1.7 Active transport : definition and description


As outlined in the historical section of this chapter, early workers in the field of ion transport believed that the finding of a higher concentration of ions on the inside of a biological membrane than outside in the medium was evidence of an active accumulation of those ions. However, the inadequacy of this definition and its lack of specificity eventually became apparent and it was replaced by the more rigorous definition of Ussing (1949) who defined active transport as the process by which an ion is moved against a gradient of electrochemical potential. Movement of the ion is therefore dependent on the decrease in free energy of a metabolic process. When the membrane conditions are known in sufficient detail the Ussing–Teorell flux-ratio equation (1.15) can be used to predict whether the net flux of an ion is passive or active. As the magnitude, but not the sign, of the logarithm of the flux ratio may be changed by non-independence and interaction of ions in the membrane, active transport is identified by a flux ratio whose logarithm has the wrong sign provided there is no obvious flow component to which the transport could be coupled.

Kedem (1961) has defined active transport in terms of irreversible thermodynamics as an entrainment between a transport flux and a metabolic reaction. This requires the inclusion of a reaction flux, $\phi_r$, driven by its affinity, $A_r$, in the phenomenonological equations for the system. Active transport is then characterised by a non-zero coupling coefficient, $R_{jr}$, between the flow of substance j and the reaction. MacRobbie (1970) has discussed the usefulness of this definition of active transport, concluding that the difficulties of applying the thermodynamic conditions correctly imposes limits upon this approach and that the usual practical test for active transport remains that of the older Ussing definition.

A wide variety of studies have revealed that the distribution of ions across plant cell membranes is the result of both passive and active ion transport. The most striking feature of ion transport in plant cells is the ability to accumulate ions from a low external concentration to a very high internal level. Epstein (1972) describes this property of plants as 'mining' of the environment and points out that only green plants and some microorganisms possess the ability to efficiently extract simple inorganic compounds and ions from the environment. A large proportion of these ions are accumulated in the central vacuole, which generally occupies 90% or more of the cell volume. The vacuolar contents vary from the complex mixture of inorganic and organic solutes found in higher plant cells to the simple salt solution found in many algae.

A feature of many plant cells is the low level of $Na^+$ in comparison with $K^+$ and an active $Na^+$ extrusion pump has been found in those cells which have so far been studied in sufficient detail. An active $K^+$ influx often accompanies the $Na^+$ efflux although in many cases the vacuolar $K^+$ is found to be near or at passive equilibrium with the level in the bathing medium. The situation in halophytes is not always in complete agreement with the above general picture and high $Na^+/K^+$ ratios are often found (see Ch. 10). Other ion species are found to be dependent upon active transport for their accumulation, an influx pump for anions and in particular $Cl^-$ being a feature of all plant cells so far studied.

In a recent review (Slayman, 1974) three distinct mechanisms have been outlined which can supply energy to active transport systems. These are redox transport, covalent bond splitting and co-transport.

Redox transport brings about the separation of protons from water and is postulated to reverse the action of an associated ATP-dependent proton transport system. Redox transport is therefore electrogenic, contributing to the potential of the membrane. It is only found to occur across the 'energy-conserving' membranes of mitochondria, chloroplasts and some bacteria.

The direct enzymatic splitting of covalent bonds is known to supply energy for $Na^+/K^+$ transport in animal systems, utilising the transport-ATPase (see Ch. 2). This mechanism may also operate in plant, bacterial and fungal cell membranes for $Na^+/K^+$ transport, although it is postulated that the major function may be the extrusion of protons across the plasma membrane.

Co-transport is the utilization of the electrochemical gradient for one species to drive the uphill transport of another. In animal systems the $Na^+$-dependent transport of $Ca^{2+}$, sugars and amino acids are well-established examples of co-transport. Recent evidence suggests that proton gradients operate in a similar manner in plant cells, bacteria and fungi.

Both covalent bond splitting and co-transport can be reversible in that they can drive either ATP synthesis or the uphill transport of the primary ions. Unlike redox transport these other mechanisms may be electroneutral or electrogenic depending on the closeness of the ionic coupling and the charge, if present, on the co-transported substrate. When these systems are electrogenic their interaction with the membrane potential may be of fundamental importance in that they may drive ATP synthesis or distribute energy to other co-transport systems.

## 1.8 Current concepts and conclusions

In order to fully describe and understand ion transport processes there are a number of facts which need to be established for any system under consideration. These are, in the order in which they are currently studied:

(1) The concentrations of ions and the electrical potentials in the various phases of the system under consideration.

(2) The influx and efflux of specific ions across the membrane boundaries between the phases, enabling the direction and location of specific ion pumps to be identified.

(3) The source (or sources) of energy for specific ion pumps.

(4) The nature of the coupling of the ion pump to the energy source.

(5) The biochemical nature of the ion pump.

Of the above five requirements only the first three have been partially met in studies on giant algal cells (Ch. 5). In higher plants, because of experimental difficulties, only the first two features have been partially described and there is a considerable gap in our knowledge of the fundamental processes of active transport in these cells.

The biophysical approach to the problems of ion transport, as outlined in this chapter, has contributed a great deal to the study of ion transport but has distinct drawbacks. It attempts to explain the transport properties of the living cell in terms of a static system of diffusion barriers and fixed compartments. During the last two decades investigators have characterized ion pumps in these biophysical terms. Future investigators should extend this research to include not only a quantification of fluxes but also move towards a conceptual understanding of the ion transport process in terms of the dynamic nature of the cell and of its membranes. The advances which have been made in this direction are discussed in the next chapter.


# *References*


W.E. BERRY and F.C. STEWARD, Ann. Bot., 48 (1934) 395.

D.T. CLARKSON, Ion Transport and Cell Structure in Plants (1974), McGraw–Hill, London.

J. DAINTY, Annu. Rev. Plant Physiol., 16 (1962) 379.

G.S. EADIE, Science, 116 (1952) 688.

E. EPSTEIN, Mineral Nutrition of Plants : Principles and Perspectives (1972) Wiley, London.

E. EPSTEIN and C.E. HAGEN, Plant Physiol., 27 (1952) 457.

N. HIGINBOTHAM, Bot. Rev., 39 (1973) 15.

D.R. HOAGLAND, Bot. Gaz., 68 (1919) 297.

D.R. HOAGLAND and A.R. DAVIS, J. Gen. Physiol., 5 (1923a) 629.

D.R. HOAGLAND and A.R. DAVIS, J. Gen. Physiol., 6 (1923b) 47.

D.R. HOAGLAND and A.R. DAVIS, Protoplasma, 6 (1929) 610.

W. HOFMEISTER, Handbuch der Physiologischen Botanik (1867) Vol. 1, Engelmann, Leipzig.

B.H.J. HOFSTEE, Science, 116 (1952) 329.

A.B. HOPE and N.A. WALKER, Aust. J. Biol. Sci., 13 (1960) 277.

A.B. HOPE and N.A. WALKER, Aust. J. Biol. Sci., 14 (1961) 26.

H.M. HOWE, Isis, 56 (1965) 408.

O. KEDEM, in A. Kleinzeller and A. Kotyk (Eds.) Membrane Transport and Metabolism (1961) Academic Press, London, p. 87.

O. KEDEM and A. KATCHALSKY, Biochim. Biophys. Acta, 27 (1958) 229.

O. KEDEM and A. KATCHALSKY, Trans. Farad. Soc., 59 (1963) 1931.

G. KLEBS, Ber. Dtsch. Bot. Ges., 5 (1887) 181.

A.D. KRIKORIAN and F.C. STEWARD, Bioscience, 18 (1968) 286.

G.G. LATIES, Annu. Rev. Plant Physiol., 20 (1969) 89.

A. LÄUCHLI, in W.P. Anderson (Ed.) Ion Transport in Plants (1973) Academic Press, London, p. 1.

H. LINEWEAVER and D. BURK, J. Am. Chem. Soc., 56 (1934) 658.

H. LUNDEGÅRDH and H. BURSTRÖM, Planta, 18 (1933) 683.

H. LUNDEGÅRDH and H. BURSTRÖM, Biochem. Z., 277 (1935) 223.

E.A.C. MACROBBIE, Quart. Rev. Biophys., 3 (1970) 251.

E.A.C. MACROBBIE, Aust. J. Biol. Sci., 19 (1966) 363.

E.A.C. MACROBBIE and J. DAINTY, Physiol. Plant., 11 (1958a) 782.

E.A.C. MACROBBIE and J. DAINTY, J. Gen. Physiol., 42 (1958b) 335.

R. MEURER, Jb. Wiss. Bot., 46 (1909) 503.

W. MULDER, Physiologie Chemie (1851).

C. NAGELI and C. CRAMER, Pflanzenphysiologische Untersuchungen (1855) Schulthess, Zürich.

P. NISSEN, Annu. Rev. Plant Physiol., 25 (1974) 53.

L. ONSÄGER, Phys. Rev., 37 (1931) 405.

E. PANTANELLI, Jb. Wiss. Bot., 56 (1915) 689.

W. PFEFFER, Osmotische Untersuchungen : Studien zur Zellmechanik (1877) Engelmann, Leipzig.

W. PFEFFER, Untersuch. Bot. Inst. Tübingen, 2 (1886) 179.

W. PFEFFER, in A.J. Ewart (Ed.) The Physiology of Plants (1900) Vol. 1. Oxford University Press, p. 86.

W. RUHLAND, Z. Wiss. Bot., 1 (1909) 747.

E.J. RUSSELL, Soil Conditions and Plant Growth (1974) 10th Edn. (revised by E.W. Russell), Longmans, London.

C.L. SLAYMAN, in J. Dainty and U. Zimmerman (Eds.) Membrane Transport in Plants and Plant Organelles (1974) Springer Verlag, Berlin.

A.J. STAVERMAN, Recl. Trav. Chim. Pays-Bas Belg., 70 (1951) 344.

A.J. STAVERMAN, Recl. Trav. Chim. Pays-Bas Belg., 71 (1952) 623.

F.C. STEWARD, Trans. Farad. Soc., 33 (1937) 1006.

F.C. STEWARD and H.E. STREET, Annu. Rev. Biochem., 16 (1947) 471.

W. STILES and F. KIDD, Proc. R. Soc. B 90 (1919) 487.

J.F. SUTCUFFE and D.A. BAKER, Plants and Mineral Salts, Arnold, London.

T. TEORELL, Arch. Sci. Physiol., 3 (1949) 205.

M. THELLIER, Ann. Bot., 34 (1970) 983.

M. TRAUBE, Arch. Anat. Physiol. Wiss. Med. (1867) 87.

H.H. USSING, Acta Physiol. Scand., 19 (1949) 43.

R.F.M. VAN STEVENINCK, in W.P. Anderson (Ed.) Ion Transport in Plants (1973) Academic Press, London, p. 25.

H. DE VRIES, Arch. Neerl. Sci., 6 (1871) 117.

H. DE VRIES, Bot. Z., 46 (1888) 229.

*Ion transport in plant cells and tissues*
*edited by D.A. Baker and J.L. Hall*
© *North-Holland Publishing Company, 1975*

CHAPTER 2

# Cell membranes

J.L. HALL and D.A. BAKER

## Contents



## 2.1 Introduction

The past few years have seen a great increase in our knowledge of the nature, properties and structural relationships of the components of cell membranes

39

and this has been partly reflected by the considerable number of useful reviews that have appeared over this period (for example, see Cook, 1971; Hendler, 1971; Manson, 1971; Bangham, 1972; Fox and Keith, 1972; Guidotti, 1972; De Pierre and Karnovsky, 1973; Oseroff, Robbins and Burger, 1973; Singer, 1974). This expansion of membrane research is partly a reflection of the fundamental importance of this subject to a wide range of problems in cell biology and also a result of the wide variety of biochemical, physical, and physiological techniques that may now be used for the analysis of membranes. Considerable improvements have been made in techniques used for the isolation of membranes and it is now possible to prepare relatively pure fractions of specific cell membranes such as the plasmalemma. These techniques have been employed in a variety of biological investigations including, of course, those relating to the mechanism of ion transport as well as other membrane functions such as the transport of uncharged solutes, macromolecules and particles, cell–cell interactions, cell growth and differentiation, and electron transport and phosphorylation. However, only in a very few instances have specific membrane components been related to specific functions. Little is known of the relationship between a particular membrane function and the nature and arrangement of particular components within that membrane. Epstein (1972) has commented on the minor contribution that membrane studies have made to our knowledge of ion transport, particularly in plants. This is perhaps a reflection of the difficulties encountered in membrane biochemistry, an over-emphasis on membrane models, and of the unwillingness of many plant physiologists to become involved in investigations at this level. Ultimately an understanding of the mechanism of ion transport must be answered in terms of the properties and arrangement of the components of the plasmalemma and other cell membranes which can only come when a full biochemical and biophysical characterization of specific membranes has been achieved.

## 2.2 *General properties of membranes*

Some general properties of biological membranes can now be considered in relation to later sections on membrane composition and structure, although it should be remembered that there is considerable evidence that specific membranes may differ widely in relation to some of these properties.

(1) Isolated membranes are composed largely of lipids and proteins usually in a ratio of about 40 : 60 although these proportions may vary considerably (see Table 2.1). There is, in addition, a wide diversity in such factors as the

range of phospholipids and enzymic activities present. These differences occur both between different membrane types within a cell and between the same membrane e.g. the plasmalemma, from different cells (see Korn, 1968; Guidotti, 1972).

(2) Cell membranes have characteristic electrical properties. The capacitance is usually about $1 \times 10^4$ $\mu$F m$^{-2}$ and the electrical resistance in the range $10^{-7}$–$10^{-10}$ ohms m$^{-2}$. The latter is considerably higher in synthetic lipid bilayers (see Table 2.3).

(3) Natural membranes have a surface tension in the range of 0.003–0.3 $\mu$N m$^{-1}$ which is considerably lower than that of neutral lipid/water interfaces and is usually explained by the presence of protein or phospholipids at the membrane surface. A similar low surface tension is usually found in synthetic lipid bilayers.

(4) Membranes have characteristic permeability properties: they are generally freely permeable to water; small molecules penetrate more easily than larger molecules and usually at rates directly related to their lipid solubility since molecular size becomes a more important factor; membranes are relatively impermeable to polar molecules and only slightly permeable to ions.

(5) When thin sections of fixed and dehydrated material stained with heavy metals are examined under the electron microscope, cellular membranes appear as two electron-opaque layers separated by a clear zone with a total width ranging from 6.0 to 10.0 nm. This image will be discussed in relation to the 'unit membrane' concept in a later section. Some of these properties are summarised in Table 2.2 and compared with those of synthetic lipid bilayer membranes.


## 2.3 Techniques for studying membranes

There has been a great increase in the range of techniques used in the study of membranes although, unfortunately, these have been largely applied to animal systems. Advances have been made in (i) the isolation of membranes, (ii) the analysis of membrane preparations, and (iii) the microscopic examination of membranes. These particular aspects will be dealt with in turn.

### 2.3.1 Isolation of membranes

The isolation of membranes and their subsequent characterization is an essential step in the understanding of membrane function. At present we rely almost entirely on centrifugation. Very simply, cells are disrupted and particles separated in a centrifuge either by the differences in their rates of sedimentation or by differences in their buoyant densities. Ideally, particular membrane types should be isolated free of contamination by other membrane and cytoplasmic material and have retained their full biochemical activity. However, current methodology frequently falls short of the perfect, difficulties arising for a variety of reasons.

(1) The wide range in size, shape and density of various organelles and membranes, often produced as a result of the homogenization procedure, makes it extremely difficult to isolate uncontaminated preparations.

(2) The lack of rigorous membrane markers in many cases makes it difficult to characterize a preparation fully and determine its purity. Even when available, such markers are not always applied.

(3) On isolation, membranes may lose or gain components, become structurally altered or lose the activity of unstable groups.

Thus, the final preparation may well be an artifact to some greater or lesser extent whose purity is extremely difficult to assess.

The subject of membrane isolation is clearly very extensive and full coverage is beyond the scope of this book. Membranes vary both physically and chemically between species, between tissue and cell types, and between and within organelles. Thus clearly no hard and fast rules can be laid down as to the correct procedure for the isolation of a particular membrane fraction. Each problem must be treated individually and care taken to optimise conditions, particularly if the tissue has not been studied previously. There have been a number of useful reviews written on the problems of membrane isolation (Steck, 1972) and more specifically on the isolation of plasma membranes (Steck and Wallach, 1970; Warren and Glick, 1971; De Pierre and Karnovsky, 1973; Oseroff et al., 1973; Wallach and Lin, 1973), mitochondrial and chloroplast membranes (Ernster and Kuylenstiernia, 1969), and microsomes (Dallner and Ernster, 1968). In relation to this text, developments in the isolation of the plasma membrane are of most interest although this has very largely been achieved with animal cells; little progress has been made in the isolation of the tonoplast. We will concentrate on recent advances made in the isolation of the plasma membrane, particularly in relation to studies on plant cells.

*Isolation of plasma membranes*

The plasma membrane is probably the most difficult cell membrane to isolate because it must be disrupted to be isolated. It is essential to keep disruption of other cell particles to a minimum since fragmentation of cell organelles can lead to contamination of plasma membrane fractions by other membrane fragments. A variety of techniques have been used for the disruption of various animal cells for this purpose including a Dounce or Potter–Elvehjem homogenizer, Waring blender, rapid decompression in an inert gas, mincing, sonication and osmotic stress (see Steck and Wallach, 1970; Warren and Glick, 1971). Depending on the tissue and conditions used, the plasma membrane of animal cells may be recovered as whole ghosts, or as fragments in the form of sheets or small vesicles. Usually the larger the fragments the easier the separation from other particles in the homogenate. The aim should be to produce large ghosts although this may not be possible with plant cells due to the close association between plasmalemma and cell wall; this might be overcome by the use of isolated, wall-less protoplasts.

An interesting technique which has been applied to cultured animal cells is that of hardening or stabilizing the plasma membrane before homogenization by the use of heavy metal ions such as $Zn^{2+}$, tris base, or sulphydryl blocking reagents such as fluorescein mercuric acetate (FMA) (see Warren and Glick, 1971). Although these methods may produce some denaturation of the membrane, it would be interesting to see if they have any widespread application in the isolation of plant surface membranes.

The media and conditions of fractionation used in the isolation of animal plasma membranes vary widely and have been thoroughly reviewed by Steck and Wallach (1970) and De Pierre and Karnovsky (1973) and so will only be discussed briefly here. There appears to be no generally accepted procedure, media varying in relation to the nature and concentration of the major solute (usually sucrose), the pH and the presence of substances such as divalent cations or EDTA. There is no agreement as to whether the medium should be isotonic or hypotonic, although many investigators use isotonic sucrose solutions. The hypotonic medium developed by Neville (1960) which consists of ice-cold water buffered with 1 Mol m$^{-3}$ $NaHCO_3$ has also been widely employed. This medium causes cell disruption by osmotic means and although reducing the mechanical force required may also lead to the lysis of cell organelles, thus increasing the possible contamination of the plasma membrane fraction. Certainly marked differences in the patterns of proteins present in plasma membrane fractions have been observed when such an hypotonic homogenization medium has been compared with isotonic sucrose (Berman et al., 1969).

The means of centrifugation also varies, the most widely used procedure being to commence with differential centrifugation followed by further purification of the plasma membrane-enriched fraction by some form of density gradient centrifugation, thus relying on differences in sedimentation or density to separate the plasmalemma from other cellular particles. These methods are discussed thoroughly in the reviews mentioned above and by Wallach (1972) and will be referred to later in relation to the isolation of plant plasma membranes. An interesting innovation is the aqueous two-phase polymer system developed by Brunette and Till (1971) which attempts to separate particles on the basis of differences in their surface properties. After homogenization and removal of cell debris, membranes are purified by repeated centrifugal partition into an interface formed between dextran and polyethylene glycol. This is a rapid procedure which appears to give high yields of surface membranes and it would be interesting if this method could be applied to plant systems. Initial investigations, however, suggest that it is not suitable since it probably relies on the large fragments of plasma membranes produced by the homogenization of mammalian cells (Hall and Roberts, 1975).

*Membrane markers*

If a membrane is to be isolated and studied in relation to a particular biochemical activity, it is clearly important to have an assay for that membrane and for other particles that may be present as contaminants. Ideally such markers for specific cellular membranes should be present in that membrane and in no others. Markers used in membrane studies are of two basic types, morphological and biochemical.

*Morphological markers*   Phase contrast and electron microscopy are useful but limited as membrane markers. They can be used to identify impurities with distinct morphologies such as chloroplasts and mitochondria but many membranes form similar smooth vesicles on cell fractionation and so are impossible to identify. In addition, morphological methods are difficult to make quantitative. However, an interesting development in the isolation of plant surface membranes has been the use of a PTA-chromic acid stain which appears to be specific for plant plasma membranes. This will be discussed in more detail below.

*Biochemical markers*   Biochemical markers make it possible to obtain both quantitative estimates of membrane purity and the degree of contamination. They include membrane-bound enzymes, antigens, viral receptors, covalent labels and biochemical composition. Thus, in relation to the isolation of animal plasma membranes, such criteria as the enzymes $(Na^+, K^+)$-ATPase, 5′-nucleotidase and adenyl cyclase, the lactoperoxidase-catalysed iodination of exposed tyrosine residues, and a high cholesterol/phospholipid ratio have been used as markers. The use of such markers has been discussed thoroughly by De Pierre and Karnovsky (1973) and Wallach and Lin (1973) and will be assessed in relation to the isolation of plant surface membranes in the next section.

*Isolation of plant plasma membranes*
Since 1970 a number of reports have appeared concerned with the isolation of membrane fractions enriched in plasma membranes from plant cells. Within these relatively few reports, there is considerable variation in the methods employed. For example, Lai and Thompson (1971, 1972a, b) use an homogenization technique based on that of Neville (1960) which involves disruption of the tissue in 50 Mol m$^{-3}$ NaHCO$_2$ solution. In contrast, other workers (e.g. Lembi et al., 1971; Leonard et al., 1973) employ more complex media which contain a high concentration of sucrose. All report the successful isolation of plasmalemma-enriched fractions. After differential centrifugation to prepare a general membrane fraction, these procedures employ density gradient centrifugation of various types to purify the plasma membrane fraction. In many of these investigations, the plasma membrane fractions have been used to study the relationship between ATPase activity and ion transport; this will be discussed in detail in section 2.7.

As with the isolation of animal plasma membranes, the question of suitable markers for the detection of this fraction is all important. Unfortunately, at present there appear to be no satisfactory enzymic markers, the marker enzymes used in animal procedures not being applicable to plant systems (see Hall and Roberts, 1975). For example, although it is clear that plant cells contain a number of membrane-bound ATPases (Edwards and Hall, 1973; Leonard et al., 1973) no characteristic surface ATPase such as the ouabain-sensitive Na$^+$, K$^+$-activated ATPase of animal cells has been described for plants. Again, 5'-nucleotidase which is generally accepted as a specific plasma membrane marker in animal cells, appears to be present in very low or undectectable levels in plant membrane fractions (Lai and Thompson, 1971; Leonard et al., 1973). Similarly there is no evidence that adenyl cyclase activity is restricted to the plasmalemma of plant cells; cytochemical studies suggest that activity is present in a number of cellular membranes and is particularly active in the endoplasmic reticulum (J.L. Hall, unpublished data).

However, a number of potentially useful criteria for plant plasma membranes have been reported recently. It seems well established that the PTA-chromic acid post-stain for thin sections is specific for the plasma membrane in some tissues (Fig. 2.1) and so is a useful aid in the microscopic examination of membrane fractions (Lembi et al., 1971; Hodges et al., 1972; Littlefield and Bracker, 1972). However work in this laboratory suggests that the stain does.

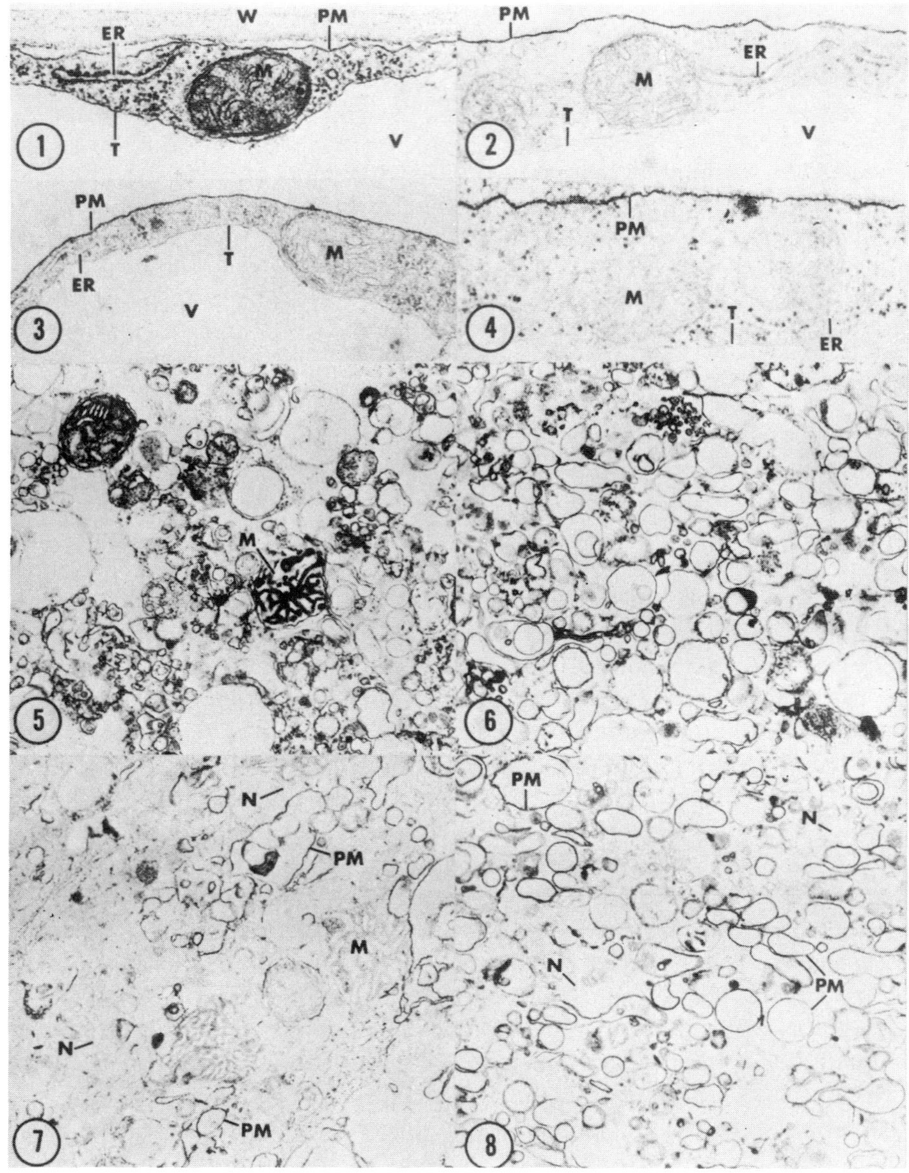

Fig. 2.1. Electron micrographs of sections of oat-root tissues and pellets of isolated membrane fractions: (1) Oat-root cortical cell stained with lead citrate ($\times$ 19 500) (2) Oat-root cortical cell stained with the periodic acid/PTA–chromic acid stain (PACP) ($\times$ 19 500) (3) Oat-root epidermal cell stained with PACP ($\times$ 28 500) (4) Oat-root stelar parenchyma cell stained with PACP ($\times$ 28 500) (5, 6) Membrane fractions from oat roots stained with lead citrate ($\times$ 11 775) (7, 8) Similar membrane fractions stained with PACP ($\times$ 11 775). W, cell wall; PM, plasma membrane; ER, endoplasmic reticulum; T, tonoplast; M, mitochondria; V, vacuole; N, unidentified membranes that do not stain with PACP. Reproduced from Hodges et al. (1972).

not give a positive reaction in all plant tissues and so should be carefully checked in each case. Another possibility is the use of glycosyl transferase (glucan synthetase) activity which appears to be associated with the plasma-lemma (Hodges et al., 1972; Van Der Woude et al., 1972) although there is evidence that the enzyme is also associated with the dictyosomes of the Golgi apparatus (Ray et al., 1969). Again, Lembi et al. (1971) have reported that $N$-1-[$^3$H]-naphthylphthalamic acid binds specifically with plasma membranes isolated from maize coleoptiles; this and similar binding reactions could provide useful markers for the identification of the surface membrane. Other criteria that have been used include a high sterol to phospholipid ratio (Hodges et al., 1972) which is characteristic of animal plasma membranes, and the general absence of specific markers for other cellular membranes. Thus although these reports represent a useful advance in our study of plant membranes there is clearly much to be done. The tonoplast, of course, is still an unexplored area as regards biochemical activity.

### 2.3.2 Analysis of membranes

The preparation of membrane fractions is followed by the analyses of the properties of these fractions. There have been two basic approaches to this problem which have been termed the preparative and analytical approach. The former concentrates on a single fraction e.g. that believed to be enriched in plasma membranes, while the latter involves the analysis of all fractions obtained. The analytical approach is considered to be preferable to the preparative method and should be routinely used in fractionation studies (De Duve, 1967; De Pierre and Karnovsky, 1973) as it provides a balance sheet for all fractions. For example, it allows an estimate to be made of the percentage recovery of an enzyme; the mere absence of a particular marker enzyme may not mean that a particular membrane or organelle is missing but simply that the enzyme has been inactivated.

The range of analytical techniques that are now applied to the study of membranes is vast and it is beyond the scope of this chapter to review these methods in detail. We will simply indicate the range of techniques and cite some useful references; the results of many of these techniques will be discussed in later sections in relation to the chemical composition and structure of membranes. The analysis of membranes can conveniently be divided into biochemical and biophysical approaches. The former include the 'solubiliza-tion' of membranes by detergents such as sodium dodecylsulphate (SDS) followed by the separation of the component proteins and lipids by gel permeation chromatography or electrophoresis (see Steck and Fox, 1972), the use of organic solvents to extract lipids and proteins (see Law and Snyder, 1972; Steck and Fox, 1972), and the use of labelling techniques (see Wallach, 1972; Oseroff et al., 1973). Biophysical procedures include X-ray diffraction (see Metcalfe, 1971; Keith and Melhourn, 1972; Oseroff et al., 1973), electron spin resonance (see Melhourn and Keith, 1972; Keith et al., 1973), nuclear magnetic resonance (see Horwitz, 1972; Oseroff et al., 1973), ultraviolet spectroscopy (see Holzwarth, 1972) and fluorescent probes (see Radda and Vanderkooi, 1972; Oseroff et al., 1973).

### 2.3.3 Microscopy of membranes

The value of electron microscopy in the study of membranes has been discussed by Branton (1969, 1971). Unlike the biochemical and biophysical techniques discussed earlier, which are essentially averaging techniques, electron microscopy permits the examination of specific regions within a membrane. Thus it has been shown that biological membranes are not homogenous structures; the hydrophobic centres containing structurally differentiated regions which may have great functional importance.

Essentially two procedures have been followed in the examination of membranes by electron microscopy. In one, tissues are fixed by chemical fixatives, dehydrated, embedded in resins and the final sections are stained with heavy metals. The second procedure, freeze-etching, does not require the use of chemical fixation. Tissues are frozen, fractured under vacuum, etched by sublimation of some of the surface ice, and finally replicas made of the surface. Neither of these procedures is free from artifacts. Chemical fixation and dehydration may produce drastic changes in the structure of the membrane and, as discussed by Korn (1968), little is known of the chemistry of either fixation or heavy-metal staining. It is not known which groups bind electron-dense atoms such as Os, U or Pb and what their localization indicates in relation to the living state. Again, with the freezing technique, there are a number of potential causes of artifact and misinterpretation. These have been thoroughly reviewed by Zingsheim (1972). Little is known of the structural changes or damage that may result from freezing, cleaving or etching, it being difficult to obtain useful criteria to assess these changes. Problems also arise in the preparation of replicas. The shadowing process may contaminate the specimen and introduce artifacts if heating occurs. Any projections may be accentuated and any depressions filled. A major disadvantage of freeze-etching is that it lacks any chemical specificity. However, accepting the limitations of these microscopial techniques and the difficulties of interpretation of the observations, electron microscopy has proved extremely useful in the study of biological membranes. The significance of the results will be discussed in the following sections in relation to the composition and structure of membranes.

## 2.4 Composition of membranes

Isolated membrane preparations are composed largely of protein and lipid with some carbohydrate; this latter is in the form of sugar residues attached to proteins and lipids to form glycoprotein and glycolipid. It can be seen from Table 2.1 that there is considerable variation in the chemical composition of different cell membranes and as we shall see later considerable

Table 2.1.

Chemical composition of cell membranes. Reproduced, with permission from Guidotti (1972) Annual Review of Biochemistry, vol. 41. Copyright © 1972 by Annual Reviews Inc. All rights reserved.

| Membrane | Protein (%) | Lipid (%) | Carbohydrate (%) | Weight fraction of protein | Ratio of protein to lipid |
|---|---|---|---|---|---|
| Myelin | 18 | 79 | 3 | 0.18 | 0.23 |
| Plasma membranes | | | | | |
|   Blood platelets | 33–42 | 58–51 | 7.5 | 0.4 | 0.7 |
|   Mouse liver cells | 46 | 54 | 2–4 | 0.46 | 0.85 |
|   Human erythrocyte | 49 | 43 | 8 | 0.49 | 1.1 |
|   Amoeba | 54 | 42 | 4 | 0.54 | 1.3 |
|   Rat liver cells | 58 | 42 | (5–10)[a] | 0.58 | 1.4 |
|   L cells | 60 | 40 | (5–10)[a] | 0.6 | 1.5 |
|   HeLa cells | 60 | 40 | 2.4 | 0.6 | 1.5 |
| Nuclear membrane of rat liver cells | 59 | 35 | 2.9 | 0.59 | 1.6 |
| Retinal rods, bovine | 51 | 49 | 4 | 0.51 | 1.0 |
| Mitochondrial outer membrane | 52 | 48 | (2–4)[a] | 0.52 | 1.1 |
| Sarcoplasmic recticulum | 67 | 33 | | 0.67 | 2.0 |
| Chloroplast lamellae, spinach | 70 | 30 | (6)[a] | 0.7 | 2.3 |
| Mitochondrial inner membrane | 76 | 24 | (1–2)[a] | 0.76 | 3.2 |
| Gram-positive bacteria | 75 | 25 | (10)[a] | 0.75 | 3.0 |
| Halobacterium purple membrane | 75 | 25 | | 0.75 | 3.0 |
| Mycoplasma | 58 | 37 | 1.5 | 0.58 | 1.6 |

[a] Deduced from the analyses.

variation is also exhibited within these groups, such as in the type and range of phospholipids found in membranes. Guidotti (1972) suggests that membranes can be classified into three groups according to their gross chemical composition. The first is a simple membrane system such as myelin which is composed largely of lipid and functions mainly as an insulator. The second group is characterized by animal plasma membranes which are composed of about 50% protein; this is presumably correlated with the increased enzymatic and transport activities of these membranes. The third group is represented by bacterial plasma membranes and the inner membranes of mitochondria which have protein as their major component (about 75%) and have additional functions such as oxidative phosphorylation. Thus, as the enzymatic functions of the membrane increase, so does the overall protein content. Until recently lipids were the most widely studied of the membrane components since they could be easily extracted by organic solvents, purified and analysed. Proteins have proved to be more difficult to study because of their insolubility, ready loss of biological activity and the serious contamination of membrane preparations by non-membranous proteins. However, as we shall see, new techniques have made the study of membrane proteins more tractable and they are currently the focus of extensive investigation. The important characteristics of membrane lipids and proteins which are largely related to animal and bacterial cell membranes may now be conveniently summarised.

### 2.4.1 Lipids

As we have seen earlier, lipids may form up to 80% of the membrane composition, as in myelin, or as little as 25% in the membranes of bacteria or the inner mitochondrial membranes. These lipids are frequently amphipathic, i.e. they contain both hydrophobic and hydrophilic groups, and so are ideally suited for the formation of interfaces between a hydrophobic region and a polar solvent. The most common lipids found in membranes are the phospholipids, glycolipids, sphingolipids and sterols. Just as the protein/lipid ratio shows considerable variation between different membranes, so does the lipid composition (Table 2.2). Simon (1974) has recently reviewed the range and activity of lipids in plant cell membranes, particularly in relation to water content. It appears that when the water content is reduced below 20% the lipid bilayer arrangement is lost and the membranes become extremely leaky.

### Phospholipids

These lipids, as their name implies, contain phosphorus in the form of phosphoric acid. Most are based on 3-phosphoglycerol and contain fatty acid

Table 2.2.
Protein and lipid content of membranes. The abbreviations are: Cer, cerebrosides; DPG, diphosphatidylglycerol; GalDG, galactosyldiglyceride; PA, phosphatidic acid; PC, phosphatidylcholine; PE, phosphatidylethanolamine; PGaa, amino acyl esters of phosphatidylglycerol; Plas, plasmalogen; SL, sulpholipid; Sph, sphingomyelin. Reproduced with permission, from Korn (1969) Annual Review of Biochemistry, vol. 38. Copyright © 1969 by Annual Reviews Inc. All rights reserved.

| Membrane | Protein/lipid (w/w) | Cholesterol/polar lipid (Mol/Mol) | Major polar lipids |
|---|---|---|---|
| Myelin | 0.25 | 0.7–1.2 | Cer, PE, PC |
| Plasma membranes | | | |
|    Liver cell | 1.0–1.4 | 0.3–0.5 | PC, PE, PS, Sph |
|    Ehrlich ascites | 2.2 | | |
|    Intestinal villi | 4.6 | 0.5–1.2 | |
| Erythrocyte ghost | 1.5–4.0 | 0.9–1.0 | Sph, PE, PC, PS |
| Endoplasmic reticulum | 0.7–1.2 | 0.03–0.08 | PC, PE, Sph |
| Mitochondrion | | | DPG, PC, PE, Plas |
|    Outer membrane | 1.2 | 0.03–0.09 | |
|    Inner membrane | 3.6 | 0.02–0.04 | |
| Retinal rods | 1.5 | 0.13 | PC, PE, PS |
| Chloroplast lamellae | 0.8 | 0 | GalDG, SL, PS |
| Bacteria | | | |
|    Gram-positive | 2.0–4.0 | 0 | DPG, PG, PE, PGaa |
|    Gram-negative | | 0 | PE, PG, DPG, PA |
|    PPLO | 2.3 | 0 | |
|    Halophilic | 1.8 | 0 | Ether analogue PGP |

molecules esterified to the first and second hydroxyl groups of glycerol, while the third hydroxyl group is esterified with phosphoric acid which in turn is esterified with another alcohol group; this phosphodiester forms the polar head of the molecule. The lipids are named according to the second alcohol at their polar head. Thus phosphatidylcholine and phosphatidylserine contain choline and serine, respectively, at their polar heads. Phospholipids are largely found in membranes and, in fact, may occur only in membranes. As can be seen from Table 2.2, the spectrum of phospholipids can vary widely from membrane to membrane (see Korn, 1968; McMurray and Magee, 1972). It appears that they may play a major role in ion selectivity. For example, Papahadjopoulos (1971) has produced vesicles from phospholipids and shown that, when prepared from phosphatidylserine and phosphatidylglycerol, they show a marked selectivity for $K^+$ permeation while there is good evidence that the membrane protein $Na^+$, $K^+$-stimulated ATPase has an absolute requirement for phosphatidylserine (Wheeler and Whittam, 1970). More recent studies have shown that the ATPase is activated by other phospholipids that

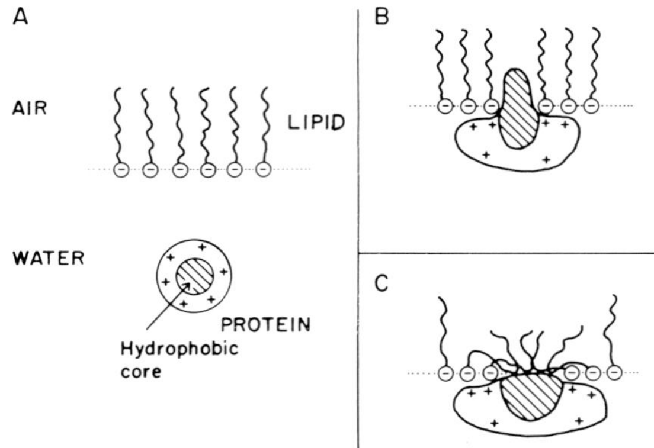



Fig. 2.2. Possible interactions of hydrophobic proteins and a phosphatidylserine monolayer at an air–water interface according to Kimelburg and Papahadjopoulos (1971)(A). Two possible conformational changes in the protein are envisaged. Either a hydrophobic portion of the protein penetrates into the hydrocarbon region of the monolayer (B), or a hydrophobic surface of the protein is exposed at the interface (C). Bangham (1972) has speculated that the ATPase protein may become operationally active in such a manner.

bear a net negative charge (K.P. Wheeler, personal communication). Kimelberg and Papahadjopoulos (1971) have suggested that proteins may react with a phosphatidylserine monolayer as shown in Fig. 2.2 and Bangham (1972) has speculated that the ATPase protein may be activated in this way. There is also evidence that exchange of phospholipids between organelles can occur and that phospholipids can be transformed in the membrane in situ (see Law and Snyder, 1972).

*Sphingolipids*
These lipids may also contain phosphoric acid but have a long chain amino alcohol as a structural backbone rather than glycerol.

*Glycolipids*
Glycolipids contain a sugar group and have a structural backbone of glycerol or of the amino alcohol, sphingosine; these latter may be classed as sphingolipids. They are generally assumed to occur mainly in the plasma membrane of animal cells and there is evidence that changes in the glycolipid pattern occur with cell transformation by tumour viruses (see Oseroff et al., 1973).

*Sterols*
These lipids are based on the perhydrocyclopentanophenanthrene nucleus but show various degrees of unsaturation and substitution. One of these,

cholesterol, is a common lipid constituent of membranes. Cholesterol appears to be able to attain two possible extreme viscosities of the hydrocarbon region, namely solid and liquid (Bangham, 1972) and the cholesterol content appears to have a marked effect on the permeability of derived liposomes to glycerol (De Kruyff et al., 1972). There is a general absence of sterols from bacterial membranes.

*Analysis of lipids*

The nature, arrangement and function of membrane lipids have been examined in a number of ways. Lipids can be extracted in organic solvents and separated by a variety of methods including Sephadex chromatography, solvent partition and thin-layer chromatography; this latter procedure can be made particularly sensitive (see Law and Snyder, 1972). The arrangement of lipids in membranes has been studied by X-ray diffraction (see Keith and Melhourn, 1972; Oseroff et al., 1973), electronic spin resonance (see Melhourn and Keith, 1972; Oseroff et al., 1972) and nuclear magnetic resonance (see Horwitz, 1972; Oseroff et al., 1973). The results of these studies will be discussed later in relation to the structure of biological membranes.

The role of lipids in membranes has also been examined by the study of lipid monolayers (see Kézdy, 1972) and lipid bilayer membranes (see Szabo, 1972). Artificial lipid bilayers and natural membranes both contain a thin hydrophobic interior. They contain many other features in common as shown in Table 2.3. However, as bilayers lack many of the functional characteristics of membranes it is clear that biological membranes are not simple continuous lipid bilayers.

Table 2.3.
Physical properties of natural and model membranes. Reproduced from Wallach (1972).

| | Natural membrane | Model lipid membrane |
|---|---|---|
| Thickness (nm) | 5–12 | 6.8–7.3 |
| Surface tension ($\mu$N m$^{-1}$) | 0.003–0.3 | 0.05–0.1 |
| Water permeability (m sec$^{-1}$) | 0.3–33 | 5–10 |
| Electrical capacitance ($\mu$F m$^{-2}$) | $0.5$–$1.3 \times 10^4$ | $0.33$–$1.3 \times 10^4$ |
| Electrical resistance ($\Omega$ m$^{-2}$) | $10^7$–$10^{10}$ | $10^{10}$–$10^{13}$ |
| Breakdown potential (V) | 0.1–3.0 | 0.15–0.20 |

## 2.4.2 Proteins

Increasing attention is now being focused on the membrane proteins although, again, this is largely in relation to animal and bacterial membranes; we know little of the nature of plant membrane proteins. Membrane proteins have been examined by such techniques as optical rotatory dispersion and infrared spectroscopy which provide information concerning the configuration of the peptide backbone. In addition, great improvements have been made in the techniques for the extraction and solubilization of proteins by organic solvents and detergents such as SDS, and in the subsequent fractionation of these proteins by centrifugation, chromatography and electrophoresis. These procedures have been thoroughly discussed by Steck and Fox (1972).

There are a variety of questions that need to be answered in relation to membrane proteins. These include the characteristics of these proteins, the range of proteins occurring in a specific cellular membrane, how the proteins are arranged in relation to each other and to the lipids and whether the proteins fulfil an inert structural role or have catalytic properties related to specific membrane functions. This field is very large and only some of the important findings will be summarised here.

According to Guidotti (1972), there appear to be two classes of membrane proteins. One group is weakly bound and can be removed by treatments such as strong salt solutions or mild detergents. The second group is tightly bound to the lipid bilayer and may have important structural or enzymatic roles. The amino acid composition of the first class is similar to other 'soluble' proteins found in the cell whereas the more tightly bound proteins appear to contain a greater number of non-polar residues which presumably facilitates the solubilization of these proteins in the lipid bilayer. There appears to be considerable heterogeneity in the molecular size of membrane proteins, molecular weights varying from about 20,000 to over 200,000. For example, a detailed examination of mammalian membrane proteins by Kiehn and Holland (1970a, b) showed a wide range of proteins of varying molecular weights (based on gel electrophoresis after solubilization in SDS), with no single protein or group accounting for the bulk of the membrane. There was no evidence for a special structural protein which would constitute a high proportion of the membrane protein. This heterogeneity appears to be the case with the great majority of cellular membranes (see Guidotti, 1972). Perhaps the most widely studied membrane in relation to proteins is that of the erythrocyte. Juliano (1973) has recently reviewed our knowledge of these proteins and presented a model for their arrangement in the membrane (Fig. 2.3).

It is often assumed that glycoproteins, which are bound to a variety of oligosaccharides, are located specifically in the plasmalemma, but this may not be the case (Oseroff et al., 1973). Bretscher (1971) has suggested that glycoproteins may transverse the lipid bilayer and so expose hydrophilic

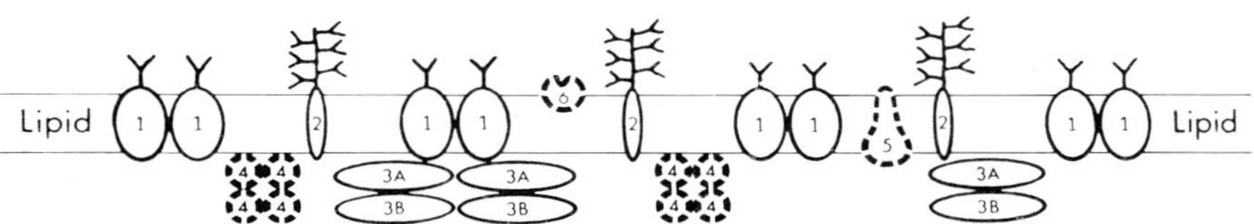


Fig. 2.3. A model of the erythrocyte membrane. (1) 100,000 molecular weight protein. (2) Sialoglycoprotein. (3) Spectrin. (4) Glyceraldehyde-3-phosphate dehydrogenase. (5) ($Na^+$, $K^+$)-ATPase. (6) Acetylcholinesterase. Only some of the membrane polypeptides are shown. There is no correlation intended between the number of copies depicted here and the number of polypeptide chains present in the membrane. The grouping of polypeptide chains into oligomeric units is based on tentative evidence. The stalk-like structures include carbohydrate (Y). Reproduced from Juliano (1973).

groups at both external and cytoplasmic surfaces of the membrane, while Gingell (1973) has proposed that membrane permeability may be related to the aggregation of mobile glycoprotein units. Glycoproteins may also have important roles in cell–cell adhesion and cell surface transformation (Oseroff et al., 1973).

Little is known of protein–protein interactions and protein–lipid interactions within membranes. Membranes may contain similar multi-protein interactions as found in viruses, microtubules and ribosomes which may be important in membrane organization. Many membrane-associated enzymes are inactivated by the removal of lipid (Steck and Fox, 1972) and there is evidence for bonding between phospholipids and membrane proteins (Korn, 1968). The association between proteins and lipids seems to play an important role in the assembly and stability of cellular membranes.

## 2.5 Membrane structure

As we have seen, membranes are composed largely of proteins and lipids. What we need to know is how these molecules are arranged within the plane of the membrane which is 7.5 ± 2.5 nm thick. Over the past 40 years or so, our ideas of membrane structure have been dominated by the model proposed by Danielli and Davson in 1935. This considers that membrane lipids are arranged in a bilayer with their polar heads arranged to the outside and covered on both surfaces with membrane proteins which are bound to the

                    *J.L. Hall and D.A. Baker*

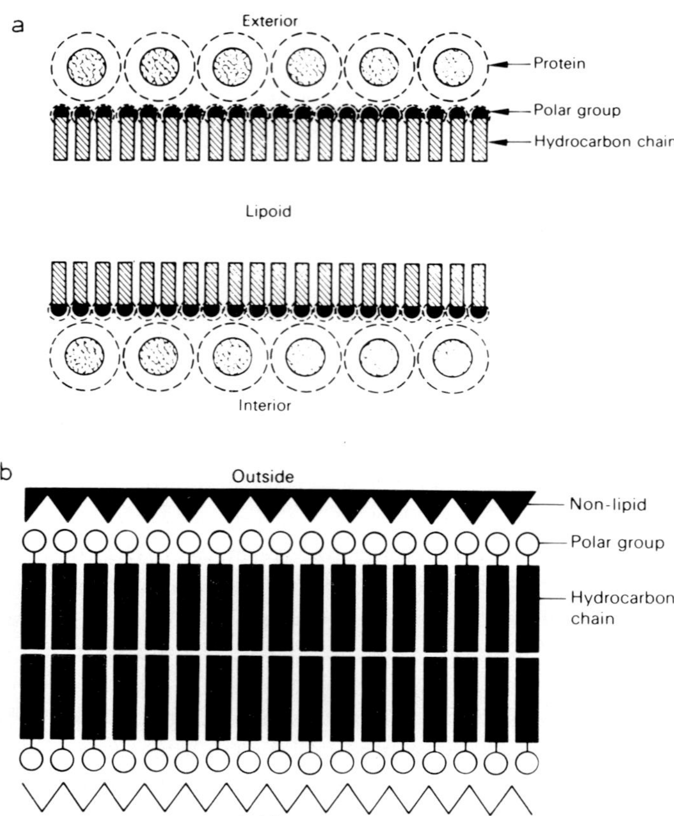


Fig. 2.4. Models of membrane structure. (a) According to Danielli and Davson (1935). (b) Unit membrane model of Robertson (1967).

lipids by electrostatic interactions (Fig. 2.4). This model was based on evidence which included observations on the permeability and surface properties of certain cells which suggested that surface membranes had a high lipid content, the calculation of Gorter and Grendel in 1925 which showed that erythrocytes contained enough lipid to form a bilayer at the surface, and the low surface tension of cells which indicated the presence of protein at the membrane surface.

This model formed the basis of the unit membrane hypothesis proposed by Robertson in 1957 which was based largely on recent electron microscope evidence and suggested a universal structure for biological membranes. It was based on the consistent appearance of biological membranes in thin sections of osmium – or permangate – fixed tissue as a three-layered structure of two dense lines separated by an unstained space. Robertson's model (Fig. 2.4) showed some differences from that of Danielli and Davson. The proteins are considered to be in the extended $\beta$-form rather than globular and it is suggested that there may be some differences between the two protein layers. The evidence leading to the proposal of the Danielli–Davson and Robertson models has been thoroughly covered by Robertson (1967), Branton (1969) and Hendler (1971).

Since the formulation of these models other pieces of evidence have arisen which may be cited in support of this general structure. These have been thoroughly reviewed and assessed by Branton (1969) and, in particular, Hendler (1971) and will only be mentioned briefly here. The classical experiments of Gorter and Grendel have been repeated using modern methods and it is clear that there is adequate lipid to provide a bilayer within the erythrocyte membrane (Bar et al., 1966). Finean and co-workers have examined packed, isolated membranes by X-ray diffraction and electron microscopy and the results are consistent with the presence of predominantly lipid bilayer (see Finean et al., 1968). However, it should be mentioned that X-ray diffraction patterns of chloroplast lamellae appear to differ from those of most other membranes (see Branton, 1969). Freeze-cleavage studies appear to be consistent with a split of the membrane down a natural plane of cleavage in the centre of a bimolecular leaflet (Branton, 1969; Zingsheim, 1972). The similarities in many physical properties between synthetic lipid bilayer membranes and biological membranes (see Table 2.3) strongly suggest that similar structures form at least part of biological membranes (Henn and Thompson, 1969; Bangham, 1972). However these different pieces of evidence largely relate to the lipid bilayer and say less about the arrangement of the protein in membranes.

The Danielli–Davson and unit membrane concepts have been criticised on numerous occasions (see Korn, 1968; Hendler, 1971; Wallach, 1972; for a detailed discussion of these criticisms). Among the criticisms listed are the wide variation found in the activity, thickness and chemical composition of different membranes, the fact that much is based on the appearance of membranes in thin sections under the electron microscope which lacks any sound chemical interpretation, and the observation that much of the X-ray diffraction data are based on studies on myelin which is an unusual membrane with a very low protein content.

In addition, the techniques of freeze-etching and -cleaving have indicated that a considerable degree of differentiation may exist within the hydrophobic interior of membranes though, again, this microscopic technique is limited

                    *J.L. Hall and D.A. Baker*

Fig. 2.5. Electron micrograph of *Avena* plasmalemma as shown by freeze-etching. The particles shown are on fracture face **B** (i.e. outer half of membrane abutting the cell wall). The large 'lump' on the membrane is where a cytoplasmic vesicle (probably a Golgi vesicle) has fused with the plasmalemma. Provided by Dr. B.A. Fineran, University of Canterbury, New Zealand.

by its lack of chemical specificity. However this technique has shown the presence of an array of particles within the matrix of most membranes (Fig. 2.5). These particles range from about 6 nm to 19 nm in diameter and vary considerably in number in different membranes. There appears to be a correlation between the metabolic activity of a membrane and the particle density, this latter being high in chloroplast lamellae and mitochondrial membranes and low or absent in myelin (Branton and Deamer, 1972). More recently, Orci and Perrelet (1973) have demonstrated an increased number of particles in pinocytotic regions of the cell membrane in intestinal smooth muscle cells. The exact chemical nature of these particles is unknown but the evidence available points to their being proteins intercalated into the lipid bilayer.

Basically different in its concept of membrane structure is the subunit model which considers lipoprotein subunits to be the basic units of structure and function in membranes (Green and Perdue, 1966; Benson, 1966; Fig. 2.6).

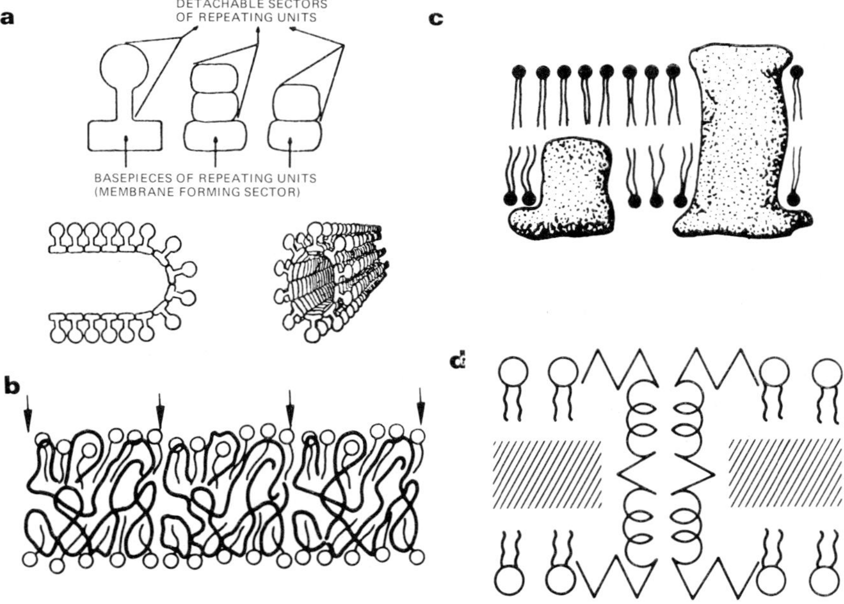


Fig. 2.6. Models of membrane structure. (a) Green and Perdue (1966) showing different forms of repeating units and the arrangement of these units in the membrane. The subunits are lipoproteins. (b) Benson (1966) consisting of a series of lipoprotein subunits each with the hydrocarbon chains of the lipid in a folded protein chain. (c) Wallach and Zahler (1966) showing membrane proteins in which two hydrophilic peptide regions at the membrane surface are separated by a hydrophobic zone. (d) Lenard and Singer (1966), suggesting that proteins with helical and random-coil portions transverse the entire membrane.

Much of the evidence for the presence of subunits has come from electron microscopy and X-ray diffraction; these are open to pitfalls of interpretation as discussed by Robertson (1966), Hendler (1971) and Zingsheim (1972). The particles seen by freeze-cleavage have not been interpreted in terms of subunits (Branton, 1969).

There have been a variety of other membrane models proposed in the last 10 years or so and many of these have been thoroughly assessed by Hendler (1971). Wallach and Zahler (1966; Fig. 2.6) suggest that the hydrophilic regions of membrane polypeptides are located on both surfaces and connected by hydrophobic rods which cross the apolar membrane core, while Lenard and Singer (1966) envisage polypeptides traversing the entire membrane (Fig. 2.6). Both these models show lipids on the outside of the membrane.

Although there have been a variety of proposals to explain the structure of biological membranes, a number of recent reviews have concluded that a lipid bilayer forms a substantial part of a membrane and is the best explanation of the unspecialized properties of the membrane (Branton, 1969; Hendler, 1971; Bangham, 1972; Zingsheim, 1972). It probably forms a structural framework to which the proteins are attached and embedded to provide functional centres in the membrane. The various models mentioned above differ largely in emphasis, with differences occurring mainly in relation to the extent of coverage of the lipid bilayer with protein which is envisaged, the degree of penetration of the membrane by protein, and the symmetry of the membrane. There is now good evidence for the occurrence of nonpolar associations between proteins and lipids which must mean that proteins penetrate into the interior of the membrane to interact with the hydrophobic lipid groups. Many of these ideas have been brought together by Singer and Nicolson (1972) in the fluid mosaic model which is perhaps the most recent attempt to provide a sound model of membrane structure (Fig. 2.7). They consider that the bulk of the phospholipid is organized as a discontinuous fluid bilayer. The proteins that are integral to the membrane are a heterogeneous set of globular molecules with ionic and polar groups protruding from the membrane and the nonpolar groups largely buried in the hydrophobic interior of the membrane.

## 2.6 The carrier concept

From what we have seen above, it is clear that the membrane is no longer considered to be a static barrier between the cell and its environment. Solute movement through membranes is not determined simply by the lipid solubility, electrical charge and molecular size of the solute but involves a specific

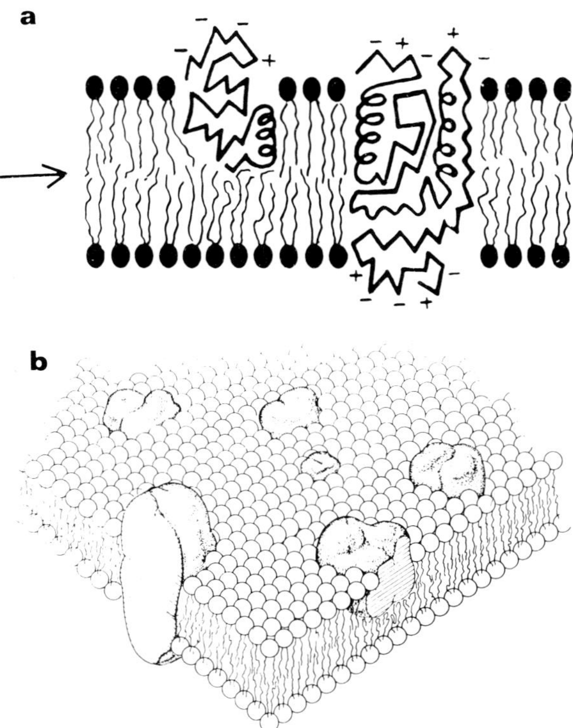

Fig. 2.7 The lipid–globular protein mosaic model of membrane structure according to Singer and Nicolson (1972). (a) Cross-sectional view. The phospholipids are arranged in a discontinuous bilayer. The proteins are shown as globular molecules, partially embedded in and partially protruding from the membrane. The arrow marks the plane of cleavage to be expected in freeze etching. (b) Three dimensional view. The solid bodies with stippled surfaces represent the globular integral proteins.

membrane component which binds and facilitates the movement of the solute through the membrane. These membrane components are known as carriers. The nature of these carriers is not well-defined although carrier-mediated transport is generally considered to involve three steps: solute recognition which involves binding of a solute to a receptor site, translocation of this complex across the membrane, and release of the solute into the cell. In the case of facilitated diffusion, movement takes place down a concentration gradient while in exchange diffusion there must be an equal movement in the opposite direction. In active transport, these processes are coupled with energy utilization. Clearly the important questions which have to be answered are: what is the nature of the carrier and how is translocation coupled to energy conservation? The answer to these questions must involve the isolation and characterization of carriers. Although we know nothing in relation to these questions in plants, some important advances have been made in the past few years, particularly in bacterial systems, and we will briefly summarise these findings here. More extensive discussions can be found by reference to Pardee (1968), Kaback (1970), Oxender (1972), Simoni (1972) and Boos (1974).

A large number of transport systems have been studied in bacteria, including those of amino acids, sugars and ions, by a combination of genetic and biochemical methods. It has been shown that transport systems, like enzymes, are quite specific which immediately suggests an important role for proteins in this process. In this respect, one of the most important developments of recent years has been the isolation of specific binding proteins for a number of amino acids, sugars and ions. These studies have largely been carried out with Gram-negative bacteria and began with the isolation of a $SO_4^{2-}$-binding protein by Pardee and Prestidge (1966). The evidence implicating these proteins in the transport process has been presented by Pardee (1968) and Oxender (1972). Since we are primarily concerned with ion transport in this text, we will only mention ion-binding proteins here. Proteins which bind $SO_4^{2-}$, phosphate and $Ca^{2+}$ have now been isolated (see Oxender, 1972; Simoni, 1972). These have molecular weights of approx. 32,000, 42,000 and 25,000 to 28,000 respectively. Like the other binding proteins they can be released by osmotic shock; in the case of the phosphate system, some transport ability can be restored to shocked cells by addition of the pure phosphate-binding protein. It seems most likely that these proteins are involved in solute recognition at the membrane surface. Whether these findings have any relevance to ion transport by plants is not known. It is also not clear what relationships these proteins have in bacteria to the second stage of carrier-mediated transport, that of translocation. This step is thought to involve some conformational change of the carrier-complex and it is interesting that isolated binding-proteins can show conformational changes when substrate is added (see Oxender, 1972; Boos, 1974).


## 2.7  Energy coupling and transport

Metabolic energy is needed for active ion transport and so there must be a mechanism to couple energy production to the transport process. This coupling is clearly a very basic biological process. The energy sources for active transport vary with different cellular systems. In most mammalian transport processes (except mitochondrial transport), the energy comes from the hydrolysis of ATP, while in many bacteria and in mitochondria it seems that membrane-associated oxidations may be directly linked to transport. The source of energy for ion accumulation in plant cells will be discussed in detail in the next section; ATP cleavage is the most highly favoured source at the present time.

How is this energy conservation linked to active transport? This question has been discussed in detail by Boyer and Klein (1972) in relation to our present knowledge of bacterial and mammalian transport systems. These authors suggest that either membrane-bound ATPases or the oxidative chain enzymes serve to utilise energy sources for transport. Solute accumulation occurs in cells lacking one or other of these systems. Therefore they suggest that either system may operate to give rise to an energised state in the membrane (designated ∼) which conserves energy to drive transport:

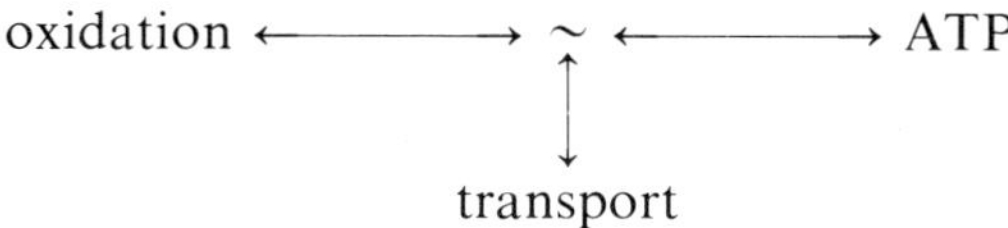

The energy may be conserved by processes such as protein conformational changes, covalent bond formation in membrane components, by establishing electro-chemical gradients, or by a combination of these processes.

How this conserved energy is used to drive transport is not understood. One possibility currently being carefully considered is that energy is used to decrease the affinity of the transport proteins for the solute (Fox and Kennedy, 1965). Such a scheme is illustrated in Fig. 2.8a. This is a simple model and does not account for membrane recognition sites. Boyer and Klein (1972) have proposed a membrane conformational change hypothesis in which the affinity for the solute remains unchanged during the transport process (Fig. 2.8b). The inner membrane entry site is 'opened' upon recognition of a loaded carrier and is then rapidly 'closed' by energy use after the solute departs. A simple competition between the rate of combination of solute with binding protein and the rate of energy input determines whether a given solute molecule is retained.

### 2.7.1 Metabolism and ion transport in plants

Although it is well established that active ion transport in plants is linked to respiratory metabolism, the actual source of energy and the way it is linked to the transport process are unknown. As regards the energy source, arguments have largely centred around two possibilities: one postulates a direct linkage

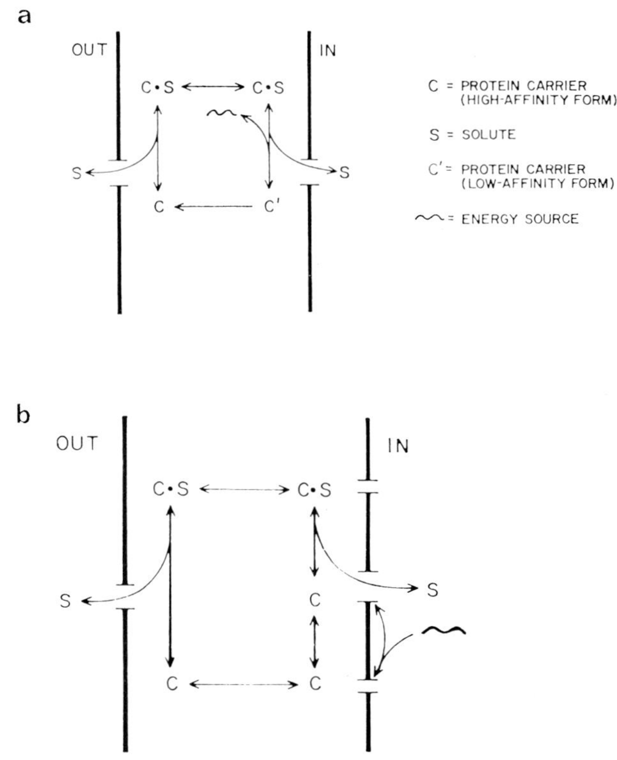



Fig. 2.8 Schemes for carrier-coupled conformational transport according to Boyer and Klein (1972). (a) Oversimplified scheme which does not account for membrane recognition sites. At the inner membrane, energy in some manner causes a decrease in solute binding. (b) In this scheme, the affinity for the solute remains unchanged. An inner membrane site 'opens' upon recognition of a loaded carrier and is 'closed' by energy input after the solute departs. This allows for a common energization mechanism for various protein carriers.

of transport to electron transfer processes while the alternative involves a coupling to ATP breakdown. At present there is no conclusive evidence in support of either of these possibilities although the weight of evidence probably favours a linkage to ATP hydrolysis (see Epstein, 1972; Hall, 1973) and this has led to a variety of investigations into ATPase activity from higher plants (see Baker and Hall, 1973).

A linkage of active ion transport to electron transfer in the respiratory chain was first proposed by Lundegårdh in 1935. Although this hypothesis is no longer tenable (see Epstein, 1972), it has led Robertson (see Robertson, 1968) to propose a linkage between ion transport and charge separation

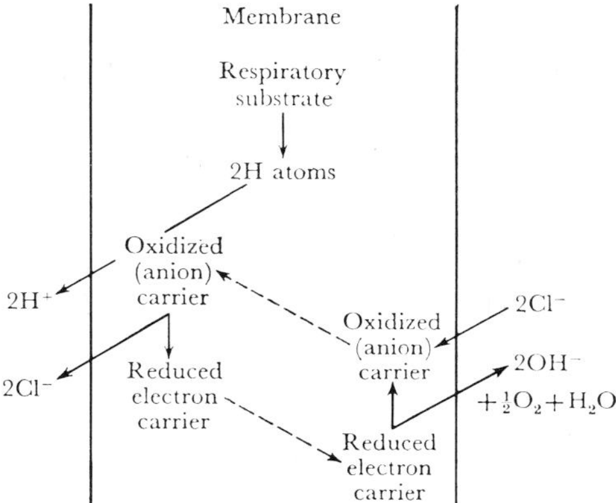


Fig. 2.9. Principle of an electron carrier in one direction acting as an anion carrier in the opposite direction; active transport of an anion or an anion pump. Reproduced from Robertson (1968).

(Fig. 2.9). This means that ion transport at the plasmalemma must be linked to the mitochondria or, in green cells, to the chloroplasts. These arguments have been thoroughly and elegantly presented by Robertson (1968) and so will not be repeated here. Perhaps the most serious drawback to this concept is the spatial one, i.e. linkage of ion transport at the plasmalemma to electron transfer reactions in the mitochondria or chloroplasts.

Evidence in favour of a phosphorylation mechanism comes from a variety of sources. It can be argued that this is a more unifying concept since it can also explain the transport of neutral molecules and transport under anaerobic conditions. The effect of uncoupling agents such as 2, 4-DNP and CCCP on active ion uptake can be most easily explained in terms of an inhibition of ATP synthesis. In most mammalian systems, it is well established that ATP hydrolysis at the cell membrane provides energy for active ion transport; this could well be a very basic biological process. In addition, a number of more specific pieces of evidence can be quoted in favour of a phosphorylation mechanism. It has been demonstrated that ATP can provide the energy for $Ca^{2+}$ transport in plant mitochondria (see Hanson and Hodges, 1967 and Ch. 3). In giant algal cells, it seems that ATP is the energy source for active cation transport though this is probably not so for $Cl^-$ transport (see Mac-Robbie, 1971 and Ch. 5). Bledsoe et al. (1969) reported that oligomycin leads to a reduction in both phosphate uptake and ATP synthesis in maize roots and concluded that ATP production in the mitochondria was reduced, thus depleting the supply of ATP at the plasmalemma for ion transport. However, Polya and Atkinson (1969) reported no correlation between ion influx and

ATP levels in carrot discs, although other studies on storage tissues are consistent with the utilization of ATP or some other high energy intermediate for active transport (MacDonald et al., 1966; Lüttge et al., 1971).

Other evidence has come from the measurement of membrane potentials. The strong depolarization induced by inhibitors in etiolated seedling tissue and *Elodea* leaves is consistent with the idea that ATP is the energy source (Higinbotham, 1973). With *Neurospora*, Slayman et al. (1973) have shown that the time course of ATP decay in the presence of respiratory inhibitors coincides with the loss of membrane potential whereas inhibition of electron transfer occured much faster; these data strongly suggest that an electrogenic ion pump in the plasmalemma is fuelled by ATP.

Finally a number of recent investigations have reported correlations between ion transport and ATPase activity in higher plants (see Fisher et al., 1970; Lai and Thompson, 1972; Leonard and Hanson, 1972a, b; Hill and Hill, 1973). These findings will be discussed more fully in a later section. Since much of this work has been stimulated by studies on mammalian systems it is pertinent at this point to include a brief discussion of the mammalian ATPase system.

### 2.7.2 Mammalian ATPase

The ($Na^+$, $K^+$)-ATPase of mammalian cells is the most extensively studied transport ATPase and has been thoroughly reviewed by Whittam and Wheeler (1970), Schwartz et al. (1972) and Dahl and Hokin (1974). The enzyme system is orientated within the membrane and is directly related to the activity of the pump which moves $Na^+$ out of and $K^+$ into the cell. Both pump and enzyme require $Na^+$ and $K^+$ together for maximal activity and both are specifically inhibited by cardiac glycosides such as ouabain (strophanthin G). The enzyme has an estimated molecular weight of between 190,000 and 500,000 and must be exposed to both sides of the membrane. The major features of this enzyme activity are summarised in Fig. 2.10 and involve three major

$$E_1 + ATP \xrightarrow{\phantom{xx} Mg^{2+}, Na^+ \phantom{xx}} E_1 - P + ADP \qquad 1$$

$$E_1 - P \xrightarrow{\phantom{xx} Mg^{2+} \phantom{xx}} E_2 - P \qquad 2$$

$$E_2 - P + H_2O \xrightarrow{\phantom{xx} K^+ \phantom{xx}} E_2 + P \qquad 3$$

$$E_2 \rightleftharpoons E_1 \qquad 4$$

Fig. 2.10. Probable reaction sequence of mammalian ATPase activity. The reactions are described in the text. $E_1$ and $E_2$ represent different conformational forms of the enzyme.

steps. The first involves a reversible phosphorylation of the enzyme by the transfer of phosphate from the terminal position of ATP. This is dependent on $Na^+$ and $Mg^{2+}$ and is inhibited by ouabain. The second stage involves a change in the phosphorylated enzyme intermediate which may be an acyl phosphate. However, there is considerable controversy concerning the exact number and nature of phosphorylated intermediates (see Fukushima and Tonomura, 1973). Finally the intermediate is dephosphorylated by a process which is dependent on $K^+$ and $Mg^{2+}$ and inhibited by ouabain. Ouabain can combine either with the free enzyme or with the phosphorylated state to produce an inactive complex. Once dephosphorylated the enzyme is ready to repeat the cycle and the scheme shown in Fig. 2.11 has been proposed by

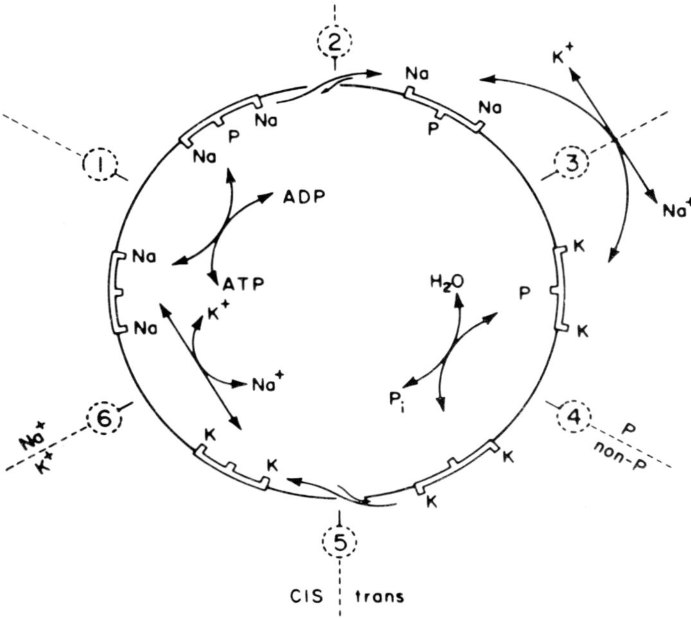


Fig. 2.11. The turnover cycle for the $(Na^+, K^+)$-ATPase membrane transport system as proposed by Albers, Koval and Siegal (1968). The cycle may include states of E–P as yet undiscovered. The circle denotes a plasma membrane of which the enzyme is a component. The phosphorylation site is on the cytoplasmic surface of the membrane. The cation effector sites of the enzyme are considered identical with the active transport carrier. The enzyme is denoted as 'cis' or 'trans' when the carriers are adjacent or opposite the phosphorylation site. Reaction 1 is the $Na^+$-dependent ATP–ADP exchange (4–6). In the native enzyme, the phosphorylated *cis* enzyme is rapidly converted to the *trans* form (reaction 2). In the presence of $K^+$, reaction 3 activates the hydrolic step (reaction 4). The *trans* enzyme becomes less stable as a result of the dephosphorylation and reverts to the *cis* form (reaction 5). The cycle is completed when $Na^+$ displaces $K^+$ from the *cis* enzyme (reaction 6).

Albers et al. (1968) to explain the sequence of events. It is not clear if the ATPase itself acts as a $Na^+$ carrier or whether the ATPase generates an energy-rich state which is then used to drive transport. Little is known of the composition or structure of the ATPase although, as we have seen in a previous section, there are indications that phospholipids play an important role in the transport activity.

### 2.7.3 Plant ATPases

As pointed out by Boyer and Klein (1972), any hydrolytic capacity for ATP exhibited by a membrane should be considered for possible involvement in active transport. In the earliest studies, plant biochemists studying ATPases may have been searching for too close a similarity with the mammalian enzyme system. However, as pointed out previously (Dodds and Ellis, 1966; Hall, 1973), we may expect a plant transport ATPase to show a number of differences from that found in mammalian cells which will be related to the different characteristics of the active transport processes. In particular, this may relate to synergistic effects of monovalent cations, the effect of $Ca^{2+}$ and inhibition by the cardiac glycoside ouabain.

ATPases have now been characterized from a variety of higher plant sources and in some cases correlations with active ion transport have been reported. However, there appears to be no characteristic transport ATPase as found in mammalian systems; plant ATPases differ widely in relation to pH effects, activation by $Mg^{2+}$ and other divalent cations, and activation by monovalent cations. In general, the ATPases are stimulated by $Mg^{2+}$ and other divalent cations but the pH optimum for this activity varies from below neutrality to pH 8.5. Many of these enzymes show further stimulation by monovalent cations applied singly, although the pH optimum for this activation varies from pH 5.0 or below (Dodds and Ellis, 1966; Williamson and Wyn Jones, 1972), to pH 6.0 (Leonard et al., 1973), to pH 8.0 or above (Greuner and Neumann, 1966; Hall and Butt, 1969; Lai and Thompson, 1971, 1972; Hall, 1971). In some cases, synergistic effects of $Na^+$ and $K^+$ have been reported. Lai and Thompson (1971) in a study of an ATPase associated with a membrane preparation from *Phaseolus* cotyledons obtained highest activity in the presence of $Na^+$ and $K^+$ together although without $Mg^{2+}$. Similarly, Hanssen and Kylin (1969) and Kylin and Gee (1970) have described ATPase activity from sugar-beet and mangrove leaves which show optimal activity in the presence of $Na^+$ and $K^+$ alone. Recently, Rungie and Wiskich (1973) have described an ATPase in a microsomal fraction from turnip which is specifically stimulated by anions and may be involved in anion accumulation across the tonoplast.

More positive evidence for a relationship between ATPase activity and ion transport has come from a number of sources. Fisher and Hodges (1969) and Fisher et al. (1970) have reported a constant relationship between the amount of cation absorbed by the roots of a number of different species and cation-stimulated ATPase activity in those roots. This result has been proposed as evidence that enough ATPase is present to account for the measured uptake rates, but this interpretation must be treated with some caution as the level of ATPase activity measured depends on a number of factors and nothing is known of the pH or concentration of $Mg^{2+}$ and ATP at the plasmalemma, nor indeed of the relationship between ATP hydrolysed and the number of ions transported. In a somewhat similar study, Williamson and Wyn Jones (1972) obtained a correlation between the $K_m$ for $K^+$ stimulation of a membrane-bound ATPase and the $K_m$ of the system II isotherm for $K^+$ uptake by maize roots.

A different approach was reported by Leonard and Hanson (1972a, b) who observed an increase in both ion absorption and ATPase activity in washed maize roots. The absorption studies were in the low system I concentration range while the ATPase was only stimulated by system II concentrations. The increased ATPase activity was only observed in the microsomal fraction and was relatively small, although, as the authors point out, it is difficult to assess the contribution of the plasma membrane to the total microsomal fraction. A further correlation has been reported by Lai and Thompson (1972a, b) who found a decrease in both ATPase activity and the export of ions from *Phaseolus* cotyledons during germination and isolated what is thought to be a partially purified plasma membrane fraction from the same tissue which formed vesicles capable of extruding $Na^+$ and $K^+$ in the presence of ATP. Again, Hill and Hill (1973) reported an increase in $Cl^-$-stimulated ATPase activity in *Limonium* when plants were transferred from a low-salt to a high-salt medium; it is thought that the enzyme is largely located in the gland cells.

Thus, there is some tentative evidence for a relationship between ATP hydrolysis and active ion transport by plant cells, although at present this lacks the clear correlations observed in a number of animal cells. This may be a reflection of the greater complexity of the plant cell in relation to this problem. The cell wall clearly makes membrane isolation more difficult and the situation is complicated by the presence of an additional transporting membrane, the tonoplast. Both gradient centrifugation (Leonard et al., 1973) and disc electrophoresis (Edwards and Hall, 1973) clearly show that higher plant cells contain a variety of membrane-bound ATPases which differ in their kinetic properties. Different membrane fractions may contain a number of these activities. This may be either a true description of the in vivo situation or could reflect the difficulty of obtaining pure preparations of membranes and organelles free from contaminants. Whatever the case, it undoubtedly confuses the characterization of transport-associated ATPase activity.

## 2.8 Membrane permeability and selectivity

### 2.8.1 Membrane permeability

The term permeability is a property of membranes and is a measure of the ease with which a solute moves passively through a particular membrane. The permeability coefficient may be defined as the number of molecules crossing a unit area of membrane in unit time when a concentration difference is applied across that membrane. Alternatively the permeability coefficient of a species j, $P_j$ may be defined as

$$P_j = \frac{D_j K_j}{\delta}$$

where $D_j$ = diffusion coefficient of species j, $K_j$ = partition coefficient of species j between the membrane phase and water and $\delta$ = the thickness of the membrane.

The value of $K_j$ which is a measure of the concentration of j in the membrane is usually obtained by measuring the partition coefficient between oil and an aqueous phase, based on the high lipid content of membranes. The problems of measuring $P_j$ are fully discussed by Nobel (1974). Generally, cells are less permeable to electrolytes than non-electrolytes, the permeability of membranes to ions is small. Cells are usually much less permeable to divalent cations and anions than to monovalent cations. This low permeability is probably due to the lipid bilayer since it has been shown that small ions do not readily penetrate across uninterrupted bimolecular phospholipid bilayers (Bangham, 1972).

There are at least two ways in which ions can pass through a membrane: by going out of solution on one side, into solution in the membrane and redissolving in the solution on the other side; or by passing through holes or pores which are part of the membrane structure (Hope, 1971). This movement will be influenced by both the properties of the membrane and of the outside solution. In relation to the membrane, permeability will be governed by its thickness, solubility of the ions in the membrane, the nature and size of the charge at the membrane surface and the properties of membrane pores. The presence of pores has been suggested to account for the very high permeability of water, although they seem unlikely to be an important factor in solute permeability. For example, although the rate of permeation to monovalent cations frequently correlates with their hydrated ion diameters, exceptions occur. In most cases, $K^+$ is usually preferred to $Na^+$ but in a few cases, such as the outside of the frog skin, discrimination in favour of $Na^+$ occurs. This suggests that a sieving action due to pores is unlikely to play an important role, since $K^+$ would always be preferred to $Na^+$.

It has become clear that permeability of solutes cannot be predicted simply by a consideration of their size, shape, charge and polarity. It is now generally considered that transport across membranes may be mediated by the presence of a carrier molecule in the membrane. When this mediation is passive, it is called facilitated diffusion.

In recent years, a clue to the molecular basis of this passive mediated transport has come from studies with a group of molecules known as iono-phores; these are surface-acting and hydrocarbon-soluble carriers. Perhaps the most widely studied of these is the antibiotic valinomycin. This and other macrocyclic compounds have been shown to have remarkable effects on the ionic permeabilities of mitochondria (see Ch. 3), synthetic bilayers and red cell membranes; $K^+$ permeability may be increased several thousand-fold, often with little effect on $Na^+$ permeability. As yet, their effects appear to be largely restricted to cations and protons although anion selective molecules have also been found. The range of these compounds and their effects on the permeability of various membrane systems are thoroughly described elsewhere (see Tosteson, 1968; Shemyakin et al., 1969; Kinsky, 1970; Clarkson, 1974). We will limit ourselves here to a brief discussion of the implication of these findings to the problem of membrane permability. It seems well established that these cyclic compounds form complexes with monovalent cations which completely surround the ion; the size of the compound may determine the nature of the cation bound. There seem to be two ways by which they may produce the marked change in membrane permeability (Kinsky, 1970). One possibility is that several of these molecules aggregate to form a physical channel in the membrane, a transient pore which admits cations and small uncharged molecules. Anions would be excluded by electrostatic repulsion due to the presence of carbonyl and ether oxygen groups in the interior making it considerably electronegative. An alternative explanation is that the complex formed with the cation diffuses through the non-polar interior of the lipid bilayer much faster than the ion alone. At present, the weight of evidence strongly favours the first explanation (Kinsky, 1970), although it is not clear how these pores are created. It is interesting to speculate that, since these compounds are structurally somewhat similar to the peptide chains of proteins, they could be similar to the sites in biological membranes responsible for their ionic selectivity (Shemyakin et al., 1969). Finally it should be mentioned that the principles on which ionophores mobilize and select ions may yet be based on Eisenman's general theory of membrane selectivity (Bangham, 1972) which is discussed in the next section.

### 2.8.2 Membrane selectivity

We have seen that biological membranes discriminate between different cations and this leads us to question what property of the membrane leads to this selectivity. If we consider the five alkali cations ($Li^+$, $Na^+$, $K^+$, $Rb^+$ and $Cs^+$), then 120 different sequences of selectivity could occur. However it is found that, in non-living systems such as glass electrodes and ion-exchange resins, only 11 selectivity orders are found (Diamond and Wright, 1969). These are the two extremes

$$Cs^+ > Rb^+ > K^+ > Na^+ > Li^+$$
$$Li^+ > Na^+ > K^+ > Rb^+ > Cs^+$$

and 9 intermediate sequences. Similar selectivity orders have been found in biological systems, which implies that the physical basis of this discrimination is the same.

These observed sequences are those predicted by Eisenman (1962) in his theory of the basis of selectivity in membranes or in a macroscopic phase. The application of this theory to biological systems is lucidly and thoroughly discussed by Diamond and Wright (1969). The principle factor in selectivity of monovalent cations is the field strength of the membrane negative charges. According to Eisenman, selectivity is based on the different attractive forces, mainly Coulombic, exerted on different cations by water and the membrane negative charges. When an ion becomes hydrated or adsorbed to a fixed negative charge there is a decrease in the free energy of the system. The relative affinities for two cations will be determined by the differences in free energy which occur when the ions move from water to the binding site; the ion which experiences the larger decrease in free energy will be preferentially selected. The selectivity pattern of anions and divalent cations can also be predicted by a similar Coulombic model and the theory can be expanded to explain the selectivity of the ionophores discussed earlier.


## 2.9 Membrane potentials

Plant cells typically show a resting electrical potential of about 100 mV or more with the interior negative. Some selected values obtained for higher

Table 2.4.
Some selected values of membrane potentials measured between the vacuole and the external medium. Reproduced from Clarkson (1974).

| Plant material/conditions | Potential difference (mV) | Concentrations of $K^+$ and $Ca^{2+}$ in medium (Mol m$^{-3}$) | |
|---|---|---|---|
| | | $K^+$ | $Ca^{2+}$ |
| *Pisum :* Cells of root cortex | $-110$ | 1.0 | 1.0 |
| Epicotyl | $-119$ | 1.0 | 1.0 |
| *Avena :* Coleoptiles | $-102$ to $-109$ | 1.0 | 1.0 |
| Roots | $-71$ to $-84$ | 1.0 | 1.0 |
| *Zea :* Sections cut from | $-90$ to $-104$ | 1.0 | 0 |
| intact root | $-76$ to $-84$ | 1.0 | 0.1 |
| | $-96$ to $-99$ | 0.1 | 0.1 |
| *Beta vulgaris* (red beet) : | | | |
| Aged disks of storage root | $-153$ | 0.6 | 0 |
| *Vigna sequipedalis* (bean) : | | | |
| Parenchyma cells | $-55$ to $-60$ | 0.1 | 0 |
| Hypocotyls of seedlings | | | |
| *Vicia faba* (broad bean) : | | | |
| Seedling roots | $-130$ | 1.0 | 0.5 |

plant tissues are shown in Table 2.4. These are measured using micro-capillary glass electrodes as described in Section 1.2 and represent measurements for the vacuole relative to the outside solution. It is not usually possible to obtain values across the plasmalemma alone because the cytoplasmic layer is too narrow. Similar values are obtained for algal cells (see Ch. 5). This potential is a result of the interaction of many processes and cannot be evaluated in absolute terms on a strict theoretical basis (Higinbotham, 1973). Action potentials which are induced by concentration gradients can be predicted and accurately measured. The origin of the electrical potential across membranes is probably largely a result of two processes, diffusion and active electrogenic transport. For example, if the two ions of a salt have different mobilities in membrane, then the unequal distribution of an ion across the membrane can give rise to a potential difference across that membrane. The application of the Goldman equation (1.19) in the interpretation of membrane potential differences is discussed in Ch. 1. The relative permeabilities of a membrane to various ions can be measured by observing the behaviour of the potential when various ionic species are substituted in the medium. This has been thoroughly discussed by Hope (1971). In addition to this diffusional process, potentials can arise as a result of active electrogenic transport; this can be defined as active transport in which a net charge is transferred across a membrane at the expense of metabolic energy (Higinbotham, 1973). There is increasing evidence that electrogenic ion pumps are widely distributed in plant and animal tissues. Membrane potentials probably arise largely as the sum of these two processes.


## Acknowledgement

We should like to thank Dr. K.P. Wheeler for his critical reading of the manuscript.


## References

R.W. ALBERS, G.J. KOVAL and G.T. SIEGEL, Mol. Pharmacol., 4 (1968) 324.

A.D. BANGHAM, Annu. Rev. Biochem., 41 (1972) 753.

R.S. BAR, D.W. DEAMER and D.G. CORNWELL, Science, 153 (1966) 1010.

D.A. BAKER and J.L. HALL, New Phytol., 72 (1973) 1281.

A.A. BENSON, J. Am. Oil Chem. Soc., 48 (1966) 265.

H.M. BERMAN, W. GRAM and M.A. SPIRTES, Biochim. Biophys. Acta, 183 (1969) 10.

C.C. BLEDSOE, V. COLE and C. ROSS, Plant Physiol., 44 (1969) 1040.

W. BOOS, Annu. Rev. Biochem., 43 (1974) 123.

P.D. BOYER and W.L. KLEIN, in C.F. Fox and A. Keith (Eds.) Membrane Molecular Biology (1972) Sinauer Assoc. Inc., Stamford, Conn., p. 323.

D. BRANTON, Annu. Rev. Plant Physiol., 20 (1969) 209.

D. BRANTON, in L.A. Manson (Ed.) Biomembranes (1971) Vol. 2, Plenum Press, p. 3.

D. BRANTON and D.W. DEAMER, Membrane Structure (1972) Springer Verlag, Vienna.

M.S. BRETSCHER, Nature New Biol., 231 (1971) 229.

D.M. BRUNETTE and J.E. TILL, J. Membrane Biol., 5 (1971) 215.

D.T. CLARKSON, Ion Transport and Cell Structure in Plants (1974) McGraw–Hill Co. (U.K.) Ltd., London.

G.M.W. COOK, Annu. Rev. Plant Physiol., 22 (1971) 97.

J.L. DAHL and L.E. HOKIN, Annu. Rev. Biochem., 43 (1974) 327.

G. DALLNER and L. ERNSTER, J. Histochem. Cytochem., 16 (1968) 611.

J.F. DANIELLI and H. DAVSON, J. Cell. Comp. Physiol., 5 (1935) 495.

C. DE DUVE, in D.B. Roodyn (Ed.) Enzyme Cytology (1967) Academic Press, New York, p. 1.

B. DE KRUYFF, R.A. DEMEL and L.L.M. VAN DEENAN, Biochim. Biophys. Acta, 255 (1972) 331.

J.W. DE PIERRE and M.L. KARNOVSKY, J. Cell Biol., 56 (1973) 275.

J.M. DIAMOND and E.M. WRIGHT, Annu. Rev. Physiol., 31 (1969) 581.

J.J.A. DODDS and R.J. ELLIS, Biochem. J., 100 (1966) 31 p.

M.L. EDWARDS and J.L. HALL, Protoplasma, 78 (1973) 321.

G. EISENMAN, Biophys. J., 2 (1962) 259.

E. EPSTEIN, Mineral Nutrition of Plants : Principles and Perspectives (1972) Wiley, New York.

L. ERNSTER and B. KUYLENSTIERNA, in E. Racker (Ed.) Membranes of Mitochondria and Chloroplasts (1969) Van Nostrand Reinhold, New York, p. 172.

J.B.R. FINEAN, S. COLEMAN, A. KNUTTON, R. LIMBRICK and J.E. THOMPSON, J. Gen. Physiol., 51 (1968) 198.

J.D. FISHER and T.K. HODGES, Plant Physiol., 44 (1969) 385.

J.D. FISHER, D. HANSEN and T.K. HODGES, Plant Physiol., 46 (1970) 812.

C.F. FOX and A. KEITH, Membrane Molecular Biology (1972) Sinauer Assoc. Inc., Stamford, Conn.

C.F. FOX and E.P. KENNEDY, Proc. Natl. Acad. Sci. U.S., 54 (1965) 891.

Y. FUKUSHIMA and Y. TONOMURA, J. Biochem., 74 (1973) 135.

D. GINGELL, J. Theor. Biol., 38 (1973) 677.

E. GORTER and F. GRENDEL, J. Exp. Med., 41 (1925) 439.

D.E. GREEN and J.F. PERDUE, Proc. Natl. Acad. Sci. U.S., 55 (1966) 1295.

N. GREUNER and J. NEUMANN, Physiol. Plant., 19 (1966) 678.

G. GUIDOTTI, Annu. Rev. Biochem., 41 (1972) 731.

J.L. HALL, J. Exp. Bot., 23 (1971) 800.

J.L. HALL, in W.P. Anderson (Ed.) Ion Transport in Plants (1973) Academic Press, London, p. 11.

J.L. HALL and V.S. BUTT, J. Exp. Bot., 20 (1969) 751.

J.L. HALL and R.M. ROBERTS, Ann. Bot., (1975) in press.

G. HANSSEN and A. KYLIN, Z. Pflanzenphysiol., 60 (1969) 270.

R.W. HENDLER, Physiol. Rev., 51 (1971) 66.

F.A. HENN and J.E. THOMPSON, Annu. Rev. Biochem., 38 (1969) 241.

N. HIGINBOTHAM, Annu. Rev. Plant Physiol., 24 (1973) 25.

B.S. HILL and A.E. HILL, J. Membrane Biol., 12 (1973) 145.

T.K. HODGES, R.T. LEONARD, C.E. BRACKER and T.W. KEENAN, Proc. Natl. Acad. Sci. U.S., 69 (1972) 3307.

G. HOLZWARTH, in C.F. Fox and A.D. Keith (Eds.) Membrane Molecular Biology (1972) Sinauer Assoc. Inc., Stamford, Conn., p. 228.

A.B. HOPE, Ion Transport and Membranes (1971) Butterworths, London.

A.F. HORWITZ, in C.F. Fox and A.D. Keith (Eds.) Membrane Molecular Biology (1972) Sinauer Assoc. Inc., Stamford, Conn., p. 164.

R.L. JULIANO, Biochim. Biophys. Acta, 300 (1973) 341.

H.R. KABACK, Annu. Rev. Biochem., 39 (1970) 561.

A.D. KEITH and R.J. MELHOURN, in C.F. Fox and A.D. Keith (Eds.) Membrane Molecular Biology (1972), Sinauer Assoc. Inc., Stamford, Conn., p. 117.

A.D. KEITH, M. SHARNOFF and G.E. COHN, Biochim. Biophys. Acta, 300 (1973) 379.

F.J. KÉZDY, in C.F. Fox and A.D. Keith (Eds.) Membrane Molecular Biology (1972) Sinauer Assoc. Inc., Stamford, Conn., p. 123.

E.D. KIEHN and J.J. HOLLAND, Biochemistry, 9 (1970a) 1716.

E.D. KIEHN and J.J. HOLLAND, Biochemistry, 9 (1970b) 1729.

H.K. KIMELBERG and D. PAPAHADJOPOULOS, Biochim. Biophys. Acta, 233 (1971) 805.

S.C. KINSKY, Annu. Rev. Pharmacol., 10 (1970) 119.

E.D. KORN, J. Gen. Physiol., 52 (1968) 257.

E.D. KORN, Annu. Rev. Biochem., 38 (1969) 263.

A. KYLIN and R. GEE, Plant Physiol., 45 (1970) 169.

Y.F. LAI and J.E. THOMPSON, Biochim. Biophys. Acta, 233 (1971) 84.

Y.F. LAI and J.E. THOMPSON, Can. J. Bot., 50 (1972a) 327.

Y.F. LAI and J.E. THOMPSON, Plant Physiol., 50 (1972b) 452.

J.H. LAW and W.R. SNYDER, in C.F. Fox and A.D. Keith (Eds.) Membrane Molecular Biology (1972), Sinauer Assoc. Inc., Stamford, Conn., p. 3.

C.A. LEMBI, D.J. MORRÉ, K. ST.–THOMSON and R. HERTEL, Planta, 99 (1971) 37.

J. LENARD and S.J. SINGER, Proc. Natl. Acad. Sci. U.S., 56 (1966) 1828.

R.T. LEONARD, D. HANSEN and T.K. HODGES, Plant Physiol., 51 (1973) 749.

R.T. LEONARD and J.B. HANSON, Plant Physiol., 49 (1972a) 430.

R.T. LEONARD and J.B. HANSON, Plant Physiol., 49 (1972b) 436.

L.J. LITTLEFIELD and C.E. BRACKER, Protoplasma, 74 (1972) 271.

U. LÜTTGE, W.J. CRAM and G.G. LATIES, Z. Pflanzenphysiol., 64 (1971) 418.

I.R. MACDONALD, J.S.D. BACON, D. VAUGHAN and R.J. ELLIS, J. Exp. Bot., 17 (1966) 822.

E.A.C. MACROBBIE, Annu. Rev. Plant Physiol., 22 (1971) 75.

W.C. MCMURRAY and W.L. MAGEE, Annu. Rev. Biochem., 41 (1972) 129.

L.A. MANSON, Biomembranes (1971) Vol. 1, Plenum Press, New York.

R.J. MELHOURN and A.D. KEITH, in C.F. Fox and A.D. Keith (Eds.) Membrane Molecular Biology (1972) Sinauer Assoc. Inc., Stamford, Conn., p. 192.

J.C. METCALFE, in D.F.H. Wallach and H. Fischer (Eds.) The Dynamic Structure of Cell Membranes (1971) H. Fischer, Springer Verlag, Berlin.

D.M. NEVILLE, J. Biophys. Biochem. Cytol., 8 (1960) 413.

P.S. NOBEL, An Introduction to Biophysical Plant Physiology (1974) W.H. Freeman and Co., San Francisco.

L. ORCI and A. PERRELET, Science, 181 (1973) 868.

A.R. OSEROFF, P.W. ROBBINS and M.M. BURGER, Annu. Rev. Biochem., 42 (1973) 1973.

D.L. OXENDER, Annu. Rev. Biochem., 41 (1972) 777.

D. PAPAHADJOPOULOS, Biochim. Biophys. Acta, 241 (1971) 254.

A.B. PARDEE, Science, 162 (1968) 632.

A.B. PARDEE and L.S. PRESTIDGE, Proc. Natl. Acad. Sci. U.S., 55 (1966) 189.

G.M. POLYA and M.R. ATKINSON, Aust. J. Biol. Sci., 22 (1969) 573.

G.K. RADDA and J. VANDERKOOI, Biochim. Biophys. Acta, 265 (1972) 509.

P.M. RAY, T.L. SHININGER and M.M. RAY, Proc. Natl. Acad. Sci. U.S., 64 (1969) 605.

J.D. ROBERTSON, in J.M. Allen (Ed.) Molecular Organization and Biological Function (1967) Harper and Row, New York, p. 65.

R.N. ROBERTSON, Protons, Electrons, Phosphorylation and Active Transport (1968) Cambridge University Press, U.K.

J.M. RUNGIE and J.T. WISKICH, Plant Physiol., 51 (1973) 1064.

M.M. SHEMYAKIN, et al., J. Membrane Biol., 1 (1969) 402.

A. SCHWARTZ, G.E. LINDENMAYER and J.C. ALLEN, in F. Bronner and A. Kleinzeller (Eds.) Current Topics in Membranes and Transport (1972) Academic Press, New York.

E.W. SIMON, New Phytol., 73 (1974) 377.

R.D. SIMONI, in C.F. Fox and A.D. Keith, Membrane Molecular Biology (1972) Sinauer Assoc. Inc., Stamford, Conn., p. 289.

S.J. SINGER and G.L. NICOLSON, Science, 175 (1972) 720.

S.J. SINGER, Annu. Rev. Biochem., 43 (1974) 805.

C.L. SLAYMAN, W.S. LONG and C.Y-H. LU, J. Membrane Biol., 14 (1973) 305.

T.L. STECK, in C.F. Fox and A.D. Keith (Eds.) Membrane Molecular Biology (1972) Sinauer Assoc. Inc., Stamford, Conn., p. 76.

T.L. STECK and C.F. FOX, in C.F. Fox and A.D. Keith (Eds.) Membrane Molecular Biology, Sinauer Assoc. Inc., Stamford, Conn., (1972), p. 27.

T.L. STECK and D.F.H. WALLACH, Meth. Cancer Res., 5 (1970) 93.

G. SZABO, in C.F. Fox and A.D. Keith (Eds.) Membrane Molecular Biology (1972) Sinauer Assoc. Inc., Stamford, Conn., p. 146.

D.C. TOSTESON, Fed. Proc., 27 (1968) 1269.

W.J. VAN DER WOUDE, C.A. LEMBI and D.J. MORRÉ, Biochem. Biophys. Res. Commun., 46 (1972) 245.

D.F.H. WALLACH, The Plasma Membrane Dynamic Perspectives, Genetics and Pathology (1972) English Universities Press Ltd., London.

D.F.H. WALLACH and P.S. LIN, Biochim. Biophys. Acta, 300 (1973) 211.

D.F.H. WALLACH and P.H. ZAHLER, Proc. Natl. Acad. Sci. U.S., 56 (1966) 1552.

L. WARREN and M.C. GLICK, in L.A. Manson (Ed.) Biomembranes (1971) Vol. 1, Plenum Press, New York, p. 257.

K.P. WHEELER and R. WHITTAM, J. Physiol., 207 (1970) 303.

R. WHITTAM and K.P. WHEELER, Annu. Rev. Physiol., 32 (1970) 21.

F.A. WILLIAMSON and R.G. WYN JONES, in Isotopes and Radiation in Soil–Plant Relationship including Forestry (1972) International Atomic Energy Agency, Vienna, p. 69.

H.P. ZINGSHEIM, Biochim. Biophys. Acta, 265 (1972) 339.

*Ion transport in plant cells and tissues*
*edited by D.A. Baker and J.L. Hall*
© *North-Holland Publishing Company, 1975*



# CHAPTER 3

# Mitochondria

J.B. HANSON and D.E. KOEPPE

## Contents



## 3.1  Introduction

Numerous biochemical reaction in plants rely on mitochondria as an energy source or for carbon moieties for degradative or synthetic reactions. Such a role requires transport of organic and inorganic compounds between mitochondria and the cytosol in which they are located. This transport can be divided into three components involving: (1) production of ATP, (2) biogenesis, growth, and self-maintenance of the organelle and (3) the production and/or transport of metabolites as part of intermediary metabolism. Of these, the production of ATP through oxidative phosphorylation and its subsequent transport to the cytosol is the most fundamental and unique property, but maintaining ATP production is also dependent on the other two components. Practically nothing is known of membrane transport processes in the latter categories.

Because of the primacy of the process of oxidative phosphorylation in cellular function, the mechanisms for the production of ATP have been studied more than any other mitochondrial process. However, there is still considerable uncertainty as to the actual mechanism of phosphorylation. The transport processes involved can be summed as follows:

$$\text{pyruvate}_{in} + 2.5 \; O_{2 \; in} \rightarrow 3 \; CO_{2 \; out} + 2 \; H_2O_{out} \; \text{(oxidation)}$$
$$P_{i_{in}} + ADP_{in} \rightarrow ATP_{out} + H_2O \; \text{(phosphorylation)}$$

In addition to inorganic phosphate ($P_i$), $Mg^{2+}$ is a necessary co-factor for the ATPase involved in phosphorylation. Potassium ions are also always found in high concentrations in plant mitochondria and are undoubtedly essential to oxidation or phosphorylation. There must be a transport mechanism for these ions, and probably for dissociable co-factors such as $NAD^+$.

Proton transport is now considered to be directly or indirectly essential to oxidative phosphorylation. Obtaining and maintaining a proton gradient requires that mitochondrial membranes be impermeable to passive proton diffusion, and that they possess mechanisms for controlled proton transport (Mitchell, 1972).

While a few studies of transport have been carried out with plant mitochondria, by far the majority have utilized mitochondria isolated from rat liver or beef heart. Generally speaking, mitochondria isolated from plants are functionally similar and have comparable transport properties to those from animal tissues. Plant mitochondria do seem, however, to be more permeable to salts, especially halides, and they oxidize NADH from the outer side of the inner membrane (Douce et al., 1973) which animal mitochondria do not. NADPH as well as NADH can serve as substrate (Koeppe and Miller, 1972). They frequently fail to have an uncoupler-stimulated ATPase, which appears to be due to a lesion in the adenine nucleotide transport system (Jung and Hanson, 1973). Plant mitochondria also lack the dramatic uncoupling reaction to $Ca^{2+}$ typical of vertebrate mitochondria (Hanson and Hodges, 1967).

In what follows, specific attention is given to transport in plant mitochondria, drawing on hypotheses and conclusions made for animal mitochondria.


## 3.2 *Structure relating to transport phenomena*

Mitochondria consist of inner and outer membranes, bounding an intermembrane (cristae) space as an outer compartment, and a matrix or inner compartment (Fig. 3.1). The outer membrane is permeable to most compounds

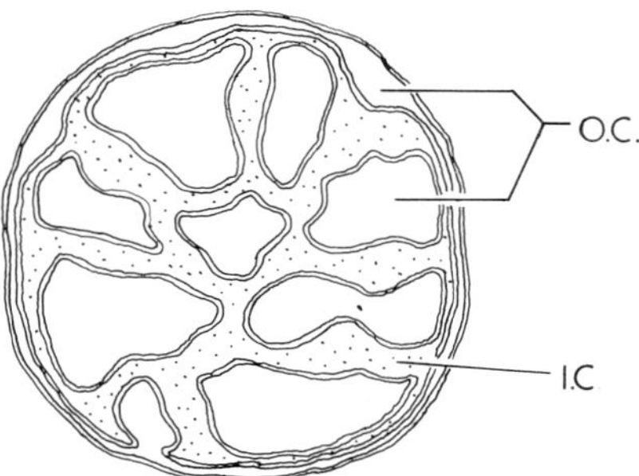


Fig. 3.1. Diagram of a cross section of a corn mitochondrion, Type III. O.C., outer compartment delimited by the outer and inner membranes (intermembrane or cristae space). I.C., inner compartment (matrix) bounded by the inner membrane. From Malone et al. (1974).

of less than 10,000 molecular weight (Ernster and Kuylenstierna, 1970) and is freely permeable to sucrose. Electron micrographs of plant outer membranes suggest that they have pits (Parsons et al., 1965), although these may not be present in all species. There have been no specific studies of transport through the outer membrane, and it is generally ignored as a solute barrier. Bean mitochondria lacking outer membranes show good respiratory control and potassium phosphate uptake (Wilson et al., 1973).

Phospholipid and fatty acid composition have been determined for the inner and outer membranes of cauliflower mitochondria (Moreau et al., 1974). Both inner and outer membranes have high proportions of phosphatidylcholine and phosphatidylethanolamine, but differ in the relative enrichment of outer membranes in phosphatidylinositol and phosphatidylglycerol, and the inner membranes in diphosphatidylglycerol. Inner membranes have a much higher proportion of unsaturated fatty acids than outer membranes, which Moreau et al. suggest may contribute to inner membrane flexibility in swelling and contraction. The inner membrane is relatively high in proteins compared to the outer.

The inner membrane is osmotically responsive and impermeable to sucrose, many ions and protons. The electron transport chain is tightly associated with the membrane as is succinate dehydrogenase. It is questionable if a transhydrogenase exists in plant mitochondria as no NADPH to NAD$^+$ electron transfer could be found in corn mitochondria (Koeppe and Miller, 1972). The coupling ATPase is readily observable as stalked particles on the inner face of the inner membrane of negatively stained mitochondria (Parsons et al., 1965), but not in freeze-etched (Tewari et al., 1972).

On the basis of transport studies (see later) the inner membrane is assumed to possess specific exchange carriers or porters. None of these have been isolated and identified, but they are presumably integral proteins embedded in the lipid phase as proposed in the fluid mosaic model of Singer and Nicolson (1972) (see section 2.5).

The matrix is enclosed by the convoluted inner membrane and contains enzymes for the complete oxidation of pyruvate as well as the mitochondrial DNA and RNA. As major solutes, the matrix has catalytic amounts of the Krebs cycle acids, $K^+$, $Mg^{2+}$, phosphate and adenine nucleotides. For freshly isolated corn mitochondria a typical analysis shows in nMol/mg protein: $K^+ = 139$; $Mg^{2+} = 44$; $P_i = 19$; adenine nucleotides $= 7$ (Jung, 1974). Insoluble salts of $Sr^{2+}$ and $Ca^{2+}$ can be identified in electron micrographs as deposited in the matrix (Mitchell et al., 1973; Peverly et al., 1974).

The proportion of matrix to intermembrane space is affected by ion fluxes across the semipermeable inner membrane driving osmotic flux of water. The magnitude of the morphological change induced by these fluxes between mitochondrial compartments is variable. Morphology both in vivo and in vitro varies not only between species, but with different tissues of the same plant as well. Malone et al. (1974) have described three mitochondrial types observed in vivo and in vitro in electron micrographs of shoots of three-day-old etiolated corn seedlings (Fig. 3.2). Meristematic tissue always contained mitochondria with a matrix density intermediate (type II) to that found in differentiated cells, which either contained a very dense matrix (type III; e.g., parenchyma cells) or a less dense matrix (type I; e.g., phloem companion cells). In vitro, the greatest expansion of the matrix in response to salt fluxes is observed in type III mitochondria, with the least in type I.


## 3.3 Passive ion permeation

As noted above, the outer mitochondrial membrane is freely permeable to most physiologically significant compounds. However, it is possible that the rapid and dramatic conformational and phosphate swelling responses to polycations (e.g., protamine, molecular weight approx. 8000; Hanson, 1972) might originate in binding to the outer membrane with indirect effects on the inner membrane. But this is not known, and we will here consider transport in terms of movement across the inner membrane only.

As with animal mitochondria, plant mitochondria have been shown to behave as nearly perfect osmometers over a range of $-1$ to $-10$ bars osmotic potential (Yoshida and Sato, 1968; Lorimer and Miller, 1969). Mitochondria isolated from several plant species swell passively when transferred into a

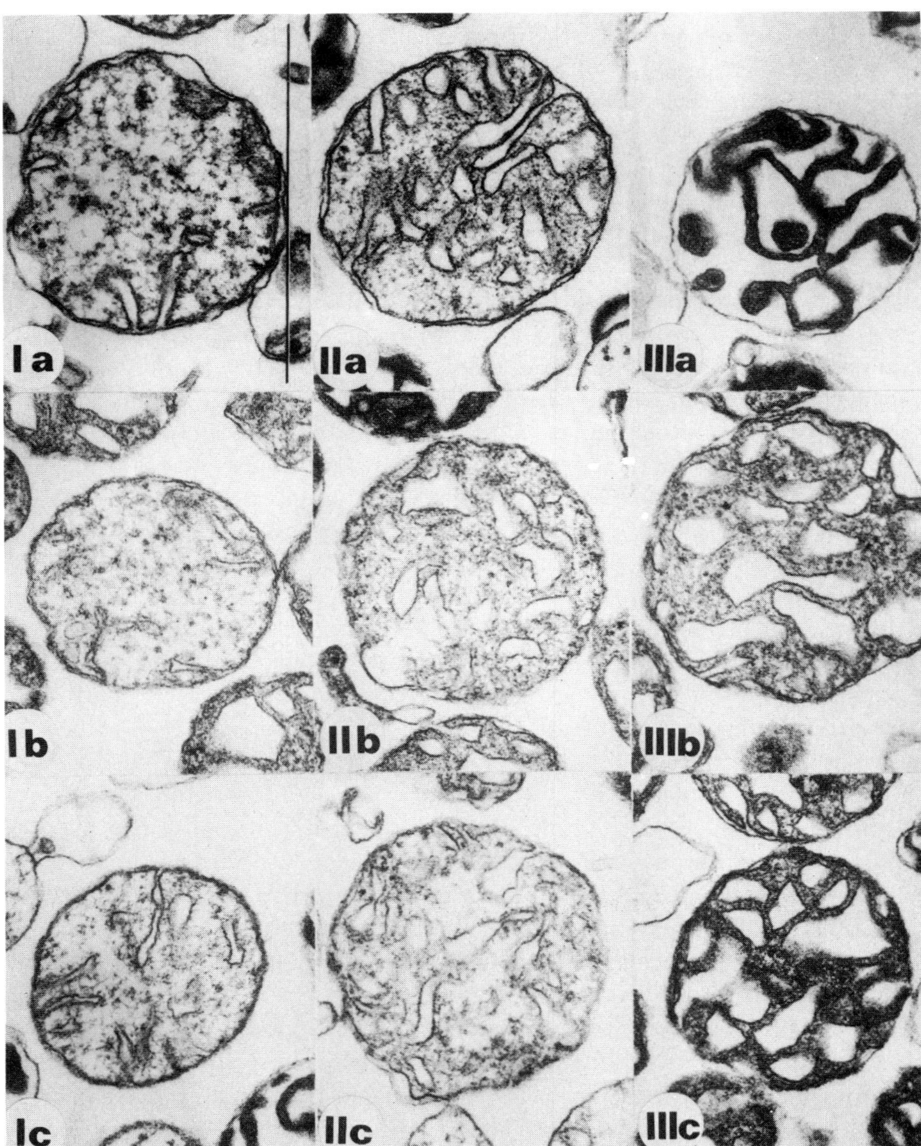



Fig. 3.2. Relative changes in the volume of outer and inner compartments of Types I, II and III corn mitochondria. Vertical files are Type I, II and III, left to right. Horizontal rows are (a) osmotically contracted in 400 Mol m$^{-3}$ sucrose, (b) passively swollen in 200 Mol m$^{-3}$ KCl, (c) actively contracted after swelling in KCl. Only Types II and III show swelling and contraction of the inner compartment (matrix space) at the expense of the outer compartment (cristae space). From Malone et al. (1974).

KCl or ribose medium from the sucrose isolation medium (Lyons et al., 1964; Longo and Arrigoni, 1964; Stoner et al., 1964; Stoner and Hanson, 1966; Earnshaw and Truelove, 1968; Yoshida and Sato, 1968; Lorimer and Miller, 1969; Overman et al., 1970). This swelling, as observed by percent transmittance or absorbancy measurements, is due to the diffusive movement of the solute into the inner (matrix) compartment accompanied by water to maintain osmotic equilibrium. Lorimer and Miller (1969) and Overman et al. (1970) have demonstrated the close correspondence between diffusion-controlled osmotic swelling as measured by percent transmission of mitochondria suspension, and osmotic theory. Given undamaged mitochondria and sufficient non-permeant solute to prevent lysis, light absorbancy is a simple and adequate measurement of solute flux in and out of the matrix.

Isolated corn mitochondria retain impermeability to sucrose unless they have been aged and treated with citrate or EDTA (Stoner and Hanson, 1966), thus indicating a role for divalent cations in maintaining the permeability barrier. If BSA is present, swelling in KCl does not appear to impair the efficiency of oxidative phosphorylation (Stoner and Hanson, 1966; Earnshaw and Truelove, 1968). The diffusive entry of KCl in swelling is highly pH sensitive, increasing rapidly with increasing pH (Stoner and Hanson, 1966; Yoshida, 1968a). In liver mitochondria this increased permeability at high pH is attributed to increased $Cl^-$ permeability (Azzi and Azzone, 1967a). At pH 9.5 corn mitochondria become permeable to ATP (Jung and Hanson, 1973).

As indicated above, it is the spontaneous influx of the anion down the concentration gradient which is affected by pH, and apparently drives the swelling. Cations penetrate down the electrical gradient established by anion influx, maintaining neutrality. Introduction of a $K^+$-binding, lipid-soluble ionophore such as gramicidin or valinomycin (Pressman, 1965) reduces resistance to $K^+$ entry and greatly accelerates the rate of swelling of plant mitochondria at neutral pH (Yoshida, 1968a; Miller et al., 1970; Wilson et al., 1972). This demonstrates the rate limiting factor is cation penetration. Yoshida and Sato (1968) using castor bean mitochondria at pH 7.4 compared swelling in chloride salts (150 Mol m$^{-3}$ monovalent, 100 Mol m$^{-3}$ divalent) and found rates to increase in the order $Ca^{2+} < Mg^{2+} \ll NH_4^+ < Li^+ < Na^+ < K^+$. With $K^+$ salts of various anions the order was $F^- < Cl^- < Br^- < NO_3^- < I^- < CNS^-$, the order of lipophilicity.

Sites of diffusive penetration of KCl and other salts are unknown, but are probably through lipid domains of the membrane. Lyons et al. (1964) reported the passive swelling of mitochondria from chilling-resistant plants was greater than that from chilling-sensitive plants; the former had a higher content of unsaturated fatty acids. Ethylene and carbon tetrachloride promote swelling of cauliflower mitochondria in KCl (Lyons and Pratt, 1964). Addition of oleic acid increases swelling of bean mitochondria in KCl and BSA is protective (Earnshaw et al., 1970). A detergent action is suspected since oleic acid and phospholipase A induce rapid swelling in sucrose (Earnshaw et al., 1970; Earnshaw and Truelove, 1970). Calcium retards and uncouplers promote swelling in KCl (Hanson et al., 1965; Yoshida, 1968a). Yoshida (1968a) attributes the action of uncoupling agents to increased permeability.

Lee and Wilson (1972) studied the swelling induced in bean mitochondria by various organic acids. The mitochondria were osmotically supported by mannitol (which penetrates castor bean mitochondria – Yoshida and Sato, 1968) and endogenous respiration was blocked with rotenone. To varying degrees passive swelling was found with 20 Mol m$^{-3}$ malate, pyruvate, $\beta$-hydroxybutyrate, malonate, maleate, glutarate, $\alpha$-ketoglutarate, citrate, succinate, acetate and propionate. Swelling increased when NADH was supplied as substrate (see next section). Lee and Wilson (1972) conclude that the poor respiration rates attained with some substrates (pyruvate, citrate) could not be due to limitations on entry.

It is not known to what degree passive transport enters into mitochondrial metabolism in vivo. It is important to note that most diffusive transport with isolated mitochondria is secured by setting up unnaturally high concentration gradients.

## 3.4 *Energy linkage for transport*

As indicated in the foregoing section, intact mitochondria possess a high degree of resistance to passive diffusion of ions down concentration gradients. However, they do exhibit high rates of water flux ($5.3 \times 10^{-5}$ m s$^{-1}$, Massari et al., 1972a) and, due to their high surface: volume ratio, come to almost instantaneous osmotic equilibrium. Hence, salt transport in or out of the matrix is reflected in osmotic volume changes which prove to be readily monitored by light absorption. After some earlier confusion as to the mechanism of energy-linked volume changes, it is now almost universally accepted that reversible swelling or contraction driven by ATP hydrolysis or respiration is due to salt transport (Chappell and Crofts, 1965; Azzi and Azzone, 1967b; Rottenberg and Solomon, 1969; Izzard and Tedeschi, 1973; Massari et al., 1972b). This may not apply to low amplitude conformation and/or volume changes associated with State 4 – State 3 transition (Hackenbrock et al., 1971); in corn mitochondria these can be seen in the light transmission trace on addition of ADP (Hanson et al., 1972).

A great deal of experimentation with animal mitochondria has also led to wide (but not universal) acceptance of the hypothesis that the energy-linked transfer of ions is based on some mechanism which augments the electro-chemical gradient of protons across the inner membrane (Mitchell, 1966, 1968; Chappell and Haaroff, 1967; Robertson, 1968; Brierley, 1970; Brierley et al., 1971, 1973; Chance and Montal, 1971; Skulachev, 1971). The proton gradient is consumed in salt transport. It is necessary to bear the essentials of this hypothesis in mind to understand the significance of observations made on ion transport in plant mitochondria.

'In simplified form, Mitchell's chemiosmotic hypothesis (1966, 1968) proposes the following (Fig. 3.3):

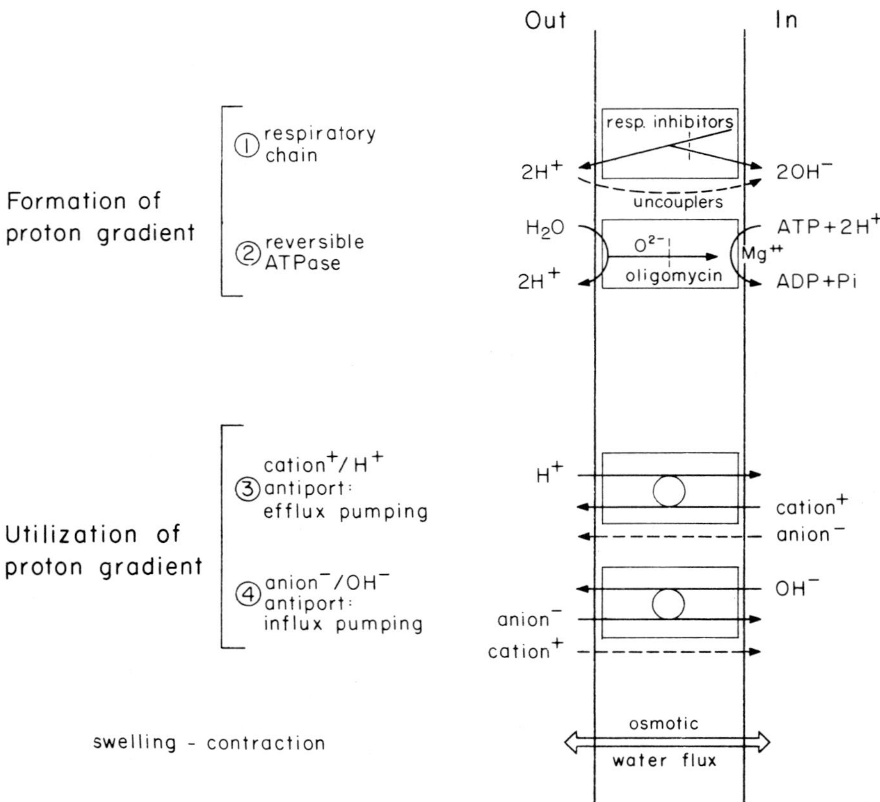


Fig. 3.3. Chemiosmotic hypothesis for formation of proton gradients across the inner mito-chondrial membrane, and utilization of the resulting protonmotive force in ion transport. See text.

1. The inner mitochondrial membrane is impermeable to protons and most ions.

2. The respiratory electron transport chain is coupled to proton extrusion (reaction 1).

3. Oxidative phosphorylation is accomplished by a reversible, proton-conducting ATPase. Hydrolysis of ATP by the ATPase can thus produce a proton gradient (reaction 2).

4. Ion transport is carried out by exchange carriers or 'antiports' which exchange $H^+$ or $OH^-$ for cations or anions (reactions 3 and 4).

5. Energy for exchange transport lies in the proton motive force (PMF), or the electrochemical gradient of protons, which sets the potential with which $H^+$ or $OH^-$ enters into exchange. In electrical units the PMF is (Mitchell, 1968, p. 24):

$$PMF = \Delta\Psi - 2.3\frac{RT}{F}\Delta pH = \Delta\Psi - 59\Delta pH \text{ (millivolts at 25 °C)} \quad (3.1)$$

It is important to this hypothesis that the dissociation constant for water is extremely low ($10^{-14}$) so that exchange transport of $H^+$ (or $OH^-$) for cations (or anions) essentially consumes the proton gradient, translating it into an electrochemical gradient of ions. Since the exchange is neutral, the electrical potential is unaltered and continued flux of an ion in exchange for protons requires collapse of the electrical potential by transport of a counter ion (gegenion) down the electrical gradient (dotted lines, Fig. 3.3). Herein lies the mechanism of salt transport. It should be noted that the hypothesis provides for both efflux (cation$^+$/$H^+$ antiporter, reaction 3, Fig. 3.3) and influx (anion$^-$/ $OH^-$ antiporter, reaction 4) transport of salt, each exchange accompanied by transport of the gegenion to maintain electrical neutrality. The membrane must therefore be permeable to the gegenion under conditions of high electrical potential difference. (Of course, an alternative to net salt transport would be back leakage of the exchange ion, wasting energy in cyclic transport.)

It should also be noted with respect to salt transport that any mechanism for producing an electrochemical gradient of protons will suffice. For example, a $K^+$ extrusion mechanism coupled to respiration or ATP hydrolysis could produce an electrochemical gradient for protons (i.e., $\Delta\Psi$ in Eq. 3.1), and part of the $K^+$ gradient might be converted into a proton gradient ($\Delta pH$ in Eq. 3.1) by a $H^+$/$K^+$ antiport.

Other models involving cations pumps (Pressman and Haynes, 1969; Moore, 1971; Izzard and Tedeschi, 1973) and $K^+$/$H^+$ exchange pumps

(Massari and Azzone, 1970) have been proposed. Rottenberg (1973) rejects these on the grounds that anion distribution in uptake studies is in accord with $\Delta pH$ across the membrane, and that $K^+$ or $Na^+$ transport are passive processes towards electrochemical equilibrium. Mitchell (1973) has discussed cation-translocating ATPase models in connection with the $(Na^+, K^+)$-ATPase of animal cell membranes.


## 3.5 *Efflux transport*


Efflux transport of salt from plant mitochondria was first observed in the energized contraction of mitochondria passively swollen in salt solutions (Lyons and Pratt, 1964; Longo and Arrigoni, 1964; Stoner et al., 1964; Stoner and Hanson, 1966; Yoshida, 1968a). The standard medium for such studies has been 100–200 Mol m$^{-3}$ KCl, with spontaneous swelling and active contraction followed by light absorbancy, generally at 520 nm. In one instance (Kirk and Hanson, 1973), it has been shown that energy-linked contraction is associated with $K^+$ and water loss from the matrix space (i.e., sucrose-inaccessible space). Simple gravimetric (Stoner and Hanson, 1966) or volume (Yoshida, 1968a) measurements confirm that there is water uptake during swelling and expulsion during contraction. Electron micrographs reveal that spontaneous swelling in KCl produces distension of the matrix compartment, and energy-linked contraction shrinks this compartment (Stoner and Hanson, 1966; Yoshida, 1968a). Of the 3 types of mitochondria obtained from entire corn seedling shoots, only Types II and III exhibit changes in cristae size or matrix density on swelling and contraction (Malone et al., 1974).

Energy for efflux pumping of KCl can be provided by ATP hydrolysis (Lyons and Pratt, 1964; Longo and Arrigoni, 1964; Stoner et al., 1964; Stoner and Hanson, 1966) or by respiration (Stoner et al., 1964; Stoner and Hanson, 1966; Yoshida, 1968a; Earnshaw and Truelove, 1968). Added $Mg^{2+}$ is required for maximum activity with ATP as substrate. Oligomycin, which inhibits the coupling ATPase complex, blocks ATP-powered contraction and permits reswelling. Respiratory inhibitors are ineffective with the ATP-system. Yoshida (1968a) had some difficulty demonstrating ATP-driven contraction in castor bean mitochondria. Subsequent investigation (Takeuchi et al., 1969) revealed these mitochondria to have little uncoupler-stimulated, oligomycin-sensitive ATPase except in the presence of respiratory substrate. Cauliflower mitochondria also lack ATP-driven contraction and uncoupler-stimulated ATPase unless 'primed' by a brief period of respiration, or 'self-primed' by prolonged exposure to ATP (Jung and Hanson, 1973; Jung, 1974). In this case the lesion appears to be in ATP/ADP exchange with the matrix, which is corrected by 'priming'.

Respiration-driven contraction is blocked by respiratory chain inhibitors, such as cyanide, but is not inhibited by the phosphorylation inhibitors oligomycin, rutamycin or aurovertin (Stoner and Hanson, 1966; Truelove and Hanson, 1966; Yoshida, 1968a). Yoshida (1968a) found an alternate electron acceptor, ferricyanide, would permit succinate oxidation and contraction in cyanide-blocked mitochondria (i.e., electron transport through a single coupling site, or 'loop', is adequate to drive contraction).

Uncouplers such as DNP or FCCP are more effective in inhibiting ATP-driven contraction than respiration-driven contraction (Stoner and Hanson, 1966). In castor bean mitochondria, Yoshida (1968a) found good uncoupling of contraction with those uncouplers which would accelerate passive swelling, a result he attributes to increased permeability. With corn mitochondria, concentrations of uncoupler optimal for releasing respiration with succinate actually slightly increased contraction (Stoner et al., 1964; Stoner and Hanson, 1966), and only with higher concentrations was inhibition obtained. More recent work (Hensley and Hanson, 1975) suggests that access to the $H^+/cation^+$ antiporter (reaction 3, Fig. 3.3) by $H^+$ and $K^+$ is restricted by a lipid layer. Facilitating penetration of $H^+$ and $K^+$ to the antiporter by DNP and valinomycin, respectively, tends to increase the rate of contraction.

Energy-linked contraction is only secured with electrolytes, never with penetrating non-polar solutes, such as ribose or mannitol (Lorimer and Miller, 1969; Yoshida, 1968a). Contraction in castor bean mitochrondria can be secured with the monovalent cations $K^+$, $Na^+$, $Li^+$, $NH_4^+$ and $Tris^+$, and the monovalent anions $Cl^-$, $Br^-$, $I^-$, $F^-$, $NO_3^-$ and $CNS^-$, the latter being highly membrane permeable and not very effective (Yoshida, 1968a). Contraction is not secured with acetate, sulphate or phosphate; these anions produce active influx pumping (see next section) and thus oppose efflux pumping and contraction (Stoner and Hanson, 1966; Truelove and Hanson, 1966; Yoshida, 1968a).

Kirk and Hanson (1973) have determined efflux rates for contraction in KCl with NADH as substrate. At an NADH oxidation rate of 174 nMol min$^{-1}$ mg protein$^{-1}$ the efflux rate was 320 nMol $K^+$ min$^{-1}$ mg protein$^{-1}$, giving a $K^+/\sim$ ratio of 0.92 ($\sim$ = ATP equivalent).

The ionophores gramicidin D and valinomycin have been used in a few studies with respiration-linked contraction. The lipid-soluble ionophores permit rapid equilibration of the electrochemical potential of $K^+$ or $Na^+$ (in the case of gramicidin) across the membrane. The ionophore–cation complex carries the charge of the cation. Gramicidin D may also transport $H^+$ (Chappell and Crofts, 1965).

Gramicidin D inhibits energy-linked contraction (Yoshida, 1968a; Miller et al., 1970). However, if corn mitochondria come to swelling equilibrium in the presence of gramicidin, it is possible to get some contraction, particularly with NADH as substrate (Miller et al., 1970). Gramicidin here acts very much like an uncoupler, releasing acceptorless respiration and eliminating respiratory control with ADP.

Valinomycin is specific for $K^+$, $Rb^+$ and $Cs^+$ (Pressman, 1965) and there is no ambiguity about possible $H^+$ transport. Valinomycin is extremely effective in producing mitochondrial swelling, release of NADH oxidation and loss of respiratory control (Wilson et al., 1972; Hanson et al., 1972). However, respiration-linked contraction of corn mitochondria swollen in KCl plus valinomycin is not very strongly inhibited, although respiration is accelerated (Kirk and Hanson, 1973). Apparently, the $H^+/K^+$ antiport (reaction 3, Fig. 3.3) functions more rapidly, probably because of lowered resistance to $K^+$ approach to the antiporter, extruding KCl at a rate which offsets the more rapid influx.

At high pH ($> 8.0$) where passive swelling is very rapid (preceding section) there is lowered energy-linked contraction in KCl, especially in the absence of BSA (Stoner and Hanson, 1966; Yoshida, 1968a). Net efflux of salt can only be realized when membrane resistance to passive influx is high.

Calcium ion slightly retards the spontaneous swelling of corn mitochondria in KCl, and enhances the rate and extent of concentration (Hanson et al., 1965). There is no satisfactory explanation of this $Ca^{2+}$ effect. Corn mitochondria lack the high affinity $Ca^{2+}$-binding component found in vertebrate mitochondria (Chen and Lehninger, 1973; Bertagnolli and Hanson, unpublished) and the extensive research on $Ca^{2+}$ reaction with vertebrate mitochondria scarcely applies. Addition of $Ca^{2+}$ causes a brief respiratory burst in corn mitochondria suspended in sucrose with succinate and lacking phosphate or other transportable anion (Hanson et al., 1965). This respiratory burst is linked to binding of about 100 nMol $Ca^{2+}$ mg protein$^{-1}$, and to a sharp contraction (Bertagnolli and Hanson, unpublished). The rate of net proton efflux increases from about 20 to about 70 nMol min$^{-1}$ mg protein$^{-1}$.

Membrane integrity is important in energy-linked efflux pumping and contraction. BSA is nearly always required for optimal activity. Earnshaw and Truelove (1968) believe the effect of BSA involves more than removal of uncoupling fatty acids by binding; BSA probably also prevents the original membrane damage leading to release of fatty acids. Binding by BSA reduces the effectiveness of certain experimental additives, especially uncouplers (Stoner and Hanson, 1966; Yoshida, 1968a).

In summary, almost all observations on efflux transport in plant mito-chondria are in accord with the model of Fig. 3.3. A proton motive force produced by respiration (reaction 1) or ATP hydrolysis (reaction 2) drives $H^+/cation^+$ exchange (reaction 3), and accumulated anions flux outward down the electrical potential by some undefined pathway. The loss of salt from the matrix causes osmotic contraction. Inhibition of contraction by ions such as phosphate is due to competing influx transport (reaction 4). Low concentrations of uncouplers, such as FCCP or DNP, and ionophores such as valinomycin, may stimulate respiration in part by enhancing antiport activity (reaction 3); this explains the failure to uncouple contraction while uncoupling phosphorylation (reaction 2 in reverse).

It must be emphasized that Fig. 3.3 is only a conceptual model, used here because it most simply and consistently explains experimental observations. It does not follow that it is correct in detail.

## 3.6 Influx transport

Energy-linked swelling in plant mitochondria was first reported for acetate, propionate, butyrate, phosphate and sulphate (Hanson and Miller, 1967; Yoshida, 1968a, b). In both reports the uptake was deduced to be in association with the postulated high energy intermediate of oxidative phosphorylation ($I \sim X$). Yoshida's (1968b) interpretation based on consumption of $I \sim X$ in proton pumping (Chappell and Crofts, 1966) was more nearly in accord with present concepts. A schematic representation of coupled transport, modified from Chappell and Haaroff (1967) and incorporating the basic concepts of Fig. 3.3 is given in Fig. 3.4. The rapid, spontaneous penetration of $Ca^{2+}$, $Sr^{2+}$ and $Mn^{2+}$ shown by Chappell and Haaroff is not included, for, as discussed above, this appears to be a characteristic of vertebrate mitochondria only. $Ca^{2+}$ is not a 'permeant cation' in plant mitochondria (for comparison of liver and cauliflower mitochondria, see Lyons and Pratt, 1964).

Rates of passive swelling of corn mitochondria in 100 Mol m$^{-3}$ potassium acetate and KCl are about the same, but introduction of a respiratory substrate causes extensive and irreversible swelling with potassium acetate, while in KCl the mitochondria contract (Hanson and Miller, 1967; Wilson et al., 1969). Oxidative phosphorylation is uncoupled in acetate, but not in Cl$^-$. Acetate and K$^+$ are accumulated in a 1 : 1 ratio. However, there is very little cation specificity shown – energy-linked acetate swelling is extensive with K$^+$, Na$^+$, Li$^+$,

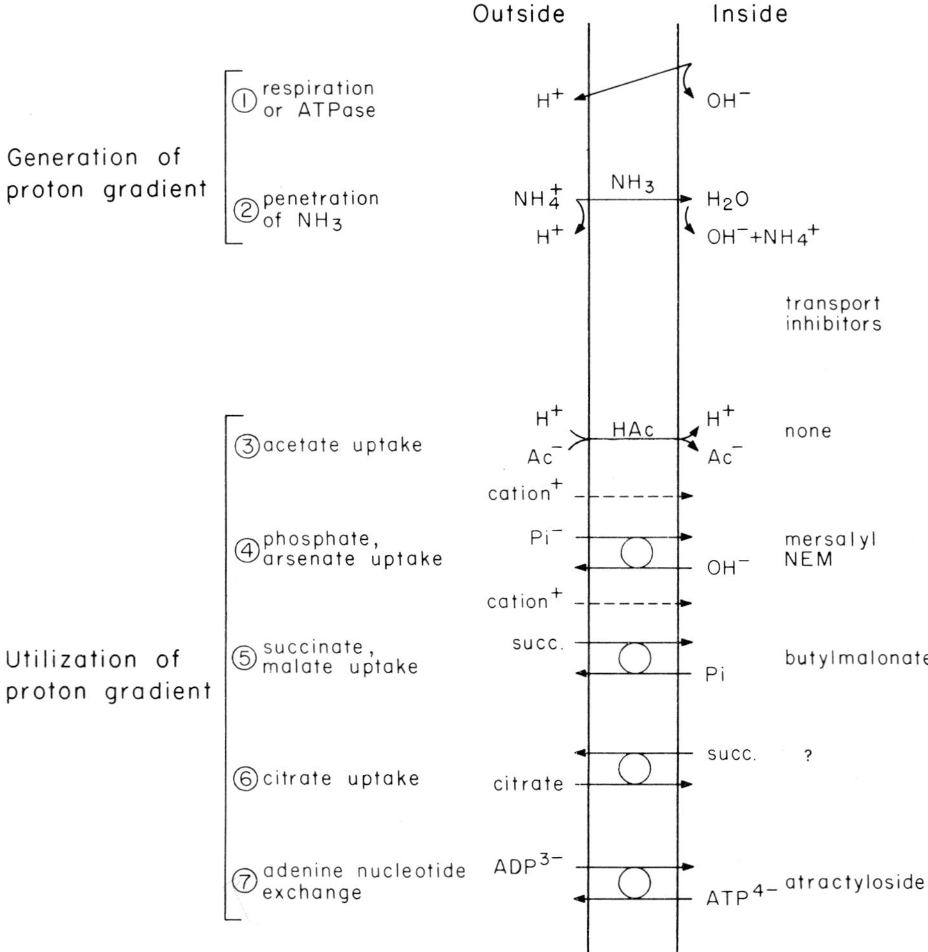


Fig. 3.4. Systems for influx transport of ions in mitochondria. Modified from Chappell and Haaroff (1967). See text.

$Mg^{2+}$, $Tris^+$, and $tetraethylammonium^+$ salts (Wilson et al., 1969). Cations must penentrate through non-discriminatory pathways. If ammonium acetate is used there is extremely rapid passive swelling and no additional swelling with substrate (Wilson et al., 1969; Yoshida, 1968b) for the reason illustrated in Fig. 3.4 (reactions 2 and 3). Penetration of $NH_4^+$ down a concentration gradient as the lipid-soluble $NH_3$ molecule produces a proton gradient and the equivalent of 'energy-linked' acetate uptake.

As an experimental technique, it proves much more useful to study the energy-linked accumulation of proton-transporting anions, such as acetate and phosphate, in media which are osmotically supported with an impermeant solute such as sucrose (Yoshida, 1968b; Hanson and Miller, 1969). The massive swelling and uncoupling are controlled, and the process is reversible; when the energy supply is exhausted or blocked the accumulated salt diffuses back down the concentration gradient established during the active uptake phase. This back diffusion is remarkably rapid but seldom complete; i.e., original volumes are not restored. The energy stored in a $K^+$ concentration gradient has been demonstrated in animal mitochondria to be translated into ATP formation when the endogenous $K^+$ is suddenly released by valinomycin (Cockrell et al., 1967; Rossi and Azzone, 1970). An explanation which conforms with the chemiosmotic scheme of Fig. 3.3 is that a membrane potential is created at the expense of the concentration gradient (Glynn, 1967).

Kirk and Hanson (1973) studied the respiratory stoichiometry of potassium acetate uptake in corn mitochondria using sucrose-supported media, following movement of $K^+$ in and out of the sucrose-inaccessible space (matrix). NADH was used as substrate; oxidation of exogenous NADH is rotenone-insensitive (Wilson and Hanson, 1969) giving a theoretical P/O ratio of 2. Assuming perfect coupling, the $K^+/\sim$ ratio approached unity (0.58–0.97). Influx rates of $K^+$ varied from 344 to 472 nMol $K^+$ $min^{-1}$ mg $protein^{-1}$. The uptake of water by the mitochondria was about that expected for osmotic equilibrium; i.e., little or no turgor pressure develops as the matrix expands.

The transport of lipid-soluble short chain fatty acids is instructive, but it is not of much physiological interest. Greater interest centres on the phosphate transport which is essential to ATP formation. The few studies made with plant mitochondria confirm the existence of the antiport mechanism visualized for animal mitochondria (Fig. 3.4, reaction 4). (Depicting the phosphate transporter as a $P_i^- - OH^-$ antiport is largely a convention; a $P_i^- - H^+$ symport would serve the same end.) Thiol reagents such as mersalyl or $N$-ethylmaleimide (NEM), which inhibit phosphate transport in animal mitochondria without inhibiting the coupling ATPase (Fonyo and Bessman, 1968; Tyler, 1969), are similarly effective with respiring plant mitochondria (Hanson, 1972; Hanson et al., 1972). Mersalyl inhibition of the $P_i^- - OH^-$ antiport can be reversed by cysteine (Hanson et al., 1972).

It has proved possible to load corn mitochondria with phosphate, block further transport with mersalyl, add ADP, and get phosphorylation to ATP at the expense of the accumulated phosphate (Hanson et al., 1972). Actractyloside, which competitively inhibits the ADP–ATP transporter (Klingenberg et al., 1969; see reaction 7, Fig. 3.4), prevents the formation of the ATP. If

                    *J.B. Hanson and D.E. Koeppe*

arsenate is substituted for phosphate, the addition of ADP withdraws arsenate from the matrix, and the rapid respiration declines (Bertagnolli and Hanson, 1973). The uncoupled respiration produced by arsenate plus ADP is blocked by adding mersalyl, illustrating the need for cyclic transport of arsenate in uncoupling.

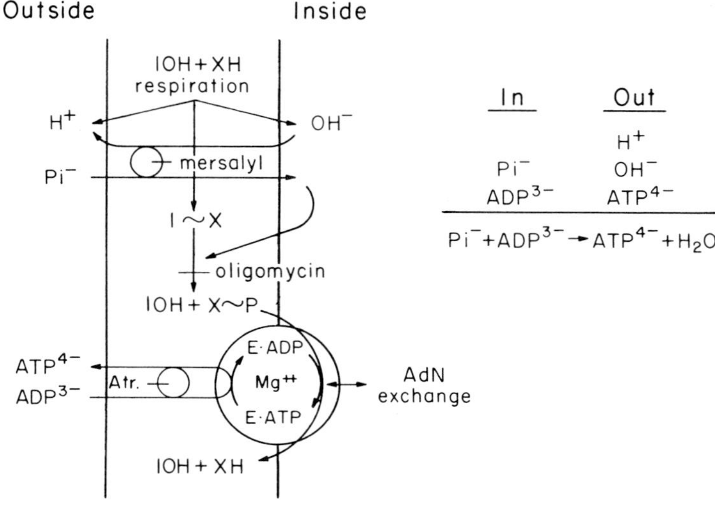


Fig. 3.5. Complex of transport reactions associated with oxidative phosphorylation in corn mitochondria. After Bertagnolli and Hanson (1973).

Fig. 3.5 is a schematic illustration of the processes associated with oxidative phosphorylation in corn mitochondria (Hanson et al., 1972; Bertagnolli and Hanson, 1973). It is deduced that respiratory coupling produces both a proton gradient for transport and a 'squiggle' for phosphorylation (shown as the conventional high energy intermediate, $I \sim X$); 'squiggle' can best be defined as a labile energized state of some membrane enzyme which is capable of reacting with phosphate or arsenate. From the chemiosmotic viewpoint (Fig. 3.3), 'squiggle' would be nothing more than the coupling ATPase activated for phosphorylation by a proton motive force. As illustrated in Fig. 3.5, the membrane potential would 'pull' $ADP^{3-}/ATP^{4-}$ exchange, thus favouring ATP formation. There is no clear understanding of the energized state, but it is manifest that the influx of phosphate can be separated from phosphorylation. Phosphorylation in turn depends on operation of the adenine nucleotide transporter (phosphorylation of endogenous nucleotides excepted), which in corn mitochondria is linked by a transfer mechanism to the $F_1$–ATPase.

Hence, the unique membrane-bound function of mitochondria – the export of ATP – is intimately associated with one active transport process (proton extrusion), two exchanges ($P_i^-$/$OH^-$ and $ADP^{3-}$/$ATP^{4-}$), one transfer reaction (AdN transporter/$F_1$-ATPase) and the donation of a phosphoryl group to ADP ($F_1$-ATPase). The net result is $P_i^-$ + $ADP^{3-}$ (in) → $ATP^{4-}$ + $H_2O$ (out). Fig. 3.5 oversimplifies many complex problems inherent in the electrogenic $ADP^{3-}$/$ATP^{4-}$ exchange, and in the higher ATP/ADP ratio for external over matrix phases (Klingenberg, 1972), but these are beyond the scope of this simple exposition. The probable participation of AMP (derived via adenylate kinase) in phosphorylation by plant mitochondria (Rusness and Still, 1973) cannot yet be fitted into the net transport picture. It is of real interest that corn, soybean and cucumber mitochondria come to phosphorylation equilibrium with approximately the following percentage of adenylates: 40 ATP, 40 ADP and 20 AMP (Rusness and Still, 1973).

A perplexing transport problem has emerged in studies of the activation of the DNP-stimulated, oligomycin-inhibited ATPase by respiration (Takeuchi et al., 1969; Carmelli and Biale, 1970; Jung and Hanson, 1973). Plant mitochondria which lack DNP–ATPase can often be activated by a brief period of respiration or by prolonged exposure to ATP ('priming'). Destruction of the membrane barrier by sonication or high pH treatment exposes an active ATPase, but the need for uncoupler is lost. Priming is absolutely dependent on added $Mg^{2+}$ and is strongly promoted by phosphate or arsenate (Jung and Hanson, 1973). Priming will last for about 2 minutes after respiration ceases, but is immediately collapsed if the DNP is added before the ATP in assay. Priming also increases the $Mg^{2+}$ content and transports ADP and ATP by an actractyloside-insensitive pathway (Jung, 1974). It appears that plant mitochondria require a transmembrane potential to maintain adequate concentrations of $Mg^{2+}$ and adenine nucleotides, and for functioning of the adenine nucleotide transporter.

Phillips and Williams (1973) and Wiskich (1974) have used the ammonium salt technique of Chappell and Haaroff (1967) to establish the existence of phosphate, dicarboxylate and tricarboxylate transporters in potato and cauliflower mitochondria (Fig. 3.4, reactions 2, 4, 5 and 6). Wiskich (1974), however, notes somewhat different results with beet root, castor bean and wheat coleoptile mitochondria. Lee and Wilson (1972) obtained respiration-linked swelling in bean mitochondria with a number of substrate acids in the absence of exogenous phosphate. Fluxes of endogenous phosphate were not determined, but the uptake appears to be a generalized transport of mono- and dicarboxylic acids by pathways not involving phosphate exchange.

$Ca^{2+}$ uptake by plant mitochondria has been difficult to understand. Unlike vertebrate mitochondria, which dramatically uncouple during active binding of small amounts of $Ca^{2+}$ (Lehninger, 1970), plant mitochondria actively absorb $Ca^{2+}$ only in connection with phosphate transport (Hanson and Hodges, 1967; Chen and Lehninger, 1973). With the possible exception of sweet potato and white potato mitochondria, plant mitochondria lack a high affinity $Ca^{2+}$ carrier (Chen and Lehninger, 1973). Corn mitochondria show energy-linked binding of the order of 100 nMol $Ca^{2+}$ mg protein$^{-1}$ in the absence of exogenous phosphate (Hanson and Miller, 1967; Kenefick and Hanson, 1967), but this proves to depend on transport of endogenous phosphate from a readily leachable phase (intermembrane?) through a mersalyl-sensitive site (Earnshaw et al., 1973). $Ca^{2+}$ actively bound in the absence of phosphate is released when respiration ceases; addition of phosphate at this point produces passive phosphate uptake and $Ca^{2+}$ retention (Earnshaw and Hanson, 1973). $Ca^{2+}$ appears to promote phosphate uptake and accompany it into the matrix where insoluble phosphates are deposited (Peverly et al., 1974).

Investigations with corn mitochondria (Bertagnolli and Hanson, unpublished) show that $Ca^{2+}$ also accelerates energy-linked acetate uptake from a sucrose medium, but $Ca^{2+}$ is not accumulated with the acetate. $Ca^{2+}$ is bound in the first 5 s after addition to the extent of 100–130 nMol mg protein$^{-1}$, just as in the absence of acetate, with no further change as acetate continues to enter. Respiration is stimulated. An old hypothesis is reinforced (Kenefick and Hanson, 1966; Hanson and Miller, 1967); $Ca^{2+}$ dissipates the energized state designated $I \sim X$ in Fig. 3.5, recycling intermediates, and thereby accelerating respiration and anion uptake.

On the whole, the information on transport processes in plant mitochondria is less than satisfactory. Most of the information on plant mitochondria is concerned with respiration and phosphorylation without regard to associated transport. Serious studies on pyruvate transport – presumably the normal respiratory substrate in vivo – are lacking, as are studies on amino acid transport. Most conditions for assaying transport in vitro are highly artificial; mitochondria in vivo are not suspended in sucrose or mannitol or KCl or ammonium succinate, but rather in 100 Mol m$^{-3}$ $K^+$ salts of macromolecules, metabolites and inorganic anions. Laties and Treffry (1969) have shown that mitochondrial morphology is drastically altered by the absence of macromolecules. Hence, some of the observations on transport may be artifactual. However, even in unnatural environments the responses observed reflect properties of the transport systems, and in proper perspective can provide valuable information.

## 3.7 Summary

Mitochondria are organized for production and extrusion of ATP at the expense of $P_i$, ADP and substrate acids taken in. There are additional transport processes associated with maintenance of the mitochondrion as a quasi-independent organelle. The available evidence on transport in plant mitochondria supports concepts developed with animal mitochondria and chloroplasts. Energy supply for transport appears to lie in an electrochemical gradient of protons (proton motive force) derived from electron transport or ATP hydrolysis in the inner membrane. The molecular mechanisms of creating the proton motive force are not resolved. Exchange transport of cations or anions with $H^+$ or $OH^-$ provides the primary element of transport, and net salt movement arises from the accompanying flux of counterions. Substrate acid transport is believed to be linked to initial transport of phosphate. Oxidative phosphorylation incorporates separate transport processes for $P_i^- - OH^-$ and $ADP^{3-} - ATP^{4-}$ exchange.

## References

A. AZZI and G.F. AZZONE, Biochim. Biophys. Acta, 131 (1967a) 468.

A. AZZI and G.F. AZZONE, Biochim. Biophys. Acta, 135 (1967b) 444.

B.L. BERTAGNOLLI and J.B. HANSON, Plant Physiol., 52 (1973) 431.

G.P. BRIERLEY, Biochemistry, 9 (1970) 697.

G.P. BRIERLEY, M. JURKOWITZ, K.M. SCOTT and A.J. MEROLA, Arch. Biochem. Biophys., 147 (1971) 545.

G.P. BRIERLEY, M. JURKOWITZ and K.M. SCOTT, Arch. Biochem. Biophys., 159 (1973) 742.

C. CARMELLI and J.B. BIALE, Plant Cell Physiol., 11 (1970) 65.

B. CHANCE and M. MONTAL, in F. Bronner and A. Kleinzeller (Eds.) Current Topics in Membrane Transport (1971) Vol. 2., Academic Press, New York, p. 99.

J.B. CHAPPELL and A.R. CROFTS, Biochem. J., 95 (1965) 393.

J.B. CHAPPELL and A.R. CROFTS, in J.M. Tager et al. (Eds.) Regulation of Metabolic Processes in Mitochondria (1966) Elsevier, Amsterdam, p. 293.

J.B. CHAPPELL and K.N. HAAROFF, in E.C. Slater, Z. Kaniuga and L. Woltczak (Eds.) Biochemistry of Mitochondria (1967) Academic Press, New York, p. 75..

C.H. CHEN and A.L. LEHNINGER, Arch. Biochem. Biophys., 157 (1973) 183.

R.S. COCKRELL, E.J. HARRIS and B.C. PRESSMAN, Nature, 216 (1967) 1487.

R. DOUCE, C.A. MANNELLA and W.D. BONNER, JR, Biochim. Biophys. Acta, 292 (1973) 105.

M.J. EARNSHAW and J.B. HANSON, Plant Physiol., 52 (1973) 403.

M.J. EARNSHAW, D.M. MADDEN and J.B. HANSON, J. Exp. Bot., 24 (1973) 828.

M.J. EARNSHAW and B. TRUELOVE, Plant Physiol., 43 (1968) 121.

M.J. EARNSHAW and B. TRUELOVE, Plant Physiol., 45 (1970) 322.

M.J. EARNSHAW, B. TRUELOVE and R.D. BUTLER, Plant Physiol., 45 (1970) 318.

L. ERNSTER and B. KUYLENSTIERNA, in R. Racker (Ed.) Membranes of Mitochondria and Chloroplasts (1970) Reinhold, New York, p. 172.

A. FONYO and S.P. BESSMAN, Biochem. Med., 2 (1968) 145.

I.M. GLYNN, Nature, 216 (1967) 1318.

C.R. HACKENBROCK, T.G. REHN, J. GAMBLE, JR, E.C. WEINBACH and J.J. LEMASTERS, in E. Quagliariello, S. Papa and C.S. Rossi (Eds.) Energy Transduction in Respiration and Photosynthesis (1971) Adriatica Editrice, Bari, p. 285.

J.B. HANSON, Plant Physiol., 49 (1972) 707.

J.B. HANSON, B.L. BERTAGNOLLI and W.D. SHEPHERD, Plant Physiol., 50 (1972) 347.

J.B. HANSON and T.K. HODGES, in D.R. Sanadi (Ed.) Current Topics in Bioenergetics (1967) Vol. 2, Academic Press, New York, p. 65.

J.B. HANSON, S.S. MALHOTRA and C.D. STONER, Plant Physiol., 40 (1965) 1033.

J.B. HANSON and R.J. MILLER, Proc. Natl. Acad. Sci. U.S., 58 (1967) 727.

J.B. HANSON and R.J. MILLER, Plant Cell Physiol., 10 (1969) 491.

J.R. HENSLEY and J.B. HANSON, Plant Physiol. (1975) in press.

S. IZZARD and H. TEDESCHI, Arch. Biochem. Biophys., 154 (1973) 527.

D.W. JUNG, The 2,4-Dinitrophenol-Stimulated Adenosine Triphosphatase Activity in Cauliflower and Corn Mitochondria (1974) PhD Thesis, Univ. of Illinois, Urbana.

D.W. JUNG and J.B. HANSON, Arch. Biochem. Biophys., 158 (1973) 139.

D.G. KENEFICK and J.B. HANSON, Plant Physiol., 41 (1966) 1601.

D.G. KENEFICK and J.B. HANSON, in M. Freid (Ed.) Isotopes in Plant Nutrition and Physiology (1967) Inter–Atomic Energy Agency, Vienna, p. 271.

B.I. KIRK and J.B. HANSON, Plant Physiol., 51 (1973) 357.

M. KLINGENBERG, in S.G. van den Bergh et al. (Eds.) Mitochondria, Biogenesis and Bioenergetics (1972) Elsevier, New York, p. 147.

M. KLINGENBERG, R. WULF, H.W. HELDT and E. PFAFF, in L. Ernster and Z. Drahota (Eds.) Mitochondria, Structure and Function (1969) Academic Press, New York, p. 59.

D.E. KOEPPE and R.J. MILLER, Plant Physiol., 49 (1972) 353.

G.G. LATIES and T. TREFFRY, Tissue and Cell, 1 (1969) 575.

D.C. LEE and R.H. WILSON, Physiol. Plant., 27 (1972) 195.

A.L. LEHNINGER, Biochem. J., 119 (1970) 129.     —

C. LONGO and O. ARRIGONI, Exp. Cell Res., 35 (1964) 572.

G.H. LORIMER and R.J. MILLER, Plant Physiol., 44 (1969) 839.

J.M. LYONS and H.K. PRATT, Arch. Biochem. Biophys., 104 (1964) 318.

J.M. LYONS, T.A. WHEATON and H.K. PRATT, Plant Physiol., 39 (1964) 262.

C.P. MALONE, D.E. KOEPPE and R.J. MILLER, Plant Physiol., 53 (1974) 918.

S. MASSARI and G.F. AZZONE, Eur. J. Biochem. 12 (1970) 301 and 310.

S. MASSARI, L. FRIGERI and G.F. AZZONE, J. Membrane Biol. 9 (1972a) 57.

S. MASSARI, L. FRIGERI and G.F. AZZONE, J. Membrane Biol. 9 (1972b) 71.

R.J. MILLER, S.W. DUMFORD and D.E. KOEPPE, Plant Physiol., 46 (1970) 471.

P. MITCHELL, Chemiosmotic Coupling in Oxidative and Photosynthetic Phosphorylation (1966) Glynn Research Ltd., Bodmin, Cornwall, England.

P. MITCHELL, Chemiosmotic Coupling and Energy Transduction (1968) Glynn Research Ltd., Bodmin, Cornwall, England.

P. MITCHELL, in van den Bergh et al. (Eds.) Mitochondria : Biogenesis and Bioenergetics, FEBS Symp. (1972) Vol. 28, Elsevier, Amsterdam, p. 353.

P. MITCHELL, FEBS Lett. 33 (1973) 267.

R. MITCHELL, H.M. JOHNSON and R.H. WILSON, Exp. Cell Res., 76 (1973) 449.

C.L. MOORE, in D.R. Sanadi (Ed.) Current Topics in Bioenergetics (1971) Vol. 4, Academic Press, New York, p. 191.

F. MOREAU, J. DUPONT and C. LANCE, Biochim. Biophys. Acta, 345 (1974) 294.

A.R. OVERMAN, G.H. LORIMER and R.J. MILLER, Plant Physiol., 45 (1970) 126.

D.F. PARSONS, W.D. BONNER, JR. and J.G. VERBOON, Can. J. Bot., 43 (1965) 647.

J.H. PEVERLY, R.J. MILLER, C.P. MALONE and D.E. KOEPPE, Plant Physiol., 54 (1974) 408.

M.L. PHILLIPS and G.R. WILLIAMS, Plant Physiol., 51 (1973) 667.

B.C. PRESSMAN, Proc. Natl. Acad. Sci. U.S., 53 (1965) 1076.

B. PRESSMAN and D.H. HAYNES, in D.C. Tosteson (Ed.) The Molecular Basis of Membrane Function (1969) Prentice–Hall, New Jersey, p. 221.

R.N. ROBERTSON, Protons, Electrons, Phosphorylation and Active Transport (1968) Cambridge Univ. Press, Cambridge.

E. ROSSI and G.F. AZZONE, Eur. J. Biochem., 12 (1970) 319.

H. ROTTENBERG, J. Membrane Biol., 11 (1973) 117.

D.G. RUSNESS and G.G. STILL, Arch. Biochem. Biophys., 159 (1973) 279.

S.J. SINGER and G.L. NICOLSON, Science, 175 (1972) 720.

V.P. SKULACHEV, in D.R. Sanadi (Ed.), Current Topics in Bioenergetics (1971) Vol. 4, Academic Press, New York, p. 127.

C.D. STONER and J.B. HANSON, Plant Physiol., 41 (1966) 255.

C.D. STONER, T.K. HODGES and J.B. HANSON, Nature, 203 (1964) 258.

Y. TAKEUCHI, K. YOSHIDA and S. SATO, Plant Cell Physiol., 10 (1969) 733.

J.P. TEWARI, J.C. TU and S.K. MALHOTRA, Cytobios, 5 (1972) 261.

B. TRUELOVE and J.B. HANSON, Plant Physiol., 41 (1966) 1004.

D.D. TYLER, Biochem. J., 111 (1969) 665.

R.H. WILSON, J. DEVER, W. HARPER and R. FRY, Plant Cell Physiol., 13 (1972) 1103.

R.H. WILSON and J.B. HANSON, Plant Physiol., 44 (1969) 1335.

R.H. WILSON, J.B. HANSON and H.H. MOLLENHAUER, Biochemistry, 8 (1969) 1203.

R.H. WILSON, E.L. THURSTON and R. MITCHELL, Plant Physiol., 51 (1973) 26.

J.T. WISKICH, Aust. J. Plant Physiol., (1974) in press.

K. YOSHIDA, J. Fac. Sci. Univ. Tokyo, 10 (1968a) 63.

K. YOSHIDA, J. Fac. Sci. Univ. Tokyo, 10 (1968b) 83.

K. YOSHIDA and S. SATO, J. Fac. Sci. Univ. Tokyo, 10 (1968) 49.

*Ion transport in plant cells and tissues*
*edited by D.A. Baker and J.L. Hall*
© *North-Holland Publishing Company, 1975*



# CHAPTER 4

# Chloroplasts

P.S. NOBEL

## Contents



## *4.1 Introduction*

Not only do chloroplasts serve as the site for photosynthesis, but they are
also intricately involved in the ionic relations of green plant cells. After a
brief description of chloroplast structure, we will consider the bioenergetics
of photosynthesis from the point of view of the energy conserving reactions in
thylakoids. We will then focus on ion and water movements into and out of
chloroplasts in vivo. For a more complete consideration of the vast literature
dealing with ion and water transport by isolated chloroplasts, the reader is

directed to the excellent reviews by Packer and Crofts (1967), Hind and McCarty (1973), Heber (1974), Murakami et al. (1974) and Walker (1974).

Chloroplasts are surrounded by two membranes, the inner one of which may invaginate to form the complex internal lamellar system (Fig. 4.1). The

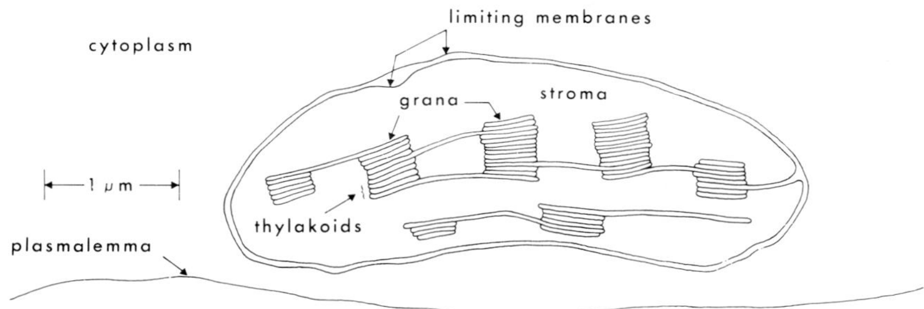



Fig. 4.1. Structural features of a chloroplast representative of those found in mesophyll cells.

photosynthetic pigments are bound to proteins in the lamellar membranes, which are about half lipid and half protein by dry weight. Each of the lamellae consists of a pair of closely apposed membranes that are 6–8 nm thick. The lamellae often form flattened sacs or discs, referred to as thylakoids (Fig. 4.1). In the red algae the lamellae are in the form of single large thylakoids traversing the whole chloroplast and separated by appreciable distances from each other. In higher plant chloroplasts ten or more thylakoids stack together to form grana. Grana are typically about 0.4 $\mu$m in diameter and 10–50 of them occur in a chloroplast of a representative mesophyll cell.

Chloroplasts in a higher plant cell generally occur immobilized in a gelatinous matrix and resemble a flattened disc slightly indented on the side facing the plasmalemma (Fig. 4.1). For leaves of plants like pea, spinach and Swiss chard (commonly used as the starting material for chloroplast isolation), the chloroplasts have a major axis of about 4 $\mu$m and a volume close to 30 $\mu$m$^3$, of which about 25% is made up of lamellar membranes and the rest is the stroma. Although there is great variation with species as well as among cells of an individual plant, about 50 chloroplasts often occur in a palisade or spongy mesophyll cell of a mesophytic leaf. The chloroplasts usually occupy about 40% of the volume of the cytoplasm and indeed form a very important ionic compartment within a mesophyll cell.

## 4.2 Bioenergetics

Photosynthesis is the largest chemical process on earth – about $5 \times 10^{16}$ g of carbon are annually fixed into organic compounds in photosynthetic organisms. We can divide the process into three stages: (1) the photochemical steps, (2) electron transfer, to which is coupled the formation of ATP, and (3) the biochemical reactions involving the incorporation of $CO_2$ into carbohydrates. For each molecule of $CO_2$ 'fixed' in chloroplasts having the Calvin–Benson pathway, 3 molecules of ATP and 2 of NADPH are required. ATP and NADPH are representatives of the two main classes of energy 'currencies' that can be used for ion transport in plants and so next we will consider their generation in chloroplasts. See Nobel (1974a) for further details and references.

### 4.2.1 Electron transfer

Chlorophylls represent the principal class of pigments responsible for light absorption in photosynthesis and they are found in all photosynthetic organisms. In 1932 Emerson and Arnold deduced that approximately 250 chlorophylls act together as a photosynthetic unit. When any one chlorophyll in a unit becomes excited by the absorption of a photon, the concomitant excitation of other chlorophyll molecules in that unit cannot be used for photosynthesis. The excitation is transferred among the chlorophylls in a photosynthetic unit until it reaches a special chlorophyll known as a 'trap', e.g. $P_{700}$ acts as the trap for Photosystem I, a terminology introduced by Duysens et al. in 1961 (P stands for pigment and the subscript 700 indicates the wavelength in nm where an absorption maximum occurs). The trap chlorophyll has its main band at longer wavelengths (lower excitation energy) than the other photosynthetic pigments, and so the excitation tends to be irreversibly funneled toward the trap, which is the site for the photochemical reactions of photosynthesis.

To convert radiant energy into a form which can be stored chemically, reducing (electron-donating) and oxidizing (electron-accepting) species must be created. An example of such a photochemical event is the transfer of an electron from an excited $P_{700}$ to a suitable electron acceptor in Photosystem I. The electron that re-reduces $P_{700}$ comes along an electron transfer pathway from Photosystem II (Fig. 4.2). Both of these photosynthetic units, which are of roughly equal size, have now been isolated and chemically characterized (Boardman, 1970). The only compositional aspect to be emphasized here is that Photosystem I contains a type of chlorophyll (Chl $a_{680}$) with an absorption maximum at longer wavelengths than the ones in Photosystem II (e.g. Chl $a_{670}$).

                                        *P.S. Nobel*

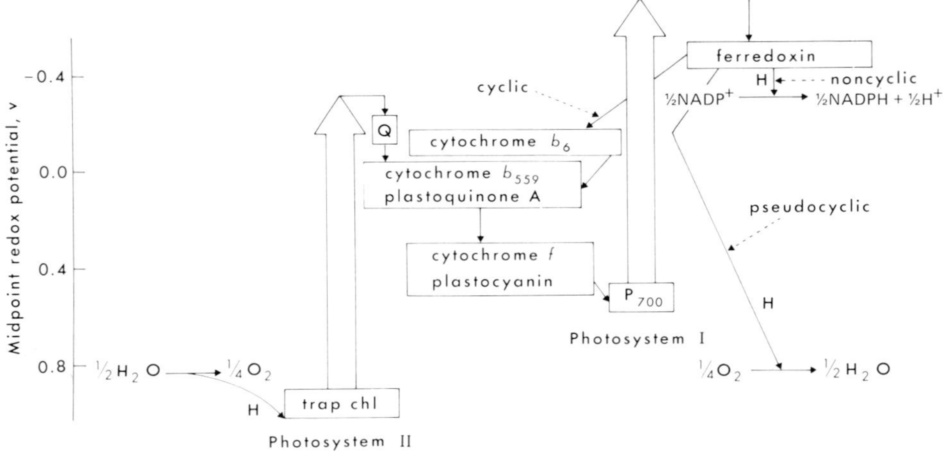



Fig. 4.2. Components involved with the three types of electron flow in chloroplasts.

Thus, when light of wavelengths greater than 690 nm is incident on a leaf, absorption by photosynthetic pigments is mainly due to Chl $a_{680}$ and $P_{700}$ in Photosystem I. As we shall see below, this preferential excitation of only one photosystem can be used to study the energy source for ion movements in green plant cells.

Studies on electron flow accompanying photosynthesis were greatly stimulated by Hill's demonstration in 1938 that isolated chloroplasts could evolve $O_2$. In 1941 Ruben and Kamen showed that the $O_2$ comes from water. We now know that the electron removed from the excited trap chlorophyll in Photosystem II is replaced by one from water in the as yet poorly understood process leading to $O_2$ evolution (Fig. 4.2). The electron from Photosystem II goes to a molecule often termed Q, since the removal of the electron from the trap chlorophyll tends to quench fluorescence, a competing pathway for the de-excitation of the photosynthetic unit. From Q the electron moves energetically downhill (toward couples with higher oxidation–reduction potentials) along an electron transfer chain that includes plastoquinone A, cytochrome $b_{559}$, cytochrome $f$, and plastocyanin before being accepted by an oxidized $P_{700}$ in Photosystem I (Fig. 4.2). The electron which had been removed from $P_{700}$ reduces ferredoxin, perhaps by way of an intermediate, and two molecules of ferredoxin then reduce one molecule of $NADP^+$ to yield NADPH. NADPH is of course utilized in $CO_2$ fixation; in addition, it or other reduced compounds could serve as energy currencies for transport processes occurring under the oxidizing conditions found in plant cells.

The overall pathway from water to Photosystem II along the electron transfer chain to Photosystem I to ferredoxin and eventually leading to the formation of NADPH is called non-cyclic electron flow (Fig. 4.2). With isolated chloroplasts it can readily be demonstrated that electrons from ferredoxin can also reduce $O_2$, which yields $H_2O_2$ and eventually $H_2O$; equal amounts of $O_2$ are evolved in Photosystem II and then consumed using reduced ferredoxin in a different reaction, and so such electron flow is termed 'pseudocyclic'. When chloroplasts are exposed to the herbicide DCMU, Photosystem II is blocked and therefore both non-cyclic and pseudocyclic electron flow are inhibited. However, another experimentally very important flow type occurs involving only Photosystem I, viz., cyclic electron flow (Fig. 4.2). Using far-red light (beyond 690 nm) or adding DCMU allows cyclic electron flow to occur while non-cyclic electron flow is suppressed, with the consequence that little NADPH is formed and so $CO_2$ fixation does not proceed at an appreciable rate.

### 4.2.2 ATP formation

The ATP required for $CO_2$ fixation in photosynthesis is produced by photophosphorylation, i.e., light absorbed by the photosynthetic pigments leads to a flow of electrons to which is coupled the phosphorylation of ADP. Photophosphorylation was first demonstrated in cell-free systems in 1954 by Frenkel, working with bacterial chromatophores and by Arnon and co-workers using spinach chloroplasts. Elucidating the mechanism has proved quite difficult, in part because the enzymes involved are located in or on the lamellar membranes. Moreover, the species involved (ADP, phosphate, ATP) cannot be readily measured quantitatively in vivo and they take part in many reactions, with the consequence that reliable data on kinetics are difficult to obtain. Here we will briefly consider the bioenergetics of ATP formation and then mention possible mechanisms.

Let us begin by considering the energy available between certain oxidation–reduction couples involved in photosynthetic electron transfer. When the chemical activity of the reduced form is the same as that of the oxidized one (an assumption we are generally forced to make, because of our ignorance of the actual activities in vivo), the Gibbs free energy change ($\Delta G$) is equal to $-nF\Delta E_o$, where $n$ is the number of moles of electrons transferred through a difference in midpoint redox potential of $\Delta E_o$. $F$, the Faraday, equals 23.06 kilocalories $Mol^{-1}$ volt$^{-1}$. Between cytochrome $b_{559}$ or plastoquinone A and cytochrome $f$ or plastocyanin the $\Delta E_o$ is 0.25 V (Fig. 4.2). Thus, each mole of electrons spontaneously moving across this part of the electron transfer pathway could release $-(1)(23.06)(0.25)$ or 5.8 kcal. How does this compare with the amount of energy required for photophosphorylation? For ionic conditions that might exist in chloroplasts, the Gibbs free energy required for the phosphorylation of one mole of ADP is approximately 10–14 kcal (see Hind and McCarty, 1973; Nobel, 1974a). Thus two moles of electrons moving along this part of the electron transfer pathway are apparently needed for the formation of one mole of ATP.

We have just used a thermodynamic argument to suggest that a coupling site for ATP formation could exist between cytochrome $b_{559}$ or plastoquinone A and cytochrome $f$ or plastocyanin. Much experimental evidence indicates that there may be two coupling sites, each requiring a pair of electrons per ATP, between the water oxidation step and the reduction of $NADP^+$, although the locus of the second site is presently uncertain (e.g., see Schwartz, 1971; Forti et al., 1972; Ort and Izawa, 1973). If there were two coupling sites, then the absorption of 4 photons by each photosystem would lead to the production of 2 NADPH plus 4 ATP. This could be used to fix 1 $CO_2$ with an ATP left over. Also both cyclic and pseudocyclic electron flow lead to the production of ATP with no accompanying reduction of $NADP^+$. We can thus surmise that chloroplasts in the light can directly or indirectly supply ATP for cellular processes in addition to that used in photosynthesis, as indeed appears to be the case (Marrê et al., 1963; Santarius and Heber, 1965; Jeschke, 1967; Keys, 1968; Raven, 1968; Stocking and Larson, 1969; Heber and Santarius, 1970; Packer et al., 1970; Simonis and Urbach, 1973; Walker, 1974).

Two hypotheses are currently being considered to explain the mechanism of coupling between electron flow and ATP formation. In the chemical hypothesis, the intermediates intervening between electron flow in the lamellar membranes and the phosphorylation of ADP are believed to be specific proteins. The electron flow could cause changes in the electronic configuration within the coupling molecules, which eventually leads to the dehydration involved in ATP formation. In the 1960s Mitchell proposed another hypothesis, now referred to as the chemiosmotic one (see Mitchell, 1966a, 1966b, and 1973; Ch. 3). Electron flow is postulated to cause an energetically uphill $H^+$ movement across the thylakoid membranes. The drop in the chemical potential of hydrogen ions from inside the thylakoids to the stroma, $\Delta\mu_{H^+}$, then acts as the Gibbs free energy source for the phosphorylation of ATP.

It can be shown that $\Delta\mu_{H^+}$ equals $F(E^i - E^o) + 1.36(pH^o - pH^i)$ kcal $Mol^{-1}$, where $E^i$ is the electrical potential within the thylakoids and $E^o$ is that outside in the stroma (see Nobel, 1974a). Mitchell proposed that two protons moved from inside a thylakoid out across the lamellar membranes into the chloroplast stroma for each ATP formed. If $E^i - E^o$ were 0.00 V and $pH^o - pH^i$

were 4.0, $\Delta\mu_{H^+}$ would be (1.36)(4.0) or 5.4 kcal Mol$^{-1}$. Two moles of protons could then release 10.8 kcal, which is approximately the free energy required for the formation of one mole of ATP in chloroplasts. In dramatic support of this proposal, Jagendorf and Uribe (1966) showed that when chloroplast lamellae are equilibrated with a solution at pH 4 (in which case pH$^i$ presumably attains a value near 4) and are then rapidly transferred to a medium with a pH of 8 containing ADP and phosphate, the lamellae are capable of forming ATP in the dark. When the $\Delta$pH is 3 or less or if the internal protons are allowed to 'leak' out of the thylakoids, very little ATP is formed.

The above calculations were made assuming that $E^i - E^o$ is negligibly small, as could occur if ions readily crossed the thylakoid membranes and thus dissipated any electrical potential difference. There is considerable evidence that $E^i - E^o$ may not be negligible, especially in vivo (see Crofts et al., 1971; Forti et al., 1972; Hind and McCarty, 1973; Larkum and Bonner, 1972; Witt, 1971). Since $F$ equals 23.06 kcal Mol$^{-1}$ V$^{-1}$, an $E^i - E^o$ of 0.23 V would make the same contribution to $\Delta\mu_{H^+}$ as a pH drop of 4. It seems likely that both the electrical and the concentration terms in the chemical potential of protons may be involved in providing the Gibbs free energy necessary for photophosphorylation, an aspect that we will next consider in a little more detail.

### 4.2.3 *Ion movements for thylakoids*

Because of obvious technical difficulties in working with organelles in plant cells, nearly all we have learned about chloroplast bioenergetics has been deduced using isolated organelles. For such studies the isolation technique has proved crucial. For measurements on osmotic responses of whole chloroplasts, a gentle 2-min isolation has been used in which the leaf material is ground in a mortar and pestle containing a buffered sucrose solution, filtered through cloth having a pore size that allows chloroplasts to pass through, and then briefly centrifuged to obtain a chloroplast-containing pellet (Nobel, 1974b). Although such chloroplasts are nearly all intact or Class I as judged by their highly refractile appearance in a phase-contrast microscope and have been shown by electron microscopy to resemble the one represented in Fig. 4.1, they do not fix $CO_2$ at an appreciable rate in the light. To obtain rates of photosynthesis by isolated intact chloroplasts comparable to rates observed in vivo, various factors are necessary in the isolation and reaction media (Jensen and Bassham, 1966; Gibbs, 1971; Walker, 1964 and 1971). Of particular importance are $MgCl_2$ (sometimes also $MnCl_2$), EDTA, iso-ascorbate, and phosphate or pyrophosphate in the isolation medium (but see

Avron and Gibbs, 1974, on the lack of requirement for $Mg^{2+}$ and EDTA). The reaction medium for $CO_2$ fixation studies also contains a fairly high concentration of bicarbonate (in order to provide $CO_2$, which acts as the actual substrate for photosynthesis) and the photosynthetic rates are often substantially increased by the addition of ATP or an intermediate of the Calvin–Benson cycle such as ribose-5-phosphate. The rate of $CO_2$ fixation of course falls as the chloroplast limiting membranes rupture and the stromal material diffuses out, which leads to a category of broken chloroplasts referred to as Class II.

Chloroplasts have proved much more difficult to maintain in an intact state once isolated than have mitochondria. Mitochondria are only a few percent of the volume of chloroplasts, which puts them in the same size range as microbodies and cellular fragments produced during the isolation procedure. Thus, considerable attention must be paid in the isolation of mitochondria to avoiding contamination by particles of similar size and density and also to maintaining enzymatic activity during the necessarily longer procedure required for their isolation (Fleischer and Packer, 1974). Most studies on isolated 'chloroplasts', especially prior to 1970, have dealt with those in the Class II category.

In addition to the Class I to Class II transition that inevitably accompanies any chloroplast isolation procedure to some extent and continues to occur once the organelles are isolated, the chloroplast-limiting membranes are often purposely broken by suspending the organelles in a solution of low osmotic pressure. Such swollen thylakoid systems form an extremely important starting material for studies of chloroplast bioenergetics. Emphasis in this field over the last ten years has been primarily on understanding the details of energy storage within the thylakoid membrane system, as ably reviewed by Hind and McCarty (1973). Neumann and Jagendorf had shown in 1964 that thylakoids could take up protons in the light. However, neither pH electrodes nor the micropipettes necessary for measuring electrical potential differences could be inserted into the lumen created within a swollen thylakoid. Thus, considerable ingenious experimentation has been directed at ways to measure $pH^i$ and $E^i - E^o$, where 'i' refers to the aqueous phase within thylakoids. Such estimates are necessary in order to determine whether the change in Gibbs free energy of protons across the membranes is sufficient to power ATP formation and we will consider both the pH and the electrical components in turn.

A key observation for estimating $pH^o - pH^i$ was made by Crofts (1967), who showed that amines penetrated thylakoid membranes rather easily in the uncharged form. Thus, uncharged forms of compounds like $NH_3$ (Crofts, 1967; Rottenberg and Grunwald, 1972), methylamine (Rottenberg et al., 1972), and ethylamine (Gaensslen and McCarty, 1971), which all have a p$K$ above 9, could be assumed to have the same concentration on the two sides of a thylakoid membrane. Based on the Henderson–Hasselbach equation, we then expect the ratio of the protonated forms of these amines between the inside and the outside of the thylakoid to be the same as the proton ratio, $[H^+]^i/[H^+]^o$ (Hind and McCarty, 1973). Although individual experiments vary, $pH^i$ in the light was estimated by such distribution ratios to be about 5 when $pH^o$ was 8. The internal pH could also affect the spectroscopic properties of pigments in the thylakoid membranes. In fact, Rumberg and Siggel in 1969 had estimated that $pH^i$ was 5 when $pH^o$ was 8 based on a calibration of the spectroscopic shifts on $P_{700}$ caused by pH. Thus, a $\Delta pH$ of at least 3 units can be caused by illuminating swollen thylakoids at an external pH of 8. A number of crucial points, however, still remain. For example, what is the pH in the stroma for chloroplasts in plant cells? Another matter deals with the internal aqueous phase within thylakoids, which often is artificially created by placing chloroplasts in a low osmotic pressure solution. In plant cells the thylakoid membranes appear to be closely appressed and do not tend to separate upon illumination (Packer et al., 1967; Miller and Nobel, 1972; Murakami et al., 1974). The actual magnitude of the proton pool created within thylakoids upon illumination in the plant cell is also unknown.

Let us now consider the electrical potential difference that can occur across thylakoid membranes. Here again spectroscopic studies have proved very useful. Light-induced absorbance changes occur in the region 515–520 nm, which presumably reflect electrochromic shifts of carotenoids and chlorophylls in the thylakoid membranes (Duysens, 1954; Junge and Witt, 1968; Witt, 1971; Strichartz, 1971; Larkum and Boardman, 1974). Schmidt et al. (1971) found that the effects of electric fields on artificial layers of carotenoids plus chlorophylls were similar to the light-induced absorbance changes of thylakoids. However, difficulty comes in trying to calibrate absorption changes in terms of $E^i - E^o$. Based on the dimensions and the electrical capacitance of the thylakoid membranes, $E^i - E^o$ was estimated to have a steady state value of about 0.10 V in the light (Schliephake et al., 1968). Also based on capacitance arguments, Junge and Witt (1968) indicated that $E^i - E^o$ could be 0.25 V. In a different approach, Jackson and Crofts (1969) assumed that the $K^+$ concentration ratio across bacterial chromatophores (when $K^+$ was the principal ion diffusing) obeyed the Nernst equation (1.7), which gave an $E^i - E^o$ of 0.29 V for the membranes of *Rhodopseudomonas spheroides*. However, Strichartz and Chance (1972) with a similar technique found a value of only 0.03 V for thylakoids from spinach chloroplasts (their calculations were based on an assumed internal $K^+$ concentration of 2 Mol m$^{-3}$). Of particular

110                                       *P.S. Nobel*

usefulness in these experiments is the antibiotic valinomycin, which apparently causes a rapid equilibration of $K^+$ chemical potential across thylakoid membranes with the consequence that $E^i - E^o$ can then be experimentally manipulated. By observing such valinomycin effects on delayed light emission, Barber (1972a and 1972b) estimated that $E^i - E^o$ was about 0.09 V upon illumination of thylakoids. Attempts are continuing to accurately determine $E^i - E^o$ by direct calibration of electrochromic shifts as well as by observing the distribution ratios of various penetrating ions.

Let us now summarize the studies with thylakoids in terms of their relevance to chloroplast bioenergetics. When $E^i - E^o$ is 0.10 V and $pH^o - pH^i$ is 3, then $\mu_{H+}$ is $(23.06)(0.10) + (1.36)(3)$ or 6.4 kcal $Mol^{-1}$ higher within the thylakoids than in the stroma (or external solution). Two moles of protons moving out of such thylakoids could supply 13 kcal, which is approximately the Gibbs free energy required to form a mole of ATP in chloroplasts assuming very efficient coupling between the processes. However, a thermodynamic argument alone cannot prove that the $H^+/ATP$ ratio is 2, and there is some evidence that it may be higher (e.g., Schröder et al., 1972). To refine our calculations we must await improvements in the measurements of $E^i - E^o$, $pH^o - pH^i$, $\Delta\mu_{H+}$, and the Gibbs free energy needed to form ATP in chloroplasts. Ionophorous antibiotics like valinomycin can be used to affect only $E^i - E^o$ and those like nigericin, which exchanges $H^+$ for $K^+$ across membranes, to influence only $pH^o - pH^i$. Thus we can study the importance of individual terms in $\Delta\mu_{H+}$. Also the molecular mechanism for $H^+$ uptake is presently unknown and there is the nagging question of whether $\Delta\mu_{H+}$ is actually the driving force for photophosphorylation or only a secondary form for energy storage. Thus, a final decision on the actual details of chloroplast bioenergetics must be left for the future.


## 4.3 Light-dependent water movements in vivo

In keeping with the general theme of this book we now turn our attention to the fluxes of substances across chloroplast limiting membranes in vivo. Although substantial movements of water, inorganic ions, and organic solutes occur into and out of chloroplasts in a plant cell, there is great experimental difficulty in actually measuring them. Very few techniques are adaptable to studies in vivo, while studies with isolated chloroplasts in artificial media are often difficult to relate to the conditions occurring in a plant cell, in part due to the lability of the organelles once outside the cell. Water movement has received far more attention than that of solutes, because water fluxes can be deduced from changes in chloroplast volume. Since there is no evidence that chloroplasts are maintained in a state far from osmotic equilibrium (e.g., see Mudrack, 1956), the movement of water presumably reflects a net change in the amount of solutes within chloroplasts, which in turn is most likely caused by solute fluxes across the limiting membranes.

Four main techniques have been used to measure chloroplast volume and its changes in vivo: (1) light microscopy, which has been employed since the end of the 19th century, (2) electron microscopy, which has been used from the 1950s, (3) gentle rapid isolation of chloroplasts so that changes occurring in the plant cell can be retained through the isolation procedure, which began in the late 1960s, and (4) monitoring of absorbance changes of leaves, which is an indirect way of studying chloroplast volume changes and also is of recent origin. Nearly all studies on chloroplast water fluxes have dealt with the effects of light. Not only is changing the illumination condition rather simple experimentally, but also the energetic basis for the fluxes can be readily identified.

### 4.3.1 Historical

Gicklhorn (1933) described a light-dependent contraction or coiling of the spiral chloroplasts in *Spirogyra*, which in fact had been reported by De Vries in 1889 and Senn in 1908. Peteler (1939) found that chloroplasts in *Melosira borreri* flattened in the light and then became round during darkness. Somewhat later Bünning (1942) observed such a light-dependent flattening of chloroplasts in *Nymphaea lotus, Nicotiana tabacum,* and *Phaseolus multiflorus.* Similar findings were reported by Busch (1953) for *Selaginella serpens* and by Vanden Driessche (1966a, 1966b) for *Acetabularia mediterranea.* Most of these studies using light microscopy were concerned with circadian rhythms and did not attempt to describe chloroplast volume changes quantitatively. However, .both Bünning and Busch made the noteworthy observation that the chloroplast shape changes were correlated with pH changes adjacent to the leaves, a subject area currently receiving much attention (e.g., see Davies, 1973; Smith and Raven, 1974).

Zurzycki (1964, 1967) found that moderate illumination caused a flattening of chloroplasts in *Mnium undulatum,* which could be reversed by strong blue light. Schorer-Mörtel (1972a, 1972b) reported rather similar responses for the chloroplast in *Mougeotia,* which also decreases in volume in the light. Hopkins (1966) observed that the chloroplast in *Surirella gemma* contracts on going from low to moderate illumination, which also occurs for *Pyrocystis lunula,* as shown by Swift and Taylor (1967). Further light-microscopic investigations by Hilgenheger and Menke (1965) on *Nitella flexilis* and Nobel et al. (1969) on *Pisum sativum* have indicated that chloroplast volume in such plants decreases about 20% in the light and these latter studies will be returned to below.

Kushida et al. (1964) using electron microscopy dramatically showed that chloroplasts in *Spinacia oleracea* flatten in the light. Similar electron microscopic results have been reported by Hilgenheger and Menke (1965) for *Nitella flexilis*, Packer et al. (1967) for *Spinacia oleracea*, and Miller and Nobel (1972) for *Pisum sativum*. The latter also found that the extent of light-dependent flattening increased in parallel with the chlorophyll in the developing chloroplasts and reached a maximum of a 23% decrease in thickness for mature plastids.

Another technique that has been successfully applied to the study of chloroplast volume changes in vivo is the measurement of leaf absorbance. Unfortunately absorbance changes relate more to alterations in refractive indices than to volume changes. Nevertheless light-dependent absorbance changes of leaves have helped show the universality of chloroplast volume changes in vivo. Packer et al. (1967) showed absorbance changes caused by chloroplast structural changes in *Spinacia oleracea*, *Beta vulgaris*, *Nicotiana tabacum*, *Senecio*, *Ulva lobata*, *Enteromorpha* and *Porphyra* (see also Murakami and Packer, 1970). Heber (1969) using absorbance changes found that the light-dependent volume decreases of chloroplasts occurred in *Spinacia oleracea*, *Chenopodium gigantospermum*, *Mimulus cardinalis*, *M. verbaneous*, *Oenothera hookeri*, *Plantago lanceolata*, and *Pisum sativum*. Thus, chloroplasts in over twenty different species, including algae as well as both primitive and advanced vascular plants, flatten in the light and then increase in volume when the illumination is removed.

### 4.3.2 Magnitude of volume changes

Let us now turn to a quantitative consideration of the effect of light on the water fluxes from chloroplasts in vivo. Most of these studies have taken advantage of the observation that many chloroplasts do not change their cross-sectional area in response to changes in illumination – rather, only the thickness (the vertical dimension in Fig. 4.1) varies. Therefore the flattening of a single chloroplast can be followed in a time sequence of photomicrographs, which preferably use the actinic light both for causing the water flux and for exposing the film (see Nobel et al., 1969). The volume change can then be readily calculated from the observed change in chloroplast thickness.

Hilgenheger and Menke (1965) showed that chloroplasts in *Nitella flexilis*

Table 4.1.
Magnitude and kinetics of the light-dependent chloroplast flattening in vivo.

| Plant | Illu-mination (lux) | Flattening half-time (min) | Maximal flattening (%) | Re-swelling half-time (min) | Technique | Reference |
|---|---|---|---|---|---|---|
| *Nitella flexilis* | 7000 | 1.9 | 18 | > 2 | light microscopy in vivo | Hilgenheger and Menke (1965) |
| *Pisum sativum* | 300 | 3.8 | 5 | | light microscopy in vivo | Nobel et al. (1969) |
| | 1000 | 2.8 | 10 | | | |
| | 4000 | 1.0 | 18 | | | |
| | 2000 | 2.5 | 15 | 4.8 | rapidly isolated chloroplasts | Nobel (1968a) |

decreased in volume about $18\%$ for an illumination of 7000 lux (Table 4.1). The decrease in chloroplast volume had a half-time of approximately 2 min and the minimal volume was reached after 20–30 min. They also observed that the extent of flattening was less at lower illuminations. Nobel et al. (1969) extended such findings using pea chloroplasts (Table 4.1) and discovered an additional feature, viz., there was a short time lag before the chloroplast flattening attained an appreciable rate. This lag amounted to 11 s at 1000 lux and 2 s at 4000 lux.

Nobel (1968b) used chloroplasts isolated in 2 min from pea leaves to study water fluxes. These chloroplasts were initially $95\%$ Class I (regarded as intact) and apparently retained the volume changes that had occurred in vivo. Once chloroplasts were isolated, their volume could be accurately determined by conventional techniques using large numbers of plastids, viz., the packing volume obtained upon centrifugation (approx. $2 \times 10^8$ chloroplasts per tube) and that obtained with a Coulter counter (approx. $2 \times 10^4$ chloroplasts per run). The half-time for the light-dependent volume decrease in vivo as measured using aqueously isolated chloroplasts was 2.5 min at 2000 lux and the extent was $15\%$ (Table 4.1). This agrees well with the light-microscopic observations made on pea chloroplasts in intact cells and tends to support the technique of using rapidly and gently isolated chloroplasts to measure water fluxes occuring in vivo. Moreover, measurements on isolated chloroplasts facilitated the study of the energetics of volume changes to be discussed below.

We will next estimate the net flux of water out of chloroplasts upon illumination of pea leaves with 2000 lux, which caused a volume decrease of about $3.2\%$ in the first minute. The volume of pea chloroplasts in the dark at approximately the osmotic potential occurring in vivo (equivalent to 260 Mol m$^{-3}$ sucrose) is 35 $\mu$m$^3$ (Nobel, 1969). The volume decrease upon illumination was thus (0.032 min$^{-1}$) (35 $\mu$m$^3$) or 1.1 $\mu$m$^3$ min$^{-1}$. Assuming that chloroplasts can be approximated by spheres 2 $\mu$m in radius, this corresponds to a volume flux of water equal to (1.1 $\times$ 10$^{-18}$ m$^3$ min$^{-1}$) (0.0167 min s$^{-1}$)/[4$\pi$ (2 $\times$ 10$^{-6}$ m)$^2$] or 3.7 $\times$ 10$^{-10}$ m s$^{-1}$. Such a volume flux of water could occur across a fairly permeable membrane with a hydraulic conductivity coefficient of 10$^{-7}$ m s$^{-1}$ bar$^{-1}$ (as might be reasonable for those surrounding chloroplasts) separating compartments that differ by less than 0.01 bar in osmotic potential.

What efflux of solutes could cause such a water flow if the chloroplasts were always close to osmotic equilibrium with the cytoplasm? The solute flux would be (3.7 $\times$ 10$^{-10}$ m s$^{-1}$) (260 Mol m$^{-3}$) or 9.6 $\times$ 10$^{-8}$ Mol m$^{-2}$ s$^{-1}$ (i.e., 9.6 picoMol cm$^{-2}$ s$^{-1}$), which is fairly large. For comparison with photosynthetic data it is convenient to express this solute flux in a rather different way, viz., it corresponds to (1.1 $\times$ 10$^{-18}$ m$^3$ min$^{-1}$) (260 Mol m$^{-3}$) [0.93 $\times$ 10$^9$ pea chloroplasts/mg chlorophyll] or 0.27 $\times$ 10$^{-6}$ Mol min$^{-1}$/mg chlorophyll. Later we will consider which solutes compose this flux.

### 4.3.3 Energetics

Light obviously plays an important role in the water efflux from chloroplasts in vivo. Using rapidly isolated chloroplasts, Nobel (1968a) showed with a rather crude action spectrum that chlorophyll was the pigment responsible. To see what aspect of photosynthesis was involved, DCMU or various uncouplers of photophosphorylation were introduced into the mesophyll cells (Nobel, 1968b). Although DCMU had very little effect, the light-dependent water efflux from chloroplasts was strongly inhibited by the uncouplers FCCP and nigericin (Table 4.2). DCMU should block non-cyclic and pseudocyclic electron flow, and so the remaining cyclic electron flow and its accompanying ATP formation could apparently support the chloroplast flattening. The inhibition caused by uncouplers suggested that ATP might be acting as the energy source. An action spectrum from 660 to 750 nm had a shoulder near 700 nm, suggesting that light absorbed by Photosystem I could lead to the water efflux. Again this is consistent with ATP or some 'high energy intermediate' related to photophosphorylation (such as $\Delta\mu_{H^+}$) being responsible for the chloroplast volume decrease.

Similar conclusions have been reached by other investigators. For instance, Heber (1969) found that when photosynthesis was suppressed by removal of $CO_2$, more ATP was apparently available for the chloroplast volume decrease.

Table 4.2.

Effects of metabolic treatments on the light-dependent water efflux from chloroplasts in vivo.

| Plant | Treatment | Effect on water efflux | Reference |
|---|---|---|---|
| *Pisum savitum* | DCMU | essentially none | Nobel (1968b) |
| | FCCP | strongly inhibited | Nobel (1968b) |
| | nigericin | completely inhibited | Nobel (1968b) |
| | far-red light | stimulated | Nobel (1968b) |
| various leaves | removal of $CO_2$ | stimulated | Heber (1969) |
| | far-red light | stimulated | Heber (1969) |
| *Dunaliella parva* | DCMU | strongly inhibited | Gimmler (1973) |
| | FCCP | strongly inhibited | Gimmler (1973) |

He also showed that far-red light absorbed by Photosystem I could support the water efflux (Table 4.2). Gimmler (1973) likewise measuring absorbance demonstrated that the chloroplast volume changes of *Dunaliella parva* responded to the ATP level in vivo – when the ATP in intact cells increased, there was a water efflux from the chloroplasts. Although the uncoupler FCCP inhibited the water efflux from the chloroplasts in the light in the same way as it had for *Pisum sativum*, DCMU also inhibited the volume decrease for *Dunaliella* (Table 4.2). Gimmler interpreted this as evidence for the lack of cyclic electron flow in this halophytic alga, which means that DCMU would then abolish all electron flow and ATP formation. In support of this hypothesis far-red light absorbed by Photosystem I did not cause a substantial increase in the ATP level in the cells.

In another study relating to the ATP level in plant cells, Nobel et al. (1969) found that the lag observed before appreciable chloroplast flattening equaled the time for ATP to increase by 10 nMol mg chlorophyll$^{-1}$. Moreover, the ATP level in the cells actually dropped before the $CO_2$ fixation rate became appreciable, suggesting that ATP was being diverted to some other process. If this process used ATP at the rate it was initially produced (the rate of photophosphorylation apparently increases with time in the light, Nobel et al., 1969), then its ATP consumption rate at 2000 lux would be 0.12 $\mu$Mol min$^{-1}$ mg chlorophyll$^{-1}$. We shall reconsider this rate after we discuss the ion fluxes that have been observed for chloroplasts in vivo.

Does the light-dependent water efflux from chloroplasts directly benefit the plant? Such a water movement can have a substantial concentrating effect on certain solutes in chloroplasts. For example, pea chloroplasts in the dark are only 51 % water by volume (Nobel, 1969 and 1974a) and so a volume decrease of 20 % upon illumination represents a loss of about 40 % of the internal water. Thus, impermeant solutes and photosynthetic enzymes would nearly double their concentration in chloroplasts as a result of the light-dependent water efflux, an aspect that may have important metabolic consequences.

## 4.4 Light-dependent ion movements in vivo

Before discussing the limited amount of data on ion fluxes for chloroplasts in vivo, we will consider their inorganic ion contents. Our emphasis here will be on $K^+$, $Na^+$, $Mg^{2+}$, $Ca^{2+}$, and $Cl^-$, which are the major ions whose concentrations have been determined in vivo.

Actually, the principal species moving into chloroplasts in the light is $CO_2$ (in the dissolved or hydrated form, not as bicarbonate, see Walker, 1974). For a pea leaf after 2 min at 2000 lux the $CO_2$ flux into chloroplasts is 0.3 $\mu$Mol min$^{-1}$ mg chlorophyll$^{-1}$ (Nobel et al., 1969), which increases to over 3 $\mu$Mol min$^{-1}$ mg chlorophyll$^{-1}$ at 50,000 lux, an illumination about half of that of full sunlight. Nearly all the entering $CO_2$ is either converted into starch or exported from the chloroplasts in the form of photosynthetic products, and so the $CO_2$ influx apparently does not lead to an osmotic water movement. However, movement of various organic solutes occurs. Heldt (1969) and Heldt and Rapley (1970a, 1970b) have described carriers that shuttle certain species involved in photosynthesis across the chloroplast limiting membranes, viz., one carrier is specific for phosphate, 3-phosphoglycerate (PGA), and dihydroxyacetonephosphate (DHAP), while another translocates certain dicarboxylic acids. It has been suggested that phosphate$^{3-}$ and $CO_2$ readily enter chloroplasts in the light, and that DHAP$^{2-}$ (or possibly PGA$^{3-}$) is an important vehicle for rapidly moving photosynthetically fixed carbon out of the plastids. Although co-transport of organic solutes and cations has received considerable attention for animal transport systems, $H^+$ is generally invoked as the charge-balancing ion for chloroplasts. For a complete discussion of the relative rate of entry of organic compounds and others directly involved in metabolism based primarily on studies with isolated chloroplasts, the reviews of Heber (1974) and Walker (1974) should be consulted.

No technique has proved entirely satisfactory for determining chloroplast contents. Much information on the ions in chloroplasts has been obtained using plastids isolated non-aqueously, a technique pioneered by Behrens and Thalacker (1957), Heber (1957), and Stocking (1959). It involves rapidly freezing the tissue, removing water by sublimation, grinding, and then isolating various fractions in organic solvents using differential centrifugation. The freezing can cause a local formation of minute ice crystals where water molecules are preferentially crystallized. The remaining solution, which is enriched in solutes, is mechanically pushed around in the cell by the developing ice crystals and the solutes may be deposited some distance from their origin. Also, the grinding and extraction of the lyophilized material may not always lead to fracture at the chloroplast surface. Thus cytoplasmic contamination is an ever present problem with the non-aqueous technique (Nobel et al., 1969; Bird et al., 1973). Chloroplasts can be isolated aqueously such that they retain most of their contents (Nobel, 1969), but an exact determination of what entered or left during the isolation procedure is difficult. The use of ion-sensitive electrodes inserted into chloroplasts offers some promise for measuring their contents, but so far this has not been technically feasible except for $H^+$ (Davis, 1974). In fact only recently have the somewhat simpler micro-pipettes necessary for measuring electrical potential differences been inserted into chloroplasts (Bulychev et al., 1972; Vrendenberg et al., 1973; Davis, 1974; Vrendenberg, 1974).

### 4.4.1 Ionic contents

Menke (1940) indicated that chloroplasts of *Spinacia oleracea* contain $K^+$, $Mg^{2+}$, and $Ca^{2+}$, the latter being rather tightly bound. Stocking and Ongun (1962) showed that chloroplasts were actually a major site for ion accumulation in green cells of *Nicotiana rustica* and *Phaseolus vulgaris*. Averaging their data for bean and tobacco leaves indicates that of the four ions considered, the amount of $K^+$ in the chloroplasts was the highest, $Mg^{2+}$ and $Ca^{2+}$ were intermediate, and $Na^+$ was the lowest. More recently Ringoet et al. (1971) have shown that substantial amounts of foliarly applied $Ca^{2+}$ accumulates in chloroplasts of *Avena sativa*.

Saltman et al. (1963) determined actual concentrations of $K^+$ and $Na^+$ in chloroplasts of *Nitella opaca* (Table 4.3). The $K^+$ concentration was about 8 Mol $m^{-3}$ higher and $Na^+$ the same amount lower than in the cytoplasm. Kishimoto and Tazawa (1965) likewise found that $K^+$ had a higher concentration than $Na^+$ in the chloroplasts of *Nitella flexilis*, and that $Cl^-$ appeared to be the predominant anion (Table 4.3). In their case the $K^+$ concentration was about 20 Mol $m^{-3}$ lower and $Na^+$ the same amount higher than in the cytoplasm. Using *Nitella translucens* MacRobbie (1964) also indicated that $Cl^-$ was the major anion in chloroplasts. Harvey and Brown (1967, 1969) found high concentrations of $K^+$ in pea chloroplasts as did Larkum (1968) for chloroplasts in *Tolypella intricata* (Table 4.3). Larkum (1968) also reported substantial amounts of $Mg^{2+}$ and $Ca^{2+}$ in chloroplasts and again indicated that $Cl^-$ was the principal anion. A similar pattern was found for *Pisum sativum*

Table 4.3.

Ionic concentrations in chloroplasts in vivo. Data are expressed as Mol m$^{-3}$ per volume of water, although much of the $Mg^{2+}$ and $Ca^{2+}$ would actually be bound.

| Plant | $K^+$ | $Na^+$ | $Mg^{2+}$ | $Ca^{2+}$ | $Cl^-$ | Reference |
|---|---|---|---|---|---|---|
| *Nitella opaca* | 84 | 4 | | | | Saltman et al. (1963) |
| *Nitella translucens* | 150 | | | | 240 | MacRobbie (1964) |
| *Nitella flexilis* | 110 | 26 | | | 136 | Kishimoto and Tazawa (1965) |
| *Pisum sativum* | c. 300 | | | | | Harvey and Brown (1967,1969) |
| *Tolypella intricata* | 340 | 36 | 14 | 21 | 340 | Larkum (1968) |
| *Pisum sativum* | 99 | 10 | 16 | 15 | 92 | Nobel (1969) |
| *Limonium vulgare* | c. 300 | c. 200 | | | c. 500 | Larkum and Hill (1970) |

by Nobel (1969), although the concentrations tended to be less, while Larkum and Hill (1970) reported even higher and variable concentrations of $Na^+$ and $Cl^-$ for chloroplasts in *Limonium vulgare* (Table 4.3). Although no completely consistent pattern emerges from these studies, we can summarize by saying that $K^+$ generally appears to be the principal cation, $Cl^-$ the principal anion and that $Mg^{2+}$ as well as $Ca^{2+}$ also occur in large amounts in chloroplasts.

### 4.4.2 *Ionic fluxes*

The difficulties of accurately determining the ionic contents of chloroplasts are compounded when we attempt to estimate fluxes occurring in vivo, since a time sequence of such determinations is then generally required. So far only the $K^+$ fluxes have been estimated using aqueously isolated pea chloroplasts under the same conditions as for the light-dependent water efflux studies (Nobel, 1969). Fig. 4.3 presents the $K^+$ efflux and the total solute efflux, the latter being calculated from the light-dependent volume decrease assuming that the volume change was caused solely by solute movement and that the chloroplasts were always close to osmotic equilibrium with the cytoplasm. Although it may be coincidental, we note that the $K^+$ efflux for the first min at 2000 lux was 0.10 $\mu$Mol min$^{-1}$ mg chlorophyll$^{-1}$, about the same as the estimated initial rate of ATP formation, 0.12 $\mu$Mol min$^{-1}$ mg chlorophyll$^{-1}$. The relative change of the net $K^+$ efflux with time was similar to that of the total solutes and equal to nearly half of it in magnitude (Fig. 4.3). Based on measurements of the chloroplast ionic contents after the water efflux had ceased, an efflux of $Cl^-$ comparable to that of $K^+$ apparently occurred. Light may thus cause effluxes of $K^+$, $Cl^-$, and a more or less osmotically equivalent amount of water from pea chloroplasts in vivo.

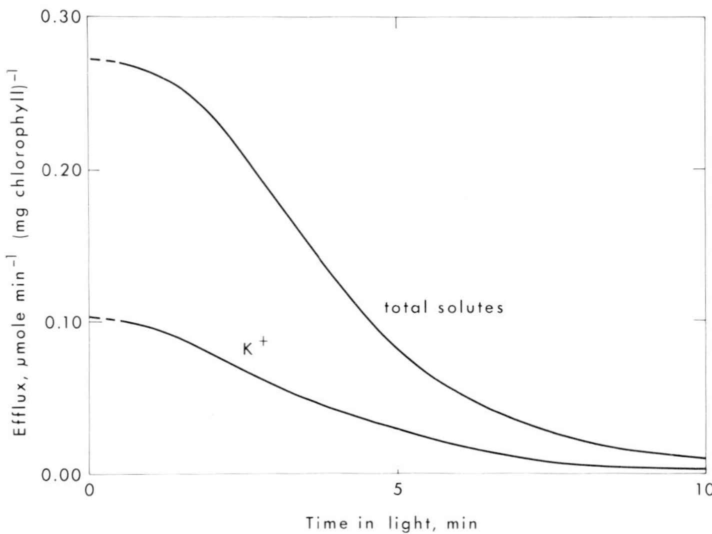



Fig. 4.3. Net light-dependent fluxes of $K^+$ and total solutes from pea chloroplasts in vivo. Illumination was 2000 lux.

What actually causes the $K^+$, $Cl^-$, and water fluxes out of chloroplasts in the light? Once $K^+$ and $Cl^-$ leave, an efflux of water would occur osmotically, and so we will focus our attention on the ionic movements. $K^+$ or $Cl^-$ could be actively transported out of chloroplasts upon illumination. Alternatively, the ion movements could be initiated by a light-dependent efflux of $H^+$ out across the chloroplast limiting membranes, for which there is some evidence (Heber and Krause, 1972; Werdan et al., 1972; Heldt et al., 1973; Davis, 1974). The $H^+$ efflux could be followed by an exchange of cytoplasmic $H^+$ for $K^+$ from the chloroplasts, while $Cl^-$ would then exit so that approximate local electroneutrality is maintained. On the other hand, the initial event might be associated with the well-documented alkalization of the stroma in the light (e.g., see Heldt et al., 1973). This could alter the permeability of the chloroplast limiting membranes, the electrical potential across them, ion binding to stromal proteins, and the concentration of various organic solutes in the chloroplasts. The light-dependent ion effluxes may then represent passive responses to one or more of such changes initiated in the stroma. Upon cessation of illumination, the pH of the stroma drops as protons move out of the thylakoids and possibly in from the cytoplasm. Because of many negatively charged proteins and membranes in chloroplasts, the stroma could have a lower electrical potential than the cytoplasm, which would encourage diffusion of $K^+$ back in from the cytoplasm in the dark. The electrical potential in the stroma should then gradually rise, as appears to be the case (Davis, 1974; Vrendenberg, 1974). The increase in positive charge may be accompanied by a nearly compensating $Cl^-$ influx and of course an osmotic movement of water inward, which would account for the observed reswelling of chloroplasts in the dark.

Movement of $Mg^{2+}$ across the chloroplast limiting membranes also occurs in vivo. Lin and Nobel (1971) found that $Mg^{2+}$ in pea chloroplasts increases by 0.10 $\mu$Mol mg chlorophyll$^{-1}$ in the light, which is in the opposite direction from the $K^+$ and $Cl^-$ fluxes. The light-dependent water efflux would concentrate any $Mg^{2+}$ dissolved in the stroma in the dark and also there appears to be a release of $Mg^{2+}$ from the thylakoids upon illumination (Lin and Nobel, 1971; Hind and McCarty, 1973). The actual concentration of non-chlorophyll $Mg^{2+}$ in chloroplasts has not been accurately determined because of ignorance of how much of such $Mg^{2+}$ is bound (the same problem prevents meaningful calculations of internal $Ca^{2+}$ concentrations). Nevertheless, because of the above three effects, the $Mg^{2+}$ concentration in the chloroplast stroma apparently increases from a rather low value in the dark up to 5 to 10 Mol m$^{-3}$ upon illumination. This may be an important regulatory mechanism for photosynthesis, since $Mg^{2+}$ concentrations in this range are necessary for optimal photophosphorylation and $CO_2$ fixation rates in vitro (e.g., see Bassham and Kirk, 1969; Bassham et al., 1970; Jensen, 1971; Hind and McCarty, 1973; Barber and Telfer, 1974; Walker, 1974). It helps account for an increasing rate of photophosphorylation in vivo with illumination time (Nobel et al., 1969). Although the kinetics of the $Mg^{2+}$ fluxes have not been measured in the plant cell, indirect evidence has been obtained by observing the changes in photophosphorylation and $CO_2$ fixation rates of rapidly isolated chloroplasts (Nobel, 1968c, 1970; Lin and Nobel, 1971). If these changes can be attributed to alterations in $Mg^{2+}$ concentrations, then the half-time for the increase in stromal $Mg^{2+}$ upon illuminating pea leaves with 2000 lux is 4–5 min, while the half-time for the loss in the dark is 5 min. Thus, the time courses of $Mg^{2+}$ changes are similar or perhaps somewhat slower than those for the light-dependent water movements and the fluxes across the chloroplast limiting membranes occur in opposite directions.

Let us now summarize the ion movements that may occur upon illumination for chloroplasts in vivo. The first species affected may be $H^+$. Protons move into the thylakoids and maybe out to the cytoplasm, leading to an increase in the stromal pH from perhaps 5 or 6 up to about 8. $Mg^{2+}$ moves in the opposite direction, leading to a substantial increase in its concentration in the stroma in the light, possibly up to 10 Mol m$^{-3}$. Meanwhile, $K^+$ and $Cl^-$ leave the chloroplasts and so there is an osmotic efflux of water, which tends to concentrate the substances left behind. The electrical potential in the thylakoids increases relative to that in the stroma, as also may the potential in the cytoplasm. The magnitude of many of these processes is still in doubt and also the role of bicarbonate and other ions, especially organic ones, is uncertain.

## 4.5 Directions for future research

As is clear from the above discussion, our knowledge of the ionic contents of chloroplasts is still meagre, especially with regard to the variation expected for different plant species. Evidence on the actual interrelationship between $H^+$, $K^+$, $Cl^-$, and water fluxes is not yet available. Fluxes of the physiologically important divalent cations have not been directly measured. Very little is known about the fractions of ions like $Mg^{2+}$ and $Ca^{2+}$ which are bound. The effect of illumination and ionic concentrations on the electrical potential differences across chloroplast membranes needs further study. We do not know the concentration of bicarbonate and organic ions in chloroplasts nor whether they contribute to osmotic movement of water. Chloroplast research is currently in a very active stage. Refinements of present experimental approaches together with relatively new techniques such as the insertion of ion-sensitive electrodes into chloroplasts, monitoring of spectral changes of pH-sensitive pigments in vivo, or use of electron and other microprobes to observe ion distributions in a plant cell will help us gain a better understanding of the ionic relations of chloroplasts in vivo.

## References

D.I. ARNON, M.B. ALLEN and F.R. WHATLEY, Nature, 174 (1954) 394.

M. AVRON and M. GIBBS, Plant Physiol., 53 (1974) 140.

J. BARBER, FEBS Lett., 20 (1972a) 251.

J. BARBER, Biochim. Biophys. Acta, 275 (1972b) 105.

J. BARBER and A. TELFER, in U. Zimmermann and J. Dainty (Eds.) Membrane Transport in Plants and Plant Organelles (1974) Springer Verlag, Heidelberg, p. 281.

J.A. BASSHAM and M. KIRK, in K. Shibata, A. Takamiya, A.T. Jagendorf and R.C. Fuller (Eds.) Comparative Biochemistry and Biophysics of Photosynthesis (1968) University of Tokyo Press, Tokyo, p. 365.

J.A. BASSHAM, A.M.EL-BADRY, M.R. KIRK, H.C.J. OTTENHEYM and H. SPRINGER-LEDERER, Biochim. Biophys. Acta, 223 (1970) 261.

M. BEHRENS and R. THALACKER, Naturwissenschaften, 44 (1957) 621.

I.F. BIRD, M.J. CORNELIUS, T.A. DYER and A.J. KEYS, J. Exp. Bot., 24 (1973) 211.

N.K. BOARDMAN, Annu. Rev. Plant Physiol., 21 (1970) 115.

A.A. BULYCHEV, V.K. ANDRIANOV, G.A. KURELLA and F.F. LITVIN, Nature 236 (1972) 175.

E. BÜNNING, Z. Bot., 37 (1942) 433.

G. BUSCH, Biol. Zentralbl., 72 (1953) 598.

A.R. CROFTS, J. Biol. Chem., 242 (1967) 3352.

A.R. CROFTS, A.C. WRAIGHT and D.E. FLEISCHMANN, FEBS Lett., 5 (1971) 89.

D.D. DAVIES, Symp. Soc. Exp. Biol., 27 (1973) 513.

R.F. DAVIS, in U. Zimmerman and J. Dainty (Eds.) Membrane Transport in Plants and Plant Organelles (1974) Springer Verlag, Heidelberg, p. 197.

L.N.M. DUYSENS, Science, 120 (1954) 353.

L.N.M. DUYSENS, J. AMESZ and B.M. KAMP, Nature, 190 (1961) 510.

S. FLEISCHER and L. PACKER, Methods in Enzymology (1974) Vol. 31, Part A, Academic Press, New York.

G. FORTI, M. AVRON and A. MELANDRI (Eds.) Proceedings of the 2nd International Congress on Photosynthesis Research (Stressa) (1972) Junk, The Hague.

A. FRENKEL, J. Am. Chem. Soc., 76 (1954) 5568.

R.E. GAENSSLEN and R.E. MCCARTY, Arch. Biochem. Biophys., 147 (1971) 55.

M. GIBBS, in M. Gibbs (Ed.) Structure and Function of Chloroplasts (1971) Springer Verlag, Heidelberg, p. 169.

J. GICKLHORN, Protoplasma, 17 (1933) 571.

H. GIMMLER, Z. Pflanzenphysiol, 68 (1973) 289.

M.J. HARVEY and A.P. BROWN, Biochem. J., 105 (1967) 30.

M.J. HARVEY and A.P. BROWN, Biochim. Biophys. Acta, 172 (1969) 116.

U. HEBER, Ber. Dtsch. Bot. Ges., 70 (1957) 371.

U. HEBER, Biochim. Biophys. Acta, 180 (1969) 302.

U. HEBER and K.A. SANTARIUS, Z. Naturforsch., 25b (1970) 718.

U. HEBER and G.H. KRAUSE, in M.D. Hatch, C.B. Osmond and R.O. Slatyer (Eds.) Photosynthesis and Photorespiration (1972) Wiley–Interscience, New York, p. 218.

U. HEBER, Annu. Rev. Plant Physiol., 25 (1974) 393.

H.W. HELDT, FEBS Lett., 5 (1969) 11.

H.W. HELDT and L. RAPLEY, FEBS Lett., 7 (1970a) 139.

H.W. HELDT and L. RAPLEY, FEBS Lett., 10 (1970b) 143.

H.W. HELDT, K. WERDAN, M. MILOVANCEV and G. GELLER, Biochim. Biophys. Acta, 314 (1973) 224.

H. HILGENHEGER and W. MENKE, Z. Naturforsch., 20b (1965) 699.

G. HIND and R.E. MCCARTY, in A.C. Giese (Ed.) Photophysiology (1973) Vol. VIII, Academic Press, New York, p. 113.

J.T. HOPKINS, J. Mar. Biol. Assoc. U.K., 46 (1966) 617.

J.B. JACKSON and A.R. CROFTS, FEBS Lett., 4 (1969) 185.

A.T. JAGENDORF and E. URIBE, Proc. Natl. Acad. Sci. U.S., 55 (1966) 170.

R.G. JENSEN and J.A. BASSHAM, Proc. Natl. Acad. Sci. U.S., 56 (1966) 1095.

R.G. JENSEN, Biochim. Biophys. Acta, 234 (1971) 360.

W.D. JESCHKE, Planta, 73 (1967) 161.

W. JUNGE and H.T. WITT, Z. Naturforsch., 23b (1968) 244.

A.J. KEYS, Biochem. J., 108 (1968) 1.

U. KISHIMOTO and M. TAZAWA, Plant Cell Physiol., 6 (1965) 507.

H. KUSHIDA, M. ITOH, S. IZAWA and K. SHIBATA, Biochim. Biophys. Acta, 79 (1964) 201.

A.W.D. LARKUM, Nature, 218 (1968) 447.

A.W.D. LARKUM and A.E. HILL, Biochim. Biophys. Acta, 203 (1970) 133.

A.W.D. LARKUM and N.K. BOARDMAN, FEBS Lett., 40 (1974) 229.

A.W.D LARKUM and W.D. BONNER, Biochim. Biophys. Acta, 256 (1972) 396.

D.C. LIN and P.S. NOBEL, Arch. Biochem. Biophys., 145 (1971) 622.

E.A.C. MACROBBIE, J. Gen. Physiol., 47 (1964) 859.

E. MARRÊ, G. FORTI, R. BIANCHETTI and B. PARISI, in La Photosynthese (1963) Centre National de la Recherche Scientifique, Gif-sur-Yvette, p. 557.

W. MENKE, Z. Physiol. Chem., 263 (1940) 104.

M.M. MILLER and P.S. NOBEL, Plant Physiol., 49 (1972) 535.

P. MITCHELL, Chemiosmotic Coupling in Oxidative and Photosynthetic Phosphorylation (1966a) Glynn Research Ltd., Bodmin, Cornwall, England.

P. MITCHELL, Biol. Rev., 41 (1966b) 445.

P. MITCHELL, in Mechanisms in Bioenergetics (1973) Academic Press, New York, p. 177.

K. MUDRACK, Protoplasma, 47 (1956) 461.

S. MURAKAMI and L. PACKER, Plant Physiol., 45 (1970) 289.

S. MURAKAMI, J. TORRES-PEREIRA and L. PACKER, in Govindjee (Ed.) Bioenergetics of Photosynthesis (1975) Academic Press, New York, p. 555.

J. NEUMANN and A.T. JAGENDORF, Arch. Biochem. Biophys., 107 (1964) 109.

P.S. NOBEL, Plant Physiol., 43 (1968a) 781.

P.S. NOBEL, Plant Cell Physiol., 9 (1968b) 499.

P.S. NOBEL, Biochim. Biophys. Acta, 153 (1968c) 170.

P.S. NOBEL, Biochim. Biophys. Acta, 172 (1969) 134.

P.S. NOBEL, D.T. CHANG, C.T. WANG, S.S. SMITH and D.E. BARCUS, Plant Physiol., 44 (1969) 655.

P.S. NOBEL, Plant Cell Physiol., 11 (1970) 467.

P.S. NOBEL, Introduction to Biophysical Plant Physiology (1974a) W.H. Freeman, San Francisco, p. 214.

P.S. NOBEL, in U. Zimmermann and J. Dainty (Eds.) Membrane Transport in Plants and Plant Organelles (1974b) Springer Verlag, Heidelberg, p. 289.

D.R. ORT and S. IZAWA, Plant Physiol., 52 (1973) 595.

L. PACKER, A.C. BARNARD and D.W. DEAMER, Plant Physiol., 42 (1967) 283.

L. PACKER and A.R. CROFTS, Current Topics in Bioenergetics, 2 (1967) 23.

L. PACKER, S. MURAKAMI and C.W. MEHARD, Annu. Rev. Plant Physiol., 21 (1970) 271.

K. PETELER, Protoplasma, 32 (1939) 9.

J.A. RAVEN, J. Exp. Bot., 18 (1968) 712.

A. RINGOET, R.V. RECHENMANN, A.J. GIELINK and E. WINNENDORP, Z. Pflanzenphysiol, 64 (1971) 80.

H. ROTTENBERG, T. GRUNWALD and M. AVRON, Eur. J. Biochem., 25 (1972) 54.

H. ROTTENBERG and T. GRUNWALD, Eur. J. Biochem., 25 (1972) 71.

B. RUMBERG and U. SIGGEL, Naturwissenschaften, 56 (1969) 130.

P. SALTMAN, J.G. FORTE and G.M. FORTE, Exp. Cell Res., 29 (1963) 504.

K.A. SANTARIUS and U. HEBER, Biochim. Biophys. Acta, 102 (1965) 39.

W. SCHLIEPHAKE, W. JUNGE and H.T. WITT, Z. Naturforsch., 23b (1968) 1571.

S. SCHMIDT, R. REICH and H.T. WITT, Naturwissenschaften, 58 (1971) 414.

G. SCHORER-MÖRTEL, Z. Pflanzenphysiol., 68 (1972a) 144.

G. SCHORER-MÖRTEL, Z. Pflanzenphysiol., 68 (1972b) 193.

H. SCHRÖDER, M. MUHLE and B. RUMBERG, in G. Forti, M. Avron and A. Melandri (Eds.) Proc. 2nd Internatl. Congr. Photosyn. Res. (1972) Junk, The Hague, p. 919.

M. SCHWARTZ, Annu. Rev. Plant Physiol., 22 (1971) 469.

W. SIMONIS and W. URBACH, Annu. Rev. Plant Physiol., 24 (1973) 89.

F.A. SMITH and J.A. RAVEN, in U. Zimmermann and J. Dainty (Eds.) Membrane Transport in Plants and Plant Organelles (1974) Springer Verlag, Heidelberg, p. 380.

        *P.S. Nobel*

C.R. STOCKING, Plant Physiol., 34 (1959) 56.

C.R. STOCKING and A. ONGUN, Am. J. Bot., 49 (1962) 284.

C.R. STOCKING and S. LARSON, Biochem. Biophys. Res. Commun., 37 (1969) 278.

G.R. STRICHARTZ, Plant Physiol., 48 (1971) 553.

G.R. STRICHARTZ and B. CHANCE, Biochim. Biophys. Acta, 256 (1972) 71.

E. SWIFT and W.R. TAYLOR, J. Phycol., 3 (1967) 77.

T. VANDEN DRIESSCHE, Exp. Cell Res., 42 (1966a) 18.

T. VANDEN DRIESSCHE, Biochim. Biophys. Acta 126 (1966b) 456.

W.J. VRENDENBERG, P.H. HOMANN and W.J. TONK, Biochim. Biophys. Acta, 314 (1973) 261.

W.J. VRENDENBERG, in U. Zimmermann and J. Dainty (Eds.) Membrane Transport in Plants and Plant Organelles (1974) Springer Verlag, Heidelberg, p. 126.

D.A. WALKER, Biochem. J., 92 (1964) 22c.

D.A. WALKER, in A. San Pietro (Ed.) Methods in Enzymology (1971) Vol. 23, Academic Press, New York, p. 211.

D.A. WALKER, in D.H. Northcote (Ed.) Biochemistry Series I, Vol. 11, M.T.P. International Review of Science (1974) Butterworth, London, p. 1.

K. WERDAN, H.W. HELDT and G. GELLER, Biochim. Biophys. Acta, 283 (1972) 430.

H.T. WITT, Quart. Rev. Biophys., 4 (1971) 365.

J. ZURZYCKI, Protoplasma, 58 (1964) 458.

J. ZURZYCKI, in T.W. Goodwin (Ed.) Biochemistry of Chloroplasts (1967) Vol. II. Academic Press, New York, p. 609.

*Ion transport in plant cells and tissues*
*edited by D.A. Baker and J.L. Hall*
© *North-Holland Publishing Company, 1975*



# Algal cells

## J.A. RAVEN

# Contents

## 5.1  Introduction

### 5.1.1  The range of algae

The algae are a diverse group of (mainly) photoautotrophic plants, defined mainly by the absence of certain vegetative and reproductive characters found in 'higher plants'. As well as the prokaryotic Cyanophyceae, they include some of the smallest eukaryotic unicells known (*Micromonas*) as well as complicated, differentiated plants such as *Macrocystis* which can be up to 50 m long (Fritsch, 1935, 1945; Dodge, 1973; Stewart, 1974). Algae are divided into some 14 classes (Dodge, 1973).

### 5.1.2  Giant algal cells

As well as being studied for their intrinsic interest, the algae have provided unique experimental material for ion-transport studies in the shape of giant, coenocytic cells. These cells have the same basic structure as the cells of higher plants, i.e. a rigid cell wall functional in volume and osmo-regulation, enclosing a thin film of cytoplasm which in turn surrounds a large aqueous vacuole which occupies most of the cell volume. While some doubts as to the comparability of the vacuole of these cells with those of smaller plant cells have been expressed (Sutcliffe, 1962), most evidence suggests that they are homologous organelles (Dainty, 1962, 1968; Hope and Walker, 1975).

These coenocytes are very large; the internodal cells of *Chara* and *Nitella* can be $10^{-3}$ m in diameter and $10^{-1}$ m long, while the spherical cells of *Valonia* can be $10^{-2}$ m in diameter (Fig. 5.1). This allows physical separation of the vacuolar sap from the cytoplasm and measurement of chemical and radio-chemical concentrations in the two phases separately. Electrical measurements

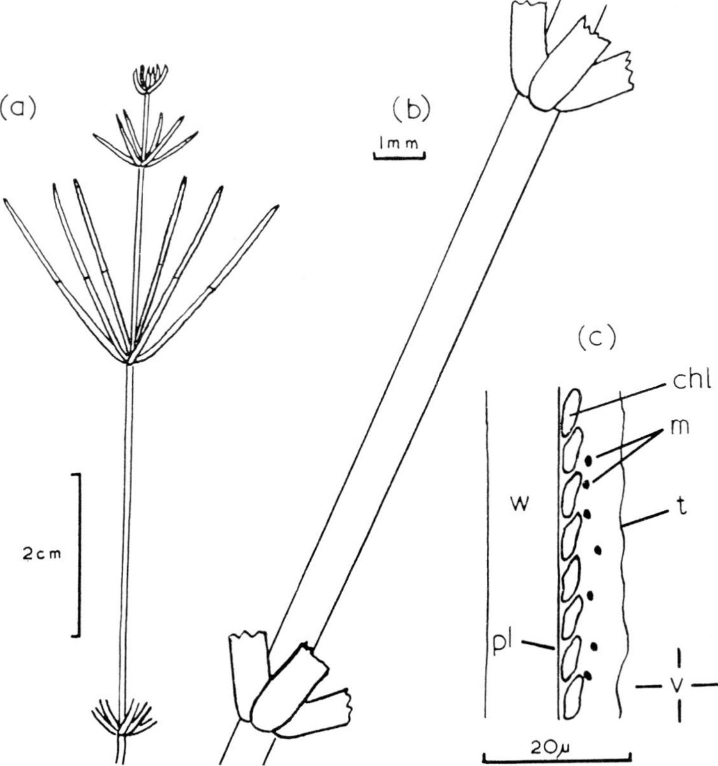

Fig. 5.1. Structure of Charophytes. (a) Apex of plant, showing the elongate internodal cells. The branches ('leaf cells') arise from the small nodal cells. Growth is apical. (b) A single internodal cell. (c) Longitudinal section of part of an internodal cell. w = cell wall; pl = plasmalemma; chl = chloroplast; m = mitochondrion; t = tonoplast; v = vacuole. The chloroplasts are embedded in stationary cytoplasm; the colourless cytoplasm between the chloroplast layer and the vacuole exhibits cytoplasmic streaming. Reproduced from Briggs et al. (1961).

with intracellular micropipettes are also made more readily on these giant cells than on smaller cells as the site of the electrode tip can be much more easily determined. It is important to bear in mind that the charophytes are multicellular plants, with a small fraction of the total number of cells (but the vast majority of their volume) being giant cells.

## 5.1.3 Microalgae

The other major kind of algal cells used in ion transport studies is the microalgal unicells such as *Chlorella* and *Scenedesmus*. These are probably as non-vacuolate as eukaryotic plant cells can ever be, although the volume of the vacuole can be an appreciable fraction (10%) of the cell volume, and the tonoplast area an appreciable fraction (35%) of the plasmalemma area (Atkinson et al., 1971). These organisms are much less amenable to electrophysiological studies than are the giant algal cells (Barber, 1968b). However, they have the advantages that they can be cultured axenically in continuous or synchronous culture, thus providing large quantities of material for relating transport phenomena to metabolism, development and nutrition. Many of the smaller unicellular algae present interesting variations on the 'plant cell norm' of structure in relation to transport. Many unicells are wall-less, and regulate their volume in freshwater by means of contractile vacuoles; many are also motile by amoeboid or, more commonly, flagellar mechanisms (see Raven, 1975).

### 5.1.4 Multicellular algae

The larger and more complex algae (e.g. many Rhodophyceae and Phaeophyceae lack the advantage of either the giant algal cells or the microalgal unicells, and present the same problems as do many higher plant tissues in the interpretation of data due to tissue inhomogeneities and symplastic transport; however, these algae have their own intrinsic interest (Raven, 1975).

### 5.1.5 Taxonomy of the algae used

Many of the algae that have been used in transport studies are in the class Chlorophyceae (green algae). These include such non-vacuolate walled unicells as *Chlorella*, *Scenedesmus* and *Ankistrodesmus*, as well as the freshwater coenocyte *Hydrodictyon* and such marine coenocytes as *Valonia*, *Halicystis*, *Codium*, *Bryopsis*, *Chaetomorpha*, *Acetabularia* and *Valoniopsis*. Among the less plant-like organisms, the flagellates *Chlamydomonas* and *Dunaliella* have also been used. The Charophyceae are closely related to the Chlorophyceae, and contain the much-used genera *Chara* and *Nitella* which have giant internodal cells.

Other algal classes have been less widely used; the Rhodophyceae have one genus (*Griffithsia*) with giant cells, and *Porphyra* which is close to being nonvacuolate. The Phaeophyceae have attracted attention in the shape of the eggs and zygotes of *Fucus* and *Pelvetia*, while the Dinophyceae include the heterotrophic giant cells of *Noctiluca*. The prokaryotic Cyanophyceae (e.g. *Anacystis*) have recently been investigated. Further discussion of algae used in ion transport work will be found in Raven (1975).

## 5.2 Electrochemical potentials

### 5.2.1 Electrical potentials

Electrical potential differences between cell compartments (cytoplasm–solution, vacuole–cytoplasm, vacuole–solution) have been measured in a number of algal cells using microelectrodes (see Ch. 1). These results show that the cytoplasm is always electrically negative with respect to the solution (Tables 5.1, 5.2 and 5.3; see also Gutknecht and Dainty, 1968; Raven, 1975). It is generally assumed that the cytoplasmic value corresponds to the 'ground cytoplasm' rather than to organelles. In most cases the vacuole is also negative with respect to the bathing solution (Tables 5.1, 5.2 and 5.3; Gutknecht and Dainty, 1968; Raven, 1975). However, some marine green algal coenocytes (*Chaetomorpha, Valonia, Valoniopsis*) have very low negative or even positive potentials in their vacuoles with respect to the bathing medium. This means that there is a large vacuole-positive P.D. across the tonoplast. Many other algae have a smaller P.D. in the same direction across the tonoplast (Table 5.1).

### 5.2.2 Chemical activities and concentrations

These are measured by standard techniques on whole cells of small algae and on separated vacuole and cytoplasm (or with further fractionation of the cytoplasm) in giant cells (MacRobbie, 1971, 1974). Care must be taken to correct for ions in the free space (Kesseler, 1964), and for metabolised ions, where there are two conflicting requirements of removal of ions from the free space and instantaneous inhibition of enzymes assimilating or regenerating the ion concerned (Raven, 1974a, 1975). Indirect arguments as to the cytoplasmic concentration of metabolised ions ($HCO_3^-$, $NO_3^-$) have also been applied, based on the observed rate of assimilation and the rate–concentration relationship for the extracted assimilating enzyme (Raven, 1970, 1975).

Generalisations as to the distribution of ions are that walled algal cells generally have a higher total salt concentration inside than outside, and a higher $K^+/Na^+$ ratio inside than outside. In vacuolate cells the major intracellular ions are $K^+$, $Na^+$ and $Cl^-$ (Tables 5.1, 5.2 and 5.3).

### 5.2.3 Electrochemical potential gradients and active transport

#### Anions
In general, active inward transport at the plasmalemma is required to account for the observed distribution of anions when the Nernst equation (Eq. 1.7) is

Table 5.1.

$K^+$, $Na^+$ and $Cl^-$ concentrations in various compartments in algal cells and their bathing medium and electrical P.Ds between compartments. Square brackets denote measurement of activity rather than of concentration. Subscripts refer to active transport, using the Nernst equation; e = electrochemical equilibrium, i = active transport inwards, o = active transport outwards. When only vacuolar and external concentrations are specified, transport is somewhere between the external solution and the vacuole. When cytoplasm concentrations are specified, subscripts to these concentrations refer to the plasmalemma, and subscripts to the corresponding vacuolar concentrations refer to the tonoplast. Cytoplasmic concentration refers to the entire cytoplasm except for *Chara* and *Nitella* where the (chloroplast-free) streaming cytoplasm was measured; ion concentrations in the chloroplasts are higher than those in the streaming cytoplasm (MacRobbie, 1970).

| Alga | Ion concentration (Mol m$^{-3}$) | | | | Electrical P.D. (mV) | | Reference |
|---|---|---|---|---|---|---|---|
| | Compartment | $K^+$ | $Na^+$ | $Cl^-$ | $E_{co}$ | $E_{vo}$ | |
| *Chara australis* | Solution | 0.1 | 1.0 | 1.6 | | | Hope and Walker (1960); Coster |
| | Cytoplasm | [115](e) | N.D. | [10](i) | −170 | −152 | (1965); Vorobiev (1965); |
| | Vacuole | 70 [48] (e) | 50(o) | 110(i) | | | Coster and Hope (1968) |
| *Nitella clavata* | Solution | 0.1 | 3.0 | 4.1 | | −106 | Barr (1965) |
| | Vacuole | 75–83(i) | 34–36(o) | 120–124(i) | | −120 | Barr and Broyer (1964) |
| *Nitella flexilis* | Solution | 0.1 | 0.2 | 1.3 | | | |
| | Cytoplasm | 125(e) | 5(e) | 36(i) | −170 | −154 | Kishimoto and Tazawa (1965) |
| | Vacuole | 80(e) | 28(i) | 136(e) | | | |
| *Nitella translucens* | Solution | 0.1 | 1.0 | 1.3 | | | MacRobbie (1962, 1964) |
| | Cytoplasm | 119(i) | 14(o) | 65(i) | −140 | −122 | Spanswick and Williams (1968) |
| | Vacuole | 75(e) | 65(i) | 150–170(e) | | | |
| *Acetabularia mediterranea* | Solution | 10 | 470 | 550 | | | |
| | Cytoplasm | 400(o) | 57(o) | 480(i) | −174 | −174 | Saddler (1970a) |
| | Vacuole | 355(e) | 65(e) | 480(e) | | | |

| | | | | | | | |
|---|---|---|---|---|---|---|---|
| *Bryopsis plumosa* | Solution | 10 | 483 | 563 | | | Gutknecht and Dainty (1968) |
| | Cytoplasm | 210(e) | 470(o) | 475(i) | −70 | −70 | Munday (1972) |
| | Vacuole | 7(o) | 529(i) | 605(i) | | | |
| *Chaetomorpha darwinii* | Solution | 13 | 500 | 523 | −72 | +8 | Dodd et al. (1966) |
| | Cytoplasm | 541(i) | 25(o) | 601(e) | −75 | −29 | Findlay et al. (1971) |
| | Vacuole | 540(i) | 25(i) | 575(e) | | | |
| *Chlorella pyrenoidosa* | Solution | 6.5 | 1.0 | 1.0 | −40 | | Barber (1968b) |
| | Cytoplasm | 108(i) | 1.1(o) | 1.3(i) | | | |
| *Halicystis ovalis* | Solution | 12 | 488 | 523 | −80 | −80 | Blount and Levendahl (1960) |
| | Vacuole | 337(e) | 257(o) | 543(i) | | | |
| *Hydrodictyon africanum* | Solution | 0.1 | 1.0 | 1.3 | | | |
| | Cytoplasm | (93)(i) | (51)(o) | (58)(i) | −116 | −90 | Raven (1967a) |
| | Vacuole | 40(e) | 17(i) | 38(e) | | | |
| *Valonia ventricosa* | Solution | 11 | 485 | 590 | | | Gutknecht and Dainty (1968) |
| | Cytoplasm | 434(i) | 40(o) | 138(e) | −71 | +17 | Gutknecht (1966) |
| | Vacuole | 625(i) | 44(i) | 643(e) | | | |
| *Anacystis nidulans* | Solution | 21 | 13 | 0.2 | (−50?) | | Dewar and Barber (1973, 1974) |
| | Cytoplasm | 149(e) | 5(o) | 2.8(i) | | | |
| *Pelvetia fastigiata* (uncleaved zygote) | Solution | 11 | 485 | 590 | −70 | | Allen et al. (1972) |
| | Cytoplasm | 400(i) | 20(o) | 322(i) | −80 | | |
| *Griffithsia* sp. | Solution | 11 | 485 | 590 | | −52 | Findlay et al. (1969) |
| | Vacuole | 500–600 (i) | 30–90(o) | 600–650 (i) | | −54 | |
| *Porphyra perforata* | Solution | 11 | 485 | 590 | −42 | | Eppley (1958) |
| | Cytoplasm | 480(i) | 51(o) | 81(e) | | | Gutknecht and Dainty (1968) |

Table 5.2.

$NO_3^-$, $H_2PO_4^-$ and $SO_4^{2-}$ concentrations in various compartments of algal cells and their bathing medium and electrical P.Ds between compartments. Subscripts referring to direction and location of active transport as in Table 5.1.

| Alga | Ion concentration (Mol m$^{-3}$) | | | | Electrical P.D. (mV) | | Reference |
|---|---|---|---|---|---|---|---|
| | Compartment | $NO_3^-$ | $H_2PO_4^-$ | $SO_4^{2-}$ | $E_{co}$ | $E_{vo}$ | |
| *Nitella clavata* | Solution | 1.8–1.9 | 0.0008 | 0.67 | | −106 | Hoagland and Davis (1923, 1929) |
| | Vacuole | 3.3–5.3(i) | 1.7–2.8(i) | 12.20(i) | | −120 | Barr and Broyer (1964) |
| *Acetabularia mediterranea* | Solution | | | 55 | −174 | −174 | Saddler (1970a) |
| | Vacuole | | | 8.8(i) | | | |
| *Chlorella* sp. | Solution | | 0.1 | 0.08 | −40 | | Barber (1968b); Bassham and |
| | Cytoplasm | | 1.0(i) | 1.0(i) | | | Krause (1969); Vallee and |
| | | | | | | | Jeanjean (1968a) |
| *Hydrodictyon africanum* | Solution | | 0.1 | 0.5 | | | |
| | Cytoplasm | | 1–2(i) | | −116 | −90 | Raven (1974a and unpublished) |
| | Vacuole | | 1–2(e) | 5–8(i) | | | |
| *Porphyra perforata* | Solution | 1–10.10$^{-3}$ | 1–10.10$^{-3}$ | | −42 | | Eppley (1958); Gutknecht and |
| | Cytoplasm | 80–140 | 2.5(i) | | | | Dainty (1968) |

Table 5.3.

H$^+$, Ca$^{2+}$ and Mg$^{2+}$ concentrations in various compartments of algal cells and their bathing medium and electrical P.Ds between compartments. Subscripts referring to direction and location of active transport as in Table 5.1.

| Alga | Ion concentration (Mol m$^{-3}$) | | | | Electrical P.D. (mV) | | Reference |
| | Compartment | H$^+$ | Ca$^{2+}$ | Mg$^{2+}$ | $E_{co}$ | $E_{vo}$ | |
| --- | --- | --- | --- | --- | --- | --- | --- |
| *Nitella clavata* | Solution | $10^{-4}$ | 1.3 | 3.0 | | $-106$ | Hoagland and Davis (1923, 1929) |
| | Vacuole | $10^{-3}$(o) | 12–19(o) | 11–22(o) | | $-120$ | Barr and Broyer (1964) |
| *Acetabularia mediterranea* | Solution | $10^{-5}$ | | | $-174$ | $-174$ | Saddler (1970a) |
| | Vacuole | 1(i) | | | | | |
| *Chlorella pyrenoidosa* | Solution | $2.10^{-4}$ | | | $-40$ | | Barber (1968b) |
| | Cytoplasm | $2.10^{-4}$(o) | | | | | Shieh and Barber (1971) |
| *Hydrodictyon africanum* | Solution | $10^{-3}$ | | | $-116$ | $-90$ | Raven (1967a and unpublished) |
| | Vacuole | $10^{-3}$(o) | | | | | |

used. The best investigated case is $Cl^-$; here active $Cl^-$ influx, probably at the plasmalemma, is ubiquitous in freshwater algae (Table 5.1). In marine algae it is also present at the plasmalemma in some species (Saddler, 1970a), but in others the electrochemical gradient for $Cl^-$ across the plasmalemma is such that a relatively small inhomogeneity of cytoplasmic $Cl^-$, with a higher concentration in the chloroplasts than the ground cytoplasm, would lead to passive $Cl^-$ distribution at the plasmalemma in *Chaetomorpha*, *Valonia* and *Porphyra* (Table 5.1). The situation at the tonoplast is unclear; in many freshwater algae $Cl^-$ is near equilibrium, while in some marine algae active influx into the vacuole is indicated (Gutknecht and Dainty, 1968; MacRobbie, 1970; Raven, 1975). In *Chaetomorpha* and *Valonia*, however, $Cl^-$ distribution at the tonoplast may, as at the plasmalemma, be passive (MacRobbie, 1970; Raven, 1975).

Less is known about other anions (Table 5.2). There is evidence for active transport of $H_2PO_4^-$, $SO_4^{2-}$ and $NO_3^-$ somewhere between the medium and the vacuole in many marine and freshwater coenocytes (Table 5.2; Raven, 1975). For *Hydrodictyon africanum*, $H_2PO_4^-$ and for green microalgae $H_2PO_4^-$ and $SO_4^{2-}$ active influx at the plasmalemma is likely (Raven, 1974a, 1975). $NO_3^-$ active transport at the plasmalemma occurs in *Porphyra* (Gutknecht and Dainty, 1968); active transport of $NO_3^-$ and of $HCO_3^-$ inwards at the plasmalemma has been suggested on indirect grounds (Raven, 1975; see Section 5.5). Thus, it appears that active influx at the plasmalemma is found for all the major transported anions. The situation at the tonoplast is less clear: $H_2PO_4^-$ in *Hydrodictyon* seems to be close to electrochemical equilibrium at that membrane (Raven, 1974a).

*Cations*

The situation here is less clear than with anions. Active $Na^+$ efflux at the plasmalemma seems to be an almost universal attribute of algae (Table 5.1). At the tonoplast there is commonly an inwardly directed $Na^+$ pump (MacRobbie, 1970; Table 5.1); this is most pronounced in *Bryopsis plumosa*, where the $K^+/Na^+$ in the cytoplasm is much higher than that in the vacuole (Rigler, quoted by Gutknecht and Dainty, 1968; Munday, 1972); this may also be the case in other marine coenocytes with low $K^+/Na^+$ in the vacuole, although in these cases the cytoplasmic $K^+/Na^+$ is not known (Gutknecht and Dainty, 1968).

$K^+$ is much more complex; here the ion can be actively transported inwards or outwards at the plasmalemma, or can be near equilibrium; at the tonoplast $K^+$ is generally at equilibrium or is actively transported inwards (Table 5.1).

$H^+$ is probably at a lower electrochemical potential in the cytoplasm than in the vacuole (Table 5.3), and there is active $H^+$ transport from cytoplasm to vacuole: an extreme case is *Desmarestia* (see Raven and Smith, 1974). At the plasmalemma there is commonly an active $H^+$ efflux under many conditions, associated with pH regulation in the face of passive $H^+$ entry and metabolic production of $H^+$ during growth with $NH_4^+$ as N source entering the cell (Raven and Smith, 1974). The converse problem is found at high external pH, where the $H^+$ electrochemical gradient is outwardly directed at the plasmalemma, and $OH^-$ production inside the cell in $NO_3^-$ and $HCO_3^-$ assimilation add to the pH stress caused by net passive $OH^-$ influx; here active $OH^-$ efflux ($H^+$ influx) at the plasmalemma must occur (Raven and Smith, 1974).

$Ca^{2+}$ and $Mg^{2+}$ are at a lower electrochemical potential in the vacuole of algal cells than in the external solution (Table 5.3), and active extrusion must be postulated if passive permeability is finite (see Robinson and Jaffe, 1973; Table 5.5).

## 5.3 *General aspects of ion fluxes*

Ion fluxes at the plasmalemma and tonoplast of algal cells have been measured both as net fluxes and as bidirectional tracer fluxes. Passive and active fluxes are considered in detail in the Sections 5.4 and 5.5. In this section a few general points will be made.

### 5.3.1 *Net fluxes associated with growth*

It is possible to compute the net fluxes associated with growth from studies of cell composition (Table 5.5). Such calculations indicate (Raven, 1975) that the major net ion fluxes during growth are $HCO_3^-$ entry and $OH^-$ efflux, for algal cells growing at high pH where $HCO_3^-$ is the inorganic carbon species entering the cell. Growth at lower pH involves entry of $CO_2$ as the inorganic carbon source; in this case the major ion fluxes are those of the inorganic N sources and associated $H^+$ fluxes. When $NH_4^+$ is the N source entering the cell at low pH values, the major net ion fluxes are $NH_4^+$ influx and an almost equal $H^+$ efflux. When $NO_3^-$ is the N source entering the cell, the major net ion fluxes during growth are $NO_3^-$ influx with an almost equivalent $OH^-$ efflux (Raven, 1975; Raven and Smith, 1974). In the large, vacuolate algal cells the next largest net influxes are of $K^+$ and $Cl^-$, followed by $H_2PO_4^-$, $SO_4^{2-}$, $Mg^{2+}$, $Na^+$, $Ca^{2+}$, etc. In the smaller, essentially non-vacuolate algae the order, after N and $H^+(OH^-)$, is $K^+$, $H_2PO_4^-$, $Mg^{2+}$, then $Na^+$, $Ca^{2+}$, $Cl^-$ (Table 5.4).

Table 5.4.
Atomic ratio of elements in algal cells, taking C = 100.
(From Raven, 1975.)

| Element | Freshwater microalgae; *Chlorella, Euglena* | Freshwater coenocytes; *Hydrodictyon*, Charophytes |
|---|---|---|
| C | 100 | 100 |
| N | 9–16 | 4.5–8.4 |
| S | 0.3–1.5 | 0.34–1.3 |
| P | 0.3–1.9 | 0.23–0.6 |
| K | 0.2–1.5 | 1.5–5.3 |
| Na | 0.025–0.10 | 0.04–1.5 |
| Ca | 0.002–0.2 | 0.12–0.51 |
| Mg | 0.08–1.4 | 0.12–0.21 |
| Cl | 0.02–0.07 | 1.7–3.0 |

### 5.3.2 Net and tracer fluxes

The tracer fluxes of $K^+$, $Na^+$ and $Cl^-$ at the plasmalemma are generally much higher than the net fluxes of these ions required for growth (e.g. Gutknecht and Dainty, 1968; Raven, 1975). However, an exception is the recovery of *Chlorella* from $K^+$ deficiency (Shieh and Barber, 1971; Barber and Shieh, 1972, 1973a). Here the net $K^+$ influx and net $Na^+$ efflux are greater than the tracer $K^+$ and $Na^+$ fluxes in either 'normal' ($K^+$-sufficient) or 'Na$^+$-rich' ($K^+$-deficient) cells.

By contrast with the normal situation with $K^+$, $Na^+$ and $Cl^-$ (and also, in at least one case, $Ca^{2+}$ and $Mg^{2+}$: Robinson and Jaffe, 1973; Table 5.5), the tracer influxes of $H_2PO_4^-$ and $SO_4^{2-}$ from 'ecological' concentrations of these ions are much closer to the net influxes, i.e. the efflux is low (e.g. Vallee and Jeanjean, 1968a; Jeanjean, 1969; Robinson, 1969a; Raven, 1974a; Table 5.5). Despite the availability of heavy nitrogen ($^{15}N$), no bidirectional flux measurements of $NH_4^+$ or $NO_3^-$ appear to have been made on algae. Tracer measurements of $H^+$ fluxes are not possible; water fluxes swamp them.

### 5.3.3 Magnitudes of fluxes at the plasmalemma of freshwater and marine algae

The plasmalemma tracer fluxes of $K^+$, $Na^+$ and $Cl^-$ are much higher in marine algae than in freshwater algae (Gutknecht and Dainty, 1968; MacRobbie, 1970, 1974; Raven, 1975), being 1–50 nMol m$^{-2}$ s$^{-1}$ for freshwater algae, and 50–5000 nMol m$^{-2}$ s$^{-1}$ in marine algae (Table 5.5). This difference could be caused by the higher ion concentration in the cytoplasm (2-fold) and

Table 5.5.
Ion fluxes at plasmalemma and tonoplast of three algae.

| Alga | Ion, concentration in solution (Mol m$^{-3}$), light or dark | Flux (n Mol m$^{-2}$ s$^{-1}$) | | | Reference |
|---|---|---|---|---|---|
| | | $\phi_{oc}$ | $\phi_{cv}$ | $\phi_{co}$ | |
| *Hydrodictyon africanum* | Cl$^-$, 1.3, Light | 10–30 | 500–1000 | 3–20 | Raven (1967a, 1969a, b, 1974a and |
| | Cl$^-$, 1.3, Dark | 2–10 | 100–400 | 2–10 | unpublished) |
| | SO$_4^{2-}$, 1.3, Light | 0.5 | 20–40 | 0.05 | Raven (unpublished) |
| | SO$_4^{2-}$, 1.3, Dark | 0.3 | 5–20 | | |
| | H$_2$PO$_4^-$, 0.2, Light | 2–3 | 100–300 | 0.2 | Raven (1974a and unpublished) |
| | H$_2$PO$_4^-$, 0.2, Dark | 1–2 | 50–200 | | |
| | K$^+$, 0.1, Light | 5–15 | 200–400 | 4–12 | Raven (1967a, 1968a and |
| | K$^+$, 0.1, Dark | 1–7 | 100–200 | 2–10 | unpublished) |
| | Na$^+$, 1.0, Light | 2–9 | 500–1000 | 4–10 | Raven (1967a, 1968a and |
| | Na$^+$, 1.0, Dark | 1–7 | 200–800 | 1–5 | unpublished) |
| *Pelvetia fastigiata* (zygote 24 h after fertilization) | Cl$^-$, 550, Light or Dark | 100 | | 90 | |
| | K$^+$, 10, Light | 330 | | 290 | |
| | Na$^+$, 550, Light | 68 | | 68 | Robinson and Jaffe (1973) |
| | Ca$^{2+}$, 7, Light | 0.74 | | 1.30 | |
| | Mg$^{2+}$, 37, Light | 4–11 | | (4–11) | |
| *Griffithsia* sp. | K$^+$, 10, Light or Dark | 500–3800 | 3600–9900 | 500–3800 | |
| | Na$^+$, 550, Light | 120–170 | 50–120 | 150 | Findlay et al. (1970) |
| | Cl$^-$, 550, Light Dark | 100–400 28–37 | $\gg \phi_{oc}$ | 40–110 | |

138    J.A. Raven

the medium (up to 500-fold) of marine algae compared with freshwater algae (see Raven, 1975).

### 5.3.4 Tonoplast fluxes

In freshwater giant algal cells the fluxes of $K^+$, $Na^+$, $Cl^-$, $H_2PO_4^-$ and $SO_4^{2-}$ at the tonoplast are apparently much higher than those at the plasmalemma, by a factor of 100 or more, based on both the analysis of tracer efflux and separation of cytoplasm and vacuole after periods of tracer influx (MacRobbie, 1964, 1966b, 1970, 1971, 1974; Raven, 1975; Table 5.5). However, in the case of $K^+$ in *Nitella axillaris* (Diamond and Solomon, 1958), $K^+$ and $Cl^-$ in *Mougeottia* sp. (Wagner and Bentrup, 1973), and $K^+$, $Na^+$ and $Cl^-$ in the brackish-water *Nitellopsis obtusa* (MacRobbie and Dainty, 1958) the tracer fluxes at the tonoplast are less than those at the plasmalemma. In marine algae the tonoplast flux can be either greater or smaller than the plasmalemma flux (Gutknecht and Dainty, 1968; Black and Weeks, 1972; MacRobbie, 1974; Raven, 1975).

### 5.3.5 Electrogenic and electroneutral fluxes

Ions can cross membranes either as charged entities (uniport in the terminology of Mitchell, 1966), or as neutral complexes in combination with some other ion. In this latter case, movement of two ions of dissimilar charge in the same direction is termed symport, while coupled movement of ions of the same charge in opposite directions is termed antiport. In giant algal cells, voltage clamp studies (Ch. 1) can be used to see if at least the passive fluxes are by uniport. There is evidence that all of these types of transport can occur in algae (see Fig. 5.2). Uniport can be either passive, or active (electrogenic pump); in both cases the fluxes contribute to the membrane electrical conductance. Antiport can be passive (e.g. exchange diffusion) or active (e.g. chemically coupled $K^+$ influx/$Na^+$ efflux); or a mixture (e.g. secondary active influx of $Cl^-$ coupled to downhill $OH^-$ efflux, as proposed by Smith, 1970). Symport appears to involve the primary active transport of one of the ions involved (e.g. $Cl^-$ in the cation–$Cl^-$ pump proposed by Raven, 1968), although the suggestion of Smith (1970) could be written as $H^+$–$Cl^-$ symport rather than $OH^-$–$Cl^-$ antiport. Electrogenic active transport of $Cl^-$ inwards leads to net salt accumulation by passive transport of a cation. (See also Ch. 8).

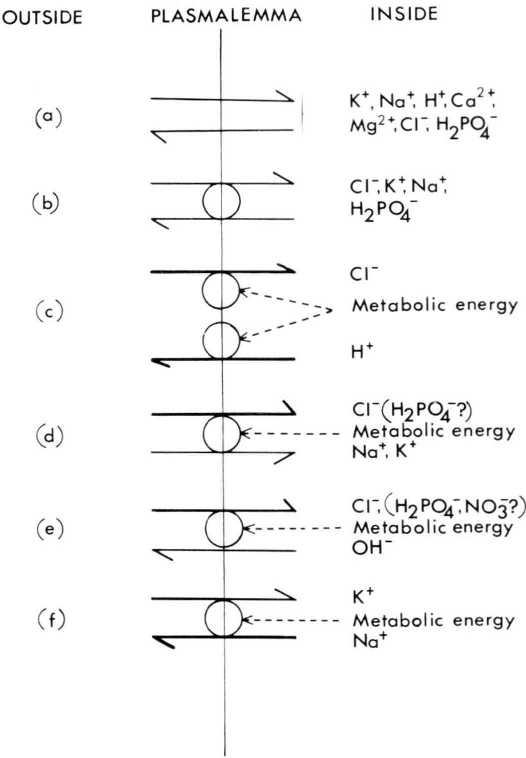



Fig. 5.2. Some possible ion transport mechanisms at the plasmalemma of algal cells. Thick arrows represent thermodynamically active fluxes, thin arrows represent thermodynamically passive fluxes. (e) represents the $Cl^-$-active influx suggested by Smith (1970) in which active transport is secondary, the energy coming from passive $OH^-$ efflux. (f) represents electroneutral antiport in which each of the fluxes is active.

## 5.4 *Passive ion fluxes*

In all algae the efflux of $Cl^-$, $H_2PO_4^-$ and $SO_4^{2-}$ at the plasmalemma are in the downhill direction. In most cases, $Na^+$, $Ca^{2+}$ and $Mg^{2+}$ influx and $K^+$ efflux at the plasmalemma are also in the downhill direction. These downhill tracer fluxes can be either by movement of the ion across the membrane as a charged entity, or as part of a neutral complex (see Section 5.3), e.g. as exchange diffusion.

### 5.4.1 *Effects of electrical potential on passive fluxes*

In order to distinguish between these alternatives, tracer fluxes have been measured under 'voltage clamp' conditions. Here current is passed between an intracellular and an extracellular electrode such as to change the trans-membrane P.D., measured using another intracellular electrode, to a value desired by the experimenter (see Ch. 2). When this has been done on giant algal cells, it has been found that $Cl^-$ efflux and $Na^+$ influx are stimulated by values of $E_{co}$ more negative than the resting potential, and decreased by more positive values of $E_{co}$. This is also true of $K^+$ influx when this flux has a substantial passive component as shown by application of the Ussing–Teorell equation (Eq. 1.15). For downhill $K^+$ efflux, by contrast, values of $E_{co}$ more negative than the resting potential decrease the flux, and more positive values increase it (e.g. Blount, 1958; Blount and Levedahl, 1960; Hope et al., 1966; Gutknecht, 1967; Kitasato 1968; Coster and Hope, 1968; Walker and Hope, 1969; Williams et al., 1972; Fig. 5.3).

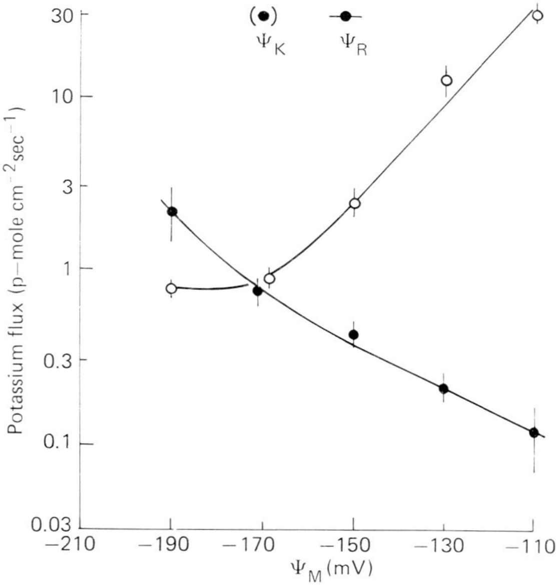


Fig. 5.3. Dependence of tracer fluxes of $K^+$ at the plasmalemma of *Chara corallina* on the trans-membrane P.D. Ordinate: Mean influx ($\bullet$) or efflux (o) of $^{42}K^+$, plotted on a logarithmic scale. Abscissa: Vacuolar P.D. Standard errors of the mean are represented by bars. The mean resting potential of the cells ($\psi_R$) is also indicated, again with standard error of the mean. The Nernst P.D. for $K^+$ in these cells was $-170$ mV. From Walker and Hope, 1969.

These results are most important in that they demonstrate unequivocally that there are true passive fluxes (i.e. other than coupled, electrically neutral (silent) fluxes) of ions at the algal cell plasmalemma (Fig. 5.2a). However, it must be borne in mind that the quantitative relationship between the changes in tracer fluxes and the imposed change in P.D. do not fit the simplest model for passive, independent ion transport (Walker and Hope, 1969).

### 5.4.2 Effects of ion concentration on passive fluxes

Further complexities are found when concentration effects on passive fluxes are examined. Since it is easier to manipulate external than internal concentration, most work has been done on passive tracer influxes. Generally, changing external concentrations of $Na^+$ or $K^+$ influences the downhill tracer influx in a more complex manner than expected from concomitant changes in the passive (i.e. chemical and electrical) driving forces (Hope and Walker, 1960, 1961; Hope, 1963; Smith, 1967; Raven, 1968). Even when contributions from the chemically coupled components of these downhill fluxes are allowed for (Smith, 1967; Raven, 1968; Findlay et al., 1969), it seems inescapable that $P_{ion}$ varies with ion concentration. In many cases 'saturation' of influx with increasing concentration has been observed which cannot be attributed to a change in the electrical driving force; this could be taken as evidence for carrier mediation of electrically detectable passive ion fluxes (but see Cram, 1974).

### 5.4.3 Ion permeability and electrical conductance

The notion that these passive fluxes involve interaction of the transported ion with a specific membrane component, other than the major lipids, is supported by a comparison of the apparent $P_{ion}$ values for algal cell plasmalemma and for model lipid membranes. In synthetic bilayers, $P_K$, $P_{Na}$ and $P_{Cl}$ are of the order of $10^{-14}$–$10^{-12}$ m s$^{-1}$, while in algal cell plasmalemma $P_K$ is $10^{-10}$–$10^{-7}$, $P_{Na}$ is $10^{-11}$–$10^{-8}$ and $P_{Cl}$ is $10^{-12}$–$10^{-9}$ m s$^{-1}$ (see Raven, 1975; Richards and Hope, 1974). The range of $P_{ion}$ values quoted above for algal cell plasmalemma is partly due to discrepancies between 'tracer' and 'electrical' determinations of $P_{ion}$ (Hope and Walker, 1961). Generally the 'electrical' $P_{ion}$ value is higher than the 'tracer' $P_{ion}$ value for freshwater algae, though the reverse can be true in marine algae, where the measured electrical conductance is less than that computed from tracer ion fluxes. In this latter case a contribution of chemically coupled ion fluxes to the flux but not to the electrical measurement has been suggested (Findlay et al., 1969; Findlay et al., 1971).

                                  *J.A. Raven*

However, correction for 'electrically silent' ion fluxes in the freshwater algae only makes the disparity there worse. One explanation is that an ion with a high $P_{ion}$ value has been neglected in the tracer measurements. It is currently thought that this ion could be $H^+$ with $P_H$ as high as $10^{-5}$ m s$^{-1}$ (Kitasato, 1968; Richards and Hope, 1974).

As might be expected from the mediated nature of these passive fluxes, they show responses (albeit non-stoichiometric) to the energetic state of the cell (see Section 5.6, Table 5.5), and also show variable and often quite large apparent activation energies (e.g. Hope and Walker, 1960; Raven, 1967a; Hope and Aschberger, 1970).

### 5.4.4 Exchange diffusion

Fluxes in the downhill direction can, as pointed out above, be electrically neutral (silent), and coupled stoichiometrically to some other ion. Exchange diffusion, e.g. of $Cl^-$, involves a 1:1 exchange of the 'cold' $Cl^-$ present on one side of the membrane with 'tracer' $Cl^-$ on the other side, with no net flux of $Cl^-$, and no requirement for energy input (Fig. 5.2b). This is detected by parallel effects on influx and efflux of such treatments as removal of the ion concerned from one side of the membrane, or of temperature, irradiance, metabolic state, etc. (e.g. Hope et al., 1966; Rybova et al., 1972; Raven, 1975). Exchange diffusion has also been invoked to explain why measured tracer ion fluxes exceed the electrically detectable fluxes (see above), and when the available energy is inadequate to support the measured tracer fluxes in the uphill direction.

### 5.4.5 Determinants of the resting potential: diffusion potentials and electrogenic pumps

Passive ion fluxes are important in determining the electrical P.D. across membranes. In a number of cases, e.g. a number of marine algae and some freshwater algae in $Ca^{2+}$-free media, there is reasonable agreement between the observed resting potential and that predicted from the Goldman equation (Eq. 1.19) over a range of external ion concentrations (see Raven, 1975). However, in the freshwater coenocytes in the presence of $Ca^{2+}$ the potential change with alterations in external $K^+$ and $Na^+$ levels is much smaller than would be expected if passive movement of these ions was the main determinant of the P.D. (e.g. Spanswick et al., 1967). However, as long as the observed P.D. stays within the range of passive equilibrium potentials (Nernst potentials) for the ions present, the potential can be explained in terms of a diffusion potential, albeit with a number of ad hoc assumptions in some cases.

Whenever the measured resting potential is outside the range of passive diffusion potentials, some additional component of the resting potential is required. This is generally thought to be an electrogenic active transport of an ion at the plasmalemma, i.e. active ion transport by a uniport mechanism. Since electrogenesis commonly involves hyperpolarisation (more negative values of $E_{co}$), active anion influx (e.g. of $Cl^-$: Saddler, 1970c) or active cation efflux (e.g. of $H^+$: Spanswick, 1972, 1974a) is invoked.

Other evidence for electrogenic pumps is more equivocal, and it is not easy to distinguish electrogenic pumps from diffusion potential contributions to the resting potential when the latter is in the range of permissible diffusion potentials. The Goldman equation predicts that changes in temperature should have relatively small effects on a diffusion potential, and that changes in metabolism should only affect the potential inasmuch as they lead to slow changes in the solute gradients. However, direct effects of temperature and of metabolism on $P_{ion}$ make such arguments rather dubious. Electrogenesis will be considered further in Section 5.6 on energy sources.

### 5.4.6 Action potentials

Characean and some other algae show action potentials. Here a transient, propagated depolarisation of the plasmalemma occurs upon electrical or other stimulation. This is slower than in animal cells, both in terms of the time course of the depolarisation and repolarisation (s rather than ms), and the velocity of propagation ($10^{-2}$ m s$^{-1}$ as compared with $10$–$10^2$ m s$^{-1}$). Tracer experiments show that the action potential is caused by a transient increase in $P_{Cl}$, which leads to an increased $Cl^-$ efflux and a decrease of the resting potential in the direction of the $Cl^-$ equilibrium potential. Recovery involves a decrease in $P_{Cl}$ relative to $P_K$. The function of the action potential in algae is not clear, but it may be involved in information transfer along and, via plasmodesmata, between cells (see Findlay, 1975; Spanswick, 1974b, 1975; Walker, 1974).

## 5.5 Active ion fluxes

When considering tracer fluxes in the downhill direction (Section 5.4) we saw that the flux could be dissected into electrogenic and electrically silent components. With fluxes in the uphill direction, we have the possibility that these two non-energy-requiring fluxes can occur, together with active components which may again be either electrogenic or electrically silent. The passive, electrogenic component can be estimated from the Ussing–Teorell equation

(Eq. 1.15) or some empirical modification of it (Walker and Hope, 1969), and the electrogenic flux in the downhill direction. The 'exchange' component can be estimated from the 'exchange' component of the corresponding downhill flux. What is left is the 'active' component, i.e. a flux against a free energy gradient for the ion, coupled to some exergonic process such as a chemical reaction (primary active transport) or a downhill solute flux (secondary active transport). Any net flux in the uphill direction represents the minimum active flux (electrogenic or electroneutral), while the tracer flux in this direction represents the maximum active flux. Unless it has been shown that influx and efflux respond differently to some treatment, active transport cannot be proved for an ion at flux equilibrium which is not at electrochemical equilibrium; the observed bidirectional fluxes could all be by exchange diffusion.

Thus, the direction of active transport can be determined from electrochemical criteria, while its magnitude can be estimated from the considerations outlined above. The energetics of the fluxes are considered in Section 5.6.

It is important to note that while there is a tendency for active fluxes to be more influenced by changes in temperature and energy supply than are passive fluxes, this is by no means always the case (Section 5.4) and these are poor criteria for distinguishing active from passive fluxes in the absence of electrochemical data. Similarly, while active fluxes often respond to changes in substrate (ion) concentration by a so-called Michaelis–Menten relationship (references in Raven, 1975), many passive fluxes also show such a relationship (Section 5.4). Another possible distinguishing feature of passive as opposed to active fluxes is response to the electrical P.D. However, there are electrogenic components of both passive and active fluxes. It is known that the electrogenic passive fluxes respond to changes in P.D.; where tested, active fluxes are not apparently altered (Gutknecht, 1967; Williams et al., 1972). This does not necessarily mean that the fluxes are not electrogenic. If the energy available from the reaction driving the pump is much more than is required by the pump at any of the imposed electrochemical potential gradients, the altered potential gradient would not necessarily alter the pumping rate. Thus, separation of active from passive fluxes by reference to such criteria as response to temperature, metabolic state, concentration dependence or response to electrical potential differences are all liable to error.

### 5.5.1 Electrogenic (uniporting) pumps

#### $Cl^-$ influx

The best authenticated examples are the $Cl^-$ influx in *Acetabularia* (Saddler, 1970c; Fig. 5.2c), and the $H^+$ efflux in a number of freshwater coenocytes

(Spanswick, 1972, 1974a; Richards and Hope, 1974). The action of such pumps involves a net charge transfer which can be balanced by either net passive transfer of an ion of the same sign in the opposite direction, or an ion of the opposite charge in the same direction. This is, formally, electrically coupled antiport or symport. In *Acetabularia*, where the active electrogenic $Cl^-$ influx appears to be a major factor in producing and maintaining turgor, it might be expected that action of the pump could involve a net passive influx of cations. This was not directly observed by Saddler (1970b, c), although the total salt concentration in the cell is high during $Cl^-$ pumping and low when the pump is only working slowly.

### $H^+$ efflux

The balancing passive movements of ions during electrogenic $H^+$ efflux (Fig. 5.2c) is not well understood. If $P_{H^+}$ is very high (see Kitasato, 1968; Richards and Hope, 1974; Spanswick, 1974a), then much of the active $H^+$ flux will be short-circuited by passive $H^+$ influx. This would make the $H^+$ electrochemical gradient set up by the $H^+$ extrusion pump an inefficient way of coupling the chemical energy used in $H^+$ extrusion to secondary active transport involving a coupling of passive $H^+$ fluxes to movement of some other solute (Smith, 1970; Komor and Tanner, 1974; see below). It must be emphasised that it is not known whether the active ion transports mentioned above as being electrogenic are completely electrogenic, or whether there is a chemically coupled, electrically silent component of these fluxes.

### 5.5.2 Electroneutral (symporting or antiporting) pumps

### $Cl^-$ influx

A possible example of chemically coupled symport is the active influx of $Cl^-$ coupled to cation influx (Fig. 5.2d) in a number of algal cells. The evidence that this pump is not electrogenic is based mainly on the slow response of $E_{co}$ to removal of $Cl^-$ from the medium (e.g. Hope and Walker, 1961; Raven, 1968; Spanswick, 1974). A net salt accumulation mechanism of this type is an alternative method of turgor generation to the electrogenic $Cl^-$ influx with passive net cation influx due to the hyperpolarisation. Evidence for linkage of cation and $Cl^-$ influx comes from reciprocal stimulation of tracer $Cl^-$ influx by $K^+$ and/or $Na^+$, and of the cation influxes by $Cl^-$. This has been found in *Nitella translucens* (Smith, 1967), *Hydrodictyon africanum* (Raven, 1968), *Tolypella intricata* (Smith, 1968), *Chara corallina* (Findlay et al., 1969) and *Enteromorpha intestinalis* (Black, 1971) and may also occur in *Chlorella pyrenoidosa* (Barber, 1969). The concentration effects of the cation on $Cl^-$

influx and of the cation on the $Cl^-$-dependent component of its own influx, and similar effects of metabolic state on the cation and $Cl^-$ components of the putatively linked influx, confirm this linkage (Smith, 1967, 1968; Raven, 1968; Black, 1971). One difficulty is that the measured cation influx is always less than the $Cl^-$ influx (see MacRobbie, 1970).

An alternative to primary active transport of $Cl^-$ chemically coupled to $K^+$ and $Na^+$ influx has been proposed by Smith (1970). His scheme (Figs. 5.2e and 5.4) links cation ($K^+$ or $Na^+$) influx to $Cl^-$ influx (net salt accu-

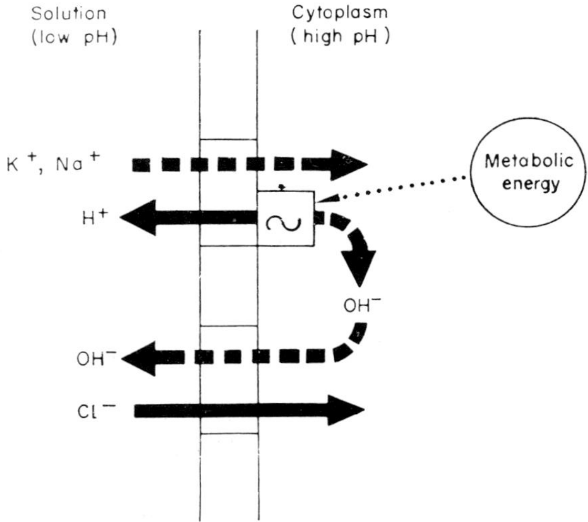



Fig. 5.4. Scheme for net salt (KCl or NaCl) transport at the plasmalemma of algal cells, from Smith (1970). The primary energy-requiring process is $H^+$ efflux. Part of this is balanced by $K^+$ and $Na^+$ influx, coupled either electroneutrally to a component of the $H^+$ efflux, or electrogenically in response to the hyperpolarisation produced by this pump. The rest of the $H^+$ efflux is balanced by downhill $OH^-$ efflux coupled to the (secondarily) active influx of $Cl^-$.

mulation) indirectly via $H^+$ fluxes. Active $H^+$ efflux is partly balanced electrically by cation ($K^+$ or $Na^+$) influx, either by chemical coupling or electrogenic coupling in a partly electrogenic $H^+$ pump. The rest of the $H^+$ efflux during net salt accumulation is balanced by effective $H^+$ passive influx in secondary active transport of $Cl^-$, where downhill $OH^-$ efflux energises active $Cl^-$ influx. Smith's hypothesis is supported by his finding that tracer $Cl^-$ influx is stimulated when a pH gradient is imposed across the plasmalemma by preincubation at high pH, followed by transition to a lower pH and addition of tracer $Cl^-$ (Smith, 1972). This work on *Chara corallina* has been confirmed for net $Cl^-$

influx in *Hydrodictyon africanum* (Raven, 1974d). However, there is no direct evidence the energy for active $Cl^-$ influx is derived from the $OH^-$ efflux rather than directly from metabolism (primary active transport: MacRobbie, 1974; Raven, 1974d). Such a scheme does, however, have teleological economy in that it relates the transport of the major turgor-generating inorganic ions to the important process of pH regulation, and also to an alternative mode of turgor generation not much favoured by algae – the accumulation of alkali cation salts of organic acids (Raven and Smith, 1973, 1974; Smith and Raven, 1974b, 1975).

## $NO_3^-$, $HCO_3^-$, $H_2PO_4^-$ and $SO_4^{2-}$ influx

What evidence is available suggests that these pumps are electroneutral, at least in the *Characeae* (e.g. Hope and Walker, 1961; Spanswick et al., 1967; Spanswick, 1970). Some coupling of influx of $NO_3^-$ or $HCO_3^-$ to $OH^-$ efflux might be expected, in view of the large net fluxes of these anions during algal growth, and the almost 1/1 ratio of $NO_3^-$ or $HCO_3^-$ assimilation to $OH^-$ production (Raven, 1970; Raven and Smith, 1973, 1974; Fig. 5.2e), and the absence of long-term stimulation of cation influx by assimilation of $HCO_3^-$ (Raven, 1970). $HCO_3^-$ influx may be related to $Cl^-$ influx (Findenegg, 1974; Raven, 1970).

$H_2PO_4^-$ influx in a number of green algae is stimulated specifically by $Na^+$ rather than $K^+$ (references in Raven, 1974a; Fig. 5.2d). Such a chemical coupling is inadequate to power the phosphate influx by the $Na^+$ gradient, although the coupling to $Na^+$ rather than $K^+$ might reduce the very large free energy gradient against which $H_2PO_4^-$ influx can occur in algae (Raven, 1974a) compared with electrogenic influx or chemical coupling to $K^+$. The $H_2PO_4^-$ influx in *Hydrodictyon africanum*, unlike the $Cl^-$ influx, is not stimulated by a pH transition treatment (Raven, unpublished), suggesting a different mechanism for influx of the two anions. This is supported by studies on energy coupling (Section 5.6). $SO_4^{2-}$ influx in *Ankistrodesmus* is specifically $K^+$-stimulated (Ullrich-Eberius, 1973a).

## $K^+$ influx and $Na^+$ efflux

An electroneutral $K^+$–$Na^+$ pump (Fig. 5.2f) occurs in *Nitella translucens* (MacRobbie, 1962), *Hydrodictyon africanum* (Raven, 1967a, 1971a) and *Enteromorpha intestinalis* (Black, 1971). Here active $Na^+$ efflux is dependent on external $K^+$; these two active fluxes show the same response to external $K^+$ concentration (Raven, 1967a), and are both inhibited by ouabain, albeit at rather high concentrations. The absence of effect of ouabain on the value of $E_{co}$ (Raven, 1967a; Spanswick et al., 1967; Black, 1971) suggests that the pump is electroneutral, although there is commonly a quantitative discrepancy between the $K^+$ influx and $Na^+$ efflux; $H^+$ may be the 'balancing' ion. In other cases the coupled $K^+$ influx and $Na^+$ efflux is not ouabain-sensitive (e.g. in *Chlorella*: Barber, 1968c, d; Shieh and Barber, 1971), while in *Chara corallina* $Na^+$ efflux is neither ouabain-inhibited nor $K^+$-stimulated (Findlay et al., 1969). The role of this pump in algal cell growth is especially important if phosphate influx is coupled to $Na^+$ influx during growth (Raven, 1974a), particularly in microalgae which contain less $Na^+$ than P (Table 5.4).

### $NH_4^+$ influx

During growth on $NH_4^+$ as N source, $NH_4^+$ influx is the major ion influx (Section 5.3). Active $H^+$ efflux of a closely similar magnitude occurs during this growth and some coupling between the fluxes might be expected. It is not clear whether this $NH_4^+$ influx is active or passive (Section 5.2).

### Tonoplast fluxes

The incisive work of MacRobbie (see MacRobbie, 1973, 1974) on charophytes (mainly *Nitella translucens*) suggests a linkage of $K^+$ and $Cl^-$ tracer influxes at the tonoplast ($\phi_{cv}$) to $Cl^-$ influx at the plasmalemma ($\phi_{oc}$). Despite the high measured tracer $Cl^-$ influx at the tonoplast (possibly confounded by uncertainty as to the specific activity of $Cl^-$ in the compartment from which the tracer $Cl^-$ in the cytoplasm is supplied to the tonoplast), perfusion of the vacuole of *Nitella flexilis* with an artifical sap in which $Cl^-$ is replaced by $NO_3^-$ leads to little net $Cl^-$ loss from the cytoplasm across the tonoplast (Kishimoto and Tazawa, 1965).

## 5.6  Energy sources for active transport

The energy for active transport at the plasmalemma and tonoplast of algal cells comes from the partial reactions of respiration and, in illuminated photosynthetic cells, of photosynthesis.

### 5.6.1  In darkness

In the dark, the evidence suggests that active ion transport requires oxidative respiratory reactions. It is inhibited by removal of $O_2$ and by inhibitors of respiratory electron transport such as cyanide and antimycin A (e.g. Raven, 1969b; Lilley and Hope, 1971a; Shieh and Barber, 1971). Eukaryotic algae are unable to grow in the complete absence of $O_2$ (Nuhrenberg et al., 1968), and appear to be unable to bring about net active transport under anaerobic conditions, although tracer exchange, possibly related to maintenance of ion levels in a pump and leak situation, can occur (Schaedle and Jacobsen, 1965, 1966, 1967).

### 5.6.2 *In light*

In the light, active ion transport can be supported by the partial reactions of photosynthesis. This is shown by light-stimulation of active transport (e.g. Table 5.5) even when respiratory reactions are inhibited (e.g. by removal of $O_2$, or the presence of cyanide or antimycin A: Raven, 1967b, 1969b, 1971d; Shieh and Barber, 1971). Even when the active transport is not light-stimulated under normal conditions, it is light-stimulated when respiration is inhibited (e.g. Robinson, 1969b). Action spectra for the light stimulation show the involvement of photosynthetic pigments (Raven, 1969a, 1971b, c; Barber and Shieh, 1973b; Wagner and Bentrup, 1973; Fig. 5.5).

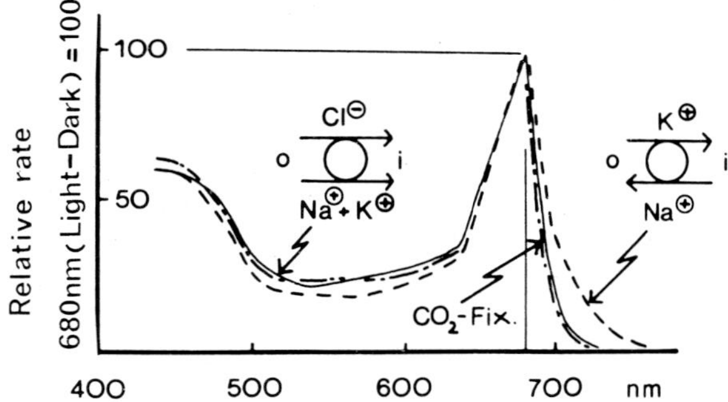


Fig. 5.5. Action spectra for photosynthetic $^{14}CO_2$ fixation, and light-stimulated $^{42}K^+$ and $^{36}Cl^-$ influx in *Hydrodictyon africanum*. Ordinate: fixation of $CO_2$ (– · – · –), influx of $K^+$ (– – –), or influx of $Cl^-$ (———). The light-stimulation at 680 nm is in each case taken as 100 %. Abscissa: wavelength in nm. From Raven (1969a), as redrawn by Lüttge (1973).

The 'light reactions' of photosynthesis are outlined in Ch. 4, and specifically algal aspects are reviewed by Raven (1974c). Experiments in which various of these partial reactions are isolated by inhibition of other partial reactions shows that light-dependent active transport can involve any of the reactions discussed by Nobel (Ch. 4).

Cyclic electron transport does not require $O_2$, $CO_2$ or $NO_3^-$ as electron acceptor, nor does it need Photosystem II as electron donor. It may thus be

150                                    *J.A. Raven*

isolated from open-chain phosphorylations by the use of DCMU or far-red
light to inhibit Photosystem II, and the absence of the electron acceptors to
prevent 'over-oxidation' of the redox components. It is generally inhibited to
a greater extent than is open-chain phosphorylation by antimycin A (Simonis
and Urbach, 1973; Hope et al., 1974; Raven, 1969a, b, 1975).

Open-chain electron transport requires both photoreactions, but is less
sensitive to inhibitors such as antimycin A than is cyclic or oxidative electron
transport. Thus, in the presence of antimycin A and the appropriate electron
acceptor, non-cyclic electron flow to $CO_2$ or $NO_3^-$, or pseudocyclic electron
flow to $O_2$ can be established. The two non-cyclic variants are inhibited by
cyanide, while pseudocyclic is not inhibited (Raven, 1969b, 1975; Ullrich, 1971,
1974; Ullrich-Eberius, 1973b).

### 5.6.3 *Involvement of Photosystem II in light stimulation*

The use of these techniques for isolating various partial reactions show two
major patterns for energy-linkage of active ion transport in algae. One type
(active $Cl^-$ and coupled cation influx at the plasmalemma of many algal cells)
has an absolute requirement in Photosystem II for light stimulation (Mac-
Robbie, 1965; Raven, 1967b, 1969a, b, 1971d; Smith, 1967, 1968; Smith and
Raven, 1974; Rybova et al., 1972; Hope et al., 1966; Lilley and Hope, 1971a;
Black, 1971; Wagner and Bentrup, 1973; Fig. 5.5). The other type can be
supported by cyclic electron transport alone, although in general open-chain
electron transport can also be used. Examples are $Cl^-$ influx in *Chlorella*
(Barber, 1968a) and *Scenedesmus* (Hope et al., 1974), coupled $K^+/Na^+$ fluxes
(MacRobbie, 1965; Raven, 1967b; Black, 1971; Barber and Shieh, 1973a, b;
Wagner and Bentrup, 1973; Fig. 5.5), $H_2PO_4^-$ influx (references in Raven,
1974b and Simonis et al., 1974) and $H^+$ efflux in *Nitella translucens* (Spanswick,
1974a). In *Chara corallina* no processes have so far been found which are
supported by Photosystem I alone in vivo (Smith and Raven, 1974a).

### 5.6.4 *Involvement of ATP*

Generally, the evidence obtained with inhibitors and uncouplers of ATP
synthesis suggests that the transport processes which can be supported by
Photoreaction I alone use ATP (e.g. MacRobbie, 1965, 1966a; Smith, 1966;
Raven, 1967b, 1969b, 1971e, 1974b; Barber and Shieh, 1973a, b; Spanswick,
1974a). Evidence on the energy-coupling of active $Cl^-$ influx in cases where
open-chain electron flow is needed is less clear; in some cases ATP is apparently required (Smith and West, 1969; Lilley and Hope, 1971a), while in

others it is not (MacRobbie, 1965; Smith, 1967, 1968; Raven, 1967b, 1969b, 1971d). The nature of the coupling in these latter cases is obscure; neither redox reactions per se nor the 'high energy intermediate' of photo- or oxidative phosphorylation are entirely acceptable as immediate energy sources (MacRobbie, 1970, 1974; Smith and Raven, 1974a, b).

### 5.6.5 *Energy and information*

Some of the complications, especially regarding the inability of Photosystem I to power $Cl^-$ influx in many algae, may be related to 'non-stoichiometric' or 'informational' effects of energy metabolism on active transport. Such effects are known for both passive and exchange diffusion fluxes of ions in algal cells; changes in metabolism alter both the permeability to ions and the rate of exchange diffusion (e.g. Gutknecht, 1965; Hope et al., 1966; Barber, 1968c, d; Raven, 1969a; Barber and Shieh, 1973a). It is thus possible that energy metabolism influences the active transport rate by altering the activity of the pump (i.e. the rate at which it operates under constant substrate supply) as well as by supplying the energy substrate (stoichiometric interaction); such a situation is common for both energy-requiring and other reactions in the carbon pathway of photosynthesis (Bassham, 1971; Preiss and Kosuge, 1970).

One possible example of such a case is that of a reaction which can be powered by ATP produced by Photosystem I alone, i.e. $K^+-Na^+$ exchange in *Chlorella* (Barber, 1968a, c, d; Barber and Shieh, 1973a). Here the ATP concentration in the cell is lower in $N_2$ in the light than in air in the dark (Strehler, 1953) yet the rate of active transport is higher in light–$N_2$ than in dark–air (Barber, 1968b, d, e; Springer-Lederer and Rosenfeld, 1968; Shieh and Barber, 1971; Barber and Shieh, 1973b). While compartmentation of ATP has not been eliminated as a reason for this anomaly, the data are consistent with some activation of the $K^+-Na^+$ transport mechanism by light other than by supplying ATP (see Meszes and Erdei, 1969). Further experiments in which ADP, AMP and $P_i$ are measured in various cell compartments are required to test this hypothesis, further discussion of which is found in Raven (1966, 1974b, 1975), Lilley and Hope (1971b), and Smith and Raven (1974 a, b). It must be emphasised that this possibility does not cast doubt on the results of experiments in which a single, isolateable source of energy (e.g. cyclic ATP synthesis) has been shown to power a pump; what is being questioned is not so much where the energy comes from as what controls its rate of use.

## 5.7  Functions of ion transport

### 5.7.1  Turgor generation and regulation in walled cells

#### Growth

Walled plant cells, including algae, require a minimum turgor pressure to bring about plastic extension (growth) of the cell wall; this has been elegantly demonstrated for *Nitella* by Green (1968). Guillard (1962) tabulates some turgor pressures for walled algal cells. While ions transported into the cell are often major contributors to turgor in algal cells (especially vacuolate algae), organic solutes are also involved (see Cram, 1975; Raven, 1975). It is not clear whether the primary factor which leads to cell expansion in algae is increased osmotic pressure of the cells or decreased wall pressure. In a normal light–dark cycle most irreversible increase in cell volume occurs in the light phase (e.g. Neeb, 1952; Tamiya, 1966; Neuscheler-Wirth, 1970; Waasland and Cleland, 1972). Intracellular osmotic pressure is often also higher in the light (light-induced net salt influx, generally related to active anion influx: Neeb, 1952; Tazawa, 1961; Nagai and Tazawa, 1962; Saddler, 1970c). Neeb (1952) investigated the two phenomena in parallel in *Hydrodictyon reticulatum*; the results suggest that a simple cause-and-effect relation between increased internal osmotic pressure and cell expansion is unlikely. Further experiments in which turgor, wall properties (e.g. Miyashi, 1972) and cell expansion are investigated in parallel are needed.

#### Osmoregulation

In many marine algae the internal osmotic pressure changes in parallel with changes in the external osmotic pressure over a considerable range, thus keeping turgor constant. This can involve both inorganic and organic (produced by photosynthesis) solutes (Gutknecht and Dainty, 1968; Raven, 1971b; Cram, 1975). Many freshwater algae also show some adaptation of their turgor pressure (Wildervank, 1932; Collander, 1936; Barr, 1965; Green, 1968), although in some cases it is the internal osmotic pressure rather than turgor which is restored after experimental perturbation (Kamiya and Kuroda, 1956). Cram (1975) summarises evidence that the signal to which the regulatory mechanism responds is indeed cell turgor pressure. The transport process which is affected is unclear; in *Chaetomorpha* $K^+$ is important (Zimmerman and Steudle, 1971), although the absence of active $Cl^-$ transport here may make it rather atypical (see Section 5.2).

### 5.7.2 *Volume regulation in wall-less systems*

*Isotonic media*
Included here are wall-less cells in sea water, and the cytoplasm vis-a-vis the vacuole in walled, vacuolate cells. Here the problem arises because the cytoplasm has a higher concentration of impermeant solutes than the sea water or the vacuole. Such a system is osmotically unstable, and the compartment with excess impermeant solutes will swell unless some permeant solute is actively extruded from it. In animal cells in isotonic media the solute is $Na^+$ (Baker, 1972); this may well also be the case in the wall-less eggs and early zygotes of *Pelvetia* where $Na^+$ is also actively extruded (Robinson and Jaffe, 1973).

Active $Na^+$ transport inwards at the tonoplast may also be involved in cytoplasmic volume regulation in walled, vacuolate cells, although active $K^+$ or $Cl^-$ transport may also be involved in some cases (Section 5.2).

Adjustment to very high external osmotic pressures in wall-less algae such as halophilic *Dunaliella* generally involves organic solutes (glycerol in this case) rather than ions (Cram, 1975; Raven, 1975). These extreme halophilic algae, unlike the corresponding bacteria, have enzymes which are intolerant of very high salt concentrations but are tolerant of organic solutes such as glycerol (e.g. Heimer, 1973).

*Hypotonic media*
Volume regulation in wall-less cells in media more dilute than their cytoplasm (which has an osmotic potential of at least 20 Mol $m^{-3}$ KCl: Raven, 1975) is by means of contractile vacuoles. They excrete a hypotonic solution which may be involved in efflux of $Na^+$ as well as water (Bruce and Marshall, 1965).

### 5.7.3 *Nutrition*

In terms of ion transport, nutrition involves not only uptake of essential elements required for growth, but also the active extrusion of solutes which are products of essential syntheses but which would be toxic if allowed to remain in the cytoplasm, e.g. $H^+$ or $OH^-$ (Raven and Smith, 1973, 1974). $K^+$ functions in both an osmoregulatory role and as an enzyme activator (Raven, 1975). As with the ions whose major function is in turgor generation, the transport of nutrient ions must be regulated in relation to cell growth. However, in the case of the nutrient ions some 'luxury accumulation' can also occur, e.g. the accumulation of P as polyphosphate in many algae under conditions of N deficiency (Healey, 1973; Raven, 1975). The increased net

$H_2PO_4^-$ or $SO_4^{2-}$ influx following readdition of these ions to P- or S-deficient cells is due to an enhanced influx rather than a decreased efflux (Vallee and Jeanjean, 1968a; Fuhs et al., 1972; Jeanjean, 1973a). Stimulation of active fluxes is also involved in restoration of a normal $K^+/Na^+$ ratio upon re-addition of $K^+$ to 'Na$^+$-rich' *Chlorella* cells (Barber and Shieh, 1972).

### 5.7.4 *Regulatory mechanisms*

Our current inability to isolate and quantify any ion-transport systems from algae makes investigation of the mechanism(s) whereby transport is controlled rather difficult. The transport system is generally assayed in vivo, and distinction between 'fine control' (alteration of activity of pre-existing transport system) and 'coarse control' (alteration in the amount of transport system present) is by means of time course (fine control can be much more rapid) and sensitivity to inhibitors of protein synthesis (coarse control should be inhibited by this treatment, fine control should not). Such experiments fall far short of the conclusive demonstration of involvement or non-involvement of protein synthesis in an in vivo process.

*Fine control*

Responses which are too rapid to be dependent on protein synthesis include light stimulation of fluxes (Barber, 1968c, d, e; Findenegg et al., 1971), and the increased capacity for $SO_4^{2-}$ influx following removal of $SO_4^{2-}$ from the medium bathing *Chlorella*. While a definite increase in the capacity is seen after some 10 min of $SO_4^{2-}$ deprivation, it must be pointed out that the increase in capacity continues for more than 8 h, so the later stages may involve protein synthesis (Vallee and Jeanjean, 1968a). The effector involved in this control is some metabolite of $SO_4^{2-}$ rather than internal $SO_4^{2-}$ itself (Vallee and Jeanjean, 1968b). The very rapid initial influx found when $SO_4^{2-}$ is added to S-starved cells declines with a half-time of about 5 min (Vallee and Jeanjean, 1968a), again by some fine control mechanism. All of these effects appear to work in *Chlorella* via effects on influx, since efflux is low (Vallee and Jeanjean, 1968a). In the case of $H_2PO_4^-$ influx in *Chlorella* (Jeanjean, 1969, 1973a), response of $H_2PO_4^-$ influx capacity to $H_2PO_4^-$ deficiency is similar, except that the deficiency effect on transport capacity takes longer to manifest and the increase in capacity is inhibited by protein synthesis inhibitors (see below).

The very close correlation between $NO_3^-$ and $NO_2^-$ influx in *Ankistrodesmus* and the rate at which they are metabolised suggests some fine control of influx of these oxy-anions by the rate at which they are metabolised, with a small intracellular pool (Ullrich-Eberius, 1973b).

*Coarse control*

In the long term (i.e. times of the order of the biomass doubling time) the increase in capacity for active transport in a growing algal cell must depend on protein synthesis, both for synthesis of the transport system and less directly. The rates of active transport during synchronous culture of algae generally follow growth (Tamiya, 1966; Meszes et al., 1967; Cook, 1968; Domanski-Kaden and Simonis, 1972; Jeanjean, 1973b). Meszes and Erdei (1968) report an increased activity of an ATPase which may be involved in $K^+$ and $Na^+$ transport in *Scenedesmus* during the cell cycle; however, fine control was not ruled out in these experiments.

Jeanjean (1973b) has shown that increased $H_2PO_4^-$ transport capacity which is normally attendant on transfer of synchronous *Chlorella* cells from dark to light does not occur if cycloheximide is present in the light, if $CO_2$ is absent in the light, or if the culture is kept in the dark. This is consistent with de novo synthesis of the $H_2PO_4^-$ transport system at this phase of the life cycle. Recovery of $H_2PO_4^-$ transport capacity in *Chlorella* following an osmotic shock is also inhibited by cycloheximide (Gaudin et al., 1973).

Muller and Paschinger (1970) and Muller et al., (1970) have shown that cycloheximide inhibits $Rb^+$ influx in *Chlorella* at concentrations which inhibit incorporation of [$^{14}$C]leucine into protein (duly corrected for effects on leucine influx) but do not decrease the ATP level. They suggest that these effects are due to turnover of the $Rb^+$ transport system rather than to a requirement for protein synthesis for ion-by-ion functioning of the transport system. Similar conclusions can be drawn from the experiments of Jeanjean (1973a, b) and Gaudin et al. (1973) on phosphate transport in *Chlorella*.

## 5.8 *Summary and prospects*

Algae have proved most valuable in distinguishing between active and passive ion transport and relating ion transport to electrical phenomena. While the presence of photosynthetic energy conversion in most algal cells can be viewed as a complication in studies of energy-coupling of transport, this additional energy source has allowed some advances to be made in the study of the enegetics of active transport. Algae have also proved useful in relating ion transport to growth and nutrition.

As regards possibilities for future research, there still remain many unsolved problems in 'membrane biophysics' of algae. The discrepancies between electrical and radiochemical measurements of electrogenic fluxes (both active and passive) require further study. Ions studied should be extended to include the major metabolic ions associated with C and N assimilation and with pH regulation, and these and previously investigated transport processes should be related to the nutrient requirements and energy availability during growth. A further very promising field is regulation of transport, both in terms of the long-term strategy of maintaining the level of transport systems at a reasonable level and the tactical aspects of short-term regulation.

## Acknowledgement

I should like to thank Dr. S.M. Glidewell for her critical reading of the manuscript and for much helpful discussion.

## References

R.D. ALLEN, L. JACOBSEN, J. JOAQUIN and L.P. JAFFE, Dev. Biol., 27 (1972) 538.

A.W. ATKINSON, JR., B.E.S. GUNNING, P.C.L. JOHN and W. MCCULLOUGH, Science, 176 (1971) 694.

P.F. BAKER, in L.E. Hokin (Ed.) Metabolic Pathways (1972) 3rd Ed., Vol. 6, Academic Press, New York, p. 243.

J. BARBER, Nature, 217 (1968a) 876.

J. BARBER, Biochim. Biophys. Acta, 150 (1968b) 618.

J. BARBER, Biochim. Biophys. Acta, 150 (1968c) 730.

J. BARBER, Biochim. Biophys. Acta, 163 (1968d) 141.

J. BARBER, Biochim. Biophys. Acta, 163 (1968e) 531.

J. BARBER, Arch. Biochem. Biophys., 130 (1969) 389.

J. BARBER and Y.J. SHIEH, J. Exp. Bot., 23 (1972) 627.

J. BARBER and Y.J. SHIEH, Planta, 111 (1973a) 13.

J. BARBER and Y.J. SHIEH, Plant. Sci. Lett., 1 (1973b) 405.

C.E. BARR, J. Gen. Physiol., 49 (1965) 181.

C.E. BARR and T.C. BROYER, Plant. Physiol., 39 (1964) 48.

J.A. BASSHAM, Science, 172 (1971) 526.

J.A. BASSHAM and G.H. KRAUSE, Biochim. Biophys. Acta, 189 (1969) 207.

D.R. BLACK, Ionic Relations of *Enteromorpha intestinalis* (L. Link) (1971) Ph. D. Thesis, University of St. Andrews, Scotland.

D.R. BLACK and D.C. WEEKS, New Phytol., 71 (1972) 119.

R.W. BLOUNT, A Quantitative Analysis of Active Ion Transport in the Single Celled Alga *Halicystis ovalis* (1958) Ph. D. Thesis, University of California, Los Angeles.

R.W. BLOUNT and B.H. LEVEDAHL, Acta Physiol. Scand., 49 (1960) 1.

D.L. BRUCE and J.M. MARSHALL, J. Gen. Physiol., 49 (1965) 151.

R. COLLANDER, Protoplasma, 25 (1936) 201.

J.R. COOK, in D.E. Buetow (Ed.) The Biology of *Euglena* (1968) Vol. 1, Academic Press, New York, p. 243.

H.G.L. COSTER, Aust. J. Biol. Sci., 19 (1965) 545.

H.G.L. COSTER and A.B. HOPE, Aust. J. Biol. Sci., 21 (1968) 243.

W.J. CRAM, in J. Dainty and U. Zimmermann (Eds.) Membrane Transport in Plants and Plant Organelles (1974) Springer Verlag, Berlin, p. 334.

W.J. CRAM, in U. Lüttge and M.G. Pitman (Eds.) Encyclopedia of Plant Physiology (New Ed.) (1975) Transport in cells and tissues, Springer Verlag, Berlin, in press.

J. DAINTY, Annu. Rev. Plant Physiol., 13 (1962) 379.

J. DAINTY, in J.B. Pridham (Ed.) Plant Cell Organelles (1968) Academic Press, London, p. 40.

M.A. DEWAR and J. BARBER, Planta, 113 (1973) 143.

M.A. DEWAR and J. BARBER, Planta, 117 (1974) 163.

J.M. DIAMOND and A.K. SOLOMON, J. Gen. Physiol., 42 (1959) 1105.

W.A. DODD, M.G. PITMAN and K.R. WEST, Aust. J. Biol. Sci., 19 (1966) 341.

J.D. DODGE, The Fine Structure of Algal Cells (1973) Academic Press, London.

J. DOMANSKI-KADEN and W. SIMONIS, Arch. Mikrobiol. 87 (1972) 11.

R.W. EPPLEY, J. Gen. Physiol., 41 (1958) 901.

G.R. FINDENEGG, Planta, 116 (1974) 123.

G.R. FINDENEGG, H. PASCHINGER and E. BRODA, Planta, 99 (1971) 163.

G.P. FINDLAY, in U. Lüttge and M.G. Pitman (Eds.) Encyclopedia of Plant Physiology (New Ed.) (1975) Transport in cells and tissues, Springer Verlag, Berlin, in press.

G.P. FINDLAY, A.B. HOPE and E.J. WILLIAMS, Aust. J. Biol. Sci., 22 (1969) 1163.

G.P. FINDLAY, A.B. HOPE and E.J. WILLIAMS, Aust. J. Biol. Sci., 23 (1970) 323.

G.P. FINDLAY, A.B. HOPE, M.G. PITMAN, F.A. SMITH and N.A. WALKER, Biochim. Biophys. Acta, 183 (1969) 565.

G.P. FINDLAY, A.B. HOPE, M.G. PITMAN, F.A. SMITH and N.A. WALKER, Aust. J. Biol. Sci., 24 (1971) 731.

F.E. FRITSCH, The Structure and Reproduction of the Algae (1935) Vol. 1, Cambridge University Press, London, 789 p.

F.E. FRITSCH, The Structure and Reproduction of the Algae (1945) Vol. 2, Cambridge University Press, London, 939 p.

G.W. FUHS, S.D. DEMMERLE, E. CARELLI and M. CHEN, Limnol. Oceanogr. Special Symposia Vol. 1 (1972) 113.

C. GAUDIN, R. JEANJEAN and F. BLASCO, C. R. Acad. Sci. Paris. D, 277 (1973) 301.

P.B. GREEN, Plant Physiol., 43 (1968) 1169.

R.T.L. GUILLARD, in R.L. Lewin (Ed.) Physiology and Biochemistry of Algae (1962) Academic Press, New York, p. 529.

J. GUTKNECHT, Biol. Bull., 129 (1965) 495.

J. GUTKNECHT, Biol. Bull., 130 (1966) 331.

J. GUTKNECHT, J. Gen. Physiol., 50 (1967) 1821.

J. GUTKNECHT and J. DAINTY, Oceanogr. Mar. Biol. Annu. Rev., 6 (1968) 163.

F.P. HEALEY, Crit. Rev. Microbiol., 3 (1973) 69.

Y.M. HEIMER, Planta, 113 (1973) 279.

D.R. HOAGLAND and A.R. DAVIS, J. Gen. Physiol., 6 (1923) 47.

D.R. HOAGLAND and A.R. DAVIS, Protoplasma, 6 (1929) 610.

A.B. HOPE, Aust. J. Biol. Sci., 16 (1963) 429.

A.B. HOPE and P.A. ASCHBERGER, Aust. J. Biol. Sci., 23 (1970) 1047.

A.B. HOPE and N.A. WALKER, Aust. J. Biol. Sci., 13 (1960) 277.

A.B. HOPE and N.A. WALKER, Aust. J. Biol. Sci., 14 (1961) 26.

A.B. HOPE and N.A. WALKER, The Physiology of Giant Algal Cells (1975) Cambridge University Press, London.

A.B. HOPE, U. LÜTTGE and E. BALL, Z. Pflanzenphysiol., 72 (1974) 1.

A.B. HOPE, A. SIMPSON and N.A. WALKER, Aust. J. Biol. Sci., 19 (1966) 355.

R. JEANJEAN, Bull. Soc. Fr. Physiol. Veg., 15 (1969) 159.

R. JEANJEAN, FEBS Lett., 32 (1973a) 149.

R. JEANJEAN, C. R. Acad. Sci. Paris D., 277 (1973b) 193.

N. KAMIYA and K. KURODA, Protoplasma, 46 (1956) 423.

H. KESSELER, Helgoländer Wiss. Meeres Unters., 11 (1964) 258.

U. KISHIMOTO and M. TAZAWA, Plant Cell Physiol., 6 (1965) 507.

H. KITASATO, J. Gen. Physiol., 52 (1968) 60.

E. KOMOR and W. TANNER, Eur. J. Biochem., 44 (1974) 219.

R. MCC. LILLEY and A.B. HOPE, Biochim. Biophys. Acta, 226 (1971a) 161.

R. MCC. LILLEY and A.B. HOPE, Aust. J. Biol. Sci., 24 (1971b) 1351.

U. LÜTTGE, Stofftransport der Pflanzen (1973) Springer Verlag, Berlin.

E.A.C. MACROBBIE, J. Gen. Physiol., 45 (1962) 861.

E.A.C. MACROBBIE, J. Gen. Physiol., 47 (1964) 859.

E.A.C. MACROBBIE, Biochim. Biophys. Acta, 94 (1965) 64.

E.A.C. MACROBBIE, Aust. J. Biol. Sci., 19 (1966a) 363.

E.A.C. MACROBBIE, Aust. J. Biol. Sci., 19 (1966b) 371.

E.A.C. MACROBBIE, Quart. Rev. Biophys., 3 (1970) 251.

E.A.C. MACROBBIE, Annu. Rev. Plant Physiol., 22 (1971) 75.

E.A.C. MACROBBIE, in W.P. Anderson (Ed.) Ion Transport in Plants (1973) Academic Press, London, p. 431.

E.A.C. MACROBBIE, in W.D.P. Stewart (Ed.) Algal Physiology and Biochemistry (1974) Blackwell, Oxford, p. 678.

E.A.C. MACROBBIE and J. DAINTY, J. Gen. Physiol., 42 (1958) 335.

G. MESZES and L. ERDEI, Acta Biochim. Biophys. Acad. Sci. Hung., 41 (1969) 131.

G. MESZES, J. KRALOVANSKY, E. CSEH and Z. BOSZORMENYI, Acta Biochim. Biophys. Acad. Sci. Hung., 2 (1967) 239.

P. MITCHELL, Chemiosmotic Coupling in Oxidative and Photosynthetic Phosphorylation (1966) Glynn Research Ltd., Bodmin, Cornwall, England.

Y. MIYASHI, Bot. Mag. Tokyo, 85 (1972) 59.

E. MULLER and M. PASCHINGER, Biochem. Physiol. Pflanz., 161 (1970) 476.

E. MULLER, M. PASCHINGER and M. LUX, Abh. Dtsch. Akad. Wiss. Berl. Nr. 4b (1970) p. 49.

J.C. MUNDAY, JR., Bot. Mar., 15 (1972) 61.

R. NAGAI and M. TAZAWA, Plant Cell Physiol., 3 (1962) 323.

O. NEEB, Flora, 139 (1952) 39.

H. NEUSCHELER-WIRTZ, Z. Pflanzenphysiol., 63 (1970) 352.

B. NUHRENBERG, D. LESEMANN and A. PIRSON, Planta, 79 (1968) 162.

J. PREISS and T. KOSUGE, Annu. Rev. Plant. Physiol., 21 (1970) 433.

J.A. RAVEN, Ionic Relations of *Hydrodictyon africanum* Yamma (1966) Ph. D. Thesis, University of Cambridge.

J.A. RAVEN, J. Gen. Physiol., 50 (1967a) 1607.

J.A. RAVEN, J. Gen. Physiol., 50 (1967b) 1625.

J.A. RAVEN, J. Exp. Bot., 19 (1968) 233.

J.A. RAVEN, New Phytol., 68 (1969a) 45.

J.A. RAVEN, New Phytol., 68 (1969b) 1089.

J.A. RAVEN, Biol. Rev., 45 (1970) 167.

J.A. RAVEN, Planta, 97 (1971a) 28.

J.A. RAVEN, Chem. Ind. (1971b) p. 859.

J.A. RAVEN, Trans. Bot. Soc. Edinb., 41 (1971c) 219.

J.A. RAVEN, J. Membrane Biol., 6 (1971d) 89.

J.A. RAVEN, J. Exp. Bot., 22 (1971e) 420.

J.A. RAVEN, New Phytol., 73 (1974a) 421.

J.A. RAVEN, J. Exp. Bot., 25 (1974b) 221.

J.A. RAVEN, in W.D.P. Stewart (Ed.) Algal Physiology and Biochemistry (1974c) Blackwell, Oxford, p. 391.

J.A. RAVEN, in J. Dainty and M. Zimmermann (Eds.) Membrane Transport in Plants and Plant Organelles (1974d) Springer Verlag, Berlin, p. 169.

J.A. RAVEN, in U. Lüttge and M.G. Pitman (Eds.) Encyclopedia of Plant Physiology (New Edn.) (1975) Transport in cells and tissues, Springer Verlag, Berlin, in press.

J.A. RAVEN and F.A. SMITH, in W.P. Anderson (Ed.) Ion Transport in Plants (1973) Academic Press, London, p. 271.

J.A. RAVEN and F.A. SMITH, Can. J. Bot., 52 (1974) 1035.

J.L. RICHARDS and A.B. HOPE, J. Membrane Biol., 16 (1974) 121.

J.B. ROBINSON, J. Exp. Bot., 20 (1969a) 201.

J.B. ROBINSON, J. Exp. Bot., 20 (1969b) 212.

K.R. ROBINSON and L.F. JAFFE, Dev. Biol., 35 (1973) 349.

R. RYBOVA, K. JANACEK and M. STAVIKOVA, Z. Pflanzenphysiol., 66 (1972) 420.

H.D.W. SADDLER, J. Exp. Bot., 21 (1970a) 345.

H.D.W. SADDLER, J. Exp. Bot., 21 (1970b) 605.

H.D.W. SADDLER, J. Gen. Physiol., 55 (1970c) 803.

M. SCHAEDLE and L. JACOBSEN, Plant Physiol., 40 (1965) 214.

M. SCHAEDLE and L. JACOBSEN, Plant Physiol., 41 (1966) 248.

M. SCHAEDLE and L. JACOBSEN, Plant Physiol., 42 (1967) 953.

Y. SHIEH and J. BARBER, Biochim. Biophys. Acta, 233 (1971) 594.

W. SIMONIS and W. URBACH, Annu. Rev. Plant. Physiol., 24 (1973) 89.

W. SIMONIS, T. BORNEFELD, J. LEE and K. MAJUMDAR, in J. Dainty and U. Zimmermann (Eds.) Membrane Transport in Plants and Plant Organelles (1974) Springer Verlag, Berlin, p. 220.

F.A. SMITH, Biochim. Biophys. Acta, 126 (1966) 94.

F.A. SMITH, J. Exp. Bot., 18 (1967) 716.

F.A. SMITH, J. Exp. Bot., 19 (1968) 442.

F.A. SMITH, New Phytol., 69 (1970) 903.

F.A. SMITH, New Phytol., 71 (1972) 595.

F.A. SMITH and J.A. RAVEN, New Phytol., 73 (1974a) 1.

F.A. SMITH and J.A. RAVEN, in J. Dainty and U. Zimmermann (Eds.) Membrane Transport in Plants and Plant Organelles (1974b) Springer Verlag, Berlin, p. 380.

F.A. SMITH and J.A. RAVEN, in U. Lüttge and M.G. Pitman (Eds.) Encyclopedia of Plant Physiology (New Edn.) (1975) Transport in cells and tissues, Springer Verlag, Berlin, in press.

F.A. SMITH and K.R. WEST, Aust. J. Biol. Sci., 22 (1969) 351.

R.M. SPANSWICK, J. Membrane Biol., 2 (1970) 59.

R.M. SPANSWICK, Biochim. Biophys. Acta, 288 (1972) 73.

R.M. SPANSWICK, Biochim. Biophys. Acta, 332 (1974a) 387.

R.M. SPANSWICK, Symp. Soc. Exp. Biol., 28 (1974b) 127.

R.M. SPANSWICK, in U. Lüttge and M.G. Pitman (Eds.) Encyclopedia of Plant Physiology (New Edn.) (1975) Springer Verlag, Berlin, in press.

R.M. SPANSWICK and E.J. WILLIAMS, J. Exp. Bot., 16 (1965) 463.

R.M. SPANSWICK, J. STOLAREK and E.J. WILLIAMS, J. Exp. Bot., 18 (1967) 1.

H. SPRINGER-LEDERER and D.L. ROSENFELD, Physiol. Plant., 21 (1968) 435.

W.D.P. STEWART (Ed.) Algal Physiology and Biochemistry (1974) Blackwell Scientific Publications, Oxford.

B.L. STREHLER, Arch. Biochem. Biophys., 43 (1953) 67.

J.F. SUTCLIFFE, Mineral Salts Absorption in Plants (1962) Pergamon Press, Oxford.

H. TAMIYA, Annu. Rev. Plant Physiol., 17 (1966) 1.

M. TAZAWA, Protoplasma, 53 (1961) 342.

W. ULLRICH, Planta, 100 (1971) 18.

W. ULLRICH, Planta, 116 (1974) 143.

C.I. ULLRICH-EBERIUS, Planta, 109 (1973a) 161.

C.I. ULLRICH-EBERIUS, Planta, 115 (1973b) 25.

M. VALLEE and R. JEANJEAN, Biochim. Biophys. Acta, 150 (1968a) 599.

M. VALLEE and R. JEANJEAN, Biochim. Biophys. Acta, 150 (1968b) 607.

L.N. VOROBIEV, Nature, 216 (1967) 1325.

S.D. WAALAND and R. CLELAND, Planta, 105 (1972) 196.

G. WAGNER and F.W. BENTRUP, Ber. Dtsch. Bot. Ges., 86 (1973) 365.

N.A. WALKER, in J. Dainty and U. Zimmermann (Eds.) Membrane Transport in Plants and Plant Organelles (1974) Springer Verlag, Berlin, p. 173.

N.A. WALKER and A.B. HOPE, Aust. J. Biol. Sci., 22 (1969) 1179.

L.S. WILDERVANK, Rec. Trav. Bot. Neerl., (1932) 227.

E.J. WILLIAMS, C. MUNRO and D.S. FENSOM, Can. J. Bot., 50 (1972) 2255.

U. ZIMMERMANN and E. STEUDLE, Mar. Biol., 11 (1971) 132.

*Ion transport in plant cells and tissues*
*edited by D.A. Baker and J.L. Hall*
© *North-Holland Publishing Company, 1975*

CHAPTER 6

# Storage tissues

W.J. CRAM

## Contents



161

## 6.1 Introduction

Excised storage tissue has only rarely been used in investigations of the ion relations of the storage organ itself. It has mostly been used as a convenient source of a representative plant cell in work on cellular ion relations. Most of the work summarised in this chapter is therefore of a general nature and unavoidably overlaps work on other tissues which has been directed to answering the same questions (see also Poole, 1975).

Storage tissue is initiated by the stimulation of non-growing stem or root tissue to growth. Cells are cut off from one or many cambia and expand. The tissue eventually stops growing and dormancy sets in. Whether or not there is a non-growing but active stage between growth and dormancy is unknown, but it seems possible that there is such a stage as can be distinguished in other organs such as root and leaf. Finally, during sprouting, reserves are mobilised, flow out of the cells and eventually move into the phloem. Growing, non-growing but active, and dormant states are, of course, not sharply distinct, but they may be discussed separately for convenience.

In practice one can only work on excised tissue because of the limitation of the long diffusion paths in the intact tissue. However, excision initiates major changes in storage tissues which continue for several days. After two to four days an indefinitely maintained capacity to accumulate ions develops and it is the properties of the tissue in this state which have been most extensively investigated.

Storage tissues are mainly composed of homogeneous, unspecialised parenchyma cells. One or many cambia occur in different tissues, but cells from non-cambial regions can be excised for experiment. In any tissue parenchyma cells vary several-fold in size. Fig. 6.1 shows the fine structure of red beet parenchyma cells. If the cytoplasm were 0.5 $\mu$m thick it would occupy 6% of the cell volume in a cell 50 $\mu$m in diameter.

## 6.2 Methods

### 6.2.1 Flux measurement

The problem is to measure fluxes in a compartmental system when there is no access to any internal compartment (see Ch. 5 on giant algal cells where cytoplasm and vacuole can be individually sampled).

A preliminary problem is to allow for diffusion through extracellular spaces. In 1 mm thick slices of tissue the extracellular spaces equilibrate with the

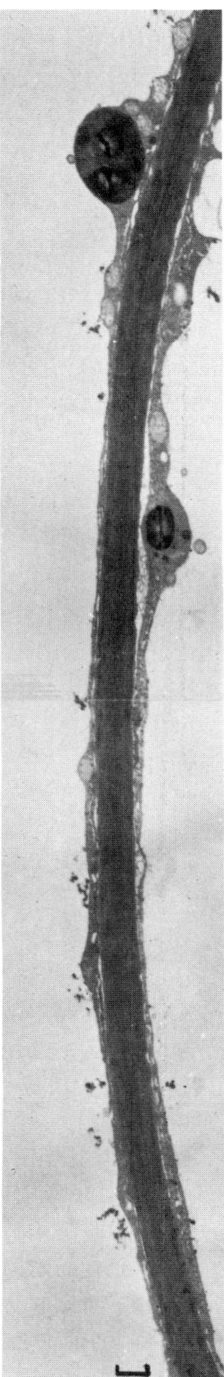

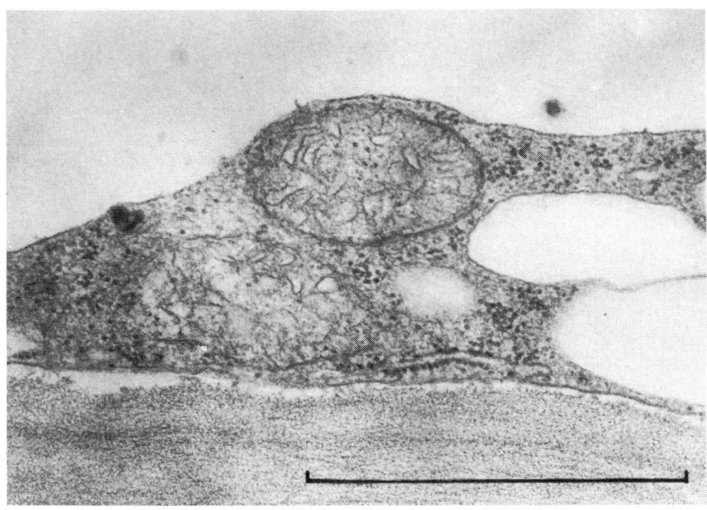

Fig. 6.1. The fine structure of parenchyma cells of red beet root tissue. Scale markers: 1 $\mu$m. Glutaraldehyde/OsO$_4$ fixation. Post-stained in lead citrate. Electron micrograph kindly provided by M.E. Van Steveninck.

external solution within about 5 minutes. Although uptake by individual cells means that the concentration at the centre of the disc is less than at the outside, the difference in concentration is generally negligible at external concentrations above 1 Mol m$^{-3}$. At lower external concentrations and in slices thicker than 1 mm the gradient of concentration within the disc may become significant, so that the concentration round the cells is not uniform and flux measurements may be in error. A further complication with cations is that they interact with fixed negative charges in cell walls and over 10–20 minutes reach a much higher concentration than in the external solution. The old terms 'water free space' and 'Donnan free space' (still occasionally used) correspond to intercellular spaces and cell walls respectively (see Section 1.3). These properties of the tissue are of little physiological significance but introduce technical difficulties to the measurement of fluxes.

### 6.2.2 Compartmental analysis

One can attempt to obtain detailed estimates of fluxes from analysis of the exchange of radioactive tracers.

For transport processes which do not distinguish between different isotopes of the same element, the movement of radioactivity $\phi^*$ (cpm kg$^{-1}$ s$^{-1}$) from compartment 1 to compartment 2 at any instant will be given by

$$\phi^*_{12} = \phi_{12} \cdot s_1 \text{ (mol kg}^{-1}\text{ s}^{-1} \cdot \text{cpm mol}^{-1})$$

In words, if there are $x$ labelled atoms per 100 atoms in compartment 1, then for every 100 atoms moving from compartment 1 to compartment 2, $x$ will be labelled.

For any particular system of compartments one can write down such equations for fluxes between each pair of compartments, and solve for the total amount of label in each compartment and thus for the total in the system. The solutions are generally complex, but for a system of two compartments in series, with the outer compartment much smaller than the inner one (Fig. 1.6) the solution is fairly simple (Pitman, 1963; Cram, 1968b; Pallaghy and Scott, 1969).

For influx, when $Q^*_T = 0$ initially and over periods short compared with the time taken to exchange in the system fully:

$$Q^*_T = \frac{\phi_{01} \cdot \phi_{10}}{(\phi_{10} + \phi_{12})^2} \cdot Q_1 \left[1 - \exp(-k_1 \cdot t)\right] + \frac{\phi_{01} \cdot \phi_{12}}{\phi_{10} + \phi_{12}} \cdot t \qquad (6.1)$$

For washing out after a loading time $t_{in}$, and writing $A$ for the first term and $B$ for the second term on the right hand side of the equation above

$$Q_T^* = A \cdot \exp(-k_1 \cdot t) + B\left[1 - \frac{\phi_{21} \cdot \phi_{10}}{\phi_{10} + \phi_{12}} \cdot \frac{1}{Q_2} \cdot t\right] \tag{6.2}$$

The rate of loss from the two compartments is therefore composed of two exponential components (one so slow that it is effectively linear with time: $(\exp(-kt) \simeq (1 - kt)$ if $kt \ll 1)$) which are characterised by their rate constants $(k)$ and their intercept sizes at the beginning of washing out $(I)$. After a loading time greater than $3.5 \times 1/k_1$:

$$I_{\text{slow}} = \frac{\phi_{01} \cdot \phi_{12}}{\phi_{10} + \phi_{12}} \cdot t_{\text{in}} \qquad k_{\text{slow}} = \frac{\phi_{21} \cdot \phi_{10}}{\phi_{10} + \phi_{12}} \cdot \frac{1}{Q_2}$$

$$I_{\text{fast}} = \frac{\phi_{01} \cdot \phi_{10}}{(\phi_{10} + \phi_{12})^2} \cdot Q_1 \qquad k_{\text{fast}} = \frac{\phi_{10} + \phi_{12}}{Q_1} \tag{6.3}$$

In addition, since the system is at a steady state,

$$\phi_{10} + \phi_{12} = \phi_{21} + \phi_{01}$$

and, since $Q_2$ is much larger than $Q_1$,

$$Q_2 \gg Q_T$$

Hence

$$\phi_{01} = k_f \cdot I_f + I_s/t_{\text{in}} \tag{6.4}$$

$$\phi_{10} = k_f \cdot I_f + k_s \cdot Q_T \tag{6.5}$$

$$\phi_{12} = \phi_{10} \cdot (I_s/t_{\text{in}})/(k_f \cdot I_f) \tag{6.6}$$

$$\begin{aligned}\phi_{21} &= \phi_{01} \cdot (k_s \cdot Q_T)/(k_f \cdot I_f) \\ &= \phi_{12} - (\phi_{01} - \phi_{10})\end{aligned} \tag{6.7}$$

$$Q_1 = (\phi_{10} + \phi_{12})/k_f \tag{6.8}$$

These equations describe the movement of tracer in a system which is completely defined as being composed of two compartments in series, the outer much smaller than the inner, at a steady state, and with fluxes between compartments which do not discriminate between isotopes.

An example of the observed time course of loss of an ion from a storage tissue is shown in Fig. 6.2. By plotting log of content against time, extrapolating, subtracting and replotting the difference, the curve can be fitted by the sum of three exponential terms. In dead tissue only the fastest is still present and appears to occupy 100% of the tissue volume. The fastest phase can therefore be taken to be the loss from extracellular spaces, and the two slower phases as loss from cells. An interpretation of cellular exchange is suggested by the presence in the cells of the thin layer of cytoplasm apparently surrounding

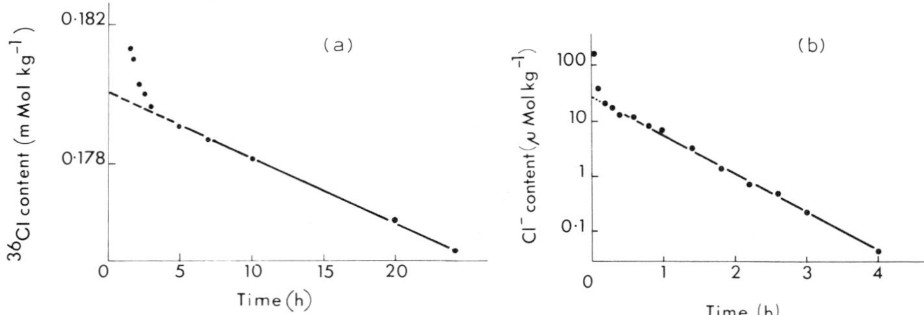

Fig. 6.2. An example of the time course of loss of $Cl^-$ from storage tissue. Potato disks, 1 mm thick, in 10 Mol m$^{-3}$ KCl, 5 °C. (a) $^{36}Cl$ activity (divided by external solution specific activity) is plotted on a log scale against time. The slowest exchanging exponential component is extrapolated to $t = 0$, and subtracted from the total. (b) The difference is plotted on a log scale against time and this plot reveals a more rapidly exchanging exponential component. From Lannoye (1970).

the large vacuole, and one can naturally interpret values of $k$ and $I$ obtained from the graph in terms of the equations describing tracer exchange in two compartments in series; this is, of course, the impetus for deriving the equations describing such a system. However, these equations only apply under certain conditions, as noted above, and one must establish that these conditions apply before using the equations.

1. Isotope discrimination. No physiological process is known to (Wagener et al., 1967) or would be expected to discriminate significantly between isotopes of the major inorganic ions.

2. Compartmentation. An implicit condition in deriving equations 6.4 to 6.8 is that the rate of movement from a compartment is limited by movement across the bounding membrane, and not by diffusion within it. With the observed rates of exchange of the compartments it is simple to show that the rate of diffusion of an ion would have to be at least $10^7$ times less than in water for exchange to be limited by internal diffusion. Biochemical reactions could not take place in the cytoplasm if this were the case for all molecules, and conductivity measurements show it is not the case for ions (Gelfan, 1928).

3. Steady state conditions. These must be established by measurement. One can depart from steady state conditions without seriously affecting the interpretation if changes in contents and fluxes are less than about 10 percent during an experiment i.e. a quasi-steady state (see Cram, 1969a).

4. The cytoplasmic compartment content is much smaller than the vacuolar. The tonoplast is not rigid since isolated protoplasts behave as osmometers (Vreugdenhill, 1957). The osmotic pressure in the cytoplasm and vacuole must

therefore be approximately the same, and the contents of the major osmotica must then be roughly proportional to volume. Cytoplasmic volume is much smaller than vacuolar in storage tissue at least.

5. Cytoplasm and vacuole in series. Appearances may be deceptive (MacRobbie, 1969). The physiological demonstration of the spatial relationship between the compartments giving rise to the two cellular efflux components is the major problem in applying compartmental analysis to small plant cells. There is no general method of solving this problem. Two investigations depending on specialized properties of tissues have been published. In beet, $Na^+$ was indirectly shown to inhibit the movement of $K^+$ from cytoplasm to vacuole, implying that $K^+$ can flow directly between the two compartments (Pitman, 1963). In carrot, $Cl^-$ was shown to move from vacuole to cytoplasm and back when the flux across the plasmalemma was inhibited by placing the tissue in water; and it was also concluded that there was no direct movement from vacuole to external solution (Cram, 1968b).

However, these experiments do not show whether there is any direct influx from the external solution to the vacuole, as appears to be the case in *Nitella* (MacRobbie, 1969) and *Chara* (Findlay et al., 1971). If there is a 'straight through' flux ($\phi_{ov}^s$) then Eqs. 6.4 to 6.8 are no longer applicable, except that Eq. 6.4 still describes the influx across the plasmalemma, which is now ($\phi_{oc} + \phi_{ov}^s$). (This can be established by considering the derivation of Eq. 6.4–6.8 with the extra flux present.) From this consideration and the effects of CCCP and oligomycin on $Cl^-$ fluxes (Cram, 1969a) one can obtain a maximum value of a straight through influx in carrot cells. In the presence of both CCCP and oligomycin ($\phi_{oc} + \phi_{ov}^s$) is unaltered, but the influx to the vacuole ($\phi_{ov}^s + (\phi_{oc} \cdot \phi_{cv})/(\phi_{co} + \phi_{cv})$) is reduced to 40%. The maximum value of $\phi_{ov}^s$ (when $\phi_{cv}/(\phi_{co} + \phi_{cv}) = 0$) is therefore 40% of the total rate of entry to the cells under these conditions.

### 6.2.3 Kinetic checks

Wash out curves may appear to be composed of exponential terms when the conditions of Eqs. 6.4 to 6.8 do not apply, for instance in a system of many compartments (Van Liew, 1962) or at a non-steady state. There are two checks for internal self-consistency in the kinetics.

1. The value of $I_c$ ($A$ in Eqs. 6.1–6.2) after various loading periods should rise with a $(1 - \exp(-kt))$ time course with the value of $k$ the same as in the wash-out curve. To a first approximation this is the case for $K^+$ in beet (Pitman, 1963), $Cl^-$ in carrot (Cram, 1968b) and $SO_4^{2-}$ in carrot (Cram, unpublished).

2. The graphs of log (content) versus time and of log (rate of loss) versus time should have the same number of components with the same $k$ for each and intercepts in the ratio of the rate constants (Cram and Laties, 1974). A curve apparently fitting the exchange of $^{14}C$ in carrot tissue was shown by this method to be erroneous (Cram and Laties, 1974).

### 6.2.4 Possible misconceptions

1. A system of two compartments in parallel also exchanges as the sum of the two exponentials. While the form of a validly fitted curve may show the number of distinguishable compartments, it does not give any information about their arrangement.

2. The cytoplasmic intercept is not equal or proportional to the amount of the substance in the cytoplasm (Eqs. 6.3 and 6.8), as has sometimes been assumed. In relation to this, the 'cytoplasmic' intercept after loading under one set of conditions and washing out under another is even less easy to interpret directly, as there is then the added complication of the non-steady state.

3. A rate constant describes the rate of turnover of particles in a compartment; it is not related to a single flux (Eq. 6.3). The ratio of two rate constants is not related to the ratio of 'fluxes from the compartments'.

4. Eqs. 6.4–6.8 describe any system in which the 5 conditions referred to apply. The mechanism of the movements is not one of these conditions and therefore has no relevance to the interpretation of the tracer kinetics. The same equations describe the kinetics of the system whether the fluxes in it are passive and/or active.

### 6.2.5 Other flux measurements

1. The influx to a cell has most frequently been estimated as the content of the tissue after loading in a labelled solution for a period, $t_{in}$, and washing in unlabelled solution for 30 minutes, divided by the loading time. This is approximately equal to $I_s/t_{in}$ in Eq. 6.3. It can be called the quasi-steady influx to the vacuole ($\phi_{ov}$). It is in general not equal to a single flux (Cram, 1969b). A change in any one of three fluxes would change $\phi_{ov}$, and hence, although this flux can easily be measured, it cannot easily be interpreted.

2. The influx of a substance across the plasmalemma in a cell can in principle be estimated from the initial influx of label to the cell after placing in a solution of the labelled substance. The practical difficulty is that after the necessarily short period in the loading solution most of the substance in a tissue will be

in the intercellular spaces and only a fraction will be in the cells. This difficulty can be avoided in three ways (Cram, 1974c).

(i) By washing out the extracellular space content after the short loading period. The substance will then, however, be lost from the cells also, and influx will be underestimated. This at least gives a minimum estimate of the plasmalemma influx (Cram, 1969b; Cram and Laties, 1971; Cram, 1973b).

(ii) By washing out the extracellular spaces and extrapolating a linear transform of the subsequent loss from the cell back to obtain the total in the cell at the beginning of washing out (i.e. the use of compartmental analysis as above).

(iii) By labelling the solution with a suitable non-penetrating substance as well as the substance under investigation. Double counting of the tissue will then allow the extracellular space content to be estimated and subtracted from the total amount of the penetrating substance in the tissue to give the amount in the cells (Cram, 1974c).

The first method is inherently inaccurate, but can be used as a quick and easy guide to minimum values of the plasmalemma influx. The second method has the advantages of being accurate and unambiguous, but has the disadvantages that it can only be applied near a steady state, is a relatively lengthy measurement, and has technical pitfalls. The third method is rapid, accurate, and applies even at a non-steady state. However it is not easy to find a substance which equilibrates with the extracellular spaces rapidly and to the same concentration as the substance under investigation, and at the same time penetrates the cell negligibly fast.

### 6.2.6 Conclusions

Not all the cellular ion fluxes can yet be experimentally distinguished, let alone measured. Three fluxes can be measured with some certainty, and these are the plasmalemma influx (which, as discussed above, does not depend on knowledge of the internal structure of the cell); the plasmalemma efflux; and the quasi-steady overall influx to the vacuole ($\phi_{ov}$). Some progress in understanding the ion relations of cells can be made using these measurements. One can investigate the properties of fluxes across the plasmalemma (at the individual membrane level), and into the vacuole (at a less defined level); and one can see whether factors which alter the overall influx to the vacuole do so via altering the plasmalemma fluxes or via effects within the cell. What cannot be done yet is to investigate the distribution and fluxes of ions between the cytoplasm and the vacuole.

## 6.3  Occurrence and mechanisms of active transport

### 6.3.1  Transport of monovalent inorganic ions

When a storage tissue is placed in a solution of KCl plus NaCl the ions
are accumulated over the course of 5 to 10 days until a steady internal concentration of about 150 Mol m$^{-3}$ Cl$^-$ salt is reached. When only these
ions are present the amount of Cl$^-$ taken up is approximately equal to
the amount of K$^+$ plus Na$^+$ taken up (Fig. 6.3a; Sutcliffe, 1952; Cram,
1968a). In most tissues K$^+$ is taken up in preference to Na$^+$, but beet accumulates Na$^+$ in preference to K$^+$ (MacDonald et al., 1960; Van Steveninck,
1970).

Concentration and electrical measurements show that the initial net influx
on transfer to salt, and the final steady state, involve active transport inwards
of Cl$^-$ in all the tissues examined and also of K$^+$ and Na$^+$ in carrot and
potato (Table 6.1). Fig. 6.4 shows the cytoplasmic and vacuolar concentrations
and plasmalemma and tonoplast fluxes calculated from tracer-exchange data,
assuming that cytoplasm and vacuole are in series. If the cytoplasm occupies
2–5% of the tissue volume, then in beet and carrot the cytoplasmic halide
concentration lies between 20 and 50 Mol m$^{-3}$ and the K$^+$ concentration
between 60 and 180 Mol m$^{-3}$. Assuming that the electrical potential difference
is largely located at the plasmalemma, Cl$^-$ must be actively transported
inwards across the plasmalemma, and also very probably inwards across the
tonoplast in carrot and beet. K$^+$ must be actively transported across the
plasmalemma in carrot and potato, but K$^+$ fluxes in beet all appear to be
down an electrochemical potential gradient. It has been suggested (Laties
et al., 1964) that the Cl$^+$ influx to potato at low temperatures is limited by
a passive influx across the plasmalemma, the active accumulation, which
occurs even at 5 °C (Lannoye, 1970), being achieved by a tonoplast pump.
The evidence for this was the shape of the influx isotherm, which is an uncertain guide (Cram, 1974c), and the results of Lannoye would argue against
both rate limitation by and solely passive movement across the plasmalemma.

Cation selectivity has only been extensively studied in beet. It is probably
not due to a cation-selective system in addition to and independent of a net
salt accumulating system (see the situation in algae, Ch. 5) since selectivity
falls as the external Cl$^-$ concentration is raised (Poole, 1973). When total
uptake is limited, external Na$^+$ depresses K$^+$ influx across the plasmalemma
(Poole, 1971b), and in addition Na$^+$ inhibits K$^+$ influx across the tonoplast
(Pitman, 1963; Poole, 1971b). The slow stimulation of K$^+$ efflux when external

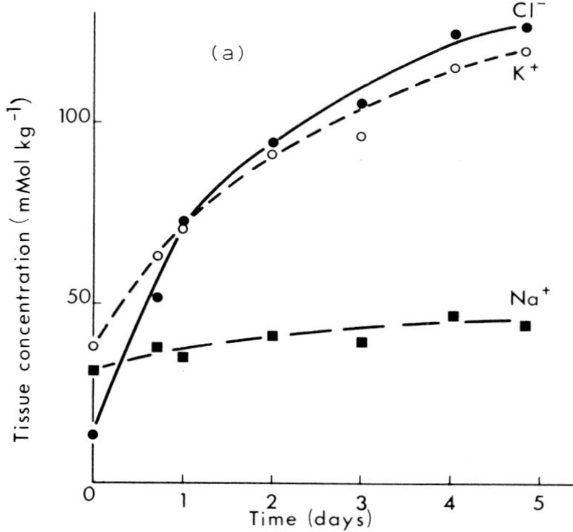


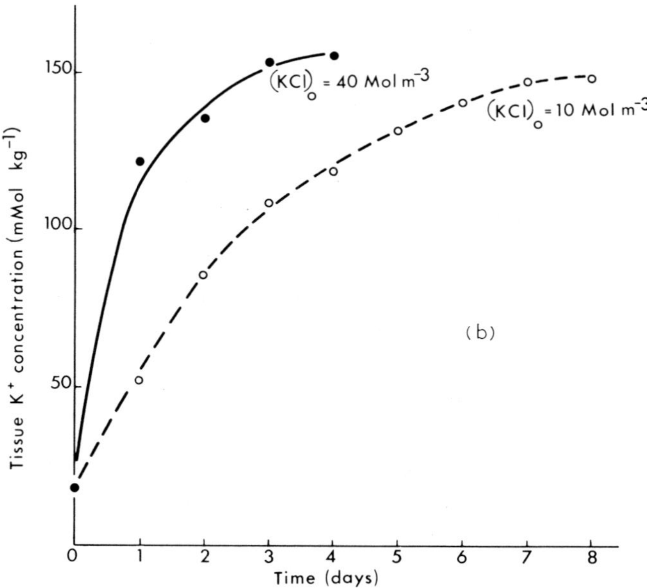


Fig. 6.3. Net accumulation of ions by washed carrot and beet tissues. (a) Carrot, in 20 Mol m$^{-3}$ Cl$^-$; 5 Mol m$^{-3}$ K$^+$, Na$^+$ and Ca$^{2+}$, 20 °C. Note that cations and anions are accumulated in equal quantities. From Cram (1968a). (b) Red beet, in 10 Mol m$^{-3}$ and 40 Mol m$^{-3}$ KCl, 25°C. Note the same final level of accumulation despite different initial influxes from the two concentrations of KCl. From Sutcliffe (1952).

W.J. Cram

Table 6.1.

The occurrence of active transport of univalent, non-metabolised inorganic ions in 4–8 day washed storage tissues.

| Species (source of data) | Ion | Concentrations $(\mathrm{Mol\,m^{-3}})$ | | $E_i^b$ (mV) | $E_{obs}^c$ (mV) | Fluxes with active components |
|---|---|---|---|---|---|---|
| | | Bathing medium[a] | Cell sap | | | |
| *Solanum tuberosum* | $K^+$ | 10 | 120 | −63 | | $\phi_{ov}$ |
| (1, 2, 3, 4) | $Na^+$ | | 2.5 | | −40 ± 2 | |
| | $Cl^-$ | 10 | 12 | +5 | | $\phi_{ov}$ |
| *Beta vulgaris* (5) | $K^+$ | 0.6 | 150 | −139 | | near equilibrium |
| | $Na^+$ | | 30 | | −155 ± 17 | |
| | $Cl^-$ | 0.6 | 8 | +65 | | $\phi_{ov}$ |
| *Daucus carota* | $K^+$ | 5 | 85 | −71 | | $\phi_{ov}$ |
| (6, 7) | $Na^+$ | 5 | 23 | −38 | −23 ± 2 | $\phi_{ov}$ |
| | $Cl^-$ | 20 | 19 | 0 | | $\phi_{ov}$ |
| *D. carota* | $K^+$ | 5 | 118 | −80 | | $\phi_{ov}$ |
| after 5 days | $Na^+$ | 5 | 45 | −55 | −26 ± 4 | $\phi_{ov}$ |
| loading (6) | $Cl^-$ | 20 | 128 | +47 | | $\phi_{ov}$ |

[a] External solution from which there is a net influx of all ions listed, except for carrot after 5 days loading in which the ions are at a steady state.

[b] The electrical potential of the vacuole relative to the external solution at which each ion would be at thermodynamic equilibrium.

[c] The observed value of the vacuolar potential relative to the external solution ± standard error of 10–20 measurements.

(1) Macklon and MacDonald (1966). (2) Macklon and DeKock (1967). (3) Lannoye (1970). (4) Hurd (1959). (5) Poole (1966). (6) Cram (1968a, and unpublished). (7) Cram (1973a).

$Na^+$ is added (Fig. 6.5) is due to $Na^+$ inhibiting $K^+$ influx to the vacuole. Hence $K^+$ flowing out of the vacuole is not pumped back in again, but raises the $K^+$ concentration in the cytoplasm and consequently the $K^+$ efflux across the plasmalemma until a new steady rate is reached. The origin of cation selectivity of other $K^+$-selective tissues has not been examined, except that no external $K^+$ stimulation of $Na^+$ efflux has been found in carrot (Cram, 1968a). This suggests that there is no $K^+/Na^+$ competition at the tonoplast, and no $K^+/Na^+$ exchange at the plasmalemma.

The distinction between active and passive components of the fluxes across the plasmalemma is difficult. In carrot increasing the external $Cl^-$ concentration stimulates $Cl^-$ efflux (Fig. 6.5). The gradual fall in $Cl^-$ efflux after the

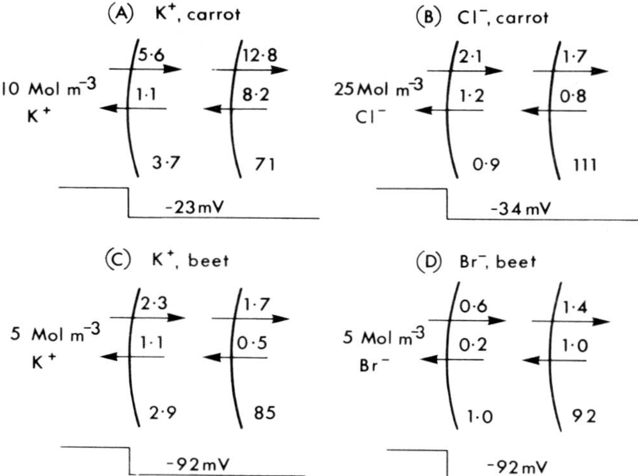



Fig. 6.4. Fluxes, concentrations and electrical potential differences in storage tissues. Fluxes at plasmalemma and tonoplast in mMol $kg^{-1}$ $h^{-1}$. Cytoplasmic and vacuolar contents in mMol $kg^{-1}$. Electrical potential differences shown relative to the external solution. Tonoplast P.D. is assumed to be small. (A) Carrot, in 10 Mol $m^{-3}$ $K^+$, 5 Mol $m^{-3}$ $Ca^{2+}$, 20 Mol $m^{-3}$ $Cl^-$, 20 °C. Cram, original observations. (B) Carrot, in 25 Mol $m^{-3}$ KCl, 0.5 Mol $m^{-3}$ $CaSO_4$, 20 °C. Cram (1968b). (C) Red beet, in 10 Mol $m^{-3}$ KCl and (D) Red beet in 10 Mol $m^{-3}$ KBr, 25 °C. Pitman (1963); Poole (1966).

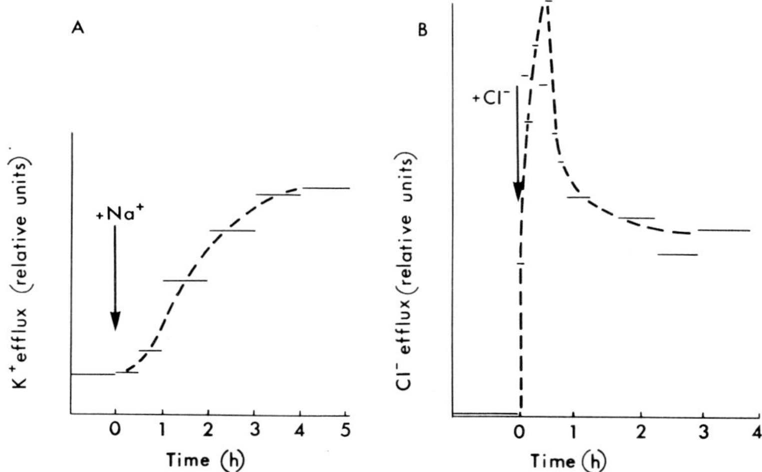



Fig. 6.5. External ion stimulation of ion efflux by interaction at the tonoplast or plasmalemma. (A) Stimulation of $K^+$ efflux from red beet due to competitive inhibition by $Na^+$ of $K^+$ influx across the tonoplast. Pitman (1963). (B) Stimulation of $Cl^-$ efflux by exchange between external and internal $Cl^-$ across the plasmalemma. Cram (1968b).

initial burst is due to the gradual fall in $Cl^-$ activity in the cytoplasm. The stimulation of $Cl^-$ efflux appears to result from a one-for-one $Cl^- - Cl^-$ exchange, which becomes the major component of the plasmalemma flux at high external $Cl^-$ concentrations (Cram, 1968b). Such a system would have no physiological significance unless the exchange is not strictly one-for-one or unless anions other than $Cl^-$ can also exchange.

If one assumes that efflux at low external concentrations is entirely passive, that the estimate of cytoplasmic content is correct, that the cytoplasmic volume is 5% of the tissue volume, and that the electrical P.D. is located across the plasmalemma, then permeability coefficients for $Cl^-$ and $K^+$ in some tissues can be calculated from the Goldman equation (Eq. 1.19). The passive component of $K^+$ flux at the plasmalemma is difficult to identify (Poole, 1969), but a maximum value is given by Poole (1966). These values are shown in Table 6.2.

Table 6.2.
Permeability coefficients of $Cl^-$ and $K^+$ in
washed storage tissue cells.

| Tissue | Temperature (°C) | $P_{Cl}$ ($\times 10^{-12}\,m\,s^{-1}$) | $P_K$ |
|---|---|---|---|
| Potato | 5 | 25[a] | — |
| Beet | 25 | 8[b,c] | 500[c] |
| | 2 | 7[b,c] | — |
| Carrot | 20 | 10[d] | 50[e] |

(a) Lannoye (1970), (b) Pitman (1963), (c) Poole (1966), (d) Cram (1968b), (e) Cram (unpublished).

The values of $P_{Cl}$ are all close to $10 \times 10^{-12}$ m s$^{-1}$. In carrot and beet $P_K$ is greater than this, which is consistent with the response of the electrical P.D. to changes in external KCl concentrations (Poole, 1966; Macklon and MacDonald, 1966; Cram, 1975b).

The nature of the linkage between cation and anion influx which gives rise to the accumulation of equal quantities of the two ions has been most extensively studied in beet (Poole, 1966, 1969, 1973). One component of the fluxes is a net influx of $K^+$ accompanied by $Cl^-$, varying independently of electrical P.D. This $K^+$ influx cannot simply be a passive flux responding to the P.D. generated by active transport of $Cl^-$. The linkage must therefore in some sense be chemical (rather than electrical) and the transport system must be electrically neutral. In the presence of external $HCO_3^-$, $K^+$ influx is stimulated and the internal potential becomes more negative. It was originally suggested that $K^+$ influx was linked via this change in P.D. to an electrogenic $HCO_3^-$ transport. However, $K^+$ influx does not respond to electrical potential changes in the manner expected. The situation can be understood better if the uptake of ions consists of a $K^+/H^+$ exchange plus a $Cl^-/OH^-$ exchange or a $Cl^- - H^+$ uptake. The internal pH would then be the equalising signal or driving force and the P.D. change would be the result, rather than the cause, of $K^+$ transport (Poole, 1973; Raven and Smith, 1973).

In several tissues $K^+$ influx changes with the nature of the external anion, though the equality of the two influxes has not been examined (e.g. Stiles and Skelding, 1940; Sutcliffe, 1952). The P.D. changes by only a few mV at the same time (Macklon and MacDonald, 1966; Lannoye, 1970; Olsen, 1972; Cram, 1975b). A flexible, but not electrical, link between $K^+$ and anion influxes can most easily be envisaged in terms similar to those discussed above in beet. The situation at individual membranes needs investigation.

Any inequality in influx of ions from simple solutions of non-metabolised ions would lead to internal pH changes, since the inequality could only be balanced by $H^+$, $OH^-$, or $HCO_3^-$ ions, which are the only ones present in unlimited quantity. In teleological terms, the origin of equality in cation and anion influx may be seen as a result of the cells' mechanism for maintaining pH (Raven and Smith, 1973). Where inequality occurs it is always accompanied by changes in the buffering systems in the cell, as discussed in the next section. One outstanding problem is why, under the same circumstances (e.g. in $K_2SO_4$ at pH 6), some cells appear to equalise fluxes (e.g. carrot, beet), while others change their buffering capacity (e.g. barley roots).

## 6.3.2 *The balance of inorganic ions and organic acid production*

$K^+$ and $Na^+$ are never accumulated stoichiometrically with $Cl^-$ in intact storage tissues (Tables 6.1, 6.3). In excised tissues $K^+$ may be accumulated in excess of inorganic anions in $SO_4^{2-}$ solutions and when $HCO_3^-$ is present in the external solution. Whether $K^+$ exchanges for internal $H^+$ or is accompanied by external $HCO_3^-$, one must expect the cytoplasmic pH to rise under these conditions. The excess $K^+$ is accumulated with organic acid anions which are produced by reactions involving $CO_2$ fixation in the cell. Various organic acids are accumulated predominantly in different storage tissues in vivo: in carrot – malate (e.g. Bryant and Overell, 1953); in beet – oxalate (MacDonald et al., 1960); in potato – citrate (Macklon and DeKock, 1967); in swede – malate (Truelove, 1962). Despite these differences, under conditions leading to organic acid synthesis and accumulation $^{14}CO_2$ is incorporated mainly into malate, and the first acid to be produced in quantity is malate in carrot (Cram and Laties, 1974), beet (Osmond and Laties, 1969) and potato (MacDonald and Laties, 1964). The origin of this selective accumulation of mainly one out of a number of organic acid anions is unknown. It is surprising that malate is not the anion accumulated in quantity over longer periods in all storage tissues.

The malate is probably formed via oxaloacetic acid from phosphoenol-pyruvate (PEP) and $HCO_3^-$, the first reaction being catalysed by PEP-carboxylase. In the presence of $K_2SO_4/KHCO_3$, when malate is accumulated, the substrates of the reaction must be present and PEP-carboxylase active, and the malate produced must move into the vacuole with $K^+$ ions. Transport of potassium malate would give the observed equality of $K^+$ and malate accumulated, excess malate being broken down in the cytoplasm. When $Cl^-$ is substituted for $SO_4^{2-}$ and $HCO_3^-$ externally, i.e. in KCl, $K^+$ is accumulated with $Cl^-$ and malate accumulation is inhibited. An elaborated compartmental analysis of the kinetics of exchange of $^{14}C$ in carrot tissue after labelling with $[^{14}C]HCO_3^-$ showed that, in KCl, malate production is directly inhibited. The fall in influx to the vacuole across the tonoplast is due to the fall in the malate concentration in the cytoplasm (Cram, 1974a). The alternative possibility, that $Cl^-$ inhibits, possibly competitively, the flow of malate into the vacuole, is thus eliminated. The kinetics of $^{14}C$ exchange after $[^{14}C]HCO_3$ labelling have also been examined in beet (Osmond and Laties, 1969), and the results are consistent with this conclusion. In confirmation that $Cl^-$ does not primarily inhibit the transport of endogenously produced malate to the vacuole, it has been found in carrot that the flow of external malate into the vacuole is not inhibited by $Cl^-$ (Cram, 1974a).

$Cl^-$ has been reported to inhibit PEP-carboxylase activity but only at much higher concentrations than are likely to occur in the cytoplasm (Osmond and Greenway, 1972; see Fig. 6.4). It seems more likely that malate production is inhibited in KCl due to a fall in cytoplasmic pH. The activity of PEP-carboxy-lase from maize roots falls 8-fold when the pH is decreased from 7.6 to 7.2 (Davies, 1973). In this pH region malate would be more than 99% in the doubly ionised form, which suggests that this is the form which is transported into the vacuole.

An as yet unresolved feature of potassium organate accumulation is its dependency on $HCO_3^-$ in the external solution. In beet, and probably in carrot, potassium organate accumulation is much faster in $KHCO_3$ than in $K_2SO_4$ (Stiles and Skelding, 1940; Sutcliffe, 1952; Hurd, 1959; Poole, 1969). More than sufficient $CO_2$ for organic acid production is produced by respiration inside the cell (e.g. Cram, 1974a) and hence the effect of $HCO_3^-$ must be to stimulate $K^+$ influx across the plasmalemma (Poole, 1969) or to alter pH in the cytoplasm.

### 6.3.3 Phosphate, sulphate and nitrogen compounds

Although there has been no systematic investigation, scattered observations of electrical potential, concentrations and fluxes suggest that the major inorganic nutrient anions are generally actively transported into storage tissues. Data suggesting active transport are available for $SO_4^{2-}$ in red beet (Poole, 1969), $SO_4^{2-}$, phosphate and $NO_3^-$ in potato (Lundegårdh, 1959; Loughman, 1960; MacDonald et al., 1960; Macklon and MacDonald, 1966; Lannoye, 1970), and for $SO_4^{2-}$, phosphate and $NO_3^-$ in carrot (Stiles and Skelding, 1940; El Shishiny, 1955; Cram, 1973a and unpublished observations). The main uncertainty is the concentration of the unmetabolised anion in the cytoplasm. However, this would have to be almost impossibly low (less than $10^{-2}$ Mol m$^{-3}$) if the influx across the plasmalemma from an external concentration of 1 Mol m$^{-3}$ against the electrical potential gradient were passive, so that it is highly likely, though not yet directly established, that these anions are actively transported across the plasmalemma.

Other major sources of N such as $NH_4^+$ and amino acids are also taken up by storage tissues. In carrot many amino acids can be accumulated in an uncombined form to a higher concentration inside than out (Birt and Hird, 1958). For the neutral amino acids at least, this could only result from active transport inwards. $NH_4^+$ ions flow into the tissues that have been examined (El Shishiny, 1955) but never reach an internal concentration higher than could be accounted for by the electrical potential difference (assuming that this is about the same in $NH_4Cl$ as in $KCl$). Hence it is not certain whether $NH_4^+$ is actively transported or not. Again, the concentration of these substances in the cytoplasm and the nature of the movements across the plasmalemma are uncertain.

The extent of reduction of $NO_3^-$ by different batches of carrot tissue varies from 0 to 80% of that taken up (El Shishiny, 1955; Cram, 1973a). Reduction of $NO_3^-$ is accompanied by accumulation of an equivalent amount of organic acid. During $KNO_3$ accumulation malate production is inhibited presumably as in $KCl$. During reduction of $NO_3^-$ an increase in the pH of the cytoplasm due to transfer of negative charges from $NO_3^-$ to $OH^-$ may be expected and this would lead to the production of malate as in $KHCO_3$ solutions.

Tissues differ in the relative rates at which $NH_4^+$ and $NO_3^-$ are metabolised (El Shishiny, 1955). Carrot incorporates more $NH_4^+$ than $NO_3^-$ and potato and radish incorporate more $NO_3^-$ than $NH_4^+$. In all three tissues the uptake of $NH_4^+$ and $NO_3^-$ is the same from the same external concentration, so that rates of net uptake do not correlate with the rate of subsequent metabolic incorporation.

## 6.4 Energy sources for active transport

When washed storage tissue is transferred from water to salt, respiration is stimulated. This classic observation is now interpreted not as a manifestation of mechanism but as the flexible adjustment of the rate of respiration (the energy-supplying process) to changes in the rate of salt transport (an energy-utilising process) (Ch. 2; Robertson, 1968). The immediate source of energy for active transport of KCl and NaCl in carrot and beet tissues has been intensively investigated. The important observations are that in both tissues the rate of transport into the vacuole and the total cell ATP concentration can change independently (Atkinson and Polya, 1968; Polya and Atkinson, 1969; Cram, 1969a); and that in carrot the plasmalemma $Cl^-$ influx is not inhibited by concentrations of CCCP which uncouple respiration and inhibit influx to the vacuole (Cram, 1969a). It thus appears that the plasmalemma influx and the influx to the vacuole have different links to energy supply, neither of which is a direct hydrolysis of ATP.

There appear to be two possibilities which would reconcile the independence of transport from total ATP level with a dependency of transport on ATP hydrolysis. One is that the ATP concentration next to the pump is not the same as the average concentration in the cell. The other is that in these experiments the effects observed are not only involved in energy source – energy sink relationships, but also in more complex control processes (see next section).

On the other hand, primary active transport of ions may not occur, the 'uphill' movement of ions being instead a secondary active transport linked for instance to the movement of $H^+$ down its electrochemical potential gradient (see Mitchell, 1970).

## 6.5 Regulation

### 6.5.1 Internal homeostasis and negative feedback

When excised, washed storage tissue is placed in a KCl solution it accumulates salt for several days until a concentration is reached above which no more is accumulated (as shown for carrot tissue in Fig. 6.3a, and for beet in Fig. 6.3b). At this steady state there is still an inflow of salt, measured using radioactive tracers, but this is balanced by an equally large efflux (Sutcliffe, 1954b; Cram, 1968c). It is simplest to suppose that as the internal salt concentration rises, passive effluxes increase until efflux equals influx. This, however, is not the case in beet and carrot, in both of which the overall influx to the vacuole decreases during accumulation (Sutcliffe, 1954b; Cram, 1968c). In carrot it has been shown in addition that the main decrease is in the tonoplast influx, the plasmalemma influx decreasing only in some batches of tissue. Efflux across the plasmalemma increases as expected during accumulation (Cram, 1968c and unpublished data). One other aspect of the simple 'pump and leak' hypothesis should be pointed out. At a steady state influx ($\phi_{ov}$) equals efflux ($C_{vac} \times P$, where $P$ is an over-simplified permeability coefficient). Hence

$$C_{vac} = \phi_{ov}/P$$

If $\phi_{ov}$ increases at higher external concentrations of KCl, then the steady state vacuolar concentration should increase proportionately. Fig. 6.3b shows that this is not the case in red beet.

The decrease in active $Cl^-$ influx during accumulation of KCl therefore appears to be a response of influx to the concentration of salt in the vacuole. Such a decrease in a process (influx) in response to an increasing value of its output (vacuolar concentration) is termed negative feedback. It should be realised that the mechanism of such a relationship is not specified by using the term. The pump and leak hypothesis in the previous paragraph may also be called a negative feedback system, net influx being related to internal concentration. In this case the relationship is simply a passive response of efflux to internal concentration.

The decrease in active influx during accumulation of KCl may be a more complex, indirect response of influx to internal concentration. It might also be a response to changes in concentrations of other substances or to changes in turgor. In order to identify the primary signal leading to a decreased $Cl^-$ influx in carrot tissue, batches of tissue were allowed to accumulate salts from a variety of solutions, so that the internal concentrations of several ions, reducing sugar, malic acid, and turgor all varied independently. $Cl^-$ influx was low in tissue which had accumulated KCl or $KNO_3$ but in tissue which had accumulated $K^+$ organic acid $Cl^-$ influx was as high as in tissue which had not accumulated any salt. $Cl^-$ influx was not related to the concentration of reducing sugars or to turgor (Cram, 1973a). Since turgor is equal to the difference between internal and external osmotic potential it can be reduced by increasing the external osmotic potential. This treatment did not stimulate $Cl^-$ influx in carrot tissue (and several other root tissues) confirming that $Cl^-$ influx does not respond to turgor changes in carrot (Cram, 1973a).

     *W.J. Cram*

Carrot cells therefore appear to have an intrinsic system for self regulating $Cl^-$ influx until a 'desired' internal $Cl^- + NO_3^-$ concentration is reached. If this desired level is constant, irrespective of external concentration and other factors which would tend to change the active influx, as appears to be the case for KCl accumulation in beet (Fig. 6.3b), then the system would be termed 'homeostatic'. The significance of this system is not yet understood (Cram, 1973a). A presumably related but equally obscure observation is that the influxes of $NO_3^-$ and of malate are also reduced in KCl-loaded carrot tissue (Cram, 1973a, 1974a).

Other experiments have suggested that active transport of other ions may be regulated in various ways. These have not been so extensively investigated, but their significance in the functioning of the cell is easier to understand.

In red beet tissue turgor increases during KCl accumulation, but, unlike carrot, KCl influx appears to be sensitive to turgor. In non-loaded tissue the internal osmotic potential is about $-1000$ kPa (10 bar) and net $K^+$ influx from 20 Mol m$^{-3}$ is 7 mMol kg$^{-1}$ h$^{-1}$. In KCl-loaded tissue the internal osmotic potential is about $-1900$ kPa and net $K^+$ influx is 0.4 mMol kg$^{-1}$ h$^{-1}$, but decreasing the external osmotic potential by 1000 kPa, and so reducing turgor to 900 kPa, increases net $K^+$ influx to 2.4 mMol kg$^{-1}$ h$^{-1}$ (Sutcliffe, 1954a). Although individual fluxes have not yet been measured it appears that in beet there exists a system of regulation of KCl transport in which the 'desired level' is a function of the KCl concentration and not simply the internal KCl concentration itself.

Such maintenance of turgor would have obvious advantages at two stages of growth of the storage organ. If turgor fell during expansion KCl influx would increase and this would tend to restore turgor. Secondly, when growth ceased the continued inflow of KCl would cause turgor to rise, but the negative feedback on KCl influx would prevent an excessive build up of turgor.

The other main type of regulation is by the uptake of metabolised substances. In certain plant tissues the uptake of some metabolites has been found to be greater in plants deficient in that metabolite, or less in plants which have accumulated that metabolite, or its metabolic products (Cram, 1975a). In *Nicotiana tabacum* tissue culture the net influx of $SO_4^{2-}$ and of $NO_3^-$ is reduced after the tissue has been allowed to accumulate the sulphur amino acid methionine (Hart and Filner, 1969) or various amino acids (Heimer and Filner, 1971), respectively. In washed carrot tissue methionine inhibits $SO_4^{2-}$ influx similarly (Cram, unpublished). The immediate products of $NO_3^-$ reduction, $NH_4^+$ and organic acids, have little or no effect on $NO_3^-$ uptake (El Shishiny, 1955; Cram, 1973a), but the effects of amino acids have not been investigated. Such negative feedback regulation appears likely to match the supply of unreduced metabolite to the demand for its utilisation in the cytoplasm.

In all the cases discussed here, whether the output of the system is turgor or the concentration of a specific substance, the negative feedback signal must in fact be pictured as the difference between the desired or reference value and the actual value of the output at any instant. Any deviation of the output from the desired value will then signal a change in the regulated active transport such that it tends to restore the output towards the desired value (Fig. 6.6). It is perhaps unnecessary to emphasise that it is not the instantaneous

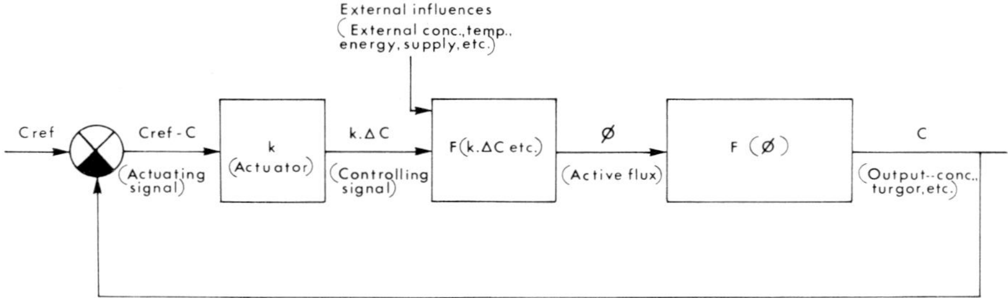


Fig. 6.6. A simplified negative feedback loop in the regulation of active transport. Electrical terms and passive flows are not explicitly included. A one-to-one correspondence between the elements of the diagram and physiological components would not be expected.

rate of active transport which is controlled at any particular value, but the long term result of the activity of active transport. A fuller discussion of negative feedback regulation of transport will be found in Cram (1975a).

The advantage of such a system is that it tends to maintain the output at the desired value in the face of natural variations in external concentration, energy supply, temperature, etc. Thus the system maintains what has been dictated by the value of $C_{ref}$ (the desired value). Long term developmental changes in the output must result from changes in the value of $C_{ref}$, and such hormonal and developmental changes will be discussed next.

### 6.5.2 Hormonal effects

Hormonal effects on ion transport in plant cells have been reviewed recently (Van Steveninck, 1975). Only two effects on washed storage tissue will be discussed.

IAA stimulates expansion growth in several plant cell types. The stimulation of growth in coleoptile and hypocotyl tissue is rapid and accompanied or followed after a short period by acidification of the medium. The latter must reflect the transport of some ion(s), $H^+$ having been suggested with an unspecified neutralising ion (Cleland, 1971; Van Steveninck, 1975). On the other hand, expansion growth of storage tissue in response to IAA is very limited, a 50% increase in 5 days being a representative value (MacDonald et al., 1961; Sperling and Laties, 1963). $H^+$ efflux has not been examined, but $K^+$ or $Rb^+$ uptake is stimulated by a fraction comparable to the stimulation of growth in potato and rutabaga (*Brassica campestris*) (Commoner and Mazia, 1942; Higinbotham et al., 1953). There is little or no effect of IAA on $K^+$ or $Na^+$ uptake in washed beet tissue (Pickles and Sutcliffe, 1955; Van Steveninck, 1965).

IAA induces much greater expansion and also cell division in tissue supplied with a full range of nutrients (e.g. Mott and Steward, 1972c). Under these circumstances concentrations of $K^+$, $Na^+$ and $Cl^-$ are maintained at lower levels than in the absence of IAA-induced growth. This is probably not simply due to dilution of cell contents due to the increasing volume, since internal osmotic potential remains constant (Mott and Steward, 1972c), but in the absence of any flux measurements on the cultures the situation cannot be analysed further. It seems possible that any changes in fluxes in response to IAA are at least in part due to the transport control system's responding to the growth of the tissue, rather than their being due to a direct effect of IAA on transport.

ABA appears to have more selective effects on ion transport (Cram and Pitman, 1972; Van Steveninck, 1972a) some of which can be understood in terms of integration of processes in the whole plant (Pitman and Cram, 1973). In storage tissue the only effect reported is a stimulation of net salt influx and an increased selectivity for $Na^+$ in washed red beet tissue. Kinetin, which has many of the properties of an ABA antagonist, inhibits the net uptake of $K^+$ and NaCl by beet tissue (Van Steveninck, 1972a, 1972b). Both hormones take effect only after a few hours and therefore do not appear to act directly on the transport process. The effects are not general, as there is no effect of kinetin on swede tissue (Van Steveninck, 1972b).

### 6.5.3 Developmental changes after excision

After excision from the intact tissue several processes change progressively for a number of days. The nature and rate of respiration change (Laties, 1967; Adams and Rowan, 1970; Jacobson et al., 1970); RNA and protein synthesis increase (e.g. Kahl, 1971); the activities of several enzymes increase (e.g. Ricardo and ap Rees, 1970; Rungie and Wiskitch, 1972); sensitivity to auxin increases (Adamson and Adamson, 1957; Sperling and Laties, 1963); and fine structure may change (Jackman and Van Steveninck, 1967; Van Steveninck, 1970).

Immediately after excision there is a loss of salt which may continue for several hours. After this there is a progressive increase in the potential rate of uptake of salt from a salt solution. In some tissues, though this is not always distinct, there is a lag before the uptake of salt begins (Asprey, 1937; Van Steveninck, 1964, 1975). During the lag phase $K^+$ influx and efflux are both high and net $K^+$ influx, presumably with organic acid, can be induced with Tris (Van Steveninck, 1961, 1964). $Cl^-$ fluxes during the lag phase are low, but the slight net influx is up an electrochemical potential gradient (Cram, unpublished) and is therefore active. At the end of the lag phase net $K^+$ influx results from a fall in $K^+$ efflux and net $Cl^-$ influx results from a rise in $Cl^-$ influx (Van Steveninck, 1964). Cation selectivity changes from lag to accumulation phase in red beet (Poole, 1971a). The change in potential linkage between the transport of $K^+$ and that of another ion (a change probably located at the tonoplast, Cram, 1967) has been discussed in relation to an $H^+$ extrusion pump by Raven and Smith (1973).

The initiation of these changes is not due to an osmotic shock (Cram, 1972). The absence of similar changes in excised maize or other root tissues (Cram, 1973b) or in potato storage tissue which is cut and then replaced in the intact tuber (Laties, 1962) strongly suggest that these are not 'wound' responses. Positive evidence as to the nature of the lag is reviewed in the next section. It seems probable that a general induction or derepression of activity takes place related to the release of dormancy.

No close parallel between respiratory or other biochemical changes and salt transport has appeared and there are no differential effects of the various treatments that have been applied during development (e.g. Rees, 1949; ap Rees, 1966). Hence, these studies have revealed little as to the nature of ion transport or its relationship to other cellular processes. For instance, one cannot tell whether $Cl^-$ transport increases because of the production of transport machinery, activation or derepression of a pre-existing pump, linkage of a pump to an energy source, or because of a change in the regulatory input to the pump.


## 6.6 *The relationship of excised storage tissue to the intact storage organ and to tissue cultures*

One can only compare enzyme activities and concentrations of substances in extracts of these tissues, since the long diffusion path in the intact tissue makes flux measurement impossible and since no flux measurements on cultures of storage tissue have been reported. Nevertheless, several conclusions can be drawn.

### 6.6.1 *The relationship of properties of the excised tissue to the stage of dormancy of the organ*

The duration of the lag in salt uptake after excision in red beet varies with the state of dormancy of the root, which suggests that the lag may be related to the time taken to lose the hormones which maintain dormancy in the intact root. An inhibitor of $MnCl_2$ transport has been isolated from red beet and found to be at a higher concentration in mid-winter dormant than in autumn growing or in incipiently sprouting spring beet (Skelding and Rees, 1952; Skelding, 1957), but the nature of the inhibitor and of its inhibiting action are not clear (Dale and Sutcliffe, 1959). In potato and Jerusalem artichoke the length of the lag in growth response to auxin also follows this pattern (Adamson and Adamson, 1957; Sperling and Laties, 1963) and similarly the rate of respiration in freshly excised carrot tissue parallels the stage of maturity of the root (Bryant and ap Rees, 1971).

Thus, as the state of the root or tuber changes, the properties of the freshly excised tissue change in parallel. This suggests that the cells in the freshly excised tissue are like those of the intact organ. There are, however, some rapid changes in glycolytic intermediates, the nature of the respiratory substrate and oxygen uptake rates within an hour of excision (Everson and Rowan, 1965; Adams and Rowan, 1970; Jacobson et al., 1970), so the state even a few hours after excision may not be that of the intact organ.

In contrast to the freshly excised tissue, washed tissue has a greater metabolic activity (previous section) and in particular the respiration rate in carrot rises to a constant rate independent of the state of maturity of the root (Bryant and ap Rees, 1971). This resurgence of activity seems most likely to reflect the release of dormancy (see Rapoport and Wolf, 1969), and suggests that washed tissue is more like the growing state of the storage organ.

### 6.6.2 *The relationship of the ionic concentrations in excised, intact and cultured tissues*

Neither the freshly excised nor the intact dormant tissue accumulate salt. This adds weight to the previous conclusion that cells in the freshly excised tissue are like cells in the intact, dormant organ, transport in both being repressed. The alternative explanation for the lack of salt accumulation in freshly excised tissue, that it is a 'wound' effect, seems unlikely since excised maize and barley root tissues do accumulate salt immediately after excision. Further, the electrical potential difference between vacuole and external solution is the same in freshly excised carrot tissue as in washed and accumulating tissue (Cram, 1975b), which appears to rule out any gross wound leakiness (but see Anderson et al., 1974).

Concentrations of ions in intact storage roots and tubers are shown in Table 6.3. The main characteristics are the high $K^+$ to $Na^+$ ratio and the

Table 6.3.
Concentrations in mature storage tissues in vivo.

| Species | $K^+$ | $Na^+$ | $Cl^-$ | Organic acids | Sugars | N | Osmotic |
| | | | | | | (mg atoms | potential |
| | | (m equiv. $kg^{-1}$) | | | (mMol $kg^{-1}$) | $kg^{-1}$) | (bars) |
|---|---|---|---|---|---|---|---|
| *Solanum* | 60–140 | 3 | 12 | 10–100 | 20 | 180–260 | – |
| *tuberosum* | | | | mainly citric | | sol : 120 | |
| (1, 2) | | | | | | insol : 80 | |
| *Beta* | 60–150 | 10–30 | 8 | 110 | 100 (+) | 140–210 | –10.2 |
| *vulgaris* | | | | mainly citric | | sol : 70 | |
| (2, 3, 4, 5, 6) | | | | | | insol : 40 | |
| *Daucus* | 70 | 14 | 10 | 100 | 200 | 70–210 | –9.1 |
| *carota* | | | | mainly malic | | sol : 40 | |
| (2, 5, 7, 8, 9) | | | | | | insol : 40 | |

(1) Macklon and DeKock (1967). (2) Duckworth (1966). (3) Poole (1966). (4) Splittstoesser and Beevers (1964). (5) MacDonald et al. (1961). (6) Sutcliffe (1954a). (7) Bernstein and Ayres (1953). (8) Ricardo and ap Rees (1970). (9) Mott and Steward (1972c).

high organic acid anion, rather than $Cl^-$, concentration. This must be the result of transport during growth. Ion concentrations in actively growing carrot tissue cultures are shown in Table 6.4. The latter is also highly $K^+$ selective (as is beet tissue in culture also, Sutcliffe and Counter, 1959), and accumulates little $Cl^-$. The same is true of growing root cells (Scott et al., 1968). Although the levels accumulated in intact growing tissues are very similar to those accumulated by growing tissue cultures, they are also very similar to concentrations in phloem (Table 6.5), which is probably a major route of supply of ions to the storage organ. Consequently one cannot tell whether the concentrations in the intact tissue result from non-selective accumulation of everything supplied in the phloem, or from selective accumulation from a soil solution. Despite this uncertainty, it seems probable (particularly if a non-growing but active stage intervenes between growth and

                           *W.J. Cram*

Table 6.4.

Concentrations in carrot root tissue cultures. Both media contained 8 Mol m$^{-3}$ K$^+$; 6 Mol m$^{-3}$ Na$^+$; 9 Mol m$^{-3}$ Cl$^-$, and 2 Mol m$^{-3}$ NO$_3^-$, together with other ions and sugar. Internal concentrations in non-growing tissue in the same external concentrations of K$^+$, Na$^+$, Cl$^-$ and NO$_3^-$ are not available. Results for clone 871 C (Mott and Steward, 1972a, b).

| | Relative growth rate (d$^{-1}$) | K$^+$ | Na$^+$ | Cl$^-$ | Osmotic potential (bars) |
|---|---|---|---|---|---|
| | | (mMol kg$^{-1}$) | | | |
| Rapidly growing[a] | 0.24 | 110 | 40 | 20 | $-7.0$ |
| Slowly growing[b] | 0.04–0.12 | 110 | 20 | 40 | $-9.5$ |

[a] Growing on basal medium plus coconut milk growth factors.

[b] Growing on basal medium with K$^+$, Na$^+$, and Cl$^-$ concentrations adjusted to be the same as in the growth medium of 1.

Table 6.5.

Concentrations of substances in phloem exudate.

| | K$^+$ | Na$^+$ | Cl$^-$ | Organic acids | Sugars (Mol m$^{-3}$) | N (Mol m$^{-3}$) | Osmotic potential (bars) |
|---|---|---|---|---|---|---|---|
| | | (equiv. m$^{-3}$) | | | | | |
| Range of species (1) | 20–85 | 1 | — | — | 300–900 | 10–100 | — |
| *Ricinus communis* (2) | 60–112 | 2–12 | 15 | 40 | 220–300 | 35 | $-14.7$ |

(1) MacRobbie (1971). (2) Hall and Baker (1972).

dormancy) that cells in the tissue culture and in the growing organ have similar transport properties.

The remaining tissue to be discussed is the well-washed, non-growing, but actively accumulating tissue used in most of the work discussed in this chapter. Washed tissue has the same ionic transport properties whether as 3 mg explants or as 80 mg discs (see Mott and Steward, 1972a, Fig. 2c; and Fig. 6.3a). Since it is capable of being stimulated to growth (Mott and Steward, 1972b) the tissue seems normal, but its activity limited by lack of external stimuli. Comparison of rapidly growing and slowly growing tissue cultures (Table 6.4) shows that, in a medium of the same K$^+$, Na$^+$, Cl$^-$ and NO$_3^-$ concentrations, the slowly growing tissue also selects K$^+$ rather than Na$^+$ and accumulates less Cl$^-$ than inorganic cations. Other data suggest that K$^+$ selectivity may be less and Cl$^-$ a larger fraction of total anions, in more slowly growing tissue (Mott and Steward, 1972a; Scott et al., 1968). The frequently observed stoichiometric uptake of $Cl^-$ with inorganic cations in non-growing tissues (e.g. Fig. 6.3a; and Mott and Steward, 1972a) is probably simply due to the absence of $NO_3^-$ from the external medium. Results to date are rather scanty, but at most suggest only quantitative differences between cells in growing tissue cultures and non-growing cells.

## 6.7 Assessment of work on storage tissue in relation to general problems of ion transport in plant cells

The general problems may be divided into those of mechanism and of regulation. The former include, for instance, the nature of the driving forces for movement and, at another level, molecular events. The latter concerns the relation of transport to internal concentrations of nutrients and hence potential syntheses and also the relation to internal osmotic potential, hydrostatic pressure and hence expansion growth. These aspects need to be considered throughout the life of the cell. Three different cell types have been obtained from storage tissue.

### 6.7.1 Non-growing but active cells

The excised, washed, non-growing tissue is a very convenient experimental material, but the transport processes of only a few ions have been investigated to any extent. (This is also true of work on roots.) While this work has revealed several major properties of the transport of these ions (reviewed in this chapter) and has allowed comparison with algae, in which KCl accumulation is a major physiological process, it may be that the early investigations of simple, non-metabolised salt transport, together with the autocatalytic nature of interest in any biological phenomenon, has led to undue emphasis on the transport of $Cl^-$ at the expense of work on those substances which are more extensively taken up as nutrients or accumulated as osmotically active particles.

Many other parenchymatous cells in the plant are also in an active but non-growing state (e.g. root cortical cells and leaf mesophyll). It is difficult to suppose that the maintenance of turgor and of cytoplasmic turnover in these cells are dependent on totally new transport systems coming into activity at the end of growth. The simplest supposition is that growing and non-growing cells have essentially the same transport processes, but that these cope with quantitatively different states. However, work on the non-growing cell is unlikely to be sufficient, for instance, to enable one to make quantitative predictions of growth from cell parameters. Quantitative differences from the growing cell may be expected to appear and it may be easier to see the way that transport systems are integrated into cellular function by investigating the growing cell.

### 6.7.2 Dormant cells

The properties of dormant cells in vivo cannot be measured. Further work on freshly excised tissue may be warranted if it is similar to the dormant tissue, but this point needs to be established.

### 6.7.3 Growing cells

The expansion of plant cells is related to turgor and so to internal solutes by an equation of the form

$$\frac{dV}{dt} = C(P - P_c) \tag{6.9}$$

where $V$ is volume, $t$ is time, $C$ is an often variable coefficient describing the elastic–plastic properties of the cell wall, $P$ is turgor pressure and $P_c$ is a threshold value of turgor above which expansion takes place (Cleland, 1971; Cram, 1974b). In growing carrot tissue cultures and possibly more generally (Cram, 1975a) the main osmoticum is potassium organic acid. One question that has yet to be answered is whether the accumulation of potassium organate in the growing cell is regulated in such a way as to maintain turgor. A second unanswered question is whether the uptake of building blocks for cytoplasmic syntheses, N compounds in particular, are regulated to match demand during growth.

These two questions are not distinct, since the interrelationship of nutrient uptake and osmotic potential generation depends on the form in which a nutrient is available. This is most striking for N, which is taken up in greatest quantity. If N is taken up as $KNO_3$ (as in roots in the soil or in leaves supplied by the xylem stream when not all the $NO_3^-$ is reduced in the root) then, for each amino acid incorporated into protein, one $K^+$ and one organic acid anion would be accumulated. The latter would be a major contribution to osmotic potential. Plant cells often take up $NO_3^-$ in exchange for an internal organic acid anion, when no decrease in internal osmotic potential accompanies N uptake (Smith, 1973; Cram, 1975a). If N were taken up as $NH_4NO_3$ then again it would not contribute to osmotic potential.

During growth of storage organs the substances accumulated in greatest quantity are $K^+$ ions, organic acid anions, N as free amino acids and protein, and sugars. If the main route of supply is the phloem, then the substances transported into the cells in greatest quantity are $K^+$, organic acid anions, amino acids, and sugars. The regulation of the transport of these substances could be investigated in tissue cultures, but this has not yet been done. However, the inter-relationship between nutrient uptake and osmotic potential suggested above has been observed in carrot tissue culture. When cultured with extra $NH_4NO_3$ the tissue grows at a higher osmotic potential (about $-5$ bars) than when cultured with extra $KNO_3$ (about $-8 \cdot 5$ bars) (Craven et al., 1972). These tissues grew at about the same rate, so that turgor depends to some extent on nutrition and does not have a particular value for a particular growth rate. (This may imply further that wall properties vary in a manner which is integrated with turgor to give a constant growth rate under different nutrient conditions.)

In order to answer basic questions about solute transport and accumulation during expansion growth it would seem easier to work on a tissue which is more readily available and reproducible than storage tissue cultures, for instance, expanding seedling tissue. It must also not be forgotten that in storage organs, as in other organs where matter is supplied in the xylem or phloem, the major regulated step may not be uptake by the cells but rather mass flow into the organ. Phloem transport may limit growth, or it may be itself controlled by signals from the organ it is supplying.

# References


P.B. ADAMS and K.S. ROWAN, Plant Physiol., 45 (1970) 490.

H. ADAMSON and D. ADAMSON, Aust. J. Biol. Sci., 10 (1957) 435.

W.P. ANDERSON, D.L. HENDRIX and H. HIGINBOTHAM, Plant Physiol., 53 (1974) 122.

T.AP REES, Aust. J. Biol. Sci., 19 (1966) 981.

G.F. ASPREY, Protoplasma, 27 (1937) 153.

M.R. ATKINSON and G.M. POLYA, Aust. J. Biol. Sci., 21 (1968) 409.

L. BERNSTEIN and A.D. AYERS, Proc. Am. Soc. Hort. Sci., 61 (1953) 360.

L.M. BIRT and F.J.R. HIRD, Biochem. J., 70 (1958) 277.

F. BRYANT and B.T. OVERELL, Biochim. Biophys. Acta, 10 (1953) 471.

J.A. BRYANT and T.AP REES, Phytochemistry, 10 (1971) 1191.

R. CLELAND, Annu. Rev. Plant Physiol., 22 (1971) 197.

B. COMMONER and D. MAZIA, Plant Physiol., 17 (1942) 682.

W.J. CRAM, PH.D. THESIS (1967) University of Cambridge, Cambridge.

W.J. CRAM, J. Exp. Bot., 19 (1968a) 611.

W.J. CRAM, Biochim. Biophys. Acta, 163 (1968b) 339.

W.J. CRAM, Abh. Dtsch. Akad. Wiss. Berl., (1968c) 117.

W.J. CRAM, Biochim. Biophys. Acta, 173 (1969a) 213.

W.J. CRAM, Plant Physiol., 44 (1969b) 1013.

W.J. CRAM, Aust. J. Biol. Sci., 25 (1972) 855.

W.J. CRAM, J. Exp. Bot., 24 (1973a) 328.

W.J. CRAM, Aust. J. Biol. Sci., 26 (1973b) 757.

W.J. CRAM, J. Exp. Bot., 25 (1974a) 253.

W.J. CRAM, Bull. R. Soc. N.Z., 12 (1974b) 183.

W.J. CRAM, in J. Dainty and U. Zimmermann (Eds.) Membrane Transport in Plants and Plant Organelles (1974c) Springer Verlag, Berlin, p. 334.

W.J. CRAM, in U. Lüttge and M.G. Pitman (Eds.) Encyclopaedia of Plant Physiology (1975a) 2nd Series, in press.

W.J. CRAM, Aust. J. Plant Physiol., 2 (1975b) in press.

W.J. CRAM and G.G. LATIES, Aust. J. Biol. Sci., 24 (1971) 633.

W.J. CRAM and G.G. LATIES, J. Exp. Bot., 25 (1974) 11.

W.J. CRAM and M.G. PITMAN, Aust. J. Biol. Sci., 25 (1972) 1125.

G.H. CRAVEN, R.L. MOTT and F.C. STEWARD, Ann. Bot., 36 (1972) 897.

J.E. DALE and J.F. SUTCLIFFE, Ann. Bot., 23 (1959) 1.

D.D. DAVIES, in B.V. Milborrow (Ed.) Biosynthesis and its Control in Plants (1973) Academic Press, London, p. 1.

R.B. DUCKWORTH, Fruit and Vegetables (1966) Pergamon, Oxford.

E.D.H. EL-SHISHINY, J. Exp. Bot., 6 (1955) 6.

R.G. EVERSON and K.S. ROWAN, Plant Physiol., 40 (1965) 1247.

G.P. FINDLAY, A.B. HOPE and N.A. WALKER, Biochim. Biophys. Acta, 233 (1971) 155.

S. GELFAN, Protoplasma, 4 (1928) 192.

S.M. HALL and D.A. BAKER, Planta, 106 (1972) 131.

J.W. HART and P. FILNER, Plant Physiol., 44 (1969) 1253.

Y.M. HEIMER and P. FILNER, Biochim. Biophys. Acta, 230 (1971) 362.

N. HIGINBOTHAM, H. LATIMER and R. EPPLEY, Science, 118 (1953) 243.

R.G. HURD, J. Exp. Bot., 10 (1959) 345.

M.E. JACKMAN and R.F.M. VAN STEVENINCK, Aust. J. Biol. Sci., 20 (1967) 1063.

B.S. JACOBSON, B.N. SMITH, S. EPSTEIN and G.G. LATIES, J. Gen. Physiol., 55 (1970) 1.

G. KAHL, Z. Naturforsch., 26b (1971) 1064.

R.J. LANNOYE, Ann. Physiol. Veg. Bruxelles, 15 (1970) 1.

G.G. LATIES, Plant Physiol., 37 (1962) 679.

G.G. LATIES, Aust. J. Sci., 30 (1967) 193.

G.G. LATIES, I.R. MACDONALD and J. DAINTY, Plant Physiol., 39 (1964) 254.

B.C. LOUGHMAN, Plant Physiol., 35 (1960) 418.

H. LUNDEGÅRDH, Physiol. Plant., 12 (1959) 336.

I.R. MACDONALD, P.C. DEKOCK and A.H. KNIGHT, Physiol. Plant., 13 (1960) 76.

I.R. MACDONALD, A.H. KNIGHT and P.C. DEKOCK, Physiol. Plant., 14 (1961) 7.

I.R. MACDONALD and G.G. LATIES, J. Exp. Bot., 15 (1964) 530.

A.E.S. MACKLON and P.C. DEKOCK, Physiol. Plant., 20 (1967) 421.

A.E.S. MACKLON and I.R. MACDONALD, J. Exp. Bot., 17 (1966) 703.

E.A.C. MACROBBIE, J. Exp. Bot., 20 (1969) 236.

E.A.C. MACROBBIE, Biol. Rev. Cambridge Phil. Soc., 46 (1971) 429.

P. MITCHELL, Symp. Soc. Gen. Microbiol., 20 (1970) 121.

R.L. MOTT and F.C. STEWARD, Ann. Bot., 36 (1972a) 621.

R.L. MOTT and F.C. STEWARD, Ann. Bot., 36 (1972b) 655.

R.L. MOTT and F.C. STEWARD, Ann. Bot., 36 (1972c) 915.

G.J. OLSEN, B. Sc. Honours Thesis (1972) University of Sydney.

C.B. OSMOND and H. GREENWAY, Plant Physiol., 49 (1972) 260.

C.B. OSMOND and G.G. LATIES, Plant Physiol., 44 (1969) 7.

C.K. PALLAGHY and B.I.H. SCOTT, Aust. J. Biol. Sci., 22 (1969) 585.

V.R. PICKLES and J.F. SUTCLIFFE, Biochim. Biophys. Acta, 17 (1955) 244.

M.G. PITMAN, Aust. J. Biol. Sci., 16 (1963) 647.

M.G. PITMAN and W.J. CRAM, in W.P. Anderson (Ed.) Ion Transport in Plants (1973) Academic Press, London, p. 465.

G.M. POLYA and M.R. ATKINSON, Aust. J. Biol. Sci., 22 (1969) 573.

R.J. POOLE, J. Gen. Physiol., 49 (1966) 551.

R.J. POOLE, Plant Physiol., 44 (1969) 485.

R.J. POOLE, Plant Physiol., 47 (1971a) 731.

R.J. POOLE, Plant Physiol., 47 (1971b) 735.

R.J. POOLE, in W.P. Anderson (Ed.) Ion Transport in Plants (1973) Academic Press, London, p. 129.

R.J. POOLE, in Encyclopaedia of Plant Physiology (1975) 2nd Series, in press.

L. RAPOPORT and N. WOLF, Symp. Soc. Exp. Biol., 23 (1969) 219.

J.A. RAVEN and F.A. SMITH, in W.P. Anderson (Ed.) Ion Transport in Plants (1973) Academic Press, London, p. 271.

W.J. REES, Ann. Bot., 13 (1949) 29.

C.P.P. RICARDO and T.AP. REES, Phytochemistry, 9 (1970) 239.

R.N. ROBERTSON, Protons, Electrons, Phosphorylation and Active Transport (1968) Cambridge University Press.

J.M. RUNGIE and J.T. WISKICH, Aust. J. Biol. Sci., 25 (1972) 103.

B.I.H. SCOTT, H. GULLINE and C.K. PALLAGHY, Aust. J. Biol. Sci., 21 (1968) 185.

A.D. SKELDING, Ann. Bot., 21 (1957) 121.

A.D. SKELDING and W.J. REES, Ann. Bot., 16 (1952) 513.

F.A. SMITH, New Phytol., 72 (1973) 769.

E. SPERLING and G.G. LATIES, Plant Physiol., 38 (1963) 546.

W.E. SPLITTSTOESSER and H. BEEVERS, Plant Physiol., 39 (1964) 163.

W. STILES and A.D. SKELDING, Ann. Bot., 4 (1940) 329.

J.F. SUTCLIFFE, J. Exp. Bot., 3 (1952) 59.

J.F. SUTCLIFFE, J. Exp. Bot., 5 (1954a) 215.

J.F. SUTCLIFFE, J. Exp. Bot., 5 (1954b) 313.

J.F. SUTCLIFFE and E.R. COUNTER, Nature, 183 (1959) 1513.

B. TRUELOVE, Ann. Bot., 26 (1962) 147.

H.D. VAN LIEW, Science, 138 (1962) 682.

M.E. VAN STEVENINCK, M. Sc. Thesis (1970) University of Adelaide.

R.F.M. VAN STEVENINCK, Nature, 190 (1961) 1072.

R.F.M. VAN STEVENINCK, Physiol. Planta 17 (1964) 757.

R.F.M. VAN STEVENINCK, Nature, 205 (1965) 83.

R.F.M. VAN STEVENINCK, Z. Pflanzenphysiol., 67 (1972a) 282.

R.F.M. VAN STEVENINCK, Physiol. Plant., 27 (1972b) 43.

R.F.M. VAN STEVENINCK, in Encyclopaedia of Plant Physiology (1975) 2nd Series, in press.

D. VREUGDENHIL, Acta Bot. Neerl., 6 (1957) 472.

K. WAGENER, U. ZIMMERMAN and H. KESSLER, Ber. Bunsenges. Phys. Chem., 71 (1967) 898.

*Ion transport in plant cells and tissues*
*edited by D.A. Baker and J.L. Hall*
© *North-Holland Publishing Company, 1975*



# CHAPTER 7

# Excised roots

R.G.WYN JONES

## Contents



## 7.1 Introduction

For the last quarter century excised plant roots, often of low internal salt status, have been a popular experimental material for investigating the absorption of ions or salts by plant tissues. This popularity is, of course, related to the fundamental biological role of roots in obtaining nutrients from the soil for the whole plant but it is also due, in part, to a methodology which has been built up around these tissues. The pedigree of this technique may be traced to the classical paper of Hoagland and Broyer in 1936. Since it must be admitted at the outset that the use of excised roots of low internal salt status has been criticised and that the interpretation of much of the data has led to strident disagreement, it is as well to reconsider briefly the advantages and pitfalls of this system as elucidated in the original paper.

Hoagland and Broyer (1936) discovered that young barley roots of low internal salt but high sugar content, which were produced by growing roots in a restricted nutrient supply, were capable of rapid net accumulation of salt. This represented a major technical advance as, in the absence of isotopes, a substantial net accumulation was essential for meaningful results to be obtained. However, while suggesting that 'many of the fundamental questions of salt absorption by root cells can be studied most effectively by eliminating during a brief experimental period, the complications of root and shoot relationships', the authors were also well aware of the difficulties. They stressed the dependence of rapid absorption on the particular state of the roots, notwithstanding their healthy appearance, and noted that, as the internal salt status rose, there was a limitation of net salt uptake even though active metabolism continued. The problems of experimental variability and reproducibility were considered very seriously. The major physiological significance of this early work was the demonstration that $KBr$ and $KNO_3$ absorption against a concentration gradient was dependent on aerobic respiration but not on $CO_2$ production in both excised roots and whole plants.

The use of low salt roots was refined by Ulrich, Machlis, Jacobson and Overstreet at Berkeley during the 1940s until a reproducible and convenient technique for studying salt uptake was achieved. The popularity of the experimental method has persisted even though its raison d'etre, namely the requirement for massive net salt accumulation prior to the use of radio isotopes, disappeared. Ironically, as the Berkeley group were also amongst the first to describe the use of radio-isotopes for measuring ion fluxes in plant tissue (Overstreet and Jacobson, 1946), further interest was aroused in their methodology. These papers are also of considerable interest as they contain the first attempts at measuring the exchange or leakage of isotope from prelabelled tissue and are therefore the precursors of efflux analysis. The authors also found that roots do not behave as perfect uniform absorption cylinders and that the highest rates of absorption were usually observed near the root tip.

During this period the idea that ion absorption was mediated by specific metabolically produced binding compounds (carriers) enjoyed great popularity in all branches of physiology (see Roberts et al., 1949; Osterhout, 1950).

In their paper on $K^+$ absorption by barley roots, Jacobson and his colleages (1950) illustrate the contemporary thinking by representing cation and anion absorption as:

$$HR + M^+ \longrightarrow MR + H^+ \tag{7.1}$$

$$R'OH + A^- \longrightarrow R'A + OH^- \tag{7.2}$$

They envisaged the complexes HR and R'OH as being formed at the external surface of the protoplasm and serving as transporting agents transferring ions from the outside to the vacuole. They produced tentative but not decisive evidence for $K^+ \rightleftharpoons H^+$ exchange. This and other papers firmly established the concept of 'a carrier' as being central to ion uptake and to a remarkable extent this position has been maintained until today (see Epstein 1973; Nissen, 1974a). In this chapter the search for the kinetic definition of these 'specific carriers' and the more recent work on flux and compartmental analysis will be discussed while some space will be devoted to a summary of the electro-chemical, metabolic and biochemical data. The first view of ion uptake will be referred to as the 'carrier-kinetic' approach and the latter as the 'thermo-dynamic approach' as they are derived from a very different view of the fundamental physical chemical basis for ion transport (see Epstein, 1973; Dainty 1962; Nobel, 1974). The anatomy of the whole root and individual root cell types are described in Ch. 8.

## 7.2 *The uptake isotherm*

In keeping with the emphasis on 'carriers' in the late 1940s, Epstein and Hagen (1952) treated ion uptake by low salt excised barley roots as being analogous to enzyme–substrate binding according to the classical Michaelis–Menten formulation of enzyme kinetics (Dixon and Webb, 1964).

$$R + M_{outside} \underset{k_2}{\overset{k_1}{\rightleftharpoons}} MR \underset{k_4}{\overset{k_3}{\rightleftharpoons}} R' + M_{inside} \tag{7.3}$$

Using this formalism Epstein and Hagen explained in a most impressive manner the interactions of the alkali metal cations, for example the competitive interactions of $Rb^+ + K^+$ (Fig. 7.1a). These data also suggested a significant difference in the kinetics of $Rb^+$ and $Na^+$ interactions above and below about 5 Mol m$^{-3}$ $Rb^+$ (Fig. 7.1b) and implied two independent sites for $Rb^+$ uptake. A logical corollary of this formalism was to define the uptake isotherm in

                                R.G.Wyn Jones

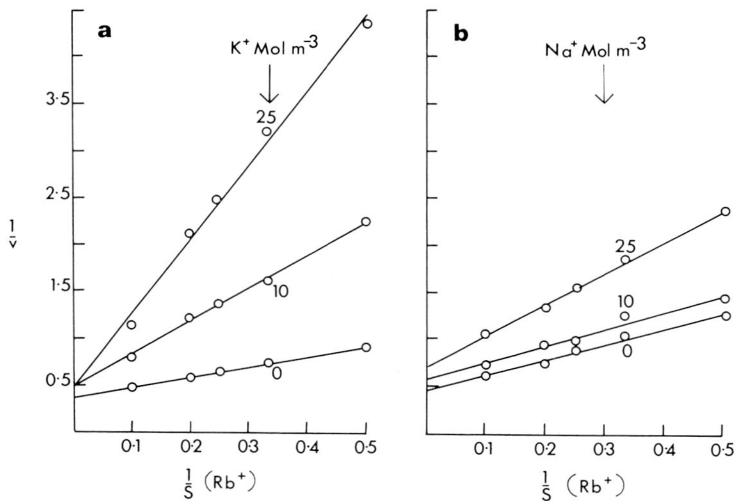

Fig. 7.1. (a) Lineweaver–Burk plot of the influence of $K^+$ at three concentrations 0, 10 and 25 Mol m$^{-3}$ on the uptake of $^{86}Rb^+$ in the range 2–10 Mol m$^{-3}$. (b) Lineweaver–Burk plot of the influence of $Na^+$ at three concentrations 0, 10 and 25 Mol m$^{-3}$ on the uptake of $^{86}Rb^+$ in the range 2–10 Mol m$^{-3}$. (From Epstein and Hagen, 1952.)

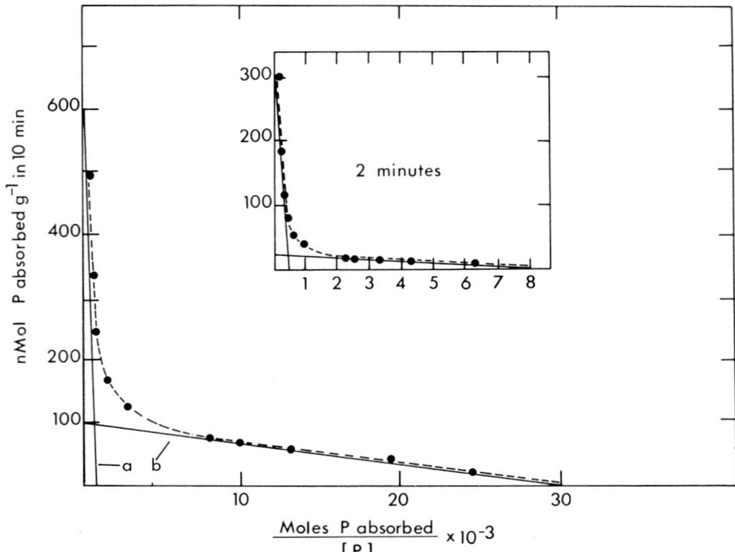

Fig. 7.2. Biphasic uptake of phosphate over wide concentration range illustrated by the Eadie–Hofstee plot of phosphate absorption isotherm for millet roots. (From Noggle and Fried, 1960.)

terms of the kinetic constants, $K_s$ ($K_m$) and $V_{max}$. Further work by Hagen, Fried, Noggle, Epstein and others suggested that two isotherms, each apparently obeying Michaelis–Menten kinetics could describe the uptake of $K^+$, $Rb^+$, $H_2PO_4^-$ and other ions over a wide concentration range (for example Fig. 7.2). One isotherm was found in the range 1–500 mMol m$^{-3}$ and the other above 1 Mol m$^{-3}$ and both were considered to describe a unique mechanism for the uptake of a particular ion as will be discussed later.

This biphasic pattern has been observed ubiquitously for many plant species, tissue types and for different ions and much of the information has been summarised by Fried and Broeshart (1967) and Epstein (1972). This pattern is usually referred to as the 'dual isotherm' and, as it appears to be a universal observation, its interpretation and explanation are major issues in plant physiology. Several conflicting hypotheses have been advanced and undoubtedly the heat generated by the adherents of some of the ideas is at least commensurate with the significance of the basic observation. The arguments have centred on the number of isotherms or phases observed, their valid kinetic characterisation and the subcellular location of the mechanism(s) considered to be responsible for the individual isotherms. Despite these differences there is substantial agreement about many of the properties of the isotherm(s).

The terms upper and lower isotherms will be used here to describe the curves relating external concentration to root uptake above and below about 1 Mol m$^{-3}$ external salt. Published $K_s$ and $V_{max}$ values will be quoted but without prejudicing later interpretation nor wishing to imply, at this stage, that each isotherm describes any one unique mechanism.

In their original work Epstein and Hagen measured uptake following a long absorption period (3 h), did not include $Ca^{2+}$ in the external medium and did not desorb ions held on the exchange sites of the cell wall. Nonetheless they obtained results capable of analysis in terms of the Michaelis–Menten formalism. It is now recognised that the large cation exchange capacity of cell walls may modify the uptake isotherm, particularly for cations, although hypotheses that ion absorption by roots may be accounted for entirely by Donnan exchange phenomena have been discounted (see Drake, 1964). External $Ca^{2+}$ is recognised as being essential to the maintenance of the integrity of the plasmalemma (see Wyn Jones and Lunt, 1967) and as $Ca^{2+}$ is more strongly absorbed on exchange sites than monovalent cations (see Winklander, 1964), this has the added advantage of minimising cell wall $K^+$-absorption in the low concentration range. A technique of a fairly short absorption period followed by an appropriate desorption is now employed (e.g. Epstein et al., 1963b; Nissen, 1971). However, the precise details of the uptake–wash regime are now known to be of paramount importance and this problem will be discussed in detail later.

The classical example of the 'carrier–kinetic' approach to ion uptake has been the investigation of alkali metal and halide ion absorption by low salt barley roots by Epstein and his colleagues (Epstein, 1973). Their data may be used to illustrate the type of information gained by this technique. In the range of the lower isotherm, $K(Rb)^+$ uptake had an apparent $K_s$ of 20 mMol $m^{-3}$ and a $V_{max}$ of 3.3 nMol $gFW^{-1}$ $s^{-1}$. In this range $Na^+$ was a very poor but competitive inhibitor of $K^+$ uptake ($K_i$ $Na^+$ 1.25 Mol $m^{-3}$) and was itself absorbed very slowly ($K_s$ 320 mMol $m^{-3}$, $V_{max}$ 2.4 nMol $gFW^{-1}$ $s^{-1}$). The substitution of $SO_4^{2-}$ for $Cl^-$ as the attendant anion had little effect on $K(Rb)^+$ uptake but did decrease $Na^+$ uptake. The rate of $Cl^-$ absorption, while exceeding that of $SO_4^{2-}$, was only a half or a quarter of the $K(Rb)^+$ rate (17 mMol $m^{-3}$, $V_{max}$ 1.46 nMol $gFW^{-1}$ $s^{-1}$) at pH 5.6 (for references, see Epstein, 1973). However there are indications that the relative rate of $K^+$ and $Cl^-$ uptake change with pH. $K^+$ uptake is more sensitive to low pH and $Cl^-$ ($Br^-$) uptake more sensitive to high pH (Jacobson et al., 1957; Jackson and Adams, 1963).

The observations on the lower isotherm are in marked contrast to the situation found in the upper range (above 1–5 Mol $m^{-3}$ external salt). $K(Rb)^+$ uptake is heavily dependent on the attendant anion; the apparent $K_s$ for $Rb^+$ increased from 14.5 Mol $m^{-3}$ with $Cl^-$ to 38 Mol $m^{-3}$ with $SO_4^{2-}$ and the $V_{max}$ was reduced five-fold (Fig. 7.3). In this range $Na^+$ is a potent competitive inhibitor of $K(Rb)^+$ uptake and by subtracting the contributions of the lower isotherm, Epstein calculated an apparent $K_s$ of 17 Mol $m^{-3}$ for $K(Rb)^+$ and a $K_i$ of 0.8 Mol $m^{-3}$ for $Na^+$. $Cl^-$ was itself absorbed very rapidly in this range but its isotherm did not describe a rectangular hyperbola. It appeared that the markedly biphasic character of the plot for $K^+$ uptake was only manifest when $Cl^-$ was the co-anion (Fig. 7.3) (Epstein et al., 1963a). A similar but not identical situation has been found in maize roots, where the anion dependence of the lower isotherm of $K^+$ uptake is somewhat more marked and the uptake rates throughout the concentration range are lower. However, as noted previously, the general pattern of a biphasic isotherm has been observed for many ions including $NH_4^+$, $Sr^{2+}$, $HPO_4^-$, $Cl^-$, $Br^-$, $SO_4^{2-}$, $NO_3^-$ etc., in many species and in other tissues. Table 7.1 summarises data on the uptake of a number of ions by roots in the lower isotherm range.

Further work by Epstein and Rains (1965), Torii and Laties (1966a) and others showed that the upper isotherm was not a single rectangular hyperbola, but was a multiple-inflection or 'lumpy' curve. In the case of $K^+$ uptake by barley, this curve was resolved by Epstein and Rains (1965) into 4 phases

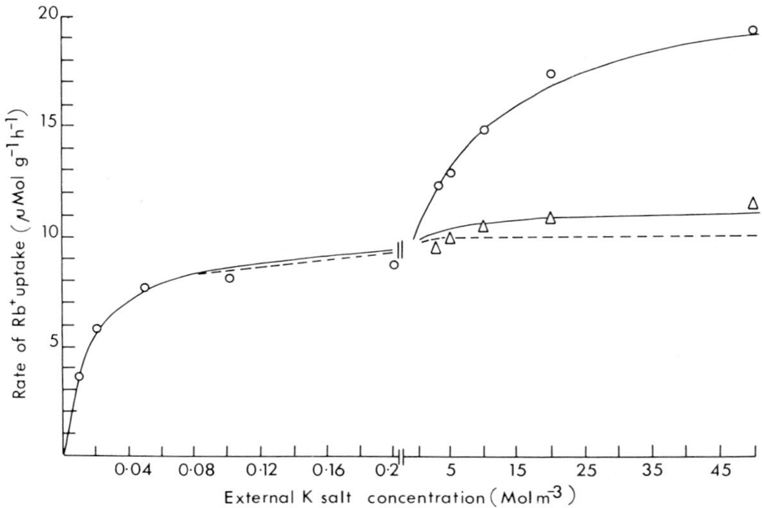

Fig. 7.3. Effect of external salt concentration and accompanying anion on $^{86}Rb^+$ uptake by barley roots. ○—○, KCl; △—△, $K_2SO_4$. Theoretical maximum rate employing kinetic constants derived from the lower isotherm (----). (From Epstein et al., 1963b.)

with inflection points at 6–8, 18–20 and 30 Mol m$^{-3}$ external KCl. Although the trends illustrated by these data are widely accepted, their formulation and interpretation are subject to wide divergences. Three hypotheses have received the greatest attention and each has been comprehensively covered by a persuasive and committed advocate in recent years.

Table 7.1.

Affinity data on lower isotherm in roots (selected). E.R., excised roots; W.S., whole seedlings.

| Ion | Species | Tissue type | Apparent affinity constant (mMol m$^{-3}$) | References |
|---|---|---|---|---|
| K$^+$ | Barley | E.R. | 21 | Epstein et al. (1963b) |
| K$^+$ | Tall wheat grass | E.R. | 8 | Elzam and Epstein (1969) |
| K$^+$ | Corn | W.S. | 26 | Claassen and Barber (1974) |
| Rb$^+$ | Corn | E.R. | 30 | Leigh (1974) |
| Rb$^+$ | Barley | E.R. | 17 | Jackman (1965) |
| Rb$^+$ | Barley | E.R. | 12 | Jackman (1965) |
| Na$^+$ | Barley | E.R. | 320 | Rains and Epstein (1967) |
| Cl$^-$ | Barley | E.R. | 14 | Elzam and Epstein (1965) |
| Cl$^-$ | Tall wheat grass | E.R. | 13 | Elzam and Epstein (1969) |
| NO$_3^-$ | Corn | W.S. | 21 | Van den Honert and Hooymans (1955) |
| H$_2$PO$_4^-$ | Corn | E.R. | 6 | Carter and Lathwell (1967) |

Therefore the major arguments advanced for each concept will only be outlined together with a brief description of a few other less-well popularised views. A more critical examination will follow in Sections 7.3 and 7.4. The major hypotheses are: (i) The parallel model: two independent mechanisms operating at the plasmalemma (Epstein, 1973). (ii) The series model: two mechanisms, one (high affinity) operating at the plasmalemma and the other at the tonoplast (Laties, 1969). (iii) The multiphasic model: a single multiphasic mechanism operating at the plasmalemma (Nissen, 1974). (iv) Other alternative schemes.

### 7.2.1 The parallel model

Epstein (1972, 1973) considers that the dual isotherm describes two independent active mechanisms for ion uptake both located on the plasmalemma of the root cortical cells, operating in parallel but not necessarily pumping ions into the same cytoplasmic compartment. The plasmalemma is regarded as an almost completely impermeable barrier to either inward or outward ion movement in the absence of a carrier at all but toxic concentrations. Independent carriers are considered to exist for $K^+$ and $Cl^-$ in both the low and high affinity systems. Uptake by the former obeys the Michaelis–Menten formalism while the latter is multiphasic. The apparent adherence of the isotherm to this formalism is regarded as evidence of a specific chemical mechanism. Epstein's principal arguments for the parallel model may be summarised as follows. Mechanism I (high affinity system) is placed at the plasmalemma of cortical cells because (a) soil $K^+$ concentrations are often low so a high affinity system will be required at the external surface and (b) the influence of extracellular $Ca^{2+}$ suggests that Mechanism I is located at the external surface. Other more convincing reasons may be added to Epstein's rationalisation. A comparison of the isotherm produced using various uptake and wash regimes (Fig. 7.4) shows the lower isotherm to be qualitatively independent of the regime used in low salt tissue. This fact, together with the analyses of efflux kinetics (see later), strongly suggests that the lower isotherm describes influx rate-limited at the plasmalemma. Membrane potential measurements (Higinbotham, 1973a) also imply that the plasmalemma is the site of the major potential difference and is therefore a membrane of limited and selective permeability.

In Epstein's concept Mechanism II is also placed at the same membrane because (a) in the presence of $Na^+$, a potent inhibitor of Mechanism II for $K^+$, $K^+$ uptake is accurately predicted by a model of two carriers operating in parallel, one sensitive to and the other insensitive to $Na^+$; (b) by comparing

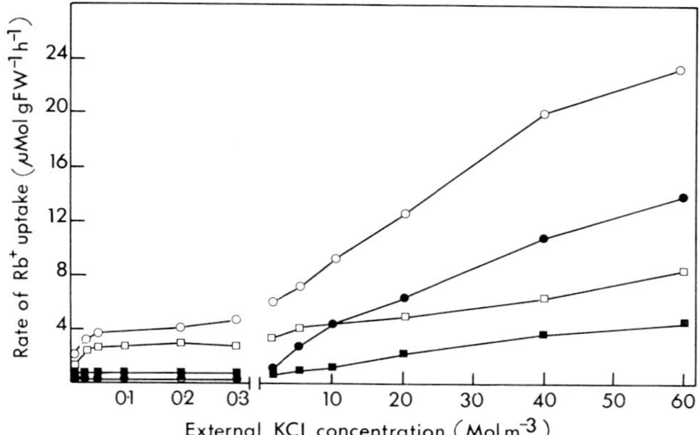

Fig. 7.4. Effect of both internal salt status and uptake–wash regime on $^{86}$Rb$^+$ uptake over the concentration range 0.01 Mol m$^{-3}$ to 60 Mol m$^{-3}$ by maize root segments. O—O, low salt, 10 min uptake, 5 min wash; □—□, low salt, 40 min uptake, 30 min wash; ●—●, high salt, 10 min uptake, 5 min wash; ■—■, high salt, 40 min uptake, 30 min wash. (From Leigh, 1974.)

the sum of short uptake pulses with a single long pulse, Welsh and Epstein (1969) found no evidence for diffusive equilibration across the plasmalemma; (c) a typical dual isotherm for Rb$^+$ absorption has been observed in the green alga, *Chlorella pyrenoidosa*, which was claimed to be non-vacuolated (Kannan, 1971) but has since been found to contain small vacuoles (Atkinson et al., 1972) and (d) evidence from long-distance transport studies are claimed to refute the idea of significant diffusion across the plasmalemma at concentrations above 1 Mol m$^{-3}$ as is required in the original series model. The different isotherms are not attributed to different cells or cell types in the root because the pattern has been observed in other tissues, e.g. storage tissue discs (where there are few variations in cell type), and even in *Chlorella*.

### 7.2.2 The series model

A contrasting hypothesis has been cogently argued by Laties (1969) who accepted that Mechanism I, as defined by the lower isotherm, is on the plasmalemma but suggested that Mechanism II is located at the tonoplast. The main pieces of evidence supporting this concept were that (a) the characteristic isotherm of Mechanism II was found in proximal sections of maize roots but not in the tips which were claimed to be non-vacuolated. (It is now realised that extensive small vacuoles are found in these cells but not the single central vacuole characteristic of mature cells; see Matile, 1968.) (b) long-distance transport was claimed to be characterised by kinetics and inhibitor sensitivities typical of the lower isotherm thereby implying that the high-affinity system lay interior to the symplasm; (c) the kinetics of experiments on beet discs in which, at low external concentrations, the apparent influx into the discs increased with time although pretreatment of the experimental tissue decreased the influx subsequently observed. Laties also pointed out that the sum of two independent systems operating in parallel when plotted according to the Lineweaver–Burke transformation of the Michaelis–Menten equation gives a curvilinear relationship not two straight lines intersecting at a point. In the model proposed by Laties, Mechanism I fills the 'empty' cytoplasm of low salt roots which, when filled, becomes the 'conduit' between the external solution and the vacuole. At higher·external concentrations passive permeation of the plasmalemma was envisaged and the rate-limiting step was thought to reside in the tonoplast. It is now realised that, for this type of model to operate, it is not essential for the influx across the plasmalemma to be passive only that it is not rate-limiting under the particular experimental procedures used to measure influx.

### 7.2.3  The single multiphasic isotherm model

The evidence for the dual absorption isotherm has been exhaustively re-examined by Nissen (1974a) who concluded that almost all the voluminous data may be accurately represented by a single multiphasic isotherm. In Fig. 7.5 the data from Epstein et al. (1963a) for $^{86}$Rb uptake have been replotted by Nissen to yield a multiphasic isotherm. In other systems, particularly $SO_4^{2-}$ uptake, an even more complex pattern with discrete breaks in the isotherm has been observed (Nissen, 1971). Nissen defines the multiphasic isotherm as follows: (i) the phases are separated by sharply defined inflection points and may be discontinuous; (ii) each phase obeys Michaelis–Menten kinetics; (iii) the kinetic constants increase in a fairly regular manner, thus, in Nissen's opinion, giving the illusion of two phases and (iv) only one phase functions at any one concentration. This is deduced from the agreement of each phase with Michaelis–Menten kinetics (see Epstein, 1973) while in Nissen's hands extrapolation and subtraction of any one phase from adjacent phases produces meaningless results. Nissen considers that the whole isotherm is located at the plasmalemma and is mediated by a single carrier or site which changes characteristically at certain discrete external salt concentrations i.e. a series of all-or-none transitions in the structure of the site. A similar multiphasic carrier may occur on the tonoplast but is only rate limiting at very high external concentrations.

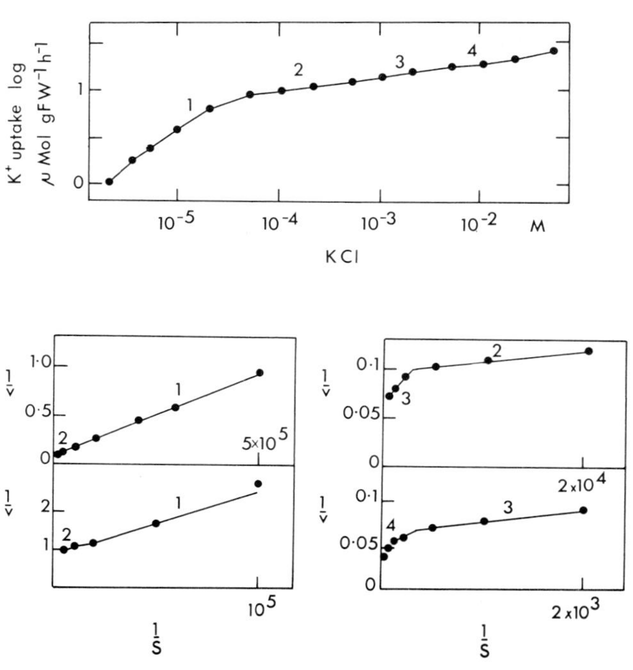

Fig. 7.5. The multiphasic isotherm for $^{86}Rb^+$, data as in Fig. 7.3 but redrawn to reveal 4 apparent individual isotherms. (From Nissen, 1973.)

### 7.2.4 Other alternative schemes

A number of other models have been suggested from time to time to explain the biphasic uptake pattern but have received less attention than those discussed above. Oertli (1967) suggested that an active uptake mechanism and a passive carrier-mediated bi-directional flux located on the plasmalemma could account for the observations – the pump–leak model. Thellier (1973) on the other hand has developed an electro-kinetic theory based on non-equilibrium thermodynamics which predicts a biphasic concentration-dependent isotherm and the observed ion-selectivities. Gerson and Poole (1971) suggested that carrier-mediated anion transport which may be modified by the membrane potential would account for the $Cl^-$ isotherm. In addition a variety of workers (e.g. Pallaghy et al., 1970; Baker and Hall, 1973; Leigh et al., 1973) have proposed various permutations on endocytosis to account for various anomalies in flux kinetics. A number of the implications of these less advertised suggestions will be reconsidered later.

## 7.3 *Problems in isotherm methodology and interpretation*

In Section 7.2 a largely uncritical account of some of the data on absorption
isotherms by low salt roots and the principal interpretations of the data has
been given. It is clear that these interpretations are mutually incompatible
and, as succinctly stated by MacRobbie (1971), all the major hypotheses are
unsatisfactory. The methodology is itself often suspect and an alternative but
difficult treatment is offered by thermodynamics and influx and efflux analysis.
This treatment, together with metabolic investigations, has in many ways
superseded the carrier–kinetic approach. However, in view of the prominence
given to carrier kinetics in current text books (Fried and Broeshart, 1967;
Epstein, 1972; Clarkson, 1974) and in recent reviews (Epstein, 1973; Nissen,
1974a), a direct examination of the problems and assumptions embodied in
the use of uptake isotherms and in the emphasis on the Michaelis–Menten
formalism is inescapable.

If it is to be deduced with validity from the observation that an isotherm
or phase thereof obeys Michaelis–Menten kinetics in that a specific, active,
carrier-mediated mechanism at a particular membrane is involved, then
certain criteria must be satisfied: (i) the influx isotherm must be capable of
exact and reproducible determination; (ii) the kinetic constants must be
calculated with known standard errors or similar; (iii) the Michaelis–Menten
constants so produced must contain worthwhile biochemical information
about such a 'carrier'; (iv) Michaelis–Menten kinetics, i.e. a rectangular hyperbola, must be characteristic of active transport processes and (v) it must be
proved that the isotherm describes a mechanism sited at a specific cytological
entity, probably a membrane.

### 7.3.1 *Some sources of variability and error*

It is readily apparent that these criteria are hard to fulfil in an extremely
heterogenous system such as a plant root. The system is heterogenous at a
gross morphological level, i.e. the root is not a uniform cylinder; at a cellular
level each of the cells of the cortex is not in the same physiological state; at a
subcellular level the cell wall and the various cell compartments must be
considered, and the root may contain substantial microbiological contamination. Some of these problems have been overcome by increased technical
sophistication. It has been shown, for instance, that bacterial contamination
is a serious hazard in studies on phosphate uptake at low concentrations
(Barber, 1968; Epstein, 1973) but, although Barber (1974) maintains that even
in the case of $Rb^+$ uptake micro-organism may contribute to the lower

isotherm, this is no explanation for the overall dual isotherm pattern. Clearly this hazard may be overcome by using axenic seedlings and roots. Similarly, problems due to unstirred layers cannot be ignored, particularly at very low concentrations, but may generally be overcome (Polle and Jenny, 1971). The variations in uptake along the length of roots first observed by Jacobson and Overstreet (1947) present an unsolved problem. Kahn and Hanson (1957) found that different values for $K_s$ and $V_{max}$ could be obtained from different segments of roots while the careful studies of Eshel and Waisel (1973) have renewed interest in this problem by showing variations of 30–50% in uptake from different segments along low salt barley roots. These workers interpret variations in Arrhenius plots along the root as suggesting qualitative and quantitative differences in the mechanism of $Na^+$ uptake (Eshel and Waisel, 1972), although other interpretations are possible.

On the other hand, the comparison of influxes into cortical half segments and whole corn root segments has proved reassuring and suggested that the stele is not a major source of error in the determination of short duration influxes (Cram, 1973a). However, a gradient of membrane potential (Fig. 7.12; Mertz and Higinbotham, 1974) and pH (Bowling, 1973) exists in the cortex and could well contribute to minor variations and irregularities in influx curves. It is now well established (Fig. 7.4) that the precise isotherm observed depends on the uptake and desorption period employed, on the internal salt status of the tissue and probably on the total ionic strength of the external medium (Higinbotham, 1974). Many of the apparent differences between the parallel and series models may well be due to technical differences and to a failure to realise the important distinction between the plasmalemma influx and the quasi-steady state influx into the vacuole or vacuole and cytoplasm combined (Cram, 1974a). Any isotherm can only at best represent the average behaviour of many cells and it has been implicitly assumed that these variations are not so great as to completely invalidate the procedure. In consequence, the replication of observations and reproducibility of the data is of great importance, particularly if small differences e.g. Nissen's phases, are to be given any credance. Unfortunately evidence of this type is frequently lacking. Epstein stated originally that his data varied by only about 5% and therefore replicates were not required (Epstein and Hagen, 1952). In more recent work (Welsh and Epstein, 1969) duplicate or triplicate samples of roots were employed. Nissen (1974b) used 4 or 5 replicates resulting in standard error of 4 to 5% but gave no information on the reproducibility of his data from experiment to experiment. Even less information is given in many studies but this is not, of course, universally true. Cram (1973a), for instance, attempted to specifically measure plasmalemma fluxes and Fig. 7.6 shows data from

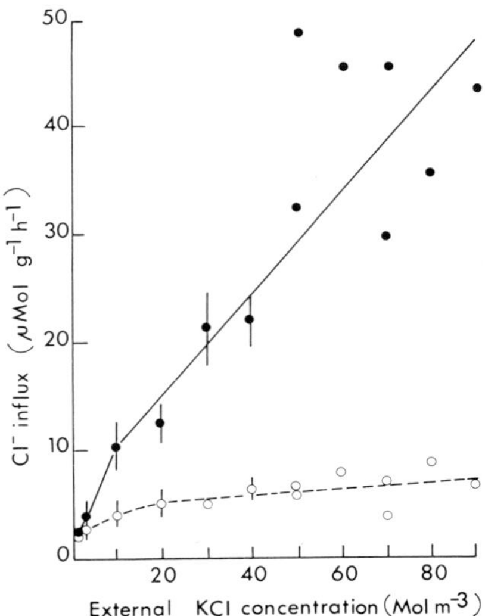

Fig. 7.6. Plasmalemma influx (●—●) and influx to the vacuole (○---○) in maize root cortical cells from 1–90 Mol m$^{-3}$ KCl and 0.5 Mol m$^{-3}$ CaSO$_4$. Values from 3 independent experiments combined. Vertical bars show ± 1 S.E.M. (From Cram, 1973.)

3 independent experiments each containing 3 replicates for the relationship between plasmalemma Cl$^-$ flux and external concentration. Quite large variations were found in influx particularly at high concentrations across the plasmalemma. Therefore the onus must be squarely on those who claim validity for small variations to prove their case beyond doubt.

### 7.3.2 Graphical presentation and mathematical calculations

In enzyme kinetics the graphical presentation of results and the calculation of kinetic constants is a major problem even in purified systems (Eisenthal and Cornish-Bowden, 1974). The conventional $v$ vs $[S]$ plot suffers from a restricted range on the $[S]$ axis and the convention in uptake studies (see Figs 7.3 and 7.4) of breaking the scale around 0.5 Mol m$^{-3}$ may be seriously misleading (see Figs 7.7 and 7.12). Furthermore, direct calculation of $K_s$ and $V_{max}$ is not possible from this presentation. The two most popular linear transformations of the Michaelis–Menten formulation are the Lineweaver–Burk plot of $1/v$ vs $1/[S]$ (Fig. 7.1) and the Eadie–Hofstee plot of $V/[S]$ vs $v$ (Fig. 7.2) (for references, see Dixon and Webb, 1964). The former is statistically very suspect leading easily to distortion of the data, while the latter, which is less often employed, has the advantage of contracting the concentration

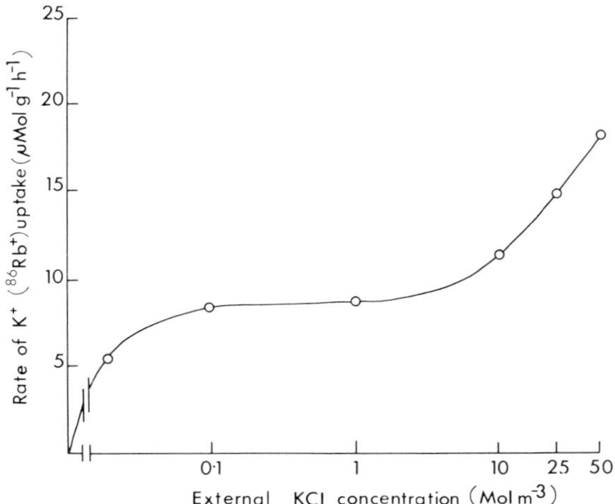



Fig. 7.7. Influence of external KCl concentration on $K^+$ ($^{86}Rb^+$) uptake by barley roots expressed on a logarithmic scale. (From Epstein, 1968.)

axis but is still not statistically satisfactory. A forthright and salutary account of the inadequacies of these techniques is given by Colquhoun (1971) who shows that the results given particularly by the Lineweaver–Burk plot when the straight line is fitted by a simple unweighted method of least squares are simply terrible.

A possible solution to the problem of calculating apparent affinity constants may be offered by the recent graphical treatment of enzyme kinetics (Eisenthal and Cornish-Bowden, 1974). In this treatment, individual values of $v$ and $[S]$ are plotted so that intersects yield individual values of $V_{max}$ and $K_s$ (Fig. 7.8). From these a median value for $V_{max}$ and $K_s$ may easily be obtained. It is claimed that this treatment is statistically more rigorous as it contains fewer assumptions about the distribution of errors and is less subject to distortion due to 'outliers'. Simple calculations yield errors at 96% confidence limits for the kinetic constants (Cornish-Bowden and Eisenthal, 1974).

The importance of this topic to the carrier–kinetic approach to ion transport is obvious. The sum of two independent mechanisms operating in parallel (both Michaelis–Menten type) is given by

$$v = \frac{V_1[S]}{K_1 + [S]} + \frac{V_2[S]}{K_2 + [S]} \tag{7.4}$$

Thus, when plotted as reciprocals, the sum of two mechanisms gives a curvilinear line (Laties, 1969) which may be contrasted with Nissen's multiphasic

     *R.G.Wyn Jones*

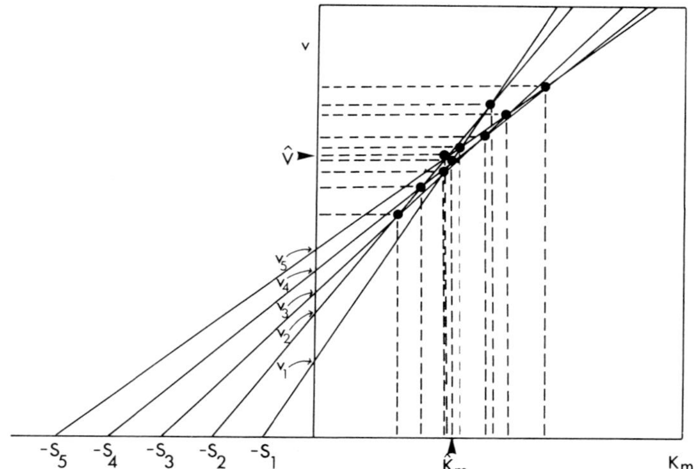



Fig. 7.8. Estimation of $K_m$ and $V_{max}$ by graphical presentation. Each intersection provides an estimate of $K_m$ and $V_{max}$. The best estimate $\hat{K}_m$ and $\hat{V}$ are taken as the medians of the two sets of estimates. (From Eisenthal and Cornish-Bowden, 1974.)

treatment of very similar data (Fig. 7.5). Clearly, extremely precise data and a rigorous statistic treatment are required to decide between a multiphasic and a biphasic system. Nissen (1974b) uses a FORTRAN programme to obtain his straight lines in which the sum of the squares of deviations in log $v$ is minimised to minimise weighting errors associated with the linear transformation of the Michaelis–Menten equation. But he makes a subjective choice of transition points, although F tests are run to see if the data fit better for 2 phases. The dangers inherent in this procedure have been emphasised by Walker (1974) who pointed out that almost anything fits two lines better than one. A solution to this problem requires a suitable statistical programme to allow the non-subjective selection of transition points and detailed evidence on the scatter of points.

It must also be pointed out that many of the relationships obtained between uptake and external concentration may equally well be interpreted in terms of negative or positive co-operativity i.e. induced changes in binding sites (Koshland, 1970), as has been suggested by Hodges (1973), and that Michaelis–Menten kinetics are now regarded as a special case of minimum interaction between bound ion and the allosteric structure of the site.

### 7.3.3 Significance of Michaelis–Menten and other formalisms

The kinetic argument would be worth pursuing if it could be shown that the fitting of a phase of an absorption isotherm to the Michaelis–Menten equation is of profound biochemical significance. As is apparent throughout this book, knowledge of animal ion-pumping systems weighs heavily on the shoulders of plant physiologists. Comparisons have their dangers but at the risk of adding fuel to this particular fire, the situation in animals will be briefly considered here.

One of the best defined animal systems is the erythrocyte, which will serve as an example. The influx of $K^+$ into this cell even over a short concentration range does not describe a rectangular hyperbola (Fig. 7.9). A sigmoidal curve

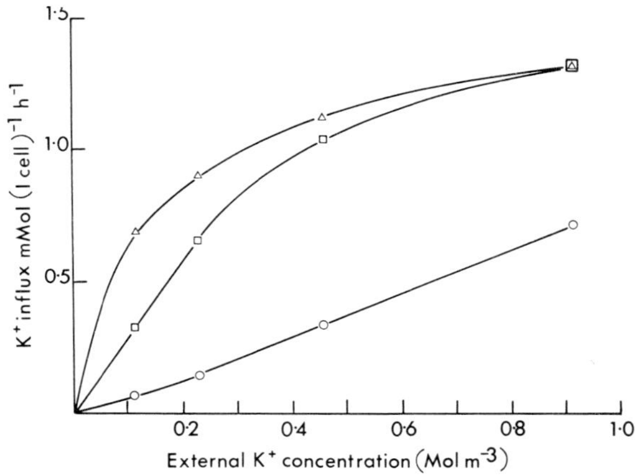

Fig. 7.9. Effect of external $Na^+$ on the foot of the $K^+$ influx curve. $\triangle$—$\triangle$, nominally $Na^+$-free but $Na^+$ concentration determined as 17 m Mol m$^{-3}$ at the end of incubation; $\square$—$\square$, 5 Mol m$^{-3}$ $Na^+$; $\bigcirc$—$\bigcirc$, 135 Mol m$^{-3}$ $Na^+$. All media made isotonic with choline. (From Garrahan and Glynn, 1967).

is clearly shown for $K^+$ influx in the presence of 5 Mol m$^{-3}$ $Na^+$ and more detailed studies showed that the $Na^+$-free curve was also sigmoidal at the lowest concentrations (Garrahan and Glynn, 1967). Uptake in this and other animal transport systems is mediated, in all probability, by a ($Na^+$, $K^+$) $Mg^{2+}$-ATPase located in the plasma membrane (Skou, 1965; Schwartz et al., 1972). Neither the $K^+$ nor the $Na^+$ activation curves of this enzyme in this or other active transport systems consistently fulfil the Michaelis–Menten equation (Robinson, 1967). The data have been interpreted in terms of allost-

                                   *R.G.Wyn Jones*

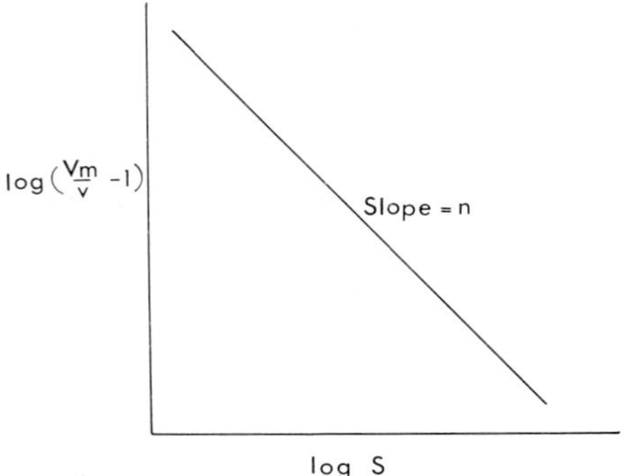



Fig. 7.10. Hill plot for the determination of co-operative interactions between substrate and enzyme. (From Hill, 1910.)

eric interactions (Koshland, 1970) and Hill plots (1910) have been used to determine the co-operativity of the interaction (Fig. 7.10).

$$v = V_{max}/(1 + K/[S]^n) \qquad (7.5)$$

where $v$ is measured initial velocity; $V_{max}$, maximum velocity; $n$, interaction co-efficient; $[S]$, activator concentration; $K$, apparent dissociation equilibrium constant, $n > 1$ indicates positive co-operativity; $n < 1$ negative co-operativity and $n = 1$ is equivalent to the Michaelis–Menten equation. It should also be noted that the degree of co-operativity in the ATPase system may be influenced by pH, total ionic strength and possibly $Ca^{2+}$ binding to the membrane (Schwartz et al., 1972).

Although comparisons of this type must be treated cautiously it is quite clear that there is no prime face biochemical reason to expect ion absorption isotherms which are the result of active processes to follow the Michaelis–Menten formalism i.e. that $n$ should equal 1. Nor can any significance be attached to the observation that a particular isotherm does so, as passive fluxes, particularly if carrier mediated, may, amongst other possibilities, display saturating kinetics and therefore produce a rectangular hyperbola and apparent Michaelis–Menten kinetics (Cram, 1974a).

Therefore, any mechanistic deduction from the shape of any isotherm is likely to be erroneous. The emphasis in the carrier–kinetic approach on the Michaelis–Menten formalism is probably misguided and has been an excessively rigid influence on this field. It might perhaps be noted that in animal ATPase work the concept of a rigid carrier site is now being questioned and the variable pores in the membrane may be a more appropriate concept to describe the method by which the ions are moved through the membrane (Singer, 1974).

It would therefore appear that the criteria outlined earlier for deducing a specific active carrier-mediated mechanism from an isotherm have not been fulfilled. The isotherms nonetheless have a descriptive and comparative value and the use of kinetic constants as a short hand to describe the shape etc. of isotherms, appears to be valuable but must not be confused with a mechanistic extrapolation. The problem of the cellular location of the isotherms measured by the various methods will be discussed in the course of the next section.


## 7.4 Flux kinetics and compartmentation

The theoretical basis for compartmental flux analysis in higher plant cells has been comprehensively discussed in the previous chapter. It is derived from a thermodynamic view of ion transport and this framework has been applied very successfully to ion relations in giant alga (MacRobbie, 1971; Ch. 5). However, even in this advantageous experimental situation, numerous technical and interpretive problems have occurred and a number of fascinating unsolved mysteries exist. Nevertheless the application of thermodynamic principles, together with metabolic studies, to alga has produced detailed evidence for a variety of influx and efflux pumps on both the plasmalemma and tonoplast. Active transport is specifically defined as transport against an (electro) chemical potential gradient of that ion and not in the looser sense used by Epstein (1973), namely transport which is dependent on metabolism.

Although the full impact of the thermodynamic approach to ion transport depends on the measurement of ion fluxes, compartmental concentrations and membrane electrical potentials and use of relationships such as the Ussing–Teorell flux ratio equation (Eq. 1.15), much valuable information has been obtained purely from flux analysis.

Efflux analysis rests on the measurement of quasi-steady state wash-out curves of ions from pre-loaded tissue. The mathematical treatment of these curves contains a number of assumptions and although the validity of the treatment is capable of being tested in a number of ways it still requires further examination (Ch. 6). The principal assumption is that the cell contains three major compartments, the cell wall, the cytoplasm and the vacuole arranged in series and although this appears to be an adequate first approximation, it cannot be assumed to be rigorously correct.

Efflux analysis allows the estimation of the permeability of the tonoplast and plasmalemma and the $t_{\frac{1}{2}}$ of efflux from the major cell compartments.

Although there are substantial variations in the published data, a number of tentative generalisation may be made about the compartmental half-times in root tissue which are consistent with the observations on other tissues. The wash-out curves, when plotted as log of tissue radioactivity against time, yield three approximately linear components by sequential subtraction of the slower component from the previous component of the curve (see Ch. 6). These phases have half-times of decay ($t_{\frac{1}{2}}$) ranging from a minute to tens of hours depending on the experimental conditions and the isotope used. In the case of $Cl^-$, $t_{\frac{1}{2}}$ for the fastest component is 1–2 min and, as this component is found in wash-out from dead tissue, it is assigned to the free space of the cell wall. The second component for $Cl^-$ efflux yielded a $t_{\frac{1}{2}}$ ranging from 12 min in fresh tissue to 5–6 min in aged tissue and, as it is absent in dead tissue, it is ascribed to efflux across the plasmalemma from the cytoplasm. The final slow component, thought to be from the vacuole, has a $t_{\frac{1}{2}}$ of up to 60 h (Cram, 1973). The equivalent figures for the wash-out of cations are, in general, greater (Table 7.2). The increased $t_{\frac{1}{2}}$ of the cell wall phase is due to the net anion

Table 7.2.
Some data on cation efflux from roots and other tissues.

| Tissue | Ion | $t_{1/2}$ cytoplasm (min) | Remarks | References |
|---|---|---|---|---|
| *Beta vulgaris* | $K^+$ | 43 | 25 °C | Pitman (1963) |
| (storage root) | $Na^+$ | 20–30 | 25 °C | Pitman (1963) |
| Barley roots | $K^+$ | 250 | about 5.0 °C | Pitman and Saddler (1967) |
| | $Na^+$ | 60 | about 5.0 °C | Pitman and Saddler (1967) |
| Corn roots | $Rb^+$ | 105 | 0.1 Mol m$^{-3}$, 2–4 °C | Leigh et al. (1973) |
| | | 118 | 10 Mol m$^{-3}$, 2–4 °C | Leigh et al. (1973) |
| *Avena sativa* | $K^+$ | 23–59 | 20–25 °C | Pierce and Higinbotham (1970) |
| (coleoptile) | $Na^+$ | 10–11 | 20 °C | Pierce and Higinbotham (1970) |
| *Pisum sativum* | $K^+$ | 74 | 20–25 °C, 1 Mol m$^{-3}$ | Macklon and Higinbotham (1968) |
| (epicotyl) | $K^+$ | 64 | 20–25 °C, 10 Mol m$^{-3}$ | Macklon and Higinbotham (1968) |

surface charge restraining cation movement, while high $t_{\frac{1}{2}}$ cyt values must indicate variation in the permeability coefficients of anions and cations. These are normally found to be in the range (0.2–10 $\times$ 10$^{-6}$ m s$^{-1}$) for the plasmalemma.

A degree of caution should be exercised over the figures as in some cases (e.g. Pitman and Saddler, 1967; Leigh et al., 1973) through–transport and leakage from the cut ends of the stele were not distinguished from cortical efflux. Further, various reports of deviations from the series of three log linear wash-out curves described above, have been observed (see later). Nevertheless efflux analysis has obvious implications for the determination of influx kinetics and the interpretation of the absorption isotherms described in Section 7.2. If efflux as well as influx is taking place then attention must be given to the precise absorption and desorption times if an absorption isotherm is to reflect influx at a given membrane. A given uptake–wash regime may measure a different flux or combination of fluxes when applied to $Cl^-$ or $K^+$ uptake if their efflux rates differ appreciably and there must be some doubt whether it is legitimate to assume that all tissues or species behave identically.

The potential importance of these considerations may be illustrated by comparing the techniques employed by Torii and Laties (1966a) and Welsh and Epstein (1969) in the papers which are, respectively, the corner stones of the series and parallel models of the dual absorption isotherm. The latter measured $^{86}Rb^+$ and $^{36}Cl^-$ uptake from 0.1 and 5 Mol m$^{-3}$ external salt over various short uptake periods (1–10 min) followed by a 30 min cold desorption. On the other hand, Torii and Laties measured absorption over a greater concentration range during periods of 1–3 h and desorbed the roots during three 1-h washing periods in water in the case of $Cl^-$ and in 1–10 Mol m$^{-3}$ RbCl in the case of $Rb^+$. This technique measures quasi-steady state influx into the vacuole and not the plasmalemma flux, while Epstein's conditions produced an isotherm akin to plasmalemma flux.

Thus, the accurate estimation of plasmalemma influx demands a knowledge of efflux kinetics and Cram (1974a and Ch. 6) has suggested three possible methods for making this measurement based on either a short uptake, short wash regime or using a non-penetrating analogue to estimate the free space content. It is apparent that many of the published uptake isotherms reflect quasi-steady state influx and not necessarily plasmalemma flux. The effect of uptake–wash regime on the isotherm is illustrated in Fig. 7.4 ($Rb^+$) and Fig. 7.6 ($Cl^-$). In the low concentration range the isotherms measured by both techniques are very similar, but there is a pronounced effect of experimental regime in the higher range (see also Cram and Laties, 1971). This suggests that plasmalemma influx is rate limiting in the low concentration range under both regimes but not in the higher range when a long uptake–wash regime is employed. This latter regime measures a mixed flux comprising plasma-lemma influx and efflux and tonoplast influx but may approximate to tono-plast flux if this process is rate-limiting and plasmalemma influx and efflux are rapid (see Figs 7.6 and 7.11). A more accurate determination of tonoplast fluxes depends on a complete compartment flux analysis and the assumption of the 3 'compartments-in-series' model for the cell.

The validity of flux analysis as outlined above is denied by Epstein (1972,

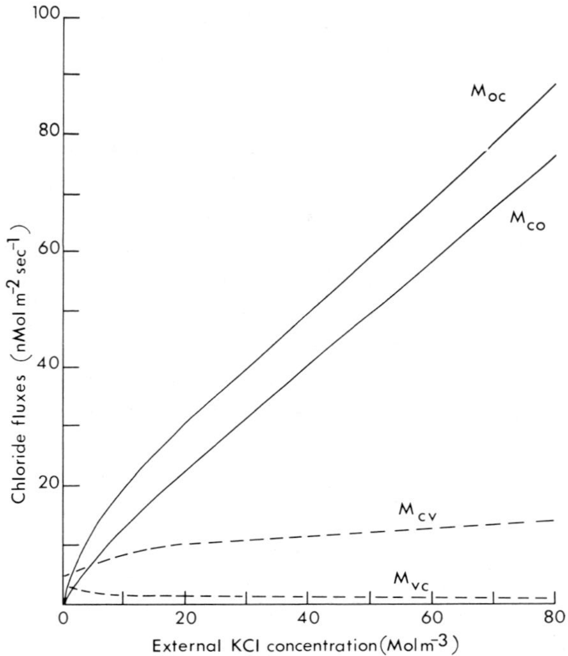

Fig. 7.11.  Calculated $Cl^-$ fluxes at the plasmalemma and tonoplast of maize root cortical cells.
(From Cram, 1973.)

1973) who regards the plasmalemma as an impermeable barrier in the absence
of a specific carrier and views carrier transport as almost exclusively inwardly
directed. He therefore considers the thermodynamic concept of movement
down a chemical potential gradient as inapplicable. He attacks efflux analysis
and the observations on outward diffusion of ions on the grounds that (a) the
excised tissue is kept so long as to produce metabolic decay, (b) some measurements were made in the absence of $Ca^{2+}$ which is required to maintain
membrane integrity (see Wyn Jones and Lunt, 1967), (c) much efflux may be
through-transport into the xylem not true efflux from the cortical cells, (d)
efflux may be carrier-mediated not simple diffusion and (e) roots are easily
damaged. Epstein also quotes a number of papers where exchange or leakage
of radio-tracers from prelabelled tissue was apparently not observed.

Most of these objections can be easily refuted by reference to recent studies
(Cram and Laties, 1971; Pitman, 1971; Jeschke, 1972, 1973, 1974; Cram,
1973a). $Ca^{2+}$ is routinely included in the bathing medium, transport and
efflux are measured separately or cortical cells are removed from the stele
and most experimenters have used barley roots which Epstein regard as being

most resistant to manipulative injury. The attribution of efflux to decay due to ageing is inconsistent with the constant through-transport found in these roots (Jeschke, 1973) while the presence or absence of a 'carrier' is irrelevant to the analysis of wash-out curves.

Thus, efflux analysis may not be lightly disregarded, although the rigid analysis of the efflux curves into three compartments in series, particularly for $Rb^+$, appears to be uncertain. Pallaghy et al. (1970) observed shoulders in the $K^+$ wash-out curves from low-salt maize root segments in the low concentration range. These abberations were confirmed by Leigh et al. (1973) but it may be objected that both these papers failed to measure through transport (Weigl, 1971). However, Baker and Hall (1973) have since observed these shoulders in $^{86}Rb^+$ efflux from isolated cortical segments in which stelar transport does not occur. Inconsistencies in influx and efflux measurements have led both Leigh and Wyn Jones (1975) and Heller et al. (1973) to suggest tentatively that there may be a fast efflux component from the cytoplasm and the uniform behaviour of the cytoplasm has yet to be proved. MacRobbie (1971) notes a number of anomalies in influx and efflux kinetics in higher plants which indicate a poor fit to the rigorous 3-compartment model and possibilities of deviations from this model must still be considered (see also Baker and Hall, 1973).

A study of the kinetics of $Cl^-$ influx and efflux in maize and barley roots (Cram and Laties, 1971; Cram, 1973a) strongly suggests that at concentrations above 1–5 Mol m$^{-3}$ the rapid plasmalemma influx of $Cl^-$ is matched by a rapid efflux and contributes little to net $Cl^-$ accumulation. This concept is reinforced by Weigl's observation (1968) that $Cl^-$ efflux is stimulated by external $Cl^-$. The plasmalemma $Cl^-$ influx was found to be active at all concentrations by Gerson and Poole (1972) but since it is less sensitive to inhibitors, and to temperature at high rather than low external $Cl^-$ concentrations, the measured flux may well comprise both an active and a passive component. It may be deduced that $K^+$ follows a similar pattern. The short uptake–wash $Rb^+$ influx rate in maize roots at 10 Mol m$^{-3}$ corrected for the cell wall component (about 8 $\mu$Mol g$^{-1}$ h$^{-1}$, Leigh and Wyn Jones, 1975) is very similar to the $Cl^-$ influx rate (about 10 $\mu$Mol g$^{-1}$ h$^{-1}$, Cram, 1973). Since Cram has also shown that the rates of $K^+$ and $Cl^-$ net accumulation in maize are identical, it is reasonable to deduce there is appreciable $K^+$ efflux at high external KCl concentrations. However at the highest concentrations (about 50 Mol m$^{-3}$) the plasmalemma $K^+$ flux rate appears to be a half or less of the $Cl^-$ rate. Inhibitor studies (Leigh and Wyn Jones, 1975) suggest that an increasing proportion of this $K^+$ influx also is passive above 20 Mol m$^{-3}$ external KCl, confirming the views of Barber (1974). The best estimates of

Cl$^-$ fluxes across both the plasmalemma and tonoplast above 1 Mol m$^{-3}$ are shown in Fig. 7.11, and clearly indicate little net accumulation as a result of uptake in the upper range.

Studies on the influence of the internal salt concentration on the fluxes have shown that, in general, influx rates decrease as the internal salt status of the root increases and that the influxes may be regarded as susceptible to some form of feed-back repression. This is, of course, only to restate in modern jargon the original observation of Hoagland and Broyer. Pitman (1970) first suggested that uptake in the upper concentration range was restricted to low salt tissue but Leigh and Wyn Jones (1973) showed clearly that Rb$^+$ and Na$^+$ uptake in the lower range was most sensitive to internal salt status (Fig. 7.4). Cram and Laties (1971) made similar observations on Cl$^-$ uptake into barley roots but curiously Cram (1973) did not observe a decrease in Cl$^-$ plasmalemma influx in maize under high salt conditions. There is general agreement that quasi-steady state influx into the vacuole is reduced by increasing internal salt.

## 7.5  Electrophysiological studies

Numerous measurements of the membrane electrical potential of root cells have been made and values of 100 to 120 mV are commonly observed. This potential probably occurs between the outside and the vacuole but the major potential difference is deduced to occur across the plasmalemma with a very small potential across the tonoplast (Pierce and Higinbotham, 1970; Higinbotham, 1973a and b). This deduction agrees with the observations on algae (Ch. 5) but its experimental verification in the small cells of higher plants is very difficult. The potential in whole barley roots decreased in response to increasing external salt in a stepwise manner with rapid depolarization being observed in the 10–100 mMol m$^{-3}$ and 1–50 Mol m$^{-3}$ KCl range (Fig. 7.2; Mertz and Higinbotham, 1974). However, Pitman et al. (1971) did not observe this stepwise pattern in root segments, but the decrease in membrane P.D. at higher external concentrations is not in doubt (Higinbotham and Anderson, 1974). In the halophyte *Triglochin maritima*, the cells grown in low NaCl but not those grown in high NaCl were depolarized by exposure to increased NaCl which suggests induced changes in the chemical composition of the plasmalemma under different growth conditions (Jefferies, 1973).

The membrane potential is made up from two major components, one due to the diffusion potential of mobile ions and calculated by the Goldman constant field equation and a second electrogenic component (see Ch. 1,

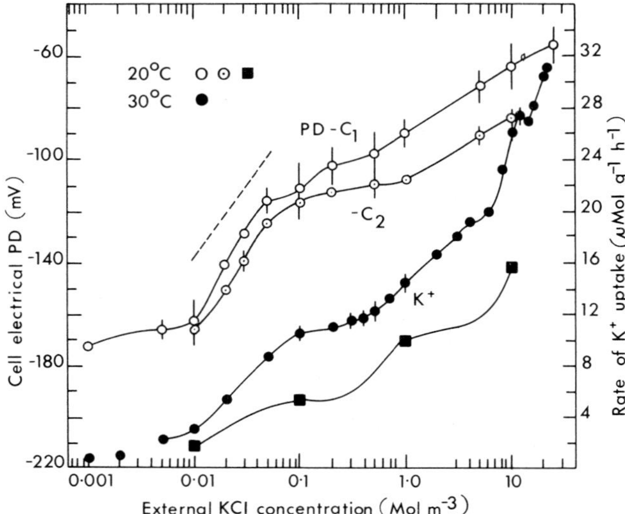



Fig. 7.12. Electrical potential difference (PD) of root cortical cells, (○—○) first and (⊙—⊙) second layer of cells and rate of $K^+$ uptake (●, ■) by excised low salt barley roots as a function of temperature, (○, ⊙, ■ : 20 °C) (● : 30 °C) and external KCl concentration. Vertical bars ± 1 S.D.M. (– – – –) represents a change in PD of 58.2 mV per 10-fold change in ion concentration. (From Mertz and Higinbotham, 1974.)

Eq. 1.19). This latter is caused by the active transport of an ion (or ions) creating a charge imbalance. A number of inhibitors ($CN^-$, DNP, CCCP) rapidly decrease the membrane potential to about $- 40$ mV which is approximately the level of the diffusion potential predicted by the Goldman equation (Higinbotham and Anderson, 1974). The reversible inhibition of the electrogenic component of the membrane potential by CO is illustrated in Fig. 7.13. It appears that the electrogenic component of the membrane potential is only significant in the higher concentration range (Table 7.3) and $K^+$ inward transport is in the wrong direction for it to be subject to an electrogenic pump. Higinbotham (1973) suggests that the apparent discontinuities or steps in the uptake isotherm may be related to sharp changes in membrane resistance rather than alterations in membrane potential.

Data from flux and membrane potential measurements have been used to estimate the active fluxes across the plasmalemma and tonoplast of root, coleoptile and epicotyl cells. For the most part, there is a reassuring agreement regarding the major active fluxes which have been deduced. Active $Na^+$ efflux at the plasmalemma has been proposed in pea, barley, maize and bean roots. These complement Jeschke's detailed evidence of an active $K^+$-stimul-

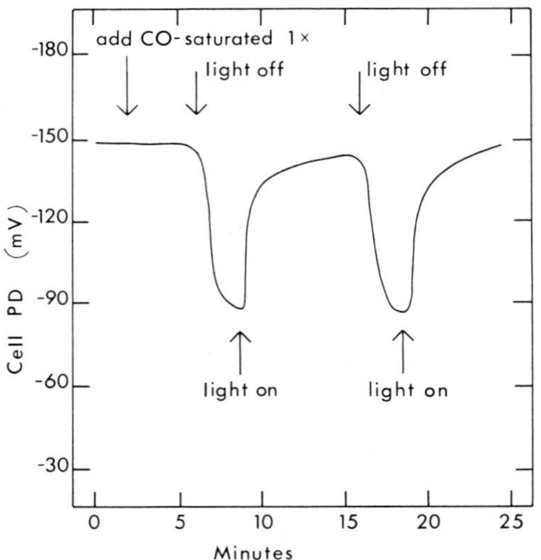


Fig. 7.13. Effect of carbon monoxide on the potential difference of pea epicotyl cells in the presence and absence of light. (From Higinbotham and Anderson, 1974.)

ated $Na^+$-efflux in barley roots (Jeschke, 1973, 1974). Similarly an active $Cl^-$ influx appears to occur at all concentrations in a number of tissues (see Higinbotham, 1973a, b for detailed references) probably because $H_2PO_4^-$ and other anions are actively concentrated in root cells (Table 7.4). However, the

Table 7.3.

Relationship in oat seedling tissue of measured and calculated cell electropotential difference to external concentration. Composition of external (1 ×) solution (Mol m$^{-3}$) KCl, 1.0; Ca(NO$_3$)$_2$, 1.0; MgSO$_4$, 0.25; NaH$_2$PO$_4$, 0.904; Na$_2$HPO$_4$, 0.048. $E_m$, measured membrane potential; $E_k$, calculated Nernst potential for $K^+$; $E_g$, membrane potential calculated from Goldman voltage equation (Eq. 1.19). From Higinbotham and Anderson (1974).

| Tissue | External solution concentration | Cell electropotential difference (mV) | | | Using Eq. 1.20 | |
|---|---|---|---|---|---|---|
| | | $E_m$ | $E_k$ | $E_m - E_k$ | $E_g$ | $E_m - E_g$ |
| Coleoptile | 0.1 × | − 109 | − 144 | + 35 | − 116 | + 7 |
| | 1 × | − 165 | − 109 | + 4 | − 89 | − 16 |
| | 3 × | − 109 | − 79 | − 30 | − 67 | − 42 |
| | 10 × | − 102 | − 55 | − 47 | − 44 | − 58 |
| Root | 1 × | − 84 | − 92 | + 8 | − 79 | − 5 |
| | 3 × | − 89 | − 71 | − 18 | − 66 | − 23 |
| | 10 × | − 71 | − 45 | − 26 | − 45 | − 26 |

situation regarding $K^+$ uptake is problematical. Pitman and Saddler (1967) produced evidence for an active $K^+$ influx at low external concentration which was masked by the $Cl^-$ pump at higher concentrations. Similarly Mertz and Higinbotham (1974) noted that the $Q_{10}$ of $K^+$ influx decreased with increasing external concentration. In pea roots equilibrated at high salt conditions and in bean root (Scott et al., 1968), $K^+$ is close to flux equilibrium, while in *Avena* coleoptile at low concentrations $K^+$ is probably pumped inwards (Table 7.4).

Table 7.4.
Comparison of observed and predicted steady-state concentrations of ions in excised roots of *Avena* (oats). Reproduced from Clarkson (1974).

| Ion | Concentration in tissue : dilute bathing solution[a] (Mol m$^{-3}$ tissue water) | | Concentration in tissue : $\times$ 10 concentrated bathing solution (Mol m$^{-3}$ tissue water) | |
|---|---|---|---|---|
| | Observed | Predicted[b] | Observed | Predicted[b] |
| $K^+$ | 66 | 27 | 73 | 159 |
| $Na^+$ | 3 | 27 | 3 | 159 |
| $Mg^{2+}$ | 8.5 | 175 | 11 | 638 |
| $Ca^{2+}$ | 1.5 | 700 | 1.5 | 2550 |
| $NO_3^-$ | 56 | 0.076 | 38 | 1.25 |
| $Cl^-$ | 3 | 0.038 | 4 | 0.62 |
| $H_2PO_4^-$ | 17 | 0.038 | 14 | 0.62 |
| $SO_4^{2-}$ | 2 | 0.0004 | 5.5 | 0.01 |
| Membrane potential | $E_{vo} = -84\,mV$ | | $E_{vo} = -71\,mV$ | |

[a] Ionic composition (Mol m$^{-3}$) $K^+$, 1.0; $Na^+$, 1.0; $Mg^{2+}$, 0.25; $Ca^{2+}$, 1.0; $NO_3^-$, 2.0; $Cl^-$, 1.0; $H_2PO_4^-$, 1.0; $SO_4^{2-}$, 0.25.
[b] Predicted from the Nernst equation (Eq. 1.7).

The minimum generalization seems to be that $K^+$ moves actively against its electrochemical potential gradient at low external concentration but approaches electrochemical equilibrium at higher external $K^+$ concentrations (Dunlop and Bowling, 1971).

There is less reliable information about active fluxes across the tonoplast of higher plants. It may be tentatively suggested that both $Na^+$ and $Cl^-$ and possibly $K^+$ are actively pumped into the vacuole although the possibility of $K^+$ being actively transported from the vacuole into the cytoplasm of halophytes cannot be discounted.

The application of electrochemical theory to ion relations and the subsequent deduction of ion pumps is only valid if the ions in the cytoplasm and vacuole are in a free state and not bound. There is little doubt that this is the case in the vacuole but as yet there are only a few studies with ion-specific electrodes and some nuclear magnetic resonance data to suggest this is true throughout plant cells (see Higinbotham, 1973b and 1974b).

## 7.6  *Metabolic and biochemical observations*

### 7.6.1  *$H^+$ and $OH^-$ fluxes and organic acids*

Two important metabolic events are directly associated with ion uptake by roots. One is the accumulation of organic anions when the uptake of organic cations exceeds that of inorganic anions, for example when barley roots are exposed to $K_2SO_4$ (Ulrich, 1941; Hiatt, 1967). The particular acid which is accumulated varies from species to species. Laties regarded organic acid synthesis in the absence of a permeant anion as closely linked to the accumulation of $K^+$ in the vacuole by Mechanism II (Torii and Laties, 1966b). However, Hiatt (1967) showed that changes in organic acid content in response to excess cation uptake were also associated with uptake in the low concentration range. The pH of the expressed root sap changed in the response to excess cation or anion uptake, increasing when excess cations were absorbed but decreasing when excess anions were taken up. Closely associated with the phenomenon of organic acid synthesis is the observation that $H^+$ is released during cation absorption and base during anion absorption. Jackson and Adams (1963) in a much neglected paper showed that both $K^+$ and $Na^+$ absorption from 10 to 30 Mol m$^{-3}$ were independent of the anion including $HCO_3^-$. It is somewhat disturbing that, in this respect, their results are at variance with these of Epstein (quoted earlier) and Hiatt (1967), both of whom found a decrease in $K^+$ uptake in the presence of $SO_4^{2-}$ compared with $Cl^-$ above 1 Mol m$^{-3}$. However, Jackson and Adams (1963) observed that $H^+$ was released at a rate which reflected the difference between the anion and cation uptake rates (Fig. 7.14). Both $H^+$ release and $K^+$ uptake were DNP sensitive. When anion uptake exceeded cation uptake, base was released, although base release exceeded anion uptake by up to 5-fold. It was concluded that the rate-limiting step for cation absorption is neither a gradient produced by simultaneous anion absorption nor an $H^+$ concentration gradient in the opposite direction because $K^+$ uptake was not sufficiently sensitive to external pH. Pitman (1970) has confirmed that $H^+$ efflux is associated with $K^+$ uptake in the higher concentration range, but was unable to detect it in the lower range, although this appears to be implied in Hiatt's data (1967). Pitman,

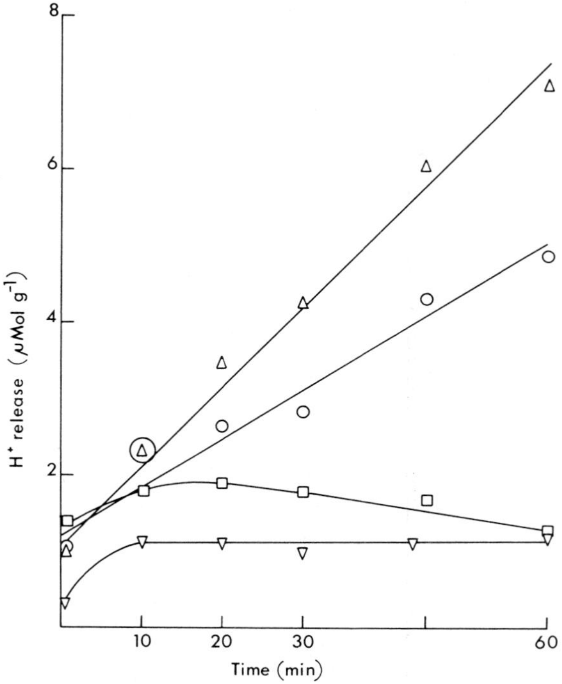

Fig. 7.14. The time course of $H^+$ release from barley roots initially at pH 7.0 in various salt solutions. O—O, KCl 10 Mol m$^{-3}$; □—□, CaCl$_2$ 10 Mol m$^{-3}$; △—△, K$_2$SO$_4$ 10 Mol m$^{-3}$; ∇—∇, CaSO$_4$ 10 Mol m$^{-3}$. (From Jackson and Adams, 1963.)

however, observed that the apparent $H^+$ release is nearly the same as $K^+$ uptake from $K_2SO_4$, but is substantially less than the $K^+$ uptake from KCl. $H^+$ release was largely associated with low salt tissue and greatly reduced in salt-saturated tissue. Pitman noted that many of the results could be explained either by postulating that $H^+$ release is related to organic acid synthesis or to a cation–$H^+$ exchange and an $OH^-$ (or $HCO_3^-$)–$Cl^-$ exchange which is suppressed by $SO_4^{2-}$. This formulation is, of course, very similar to the original concept of Jacobson et al. (1950). Thus, the possibility of exchange pumps is still of great interest in root physiology.

### 7.6.2 Ion pumps

The biochemical information regarding the energy source used to activate the various proposed ion pumps is discussed in detail elsewhere in this volume and although ATP is generally considered a likely candidate, definitive evidence is lacking (see Ch. 2).

Nonetheless, the analogy with animal systems alluded to earlier has inspired numerous studies on the relationship of plant ATPase activity to ion transport. An exhaustive review of work on root ATPases will not be attempted here as it is covered elsewhere (see Ch. 2). However, a most interesting observation was that of Fisher et al. (1970) who found an excellent correlation between $K^+$ uptake rates and the total $K^+$-stimulated ATPase activity in barley, oats, wheat and corn roots in the high concentration range. Leonard and Hanson (1972) have observed a stimulation of both $K^+$-stimulated ATPase and $K^+$ and $Cl^-$ fluxes by washing corn roots, while Leonard and Hodges (1973) have identified a $K^+$-stimulated ATPase on the plasmalemma of oat roots which gave a biphasic $K^+$-stimulation curve very similar to the classical absorption isotherm (Fig. 7.2). Drawing on Mitchell's concept of an electrogenic $H^+$ efflux pump (Mitchell, 1973), Hodges (1973) has proposed that uptake in roots is mediated by a $K^+$-activated ATPase on the plasmalemma which catalyses $K^+/H^+$ or related exchange reaction. This enzyme is thought to generate a pH gradient and an electrogenic membrane potential and to be linked to a second anion carrier system (see Fig. 7.15). The possibility of an electrogenic $H^+$ efflux pump, linked to $K^+$ influx in higher plants has been widely discussed (Higinbotham and Anderson, 1974; Slayman, 1974) but the evidence for this

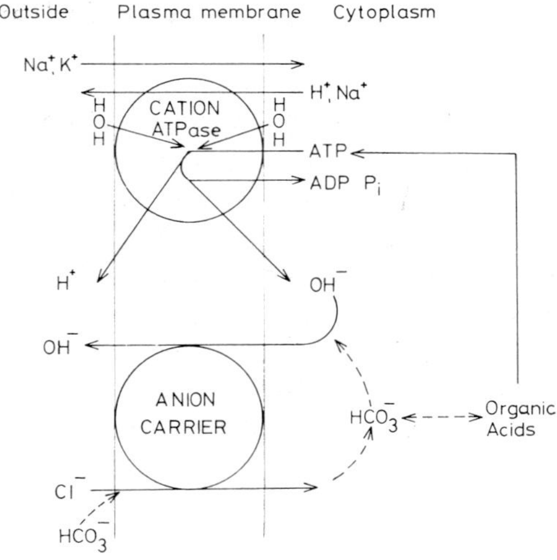


Fig. 7.15. Diagramatic representation of model for plasmalemma $H^+$ efflux pump mediated by cation-stimulated ATPase. (From Hodges, 1973.)

mechanism operating across the plasmalemma of root cortical cells is not entirely convincing. The observations on active $K^+$ influx and the electrogenicity of the membrane potential both appear incompatible with an electrogenic $H^+$ efflux pump or $K^+$ influx pump, although not with $K^+/H^+$ exchange. The electrogenic component of the potential increases as the external KCl concentration increases (Table 7.3; Mertz and Higinbotham, 1974) while, on the contrary, active $K^+$ uptake is associated with low external concentrations and $K^+$ appears to move passively, although possibly associated with a carrier, at higher concentrations. Much of the upper isotherm of $K^+$ uptake is anion-dependent. Washing of corn roots enhances the electrogenicity of the membrane potential of epidermal cells but decreases the $H^+$ extrusion (Lin and Hanson, 1974b). Thus, the operation of an electrogenic $H^+$ efflux–$K^+$ influx pump appears doubtful. It is also difficult to accept unquestioningly an active $K^+$ influx mediated by an ATPase at the plasmalemma in the high concentration range where electrogenicity is most pronounced. As stated, $K^+$ uptake is greater from KCl than from $K_2SO_4$ in this range whereas the reverse is true of $H^+$ efflux although, of course, the latter could be masked by $OH^-/Cl^-$ exchange. The evidence for a $K^+$-ATPase active in the range 1–50 Mol m$^{-3}$ rests heavily on the correlations observed by Fisher et al. (1970) between $K^+$ influx measured by a relatively long uptake – long wash technique and total KCl-stimulated activity in four cereals. These data may require reinterpretation in view of the realisation that plasmalemma fluxes may differ qualitatively and quantitatively from quasi-steady state influx measured by long uptake–long wash techniques. In addition, Leigh et al. (1974) have observed that numerous correlations may be made between root KCl–ATPase activities and fluxes measured by various techniques. Thus, the situation is probably more complex than first envisaged and further detailed studies are required on plant root ion-stimulated ATPases before clear deductions may be made.

## 7.7 *Summary*

Despite the many uncertainties which permeate the available evidence, it is possible to summarise the results to date and produce a fairly coherent interpretation of the results. Below about 1 Mol m$^{-3}$, $K^+$ is probably taken up actively at the plasmalemma of root cortical cells by a process which is rate limiting under most experimental conditions (apparent affinity constant about 20 mMol m$^{-3}$). This process is highly selective for $K^+$ over $Na^+$, is relatively insensitive to anions but extremely sensitive to inhibitors, DNP, CCCP etc.

This process is repressed as the internal salt status increases and is therefore associated with the net uptake of $K^+$. This interpretation is borne out by experiments on intact plants (e.g. Claassen and Barber, 1974) where an uptake system of similar affinity is found (apparent $K_m$ 27 mMol m$^{-3}$). In this range $K^+$ uptake may, depending on pH and an associated anion, greatly exceed the anion uptake and under these conditions $H^+$ efflux has been observed (Jackson and Adams, 1963; Hiatt, 1967). However $H^+$ efflux in this range is not confirmed by Pitman (1970) and the possibility of the cell wall exchange capacity masking $H^+$ release into the medium could be entertained. Further, both $K^+$ influx in the low range (Leigh and Wyn Jones, 1973) and $H^+$ efflux (Pitman, 1970) are decreased by high salt status. The $K^+$-stimulated $Na^+$ efflux observed by Jeschke in barley (1973, 1974) correlates well with this process as the affinity for the $K^+$ stimulation of $Na^+$ efflux is 20 mMol m$^{-3}$. Thus, a system of active but non-electrogenic $K^+/H^+$ or $K^+/Na^+$ exchange is a viable hypothesis. A similar $K^+$–$H^+$ non-electrogenic exchange pump at pH 5.5 to 6.0 has been deduced in red beet root storage tissue by Poole (1974) who also considers the possible change in the electrogenicity of the pump with pH. In this tissue there is evidence of electrogenic-linked ($K^+$–$H^+$) transport at high pH. Poole explains this change in terms of an increased dissociation of $H^+$ from the $K^+$–$H^+$ carrier at high external pH which would increase the affinity of the carrier for $K^+$. If the carrier sites were to return empty then a potential would be generated.

In the root system the exchange process would be envisaged as being fairly flexible and the same carrier could well mediate passive $K^+$–$K^+$ exchange at the plasmalemma at high external potassium salt concentrations. In this range there may also be inhibitor-insensitive passive diffusion and a chloride-dependent component to the $K^+$ flux. These processes could account for the increased $Na^+$ influx observed by Epstein (1973) and others in the upper range and the decrease in selectivity.

It is tempting to speculate that an ATPase in the plasmalemma may mediate the $K^+$–$H^+/Na^+$ exchange process. A $Mg^{2+}$-dependent ATPase stimulated by low $K^+$ concentrations has been observed in a plasmalemma-enriched fraction from oats (Leonard and Hodges, 1973), corn (Leigh et al., 1974) barley (Palacios-Alaiz, Wyn Jones and Laidman, unpublished observations). There is, as yet, no direct evidence to link these enzymes with the pump but for this postulate to hold, it may be predicted that the enzyme should be sensitive to oligomycin, CCCP and DNP and might be linked to the mersalyl-sensitive $H_2PO_4^-$-transport system investigated by Lin and Hanson (1974a, b). The plasmalemma ATPase of the three cereals is also stimulated by high concentrations of $K^+$ and could mediate $K^+/K^+$ exchange at high concentration.

Exchange reactions are frequently associated with ATPase activities in animals (Schwartz et al., 1972).

In plant roots there is also evidence, as outlined previously, for an active anion pump. Excess anion uptake is accompanied by base release (Jackson and Adams, 1963, Fig. 7.14) and anion uptake is markedly more sensitive to increased pH than $K^+$ uptake. Thus, $Cl^--OH^-$ or $HCO_3^-$ exchange as envisaged by Jacobson et al. (1950) remains a possibility. Active $Cl^-$ uptake becomes increasingly dominant (Pitman and Saddler, 1967) as the external concentration rises and this is clearly compatible with the behaviour of an electrogenic component of the membrane potential. This transporter would be envisaged as mediating various anion exchange reactions leading to $Cl^-$ or $NO_3^-$ uptake (Blevins et al., 1974) as well as $Cl^-$ exchange at high external $Cl^-$ concentrations as observed by Cram (1973a). There is evidence in other systems (*Potamogeton*, Denny and Weeks, 1970; *Acetabularia*, Saddler, 1970; *Limonium*, Hill and Hill, 1973) for electrogenic anion pumps. However, as pointed out by Higinbotham and Anderson (1974) such an electrogenic anion pump would have to be flexible, involving $OH^-$ and $HCO_3^-$. It is of interest that Cram (1974b) has shown a competitive interaction between $Cl^-$ and $HCO_3^-$ during plasmalemma influx. It is also difficult and possibly inadvisable to attempt to distinguish $H^+$ influx from $OH^-$ efflux. Thus, there are formidable technical problems to be overcome before the systems such as those discussed can be proven. It should also be noted that Poole's hypothesis of an ion pump whose electrogenicity is dependent on the pH of the medium may also be applied to any postulated electrogenic anion pump. In this discussion the role of $Cl^-$ has been emphasised although this may be misleading as in the soil $NO_3^-$ is frequently the dominant anion. However, little work on $NO_3^-$ fluxes has been undertaken with excised roots. It is interesting that Cram (1973b) observed that $Cl^-$ influx is significantly correlated with the log of the $(Cl^- + NO_3^-)$ concentration in the vacuole suggesting that either $Cl^-$ and $NO_3^-$ influxes are mediated by the same system or that a tight and common control is exercised over both influxes. The energy source for this postulated anion pump is unknown but Jacobson and Young (1971) observed that cyclopropane, which stopped cyclosis, inhibited $Cl^-$ but not $K^+$ uptake from 5 Mol $m^{-3}$ KCl into barley roots, thereby implying a different energy source. However, the kinetics of the depolarisation and repolarisation of the electrogenic potential following CO is consistent with the formation of a respiratory intermediate such as ATP (Anderson et al., 1974).

There is some electrochemical evidence for active transport from the cytoplasm to the vacuole. Precise information is sparse but both active cation and anion influxes are possible. It might be noted that Fisher et al. (1970)

correlated total KCl–ATPase in barley, oats, wheat and maize with influx measured by a 30 min uptake–30 min cold wash regime, Leigh et al. (1974) found an excellent correlation between KCl–ATPase activity on an internal membrane and quasi-steady state influx into corn roots. These observations may be ascribed tentatively to a salt transport system into the vacuole, and it might be noted that Cram (1973a) observed a precise correlation between net $K^+ + Cl^-$ influxes into corn roots. The various cross correlations observed by Leigh et al. (1974) however, necessitate a very cautious interpretation of the data at this stage.

In general, the data on uptake by roots are therefore in reasonable agreement with the better defined algal system (Ch. 5) and, as in the algae, it is probable that variations exist from species to species ready to undermine over-enthusiastic generalisations.

The inadequacies of the 3-compartment model have been alluded to at various times in this chapter. There is clearly room for endocytosis to contribute to uptake (see Baker and Hall, 1973; Robards and Robb, 1974) although the extent of this contribution is unknown. It would appear improbable from the evidence discussed that the whole of the lower isotherm is due to endocytosis. There is a clear requirement for work on membrane turnover rates in root cells so that the extent of pinocytosis may be assessed and it is important also to bear in mind the dynamic nature of the cell in our studies on ion uptake in roots.

No excuse is required for the emphasis in this chapter on technical and methodological problems as well as conceptual advances. It is apparent that every technique applied to the problem has its limitations (see the various discussions in 'Membrane Transport in Plants and Plant Organelles', 1974) and that technical difficulties are the major barrier to more rapid progress. However, both the variety and sophistication of the methods are increasing. It was singularly apt that Laties (1969) in his famous review should quote the Hindu proverb of the villagers trying to surmise the shape of an elephant in a dark room. Still more hands are being applied to our 'elephant' and one can only hope that, not too many of us are not clutching at the trunk and shouting 'Eureka, I found a straight-through pathway' or stroking the forehead and claiming it is described by a rectangular hyperbola.

## *Acknowledgement*

I would like to thank Dr. R.A. Leigh for helpful criticisms of the first half of this manuscript and for many stimulating discussions on this topic over the last several years.

# References


W.P. ANDERSON, D.L. HENDRIX and N. HIGINBOTHAM, Plant Physiol., 54 (1974) 712.

A.W. ATKINSON, B.E.S. GUNNING, P.C.L. JOHN and W.MCCULLOUGH, Science, 176 (1972) 694.

D.A. BAKER and J.L. HALL, New Phytol., 72 (1973) 1281.

D.A. BARBER, Annu. Rev. Plant Physiol., 19 (1968) 71.

D.A. BARBER, New Phytol., 73 (1974) 91.

D.G. BLEVINS, A.J. HIATT and R.H. LOWE, Plant Physiol., 54 (1974) 82.

D.J.F. BOWLING, J. Exp. Bot., 24 (1973) 1041.

O.G. CARTER and D.J. LATHWELL, Plant Physiol., 42 (1967) 1407.

N. CLAASSEN and S.A. BARBER, Plant Physiol., 54 (1974) 564.

D.T. CLARKSON, Ion Transport and Cell Structure in Plants (1974) McGraw-Hill, Maidenhead.

D. COLQUHOUN, Lectures on Biostatistics (1971) Clarendon Press, Oxford, p. 214.

A. CORNISH-BOWDEN and R. EISENTHAL, Biochem. J., 139 (1974) 721.

W.J. CRAM, Aust. J. Biol. Sci., 26 (1973a) 757.

W.J. CRAM, J. Exp. Bot., 24 (1973b) 328.

W.J. CRAM, in U. Zimmerman and J. Dainty (Eds.) Membrane Transport in Plants and Plant Organelles (1974a) Springer Verlag, Berlin, p. 334.

W.J. CRAM, J. Exp. Bot., 25 (1974b) 253.

W.J. CRAM and G.G. LATIES, Aust. J. Biol. Sci., 24 (1971) 633.

J. DAINTY, Annu. Rev. Plant Physiol., 13 (1962) 379.

P. DENNY and D.C. WEEKS, Ann. Bot., 34 (1970) 483.

M. DIXON and E.C. WEBB, Enzymes (1964) 2nd Edn., Academic Press, New York.

M. DRAKE in F.E. Bear (Ed.) The Chemistry of the Soil (1964) Van Nostrand Reinhold, New York, p. 395.

J. DUNLOP and D.J.F. BOWLING, J. Exp. Bot., 22 (1971) 445.

R. EISENTHAL and A. CORNISH-BOWDEN, Biochem. J., 139 (1974) 715.

O.E. ELZAM and E. EPSTEIN, Plant Physiol., 42 (1965) 620.

O.E. ELZAM and E. EPSTEIN, Agrochimica., 13 (1969) 196.

E. EPSTEIN, Mineral Nutrition of Plants : Principles and Perspectives (1972) Wiley, New York.

E. EPSTEIN, Int. Rev. Cytol., 34 (1973) 123.

E. EPSTEIN and C.E. HAGEN, Plant Physiol., 27 (1952) 457.

E. EPSTEIN and D.W. RAINS, Proc. Natl. Acad. Sci. U.S., 53 (1965) 1320.

E. EPSTEIN, D.W. RAINS and O.E. ELZAM, Proc. Natl. Acad. Sci. U.S., 49 (1963a) 684.

E. EPSTEIN, W.E. SCHMID and D.W. RAINS, Plant Cell Physiol., 4 (1963b) 79.

A. ESHEL and Y. WAISEL, Plant Physiol., 49 (1972) 585.

A. ESHEL and Y. WAISEL, Physiol. Plant., 28 (1973) 557.

J.D. FISHER, D. HANSEN and T.K. HODGES, Plant Physiol., 46 (1970) 812.

M. FRIED and H. BROESHART, The Soil–Plant System (1967) Academic Press, New York.

P.J. GARRAHAN and L.M. GLYNN, J. Physiol., 192 (1967) 175.

D.F. GERSON and R.J. POOLE, Plant Physiol., 48 (1971) 509.

D.F. GERSON and R.J. POOLE, Plant Physiol., 50 (1972) 603.

R. HELLER, C.L. CRIGNON and D. SCHNEIDECKER, in W.P. Anderson (Ed.) Ion Transport in Plants (1973) Academic Press, London, p. 337.

A.J. HIATT, Plant Physiol., 42 (1967) 294.

N. HIGINBOTHAM, Annu. Rev. Plant Physiol., 24 (1973a) 25.

N. HIGINBOTHAM, Bot. Rev., 39 (1973b) 15.

N. HIGINBOTHAM, in U. Zimmermann and J. Dainty (Eds.) Membrane Transport in Plants and Plant Organelles (1974a) Springer Verlag, Berlin, p. 348.

N. HIGINBOTHAM, Plant Physiol., 54 (1974b) 454.

N. HIGINBOTHAM and W.P. ANDERSON, Can. J. Bot., 52 (1974) 1011.

A.V. HILL, J. Physiol., 40 (1910) IV–VII.

B.S. HILL and A.E. HILL, in W.P. Anderson (Ed.) Ion Transport in Plants (1973) Academic Press, London, p. 379.

D.R. HOAGLAND and T.C. BROYER, Plant. Physiol., 11 (1936) 471.

T.K. HODGES, Adv. Agron., 25 (1973) 163.

R.H. JACKMAN, N.Z. J. Agric. Res., 8 (1965) 763.

P.C. JACKSON and H.R. ADAMS, J. Gen. Physiol., 46 (1963) 369.

L. JACOBSON and R. OVERSTREET, Am. J. Bot., 34 (1947) 415.

L. JACOBSON, R. OVERSTREET, R.M. CARLSON and J.A. CHASTAIN, Plant Physiol., 32 (1957) 658.

L. JACOBSON, R. OVERSTREET, H.M. KING and R. HANDLEY, Plant Physiol., 25 (1950) 639.

L. JACOBSON and L.C.T. YOUNG, Physiol. Plant., 25 (1971) 258.

R.L. JEFFERIES, in W.P. Anderson (Ed.) Ion Transport in Plants (1973) Academic Press, London, p. 297.

W.D. JESCHKE, Planta, 106 (1972) 73.

W.D. JESCHKE, in W.P. Anderson (Ed.) Ion Transport in Plants (1973) Academic Press, London, p. 285.

W.D. JESCHKE, in U. Zimmermann and J. Dainty (Eds.) Membrane Transport in Plants and Plant Organelles (1974) Springer Verlag, Berlin, p. 397.

J.S. KAHN and J.B. HANSON, Plant Physiol., 32 (1957) 312.

S. KANNAN, Science, 173 (1971) 927.

D.E. KOSHLAND, in P.D. Boyer (Ed.) The Enzymes (1970) 3rd Edn. Vol. 1, Academic Press, New York, p. 341.

G.G. LATIES, Annu. Rev. Plant Physiol., 20 (1969) 89.

R.A. LEIGH, Ph. D. Thesis (1974) University of Wales.

R.A. LEIGH and R.G. WYN JONES, J. Exp. Bot., 24 (1973) 787.

R.A. LEIGH and R.G. WYN JONES, J. Exp. Bot. (1975) in press.

R.A. LEIGH, R.G. WYN JONES and F.A. WILLIAMSON, in W.P. Anderson (Ed.) Ion Transport in Plants (1973) Academic Press, New York, p. 402.

R.A. LEIGH, R.G. WYN JONES and F.A. WILLIAMSON, in U. Zimmermann and J. Dainty (Eds.) Membrane Transport in Plants and Plant Organelles (1974) Springer Verlag, Berlin, p. 307.

R.T. LEONARD and J.B. HANSON, Plant Physiol., 49 (1972) 436.

R.T. LEONARD and T.K. HODGES, Plant Physiol., 52 (1973) 6.

W. LIN and J.B. HANSON, Plant Physiol., 54 (1974a) 250.

W. LIN and J.B. HANSON, Plant Physiol., 54 (1974b) 799.

A.E.S. MACKLON and N. HIGINBOTHAM, Plant Physiol., 43 (1968) 888.

E.A.C. MACROBBIE, Annu. Rev. Plant Physiol., 22 (1971) 75.

PH. MATILE, Planta, 79 (1968) 181.

S.M. MERTZ and N. HIGINBOTHAM, in U. Zimmermann and J. Dainty (Eds.) Membrane Transport in Plants and Plant Organelles (1974) Springer Verlag, Berlin, p. 343.

P. MITCHELL, FEBS Lett., 33 (1973) 267.

P. NISSEN, Physiol. Plant., 24 (1971) 315.

P. NISSEN, Annu. Rev. Plant Physiol., 25 (1974a) 53.

P. NISSEN, in U. Zimmermann and J. Dainty (Eds.) Membrane Transport in Plants and Plant Organelles (1974b) Springer Verlag, Berlin, p. 348.

P.S. NOBEL, Introduction to Biophysical Plant Physiology (1974) Freeman, San Francisco.

J.J. OERTLI, Physiol. Plant., 20 (1967) 1014.

W.J.Y. OSTERHOUT, Biol. Bull., 99 (1950) 308.

R. OVERSTREET and L. JACOBSON, Am. J. Bot., 33 (1946) 107.

C.K. PALLAGHY, U. LÜTTGE and K. VON WILLERT, Z. Pflanzenphysiol., 62 (1970) 51.

W.S. PIERCE and N. HIGINBOTHAM, Plant Physiol., 46 (1970) 666.

M.G. PITMAN, Aust. J. Biol. Sci., 16 (1963) 647.

M.G. PITMAN, Plant Physiol., 45 (1970) 787.

M.G. PITMAN, Aust. J. Biol. Sci., 24 (1971) 407.

M.G. PITMAN, S.M. MERTZ, J.S. GRAVES, W.S. PIERCE and N. HIGINBOTHAM, Plant Physiol., 47 (1971) 76.

M.G. PITMAN and H.D.W. SADDLER, Proc. Natl. Acad. Sci. U.S., 57 (1967) 44.

E.O. POLLE and H. JENNY, Physiol. Plant., 25 (1971) 219.

R.J. POOLE, Can. J. Bot., 52 (1974) 1023.

D.W. RAINS and E. EPSTEIN, Plant Physiol., 42 (1967) 314.

A.W. ROBARDS and M.E. ROBB, Planta, 120 (1974) 1.

R.B. ROBERTS, I.Z. ROBERTS and D.B. COWIE, J. Cell Comp. Physiol., 34 (1949) 259.

J.D. ROBINSON, Biochemistry, 6 (1967) 3250.

H.D.W. SADDLER, J. Gen. Physiol., 55 (1970) 802.

A. SCHWARTZ, G.E. LINDEMAYER, J.C. ALLEN, in F. Bronner and A. Kleinzeller (Eds.) Current Topics in Membranes and Transport (1972) Academic Press, New York, p. 1.

B.I.H. SCOTT, H. GULLINE and C.K. PALLAGHY, Aust. J. Biol. Sci., 21 (1968) 185.

S. SINGER, Annu. Rev. Biochem., 43 (1974) 805.

J.C. SKOU, Physiol. Rev., 45 (1965) 596.

C.L. SLAYMAN, in U. Zimmermann and J. Dainty (Eds.) Membrane Transport in Plants and Plant Organelles (1974) Springer Verlag, Berlin, p. 107.

M. THELLIER, in W.P. Anderson (Ed.) Ion Transport in Plants (1973) Academic Press, London, p. 47.

K. TORII and G.G. LATIES, Plant Physiol., 41 (1966a) 863.

K. TORII and G.G. LATIES, Plant Cell Physiol., 7 (1966b) 395.

A. ULRICH, Am. J. Bot., 28 (1941) 526.

T.H. VAN DEN HONERT and J.J.M. HOOYMANS, Acta Bot. Neerl., 4 (1955) 376.

N.A. WALKER, in U. Zimmermann and J. Dainty (Eds.) Membrane Transport in Plants and Plant Organelles (1974) Springer Verlag, Berlin, p. 173.

J. WEIGL, Planta, 79 (1968) 197.

J. WEIGL, Planta, 98 (1971) 315.

R.M. WELSH and E. EPSTEIN, Plant Physiol., 44 (1969) 301.

L. WINKLANDER, in F.E. Bear (Ed.) The Chemistry of the Soil (1964) Van Nostrand Reinhold, New York, p. 163.

R.G. WYN JONES and O.R. LUNT, Bot. Rev., 33 (1967) 407.

*Ion transport in plant cells and tissues*
*edited by D.A. Baker and J.L. Hall*
© *North-Holland Publishing Company, 1975*



# CHAPTER 8

# Long distance transport in roots

W.P. ANDERSON

## Contents



## *8.1 Introduction*

This chapter will cover the transport of the major nutrient ions in the higher plants from the soil solution to the root tissue and hence through the xylem to the shoot. The mechanism driving water along the same pathway will also be discussed, but in much less detail, and only in so far as it is required to understand the movement of ions.

In point of fact, very few species have been commonly used for investigative study of ion transport, whether for examination of ion accumulation into the vacuoles of root cells (see Ch. 7) or for exploring the mechanism of ion through-

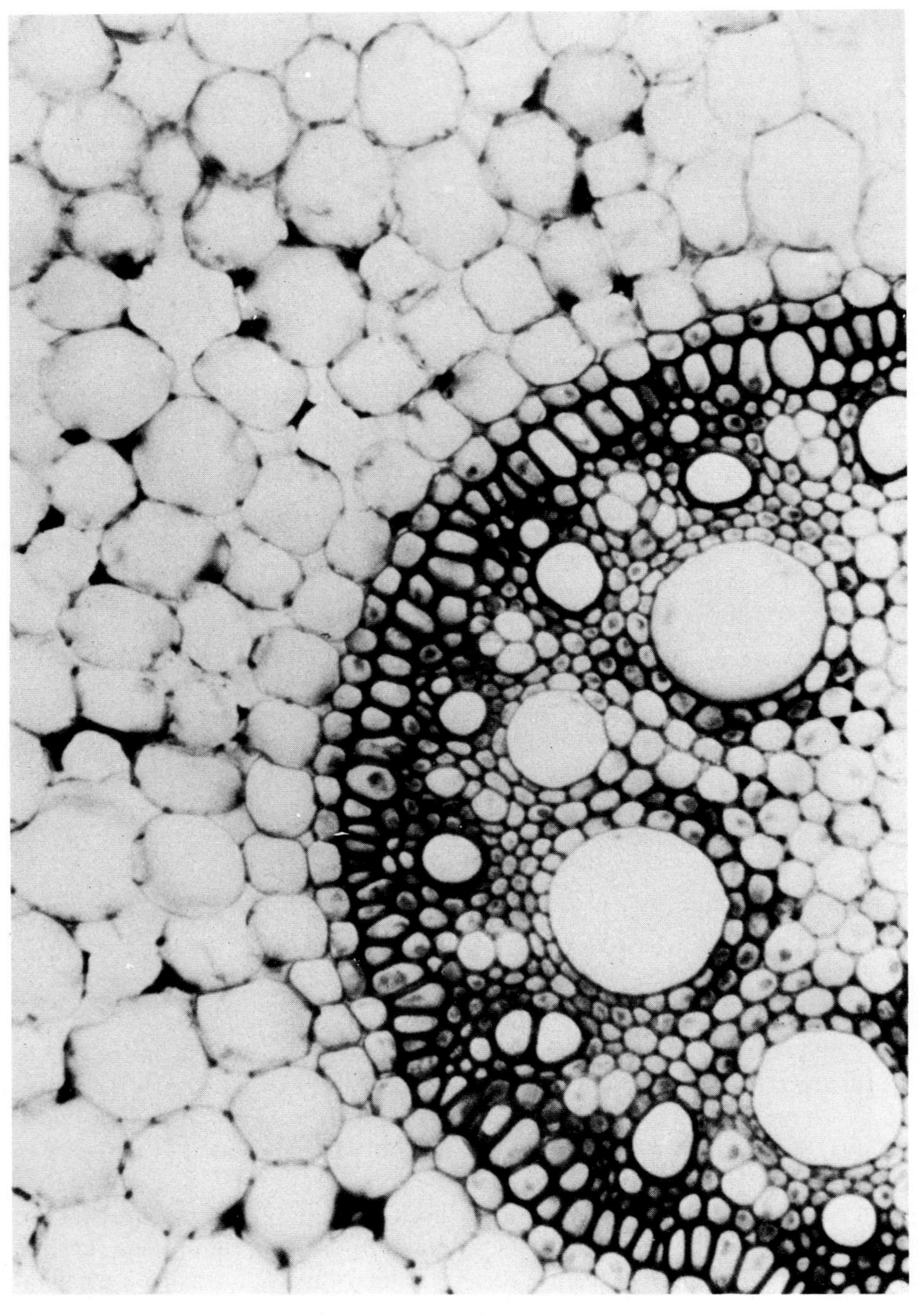

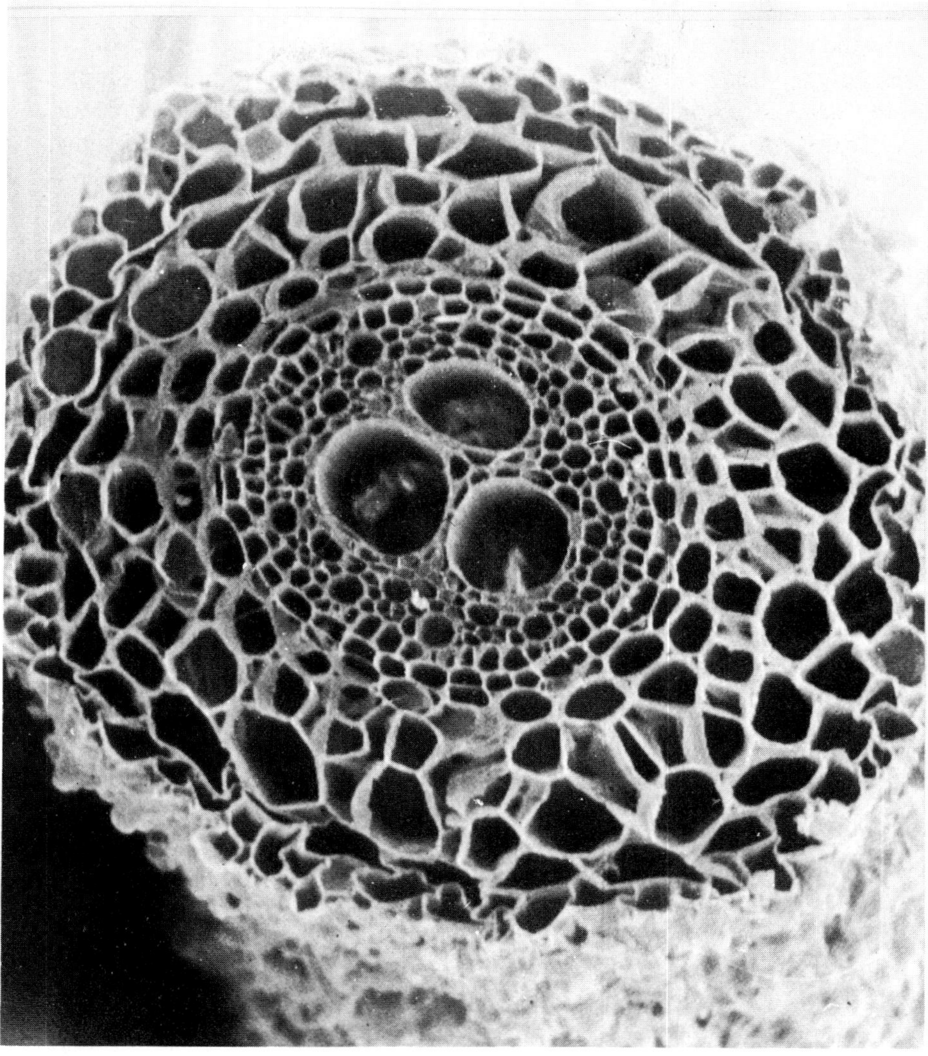

Fig. 8.1. (Left hand page) Cross-section of a *Zea mays* primary root. The cortex and stele are plainly visible, separated by the endodermis and pericycle. The large, central metaxylem vessels are thought to take no part in xylem transport at this stage. (Right hand page) Scanning electron micrograph of critical point dried barley root clearly showing the stele and cortex. Provided by Dr. A.W. Robards, Dep. of Biology, University of York.

put from the soil solution to the xylem translocation stream. Most of the reported experiments have been on the common cereals, maize, wheat, barley and oats, and to a somewhat lesser extent on pea, bean, tomato, sunflower, onion and castor-oil plants. In what follows here, the details will be drawn for the most part from work on these species, with the implication throughout that similar conclusions will apply to the root systems of all 'typical' terrestrial higher plants.

Much of the most informative work on ion transport through the root to the xylem has been done on excised root systems which exude sap from the xylem vessels under root pressure. This has been a much used experimental system since the earliest days (e.g. Priestley, 1920) and the basic mechanism of xylem exudation from excised roots has long been recognised (Priestley, 1920; Arisz et al., 1951). More recently, root pressure exudation has been described in an irreversible thermodynamic formalism (House and Findlay, 1966) and the possibilities of there being two osmotic barriers in series between the external solution and the xylem (Ginsburg, 1971) and of longitudinal variation in the transport parameters (Anderson et al., 1970) have been considered.

The long distance transport function of the root can only be understood if the anatomical structure is known. The overall secretion phenomenon of the xylem is the result of the co-ordinated operation of all the various cell types. Again, the anatomical details given below are drawn mainly from work on cereal roots, but the main features and the major comments will apply to many higher plant species.


## 8.2 *Anatomy of roots*

Fig. 8.1 shows cross-sections of two cereal roots in the young, absorbing zone of the root, which seems to extend for at least 100 mm from the root tip in hydroponically grown seedlings. There are two distinct regions in the root tissue, the cortex and the stele. At the outer surface the cortex is bounded by a sheath of epidermal cells which may develop root hairs depending upon culture conditions. At the inner surface the cortex is delimited from the stele as a single layer of cells, the endodermis. These features can be distinguished at an early stage in root ontogeny, usually within a millimetre or so from the meristem but the exact distance depends on species and growth conditions. At this stage the Casparian band forms in the endodermis; this is a suberised strip in the endodermal cell wall which completely surrounds each cell and is in contact between cells so as to form a continuous barrier in the apoplasm between the cortex and the stele. The function of this Casparian band has been much debated, but it seems pertinent to point out that similar structures are found surrounding other secretory tissues, e.g. salt glands (see Ch. 11) and the nitrogen fixing nodules on legume roots (Gunning et al., 1974). These coincidences may be taken as prima facie evidence that the Casparian bands in roots and the suberised bands in other tissues have a common function of isolating one region of apoplasmic space from another. This is, of course, the role which has been classically assumed (Crafts and Broyer, 1938; Van Fleet, 1961).

At maturity, the epidermal, cortical and endodermal cells of roots are highly vacuolated and in some cereal species the cortical cell vacuoles may comprise 90% of the cortical cell total volume, or 80% of the total root tissue volume. The outer surface of the epidermis may be cuticularised to varying degrees depending on culture conditions. In young, hydroponically grown roots there is essentially no cuticle on the epidermis.

The epidermal and cortical cells of the root are interconnected in most species by plasmodesmata, bridges across the cell walls connecting the cytoplasm of neighbouring cells, to form a symplasm. Perhaps the most complete study of plasmodesmatal structure is by Robards (1971) from whom Fig. 8.2 is taken. It is probably fair to say that there is general agreement that the structure in a large variety of species is similar to that shown here, although there may be dissent as to whether the tubular inclusion (the desmotubule) is or is not simply a strand of endoplasmic reticulum. All are agreed that there is continuity of both plasmalemma and cytoplasm from one cell to the next through the plasmodesmata.

The significance of plasmodesmata for transport across the root will be discussed in a later section of this chapter. In this connexion, the frequency of plasmodesmata is obviously of equal importance as the pore size of a single plasmodesm and Tyree (1970) has performed a useful service in searching the literature and gathering together the relevant data, which has been reproduced in Table 8.1. The fine work on plasmodesmatal frequency in the tertiary endodermis of barley by Clarkson et al. (1971) will be referred to in detail later.

In contrast to the cortex which contains only a single cell type, the stele has several distinct cell types within it. There are the pericycle cells, xylem and phloem parenchyma, xylem elements, phloem elements and, in many species, a central core of pith. In the present context, which attempts to correlate structure and function, it is the differentiation of the xylem and perhaps the xylem parenchyma which must be focused upon. The opinion of most anatomists is that the early metaxylem, through which the exudation stream

*W.P. Anderson*

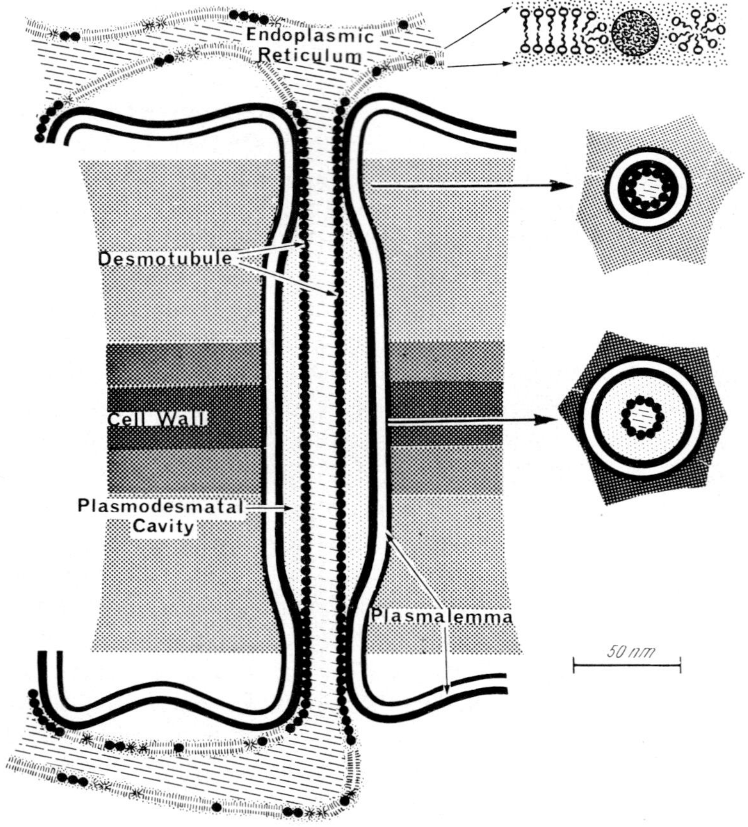


Fig. 8.2. Diagram of plasmodesmal structure. Reproduced from Robards (1971).

can be seen to flow in roots excised from young maize seedlings (e.g. by simply looking at the cut end under a microscope), is fully mature within a few millimetres from the meristem. The implications of full maturity in metaxylem are that the vessels have completed secondary wall deposition, have fully perforated end walls and no longer contain viable cytoplasm; in other words, the present consensus is that xylem translocation is conducted along dead, open vessels which play no part in that translocation other than to provide an open conduit of relatively low hydraulic resistance.

However, there does seem to be marked species variation; several reports (Scott, 1949, 1965; Anderson and House, 1967; Burley et al., 1970; Higinbotham et al., 1973) in *Ricinus communis* and *Zea mays* have suggested the presence of (at least some) cytoplasm-containing early metaxylem vessels in

Table 8.1.
Sizes and frequencies of plasmodesmata in cell walls.

| Plant | Plasmodesmal radius ($\mu$m) | Plasmodesmatal frequency ($N\ m^{-2} \times 10^{-12}$) |
|---|---|---|
| *Laminaria digitata* | | |
| trumpet cell cross wall | $3 \times 10^{-2}$ | 50–60 |
| *Tamarix aphyla* | | |
| salt gland | $3-4 \times 10^{-2}$ | 2 |
| *Allium cepa* | | |
| root meristem | $4-5 \times 10^{-2}$ | 6–7 |
| root cortex | $4-7 \times 10^{-2}$ | 1.5 |
| *Viscum album* | | |
| root cortex | $2-3 \times 10^{-2}$ | 2 |
| ectodesm | $1.5 \times 10^{-2}$ | – |
| *Avena sativa* | | |
| coleoptile cortex | $3-5 \times 10^{-2}$ | 3.6 |
| *Lycopersicum esculentum* | | |
| pollen mother cells | $1-9 \times 10^{-2}$ | – |
| *Salix fragilis* | | |
| cambium | $1.2 \times 10^{-2}$ | – |
| *Nitella translucens* | | |
| internode | $3.5 \times 10^{-2}$ | 3 |
| *Zea mays* | | |
| root cap | $2.1 \times 10^{-2}$ | 5 |
| stele | $2.1 \times 10^{-2}$ | 1.4 |

regions ranging up to 200 mm from the root tip. On the other hand, Läuchli (to be published) can find no cytoplasm-containing vessels in barley roots beyond 2–3 mm from the root tip.

The parenchyma cells of the stele are usually highly vacuolated and show the normal complement of mitochondria. In maize one frequently finds densely cytoplasmic parenchyma cells immediately adjacent to the metaxylem vessels and it might be conjectured that these are engaged in some active secretion of ions into the vessels. There is no evidence of transfer cells adjacent to xylem vessels in the roots of a large number of species (Gunning, private communication), which might argue against secretory activity by the xylem parenchyma if it is accepted that the characteristic wall in-growths of transfer cells are associated with active absorptive or secretory activity. The parenchyma of the stele are interconnected by plasmodesmata in much the manner as are the cortical cells, and the symplasm is continuous from the cortex, through the endodermis to the stele.

## 8.3 Root pressure exudation

The technique of root pressure exudation experiments is very straightforward. As is shown in Fig. 8.3, the roots or root systems are cleanly excised from the

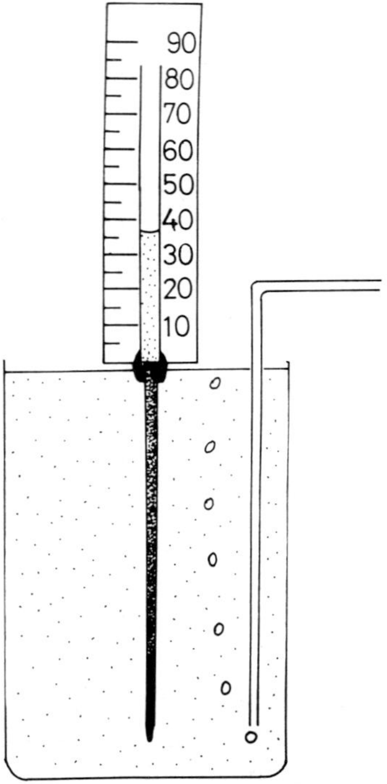


Fig. 8.3. Diagram of common method of observing excised root exudation. The excised root is sealed into a uniform bore capillary tube which is mounted on a graduated scale.

plant and the cut ends are inserted and sealed into some suitable collecting tube. The roots are bathed in aerated solution of known composition and the rate of exudation is continuously monitored by measuring the fluid column height in the collecting tube; at appropriate times the exuded xylem fluid may be removed for chemical and radioactive assay.

It is quite apparent that a root system cannot survive indefinitely after being detached from the plant and the question immediately arises as to how serious an alteration is caused to the behaviour of the root by excision. The most obvious change is that the root no longer receives a supply of photosynthate (or carbohydrate from the seed reserves in the case of young seedlings) to provide substrates for its energy metabolism; however, observations by several workers on a number of species seem to indicate that there is little diminution in respiration in excised roots for the first 24 h after excision, so that it must be assumed that the root tissue of a healthy plant has stored sufficient metabolic substrates to continue respiring normally for at least this length of time, which encompasses the usual experimental period in excised root transport experiments. Perhaps a more important and certainly more immediate effect of root excision is the disruption in the normal hormonal gradients within the plant. There is an increasing body of evidence (see later) that various plant hormones and growth regulators have marked effects on transport, both ion accumulation into root cells and long distance movement of ions through the root from the external solution to the xylem stream. Therefore these effects have to be kept in mind as possible shortcomings to the results obtained from excised root studies.

Table 8.2 contains a collection of data from several sources on the results obtained from excised root exudation experiments. Note that there is a quite obvious accumulation of the major nutrient ions from the external solution into the xylem sap. The ionic composition of the sap may be assayed by standard chemical methods on samples taken from the collection tubes at appropriate times. The reader will find details of the analytical methods in the literature citations accompanying Table 8.2.

From the primary experimental data of the rate of xylem exudate production (the exudation volume flux) and the concentration of any ion in the xylem sap, the ion flux through the xylem may be calculated as

$$J_s = J_v \cdot C_s \tag{8.1}$$

where $J_v$ is the exudation volume flux (normalised over unit area of root surface or over unit fresh weight of root for highly branched root systems) and $C_s$ is the concentration of the ion 's' in the exuded xylem sap. The quantity $J_s$ is, quite unequivocally, the ion flux emerging at the cut end of the root, but there is no necessity that the flux of ion 's' in the xylem at some other plane within the root is identical to $J_s$; there may be transport of ions either from or to the xylem stream at any point within the root, so that the ion flux translocated along the xylem may vary as one progresses along the root. An equation can be set up to describe this situation (see later). Further, there is no implication that the ion flux exuded from the xylem is exactly balanced by ion uptake into the root tissue from the external solution. Several workers (e.g. Hodges and Vaadia, 1964; Jarvis and House, 1970; Anderson et al., 1971) have produced evidence to suggest that the ion supply to the xylem stream of excised roots may, under certain circumstances, be either partly or wholly

Table 8.2.
Exudation data from excised root preparations. See Table 8.3 for composition of solution X.

| Material/preparation | Water flux | Exudate concn | | Exudate ion flux | |
|---|---|---|---|---|---|
| *Zea mays* | | | | | |
| in   1 Mol m$^{-3}$ KCl*(1) | 7.22[a] | 20.4[b] | 25.1[c] | 147.3[d] | 181.2[e] |
| 10 Mol m$^{-3}$ KCl*(1) | 6.92[a] | 28.4[b] | 30.7[c] | 196.5[d] | 212.4[e] |
| 50 Mol m$^{-3}$ KCl*(1) | 3.60[a] | 54.3[b] | 57.6[c] | 195.5[d] | 207.4[e] |
| 0.1 Mol m$^{-3}$ K$_2$SO$_4$**(2) | 1.94[a] | 13.8[b] | 2.9[f] | 26.7[d] | 5.6[g] |
| 10 Mol m$^{-3}$ K$_2$SO$_4$**(2) | 1.66[a] | 25.2[b] | 6.2[f] | 41.8[d] | 10.3[g] |
| in 0.1X solution (3) | – | 14.7[b] | 0.8[h] | | |
| | | 1.4[j] | 5.6[k] | | |
| in 1X solution (3) | – | 19.4[b] | 1.2[h] | | |
| | | 3.3[j] | 1.4[m] | | |
| | | 9.7[k] | 3.1[c] | | |
| *Ricinus communis* | | | | | |
| in 1 Mol m$^{-3}$ KNO$_3$*(4) | 6.76[n] | 12.2[b] | | 82.4[p] | |
| 1 Mol m$^{-3}$ KNO$_3$***(4) | 4.72[n] | 7.5[b] | | 35.4[p] | |
| 5 Mol m$^{-3}$ KNO$_3$*(4) | 4.33[n] | 15.6[b] | | 67.5[p] | |
| 5 Mol m$^{-3}$ KNO$_3$***(4) | 3.39[n] | 16.4[b] | | 55.6[p] | |
| *Sinapis alba* | | | | | |
| in   1 Mol m$^{-3}$ KCl*(5) | 0.2[q] | 4.4[b] | 5.1[c] | | |
| | | 0.6[j] | | | |
| in 10 Mol m$^{-3}$ KCl*(5) | 0.16[q] | 12.4[b] | 13.2[c] | | |
| | | 1.3[j] | | | |
| *Triticum aestivum* | | | | | |
| in distilled water (6) | 0.56[q] | 22.1[k] | | | |
| *Avena fatua* | | | | | |
| in   1 Mol m$^{-3}$ KCl*(7) | 1.66[a] | 5.9[b] | | 9.8[d] | |
| in 10 Mol m$^{-3}$ KCl*(7) | 1.11[a] | 10.4[b] | | 11.5[d] | |
| in 50 Mol m$^{-3}$ KCl*(7) | 0.55[a] | 63.7[b] | | 35.3[d] | |
| *Avena sativa* | | | | | |
| in   1 Mol m$^{-3}$ KCl*(7) | 3.32[a] | 5.0[b] | | 16.6[d] | |
| in 10 Mol m$^{-3}$ KCl*(7) | 1.94[a] | 8.4[b] | | 16.3[d] | |
| in 50 Mol m$^{-3}$ KCl*(7) | 1.11[a] | 45.3[b] | | 50.2[d] | |
| *Helianthus annuus* | | | | | |
| in complete nutrient solution (8) | 33.3[q] | 0.5[f] | | | |
| *Allium cepa* | | | | | |
| in   1 Mol m$^{-3}$ KCl*(9) | 3.6[a] | 12.5[b] | 5.6[c] | 45.0[d] | 20.2[e] |
| in 10 Mol m$^{-3}$ KCl*(9) | 3.3[a] | 12.3[b] | 9.1[c] | 40.6[d] | 30.0[e] |

Superscripts are as follows: (a) $\times 10^{-9}$ m$^3$ m$^{-2}$ s$^{-1}$; (b) Mol m$^{-3}$ K$^+$; (c) Mol m$^{-3}$ Cl$^-$; (d) nMol K$^+$ m$^{-2}$ s$^{-1}$; (e) nMol Cl$^-$ m$^{-2}$ s$^{-1}$; (f) Mol m$^{-3}$ SO$_4^{2-}$; (g) nMol SO$_4^{2-}$ m$^{-2}$ s$^{-1}$; (h) Mol m$^{-3}$ Na$^+$; (j) Mol m$^{-3}$ Ca$^{2+}$; (k) Mol m$^{-3}$ NO$_3^-$; (m) Mol m$^{-3}$ Mg$^{2+}$; (n) $\times 10^{-9}$ m$^3$ gFW$^{-1}$ s$^{-1}$; (p) nMol K$^+$ gFW$^{-1}$ s$^{-1}$; (q) $\times 10^{-12}$ m$^3$ plant$^{-1}$ s$^{-1}$; * 0.1 mM CaCl$_2$ added; ** 0.1 Mol m$^{-3}$ CaSO$_4$ added; *** 2.5 Mol m$^{-3}$ CaCl$_2$ added. The source references are: (1) House and Findlay (1966); (2) Anderson and Collins (1969); (3) Davis and Higinbotham (1969); (4) Minchin and Baker (1973); (5) Collins and Kerrigan (1973); (6) Lundegårdh (1953); (7) Collins, J.C. (personal communication); (8) Petterson (1966); Hay and Anderson (1972).

derived from the endogenous reservoirs within the root tissue. Presumably the cortical cell vacuoles, which constitute by far the largest volume fraction of the root, are the chief ion reserves in question.

## 8.4 Electrophysiology of roots

There are two aspects of root electrophysiology which have been commonly studied, one being the so-called exudate potential which can be measured between the xylem exudate of an excised root and the bathing solution, and the other being the electrophysiology of the individual cells of the root tissue. These studies are usually combined and proceed simultaneously in various laboratories.

Table 8.3 contains some of the experimental data on exudate potentials

Table 8.3.
Xylem exudate electropotentials of roots.

| Material and conditions | Exudate potential (mV) (exudate negative) |
|---|---|
| *Zea mays* | |
| in   0.1 Mol m$^{-3}$ KCl (1) | 50 |
| 0.3 Mol m$^{-3}$ KCl (1) | 48 |
| 1.0 Mol m$^{-3}$ KCl (1) | 30 |
| 3.0 Mol m$^{-3}$ KCl (1) | 25 |
| 10.0 Mol m$^{-3}$ KCl (1) | 18 |
| *Zea mays* | |
| in 0.2 Mol m$^{-3}$ KCl (2) | 58 |
| 0.2 Mol m$^{-3}$ KCl with 10$^{-5}$ DNP added (2) | 32 |
| *Zea mays* | |
| in 1X solution (3) | 49 |
| 1X solution with 1 Mol m$^{-3}$ CN added (3) | 32 |
| *Ricinus communis* | |
| in 1/10 Stout and Arnon's solution (4) | 58 |
| *Helianthus annuus* | |
| in 1/10 culture solution (5) | 48 |
| in full strength culture solution | 34 |

The source references are: (1) Dunlop and Bowling (1971); (2) Shone (1969); (3) Davis and Higinbotham (1969); (4) Bowling and Spanswick (1964); (5) Bowling (1966). The solution described as 1X has the following composition: 1 Mol m$^{-3}$ KCl; 1 Mol m$^{-3}$ Ca(NO$_3$)$_2$; 0.25 Mol m$^{-3}$ MgSO$_4$; NaH$_2$PO$_4$ and Na$_2$HPO$_4$ to give 1 Mol m$^{-3}$ Na$^+$ and a pH of 5.5 to 5.7

which have been obtained in a variety of species. The exudate potential is very easily measured by placing one reversible electrode (e.g. a $3 \times 10^3$ Mol m$^{-3}$ KCl salt bridge to a Ag/AgCl electrode) in the xylem exudate and a second similar electrode in the bathing solution, and recording between them with a potential measuring device of suitably high input impedance. Note that the exudate potentials in all the species given in Table 8.2 are negative with respect to the bathing solution and have values of several tens of millivolts. In general the exudate potential behaves partly as if it were a diffusion potential which can be described by a suitable form of the Goldman equation (see Ch. 1; Dainty, 1962) and partly as if it were the result of electrogenic ion pumping from the bathing solution to the xylem stream.

This is very clearly shown in the work of Davis and Higinbotham (1969) and of Shone (1969), working in each case on the excised primary roots of several days old seedlings of *Zea mays*. Both laboratories found the exudate potential was depolarised by increasing external ion concentrations, as one would expect for a diffusion potential, with the greatest effect being given by alterations in the most permeable ion, K$^+$. However, upon poisoning the root with either DNP or CN$^-$, the exudate potential was found to decrease rapidly, more rapidly than one would expect for a purely diffusion potential. The only effect of metabolic inhibition on a diffusion potential, assuming it causes no alteration to membrane ionic permeabilities, is to reduce the internal ionic concentrations by leakage (and hence by the Goldman equation the potential also is decreased), a process requiring several hours rather than the observed several minutes taken for the exudate potential to depolarise.

The suggestion from this work by Davis and Higinbotham (1969) and Shone (1969) was taken further by Higinbotham et al. (1970) who proposed that the membrane potential of a higher plant cell, as well as the exudate potential of an excised root, was a linear combination of a diffusion potential across the membrane as described by a Goldman equation, and an electrogenic ion pump potential, as given by

$$E_M = E_{eq} + E_x \qquad (8.2)$$

where $E_M$ is the membrane potential, $E_{eq}$ is the Goldman (diffusion) potential and $E_x$ is the electrogenic pump potential. More recently Higinbotham and Anderson (1974) have reviewed the evidence for this point of view and have produced new data of very rapid depolarisations of higher plant cell membrane potentials following CO inhibition of respiration. Using this inhibitory technique, Anderson et al. (1974) have recorded depolarisation times of several seconds on both root and shoot cells of *Zea* and *Pisum*; these depolarisations are completely reversible upon release of inhibition and a cell can be taken through many, rapid depolarisation–repolarisation cycles (Fig. 8.4).

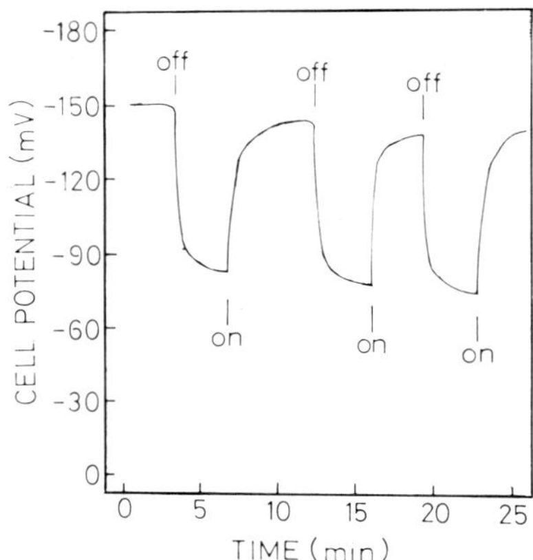


Fig. 8.4. Depolarisation–repolarisation cycles in the cell potential of a pea root epidermal cell. Resporatory inhibition is by CO attachment to haem iron in the dark, while in the light, normal $O_2$ attachment proceeds.

Similarly rapid depolarisations in the membrane potential of *Neurospora crassa* upon respiratory inhibition have been reported by Slayman (1965), who also interprets his results in terms of the observed membrane potential being comprised of two additive components, one diffusional and the other an electrogenic pump component. In the green alga *Nitella*, there is also evidence that the membrane potential is partly due to the action of an electrogenic pump (Kitasato, 1968; Spanswick, 1973) which in this case may be powered rather directly by the photosynthetic reactions (see Ch. 5). Thus, there seems to be a growing consensus that the membrane potential of plant cells widely distributed throughout the plant kingdom is composed of an addition of a diffusional and an electrogenic pump potential.

Very briefly, the distinction between an electrogenic ion pump and a neutral ion pump is as follows: a neutral ion pump is one which operates across the membrane in such a manner that on average, zero net electric charge is transferred. (Examples would be a salt pump, carrying KCl as such across the membrane, or an obligately $1:1$ coupled exchange pump, where for every $K^+$ taken inward, a $Na^+$ must be taken outward.) An electrogenic ion pump is a membrane carrier mechanism where there is no restriction of charge neutrality on its operation. (The most likely example is an exchange pump with stoichiometry other than 1 : 1; indeed it seems highly likely that one of the fine controls on ionic regulation in cells is that the ion pumps have variable stoichiometry.) The simplest way to imagine the effect of an electrogenic pump is to think of it as driving an electric current (the ion flux) through a resistor (the membrane), so that by Ohm's law a potential difference will be caused across the resistor (membrane). If such a pump stops, the potential so produced will decay to zero immediately (strictly, as soon as the membrane capacitance discharges, which for our present discussion is immediately). The cessation of a neutral pump will have no immediate effect on the observed membrane potential because no current is carried on such a pump and the membrane potential will only decay at the same rate as the internal ion concentrations leak away.

Studies on the electrophysiology of root cells as such have been in progress for many years. Early work (e.g. Etherton and Higinbotham, 1960; Etherton, 1963) was chiefly concerned with thermodynamic characterisation of the ionic situation, attempting to establish which ions were passively distributed across the membranes and which were maintained at their recorded internal concentrations by the action of active carriers. Note that for ions it is not sufficient merely to measure the concentration gradients across the membrane; ions are also driven by the electro-potential gradient (see Ch. 1). More recently work has turned to investigations of the mechanism of electrogenesis, with the results outlined above, and to establishing the mechanisms by which the root tissue as an organ accumulates ions from the bathing solution and transports them inward to the xylem stream. Bowling (1973) has published evidence of the cell potential profile across a sunflower root, reproduced here as Fig. 8.5. Although the reported values for the cell potentials shown here seem to be very low, and indeed the author has found considerable difficulty in obtaining reliable cell potential measurements beyond the first layer of cortical cells in roots because of the build up of debris on the microelectrode tip, at face value these data of Bowling's show that the root symplasm is an equipotential network, with the cells being essentially short-circuited one to another by the plasmodesmata. This idea is supported by the elegant resistance measurements made by Spanswick (1972) on maize root cortex. He was able to insert electrodes into adjacent cortical cells and by pulsing current from one to the other, to measure the cell-to-cell resistance. The voltage attenuation from one cell to the next was found to be 0.24 in maize cortex, while the attenuation within the cytoplasm of a single cell was negligible. The conclusion is therefore that the symplasm is a low resistance pathway for radial ion movement across the root and that what resistance there is occurs at the plasmodesmata between cells.

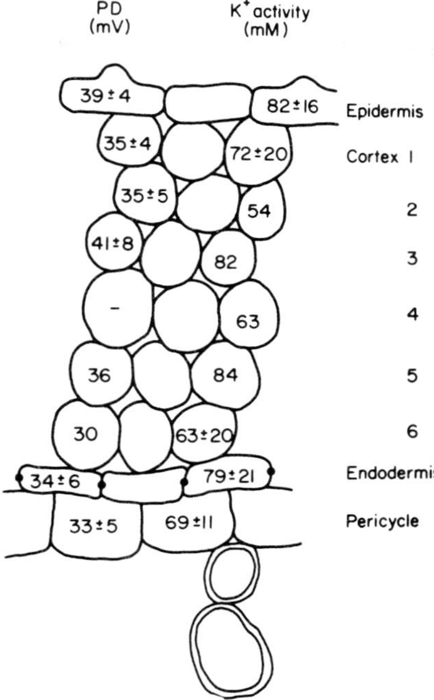



Fig. 8.5. Sketch of a radial section through a *Zea mays* primary root, detailing the cell electro-potentials and the vacuolar $K^+$ activities as obtained with a $K^+$-sensitive microelectrode. Reproduced from Bowling (1973a).

## 8.5 *Microscopic techniques*

Microscopic and ultrastructural investigations of the transport function of roots are pursued by three principal methods, autoradiography, histochemical studies and ion localisation either by precipitation techniques, electron microprobe analysis, or a combination of these two. Autoradiography was a very popular tool a few years ago but now seems to have been largely superseded by ion localisation methods which involve much less tissue preparation and are capable of higher resolution. Further, there are only a limited number of isotopes of the ions of interest which are suitably weak $\beta$-emitters to allow reasonable autoradiographic work.

Histochemical studies, particularly localisation of acid phosphatase and ATPase activity have recently been much developed (see Ch. 2 and Hall (1973)). In roots, the work of Leigh et al. (1973) has shown that phosphatase activity is associated with the cell membranes, the ER cisternae and with certain small cytoplasmic vesicles which may act, it is thought, as pinocytotic ion transport systems. One of the problems is that many membranes have associated phosphatase activity, and it is clear that the staining and bio-chemical techniques employed will demonstrate enzymes other than those involved in ion transport. Further discussion is rather outside the author's expertise and is probably redundant in view of the coverage mentioned above. There is of course no doubt that ATPase activity is associated with plant cell membranes. For example, Hodges et al. (1972) have shown the presence of a salt-stimulated ATPase activity in a fraction enriched in plasma membranes of *Avena sativa* root cells, and Kylin (1973) has reviewed his work on similar ion-stimulated ATPase activity in a variety of species (see Chs 2 and 7).

Ion localisation studies are presently perhaps the most relevant microscopic investigation which bears directly on the transport function of roots. Pre-cipitation techniques for localisation, as the name suggests, involves the rapid formation of an insoluble salt within the tissue, so that a normally soluble ion is fixed in position; e.g. $Cl^-$ is precipitated by treatment with $Ag^+$, or $Na^+$ may be localised by precipitation with pyroantimonate. The precipitates should always be checked by electron microprobe analysis, to ensure the specificity of the localisation reaction. Pyroantimonate, at one time thought to be specific for $Na^+$, is now known to also form precipitates with $Mg^{2+}$ and $Ca^{2+}$ (Tandler et al., 1970); the $Cl^-$ reaction with $Ag^+$ seems to be quite specific in normal plant tissue, but a similar reaction is produced with other halides if the tissue has been pretreated.

Electron microprobe analysis involves the micro-bombardment of a region of tissue with a finely focused electron beam, resulting in the production of characteristic X-ray spectral lines due to excitations in the inner electron shells of the atoms present. The wavelengths (or energies) of these various characteristic line spectra are well documented so that the various atoms present in the tissue may be identified and estimates of the quantities of each may be made. A recent review of this technique, particularly as it is applied to plant specimens has been given by Läuchli (1973). As was indicated above, this technique is often used in conjunction with precipitation localisation to identify the nature of the precipitate.

Fig. 8.6 shows a precipitate of AgCl in the plasmodesmata between the innermost cortical cell and the endodermis of a barley root, the implication being that we have here a visualisation of the normal pathway of $Cl^-$ move-ment from one cell to the next. Läuchli and a variety of co-workers have conducted an extensive series of studies using these techniques on plant roots (Läuchli, 1967; Läuchli et al., 1971; Läuchli et al., 1974a, b). In summary, the evidence seems to be that the Casparian band on the root endodermis effect-

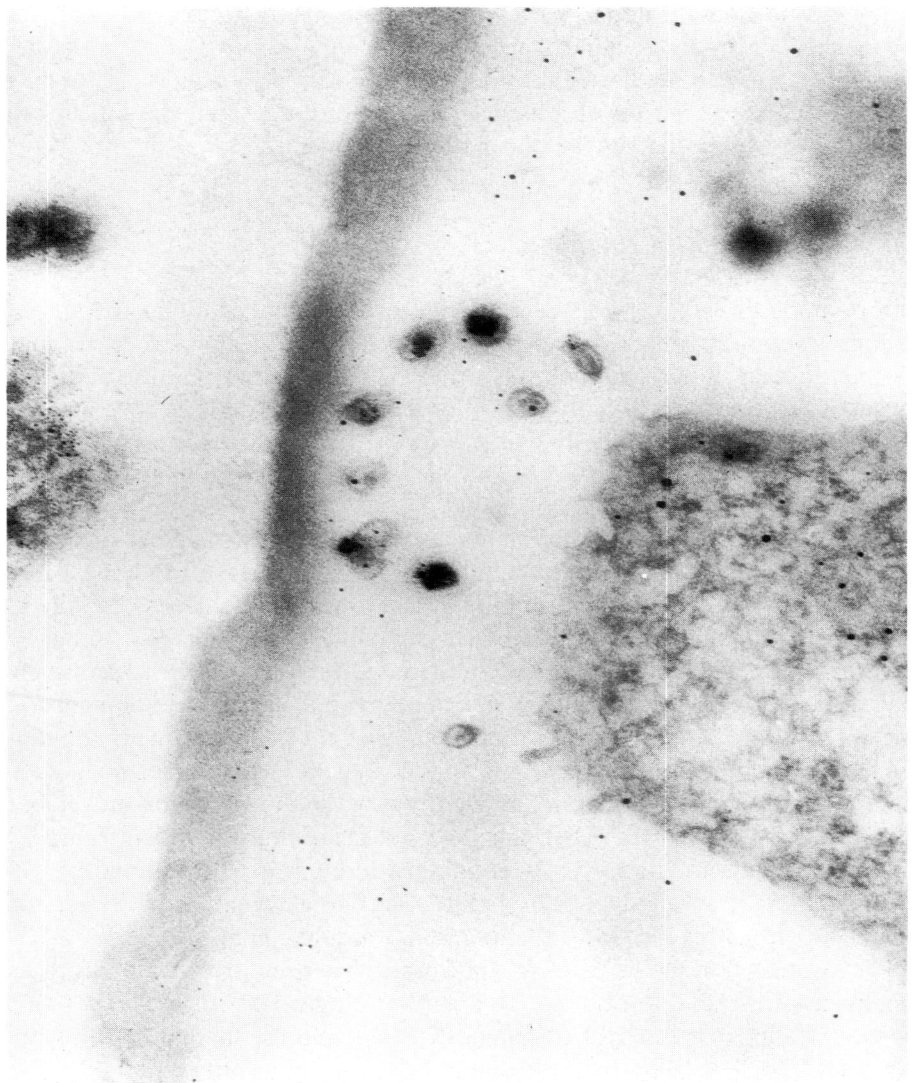

Fig. 8.6 An example of Cl⁻ localisation in barley roots, 10 mm from tip, showing plasmodes-mata in the cell wall between the endodermis and stele, in cross-section. The dark deposits in the plasmodesmata are AgCl. Magnification × 55 000. Photograph kindly supplied by Dr. A. Läuchli.

ively stops all radial apoplasmic ion movement into the stele and that ions are markedly accumulated in the xylem parenchyma, which seem to be rather directly involved in regulating ion transport to the xylem vessels. There may be some close association of ions with the ER cisternae and the cell to cell movement seems to be via the plasmodesmata.


## 8.6 *The function of the cortex*

In the young absorbing roots of most species the cortex is by far the largest volume fraction; in maize it occupies about 90% of the total root volume. Its chief functions are to absorb the ions from the external solution and then to transport these ions radially inward to supply the cells of the stele and the xylem vessels which eventually will supply the aerial parts of the plant. The mechanisms of initial ion uptake from the external solution have been dealt with in detail in Ch. 7.

There are two parallel pathways for movement across the cortex towards the stele, one through the extracellular space or apoplasm and the other through the symplasm, from the cytoplasm of one cell to the next by way of the plasmodesmata. Free space measurements in various roots (e.g. in barley by Pitman, 1965; in maize by Collins and House, 1969) show quite conclusively that diffusion in the free space is sufficiently rapid not to rate control ion uptake by even the innermost cortical cell. This is of course similar to what has been found in storage tissue (e.g. Briggs et al., 1962), and one may conclude that the apoplasm could act as a sufficiently rapid pathway for ion movement across the root cortex. However, this does not mean that the radial transport of ions to the xylem is necessarily conducted along this pathway and as will be seen later, it is highly likely that the second alternative pathway, the symplasm, in fact carries most of the radial ion movement.

The symplasm in the roots of several species has been investigated as to its effectiveness in radial ion transport. The physiological work by Arisz (1956), the electrical measurements by Spanswick (1972) and the theoretical study by Tyree (1970) all suggest that the symplasm is a sufficiently low resistance pathway that the observed ion fluxes to the xylem can be easily maintained. Briefly, there are two distinct areas for consideration in assessing the transport parameters of the symplasm. Firstly the mechanistic consideration of how ions move in the peripheral cytoplasm layer at the cortical cells and how they pass through the plasmodesmata, and secondly the anatomical considerations of what are the dimensions of a plasmodesm, how many plasmodesmata there are per cell and how they are distributed on the different walls of a cell. Further information of this nature is urgently required, but there is a certain amount available (see Table 8.1); the recent work by Gunning et al. (1974) on plasmodesmatal frequency in transfer cells could serve as a model of what is required.

The most important feature of symplasmatic ion transport in roots, stressed in the original discussion by Arisz (1956), is that cell to cell movement of ions takes place, almost without restriction, through the plasmodesmata. Tyree (1970) has demonstrated on what seem to be reasonable theoretical grounds that: (1) cell to cell movement is chiefly through plasmodesmata and only a negligible flux of all molecular species (with the possible exception of those which are very highly permeable in cell membranes, water being perhaps one such) cross the membrane bounded regions of the cell to cell junctions; (2) diffusion is the chief mechanism of transit through the plasmodesmata; (3) diffusion and perhaps cyclosis in the peripheral cytoplasmic layer of any single cell is sufficiently rapid to ensure perfect mixing, so that plasmodesmatal transit is the rate-controlling event for symplasmatic transport, once the ions are in the symplasm. Note that this last statement refers only to ions within the symplasm; overall transport from the external solution to the root xylem stream may be rate-controlled by membrane transport rates during loading or unloading of the cells in the symplasm.

The usual picture in one's mind is that the symplasm is loaded by uptake across the plasmalemma of the cortical cells from the external solution; this is probably the most common method of loading the symplasm, but it should be kept in mind that under certain circumstances the symplasm can be chiefly loaded from the ion reservoirs in the cortical cell vacuoles. For example, if radioisotope is added to the external solution bathing excised roots, the collected xylem exudate and the root tissue can be assayed at various times to demonstrate this effect.

Hodges and Vaadia (1964) added $^{36}Cl$ to the bathing media of excised onion roots of different salt status and then assayed the collected exudate at various times for $^{36}Cl$ activity; the root tissue was assayed at the end of the experiment (Fig. 8.7). Note that in all cases there is an approximately linear rise in exudate $^{36}Cl$ activity, although the rate of increase is much higher in low salt roots than in high salt roots. The main implication to be made here is that $Cl^-$ supply to the xylem stream is not simply a straightforward, direct passage from external solution to xylem, but that there is significant $Cl^-$ exchange between the symplasmatic through-put to the xylem and the vacuoles of the cortical cells. In the high salt roots there is a larger vacuolar pool of unlabelled $Cl^-$, and therefore there is a correspondingly greater isotopic dilution of the $^{36}Cl$ supply to the xylem, resulting in a lower rate of appearance of radioactivity in the xylem exudate.

                                W.P. *Anderson*

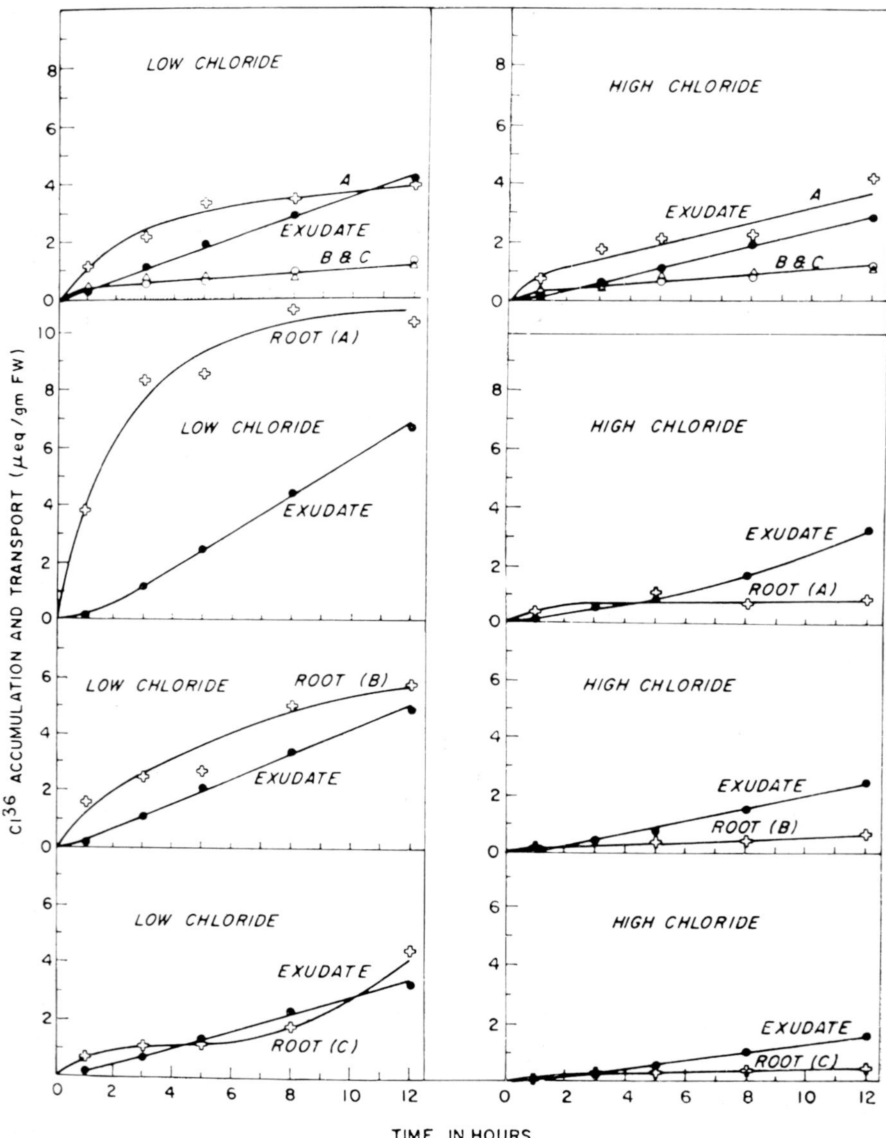



Fig. 8.7. The rate of appearance of $^{36}$Cl in root tissue and exudate, in plants grown in either high or low (Cl$^-$) prior to experimentation. Reproduced from Hodges and Vaadia (1964).

There is other evidence, again in onion but also in maize (Anderson et al., 1971), that a large proportion of the ion supply to the xylem stream may be derived from the cortical cell vacuoles. In this work, onion bulbs and maize seeds were separated into two batches and were grown in either a nitrate medium or in a chloride medium. After growth the roots were excised and transferred for exudation to the alternative solution to the one in which they had grown. In both treatments and for both species, it was found that the exudate collected for up to 24 h after excision and solution-switch, contained as the predominant anion, the anion supplied during growth, which of course was not present in the solution actually bathing the root during exudation. Again the implication is that the cortical cell vacuoles are providing anions to the symplasmatic through-put to the xylem. Unpublished data from the author's laboratory confirm essentially the same situation for $K^+$; the rate of appearance of $^{42}K$ in the exudate of excised maize roots is slower than might be expected if only the symplasm had to load and the most apparent interpretation is that there is isotopic $K^+$ exchange between the symplasm and the cortical vacuoles.

The overall flux balance in the cortical cells has been given by Pitman (1971) and the basic picture can be represented as in Fig. 8.8. In understanding this figure it should be realised that here the whole cortical symplasm is taken as

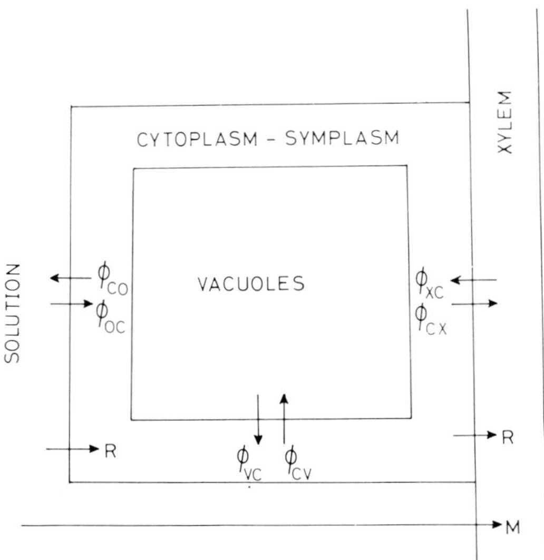


Fig. 8.8. A diagram from Pitman (1971) showing how fluxes can be considered in the root. A rather detailed analysis may be found in the original publication.

                           *W.P. Anderson*

a perfect cytoplasmic continuity; the case for this assumption rests on the work of Tyree (1970) and Spanswick (1972) as discussed earlier. Note however that the data described in the preceding paragraphs can be readily understood on the basis of Fig. 8.8; briefly, the fluxes $\phi_{cv}$ and $\phi_{vc}$ are of sufficient magnitude to significantly affect the isotopic or indeed ionic composition of the through-put to the xylem $\phi_x$.

There is further evidence of the flux situation in the root cortex to be found in the work by Läuchli et al. (1973) who have studied the effects of the protein synthesis inhibitor, cycloheximide, on the uptake and xylem transport of $K^+$ in excised barley roots. They found that cycloheximide did not affect respiration, or uptake of $K^+$ into the cortical cell vacuoles, but that it reduced to zero $K^+$ transport through the symplasm to xylem. The explanation offered by these authors is that in some manner continuing protein synthesis is required to maintain the plasmodesmata between one cell and the next in their functional state. Thus cycloheximide, by inhibiting protein synthesis, causes the plasmodesmata to become effectively closed and symplasmatic transport is thereby reduced to zero (see Ch. 9).

## 8.7  The function of the endodermis

The traditional function of the endodermis (Crafts and Broyer, 1938; Van Fleet, 1961) is assumed to be that the Casparian band in the walls of the endodermal cells acts as an effective barrier to apoplasmic movement of both ions and water between the root cortex and the stele. Transport into or from the stele must therefore occur within the endodermal cells.

There is good anatomical evidence that the Casparian band does form at a very early stage in root ontogeny and there is also indirect evidence that free space movement of ions and water is restricted by it. It is certainly hard to imagine how the absorbing region of a root could function with the selectivity it is known to have, if there were a continuous apoplasmic pathway from the external solution to the stele. Further, the evidence adduced earlier in this chapter of the occurrence of similar structures surrounding other plant secretory tissues (e.g. salt glands, nitrogen fixing nodules) strengthens the idea that the Casparian band forms an impermeable apoplasmic barrier, separating one region of free space from another.

It is most likely (see later) that ion movement across the endodermis takes place in the symplasm; the cortical symplasm is continuous with plasmodesmata in the outer tangential endodermal wall, while in the inner tangential wall there are plasmodesmatal connections to the pericycle cells and hence to the parenchyma cells of the stele. Helder and Boersma (1969) measured an average of 20,000 plasmodesmata per cell in the endodermis of young barley roots, while Clarkson et al. (1971), who conducted a similar study in the tertiary endodermis of older regions of barley roots, report an average of $1.25 \times 10^4$ pits mm$^{-2}$ of inner tangential wall, each pit containing an average 54 plasmodesmata, so that the estimated plasmodesmatal frequency is $67 \times 10^4$ mm$^{-2}$ of the total endodermal surface area. Comparison of this value of plasmodesmatal frequency in mature cells with that reported by Helder and Boersma (1969) in the young endodermis of the same species, reveals a reduction in frequency with age of about $4 \times 10^6$ mm$^{-2}$ of the total endodermal surface area. Clarkson et al. (1971) suggest explanations for this reduction and calculate from their measurements on mature cells that the xylem exudation fluxes of water and of phosphate can reasonably be expected to pass through these plasmodesmata. They further conclude that the passage cells in the mature endodermis, considered by Lundegårdh (1950), among others, to facilitate endodermal transit, occur so infrequently in barley that only small fractions of the total fluxes of ions and water will pass through them. The symplasmic pathway, even through mature endodermal cells, will carry the bulk of material in transit to the xylem.

In summary, the endodermis seems to act primarily as a barrier between the cortical free space and the free space in the stele, because the Casparian band does form an impermeable region within the endodermal cell walls and so isolates these two regions of the root tissue. Solute uptake across the membranes of the endodermis is probably no different in rate than uptake by the cortical cells; passage across the endodermis is chiefly symplasmic, with the ions arriving at the endodermal cells already in the symplasm after uptake by the epidermal and outer cortical cells.


## 8.8 The function of the stele

In contrast to discussions on the functions of the cortex and the endodermis where there are only minor areas of disagreement, the stele and in particular the xylem parenchyma are presently the subjects of active research and rather polarised discussion. There are two main topics in dispute, firstly the functioning of the xylem parenchyma cells and secondly the structure and function of the xylem vessels in the young, absorbing regions of the root.

The classic exposition of the role of the stele is due to Crafts and Broyer (1938) who assumed that the cells were 'leaky' because low oxygen tension within the stele so reduced cellular respiration rates that membrane integrity and regulatory ion-carrier mechanisms could not be maintained. Whether or not the parenchyma cells of the stele are 'leaky', it now seems clear that the reason is not anaerobiosis as was originally supposed. Hall et al. (1971) measured the respiration rate in freshly isolated maize root steles and found the rate to be on average 192 $\mu$l $O_2$ $g^{-1}$ $h^{-1}$, a value which seems quite sufficient to allow the cells to maintain themselves and conduct net transport of ions (Anderson, 1972). It is of course true that in situ respiration rates for steles are not available, but Bowling (1973) has measured the $O_2$ partial pressure in the stele of sunflower roots by a micro-electrode method and found a gradient of 151 mm Hg at the epidermis, to 130 mm Hg at the pericycle and 127 mm Hg at the protoxylem. This does not seem to be a very large gradient, certainly not one large enough to cause anaerobic suffering in the root stele and induce 'ion leakiness' in the parenchyma cells. Further, it should be remembered when thinking about the quoted value of 192 $\mu$l $O_2$ $g^{-1}$ $h^{-1}$ (Hall et al., 1971) that a gram fresh weight of stele contains proportionately far more inert material (thickened and lignified walls and inert vessels) than does a gram of cortical tissue, so that it is possible that the stele respires just as rapidly as does the cortex.

There seems therefore to be little doubt that the cells of the stele are not under great anaerobic stress, but are the parenchyma cells in fact 'leaky' to ions? Certainly ions leave the parenchyma of the stele and enter the xylem stream, but the question is whether they leak passively from the parenchyma or are secreted actively into the xylem. The first alternative has received support from Baker (1973a, b) who has demonstrated that [86]Rb efflux from isolated maize root steles is enhanced by low temperature and by the respiratory uncoupler CCCP. Such effects indicate against active secretion of $K^+$ from the parenchyma of the stele.

In contrast, other work has been interpreted as evidence in favour of active ion secretion by the parenchyma into the xylem. Läuchli and Epstein (1971) and Läuchli et al. (1971) have reached this conclusion on the basis of the results from several experimental techniques (radioactive pulse labelling, electron microprobe analysis of ion concentration profiles across the roots and inhibitor studies). They conclude that ion transport to the xylem requires two carrier-mediated membrane events, uptake at the plasmalemma of the epidermal and outer cortical cells and secretion to the xylem at the plasmalemma of the xylem parenchyma.

A similar conclusion has been reached independently by Pitman (1972) on the evidence of a comprehensive flux analysis for [36]Cl in excised barley roots after a variety of pre-treatments and with or without the addition of CCCP. Xylem exudation from the excised roots is inhibited by CCCP but the plasmalemma and tonoplast effluxes in the root cortex are not. Since the ions in the cortical cell vacuoles could therefore supply the symplasm, and hence the xylem, in pre-loaded tissue even after the application of CCCP, Pitman argues that the observed cessation of xylem exudation must be due to CCCP inhibition of an active ion secretion by the xylem parenchyma.

In considering differential inhibitor effects on uptake and exudation, it must always be borne in mind that the inhibitor may be exercising its observed effect through its action on symplasmatic transport. Although it is true that diffusion is the chief mechanism of movement in symplasmatic transport (Tyree, 1970), it is still reasonable to suppose that the continued functioning of the plasmodesmata as effective transit channels may require continuing energy supply, such as is provided by respiration. Indeed there is an increasing body of evidence that the symplasm is a transport pathway under metabolic and biochemical control; the work on the effect of the protein synthesis inhibitor, cycloheximide, by Läuchli et al. (1973) has been quoted earlier. There is also evidence that certain plant hormones (e.g. cytokinins and abscisic acid) may act on the plasmodesmata in barley roots (Pitman et al., 1974).

To be strictly accurate, the observed effect of all these hormones and inhibitors on symplasmatic transport cannot be unequivocally attributed to their action at the plasmodesmata; it is also possible that phase changes in the cytoplasm induced by these agents, cause ionic mobility in the peripheral cytoplasm of the cortical cells to be altered, and so cause alterations in symplasmatic ion transport. The most likely outcome is that inhibitors and hormones will be found to cause ion mobility alterations in both plasmodesmata and peripheral cytoplasm; the argument will probably develop into a debate on what proportion of the total observed alteration in overall symplasmatic transport is due to which effect.

Under certain conditions and in certain species, the xylem parenchyma may be important in selective reabsorption of ions from the xylem stream, usually in those regions of the root above the chief absorbing zones. $Na^+$ reabsorption may be particularly important for translocation selectivity in certain species; Läuchli et al. (1974b) have found that xylem parenchyma at the proximal ends of the roots of salt-tolerant cultivars of soy bean, develop wall ingrowths reminiscent of those found in transfer cells (Pate and Gunning, 1972).

Finally there is the function of the xylem vessels. In all the prior arguments it has been assumed that the vessels are inert conduits, but from time to time there have been proposals that the vessels are more actively involved. Hylmö (1953) suggested in his so-called test tube hypothesis that the osmotic potential to drive root exudation is developed across the living membranes of xylem vessels, which are open at one end like a test tube. Anderson and House (1967) observed a correlation between $K^+$ translocation in the xylem of young maize roots and the number of vessels still containing membrane-bounded cytoplasm, and suggested that the final step in radial ion transport was at the xylem vessel membranes. Scott (1965) reached a similar conclusion using *Ricinus communis* roots and stated that all solute movement in the root hair zone is into living, nucleated xylem vessels. Higinbotham et al. (1973) have further support for this point of view from their studies in young maize roots. They estimate that most of the outer ring metaxylem vessels are not yet open at 100 mm from the tip and suggest that root pressure in a vessel is similar to turgor pressure in a normal plant cell. However, there seems to be wide variation between species; Läuchli et al.(1974a) find no cytoplasm in xylem beyond a few millimeters of the tip in barley roots.

Perhaps the most serious criticism of all such proposals was pointed out by Anderson et al. (1970). In maize, if the exudation must cross living membranes bounding the outer ring metaxylem vessels, then these membranes must be able to conduct $K^+$ fluxes of several $\mu$Mol m$^{-2}$ s$^{-1}$, many times the 'typical' values for other root cells; the discrepancy is even larger for the anion fluxes. Although $Cl^-$ fluxes of 8 $\mu$Mol m$^{-2}$ s$^{-1}$ have been measured in *Acetabularia mediterranea* (Saddler, 1970), such values have never been observed in non-marine species; Cram (1973) has reported $Cl^-$ flux values of the order of 100 nMol m$^{-2}$ s$^{-1}$ in maize root cortex.

## 8.9 The common radial transport model

This section is essentially a summary of the preceding three sections and incorporates the commonly held view at the present time of how a root absorbs and transports salts and water from the external solution to the xylem translocation stream. The most general statement is that the root accumulates ions into the xylem and so creates an osmotic pressure gradient down which water flows from the external solution, also into the xylem vessels. Water so entering the xylem then flows away upward to the collecting tube of an excised root, or to the growing shoot of an intact seedling, carrying the accumulated ions in the flow.

Fig. 8.9 summarises the main features of the current radial transport model for young, absorbing regions of the root. The proposals are essentially similar to those given by many other workers and are assumed, at least tacitly, to apply to the root systems of all higher plants. Ions are accumulated at the epidermis and outer cortex and traverse the root radially in the symplasm, which is continuous through the endodermis, to the xylem parenchyma. At

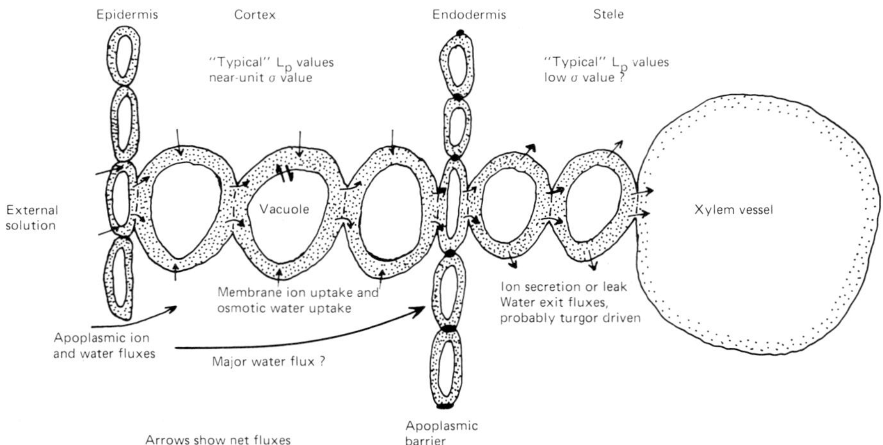


Fig. 8.9. A diagram of the most commonly accepted model for radial transport across the root.

the parenchyma, ions are either leaked or secreted into the xylem vessels (see the previous section). Apoplasmic ion movement across the cortex is unlikely to carry a very significant fraction of the total cortical ion flux, except perhaps at non-physiologically high, external ionic concentrations, for the following reasons. It seems certain that the Casparian band on the root endodermis is an effective barrier to all apoplasmic movement. Therefore all material entering the stele must pass through the endodermal cells, either by being already in the cortical symplasm, or by being taken up at the endodermal membrane. Since the calculated (Tyree, 1970) and measured (e.g. Spanswick, 1972) transport parameters for the symplasm and the measured parameters for apoplasmic ion diffusion (e.g. Pitman, 1965) indicate that these alternative pathways across the cortex are equally conductive to ions, it therefore follows that membrane penetration is the rate-controlling step for ion entry to the stele of the root. If the ion transport capability of the endodermis is no different. than that of the epidermal and cortical cells (and ultrastructural work on endodermal cells shows them to be normally vacuolated, with normal mito-chondrial numbers) then the net ion fluxes to the stele should be carried in the symplasm pathway versus the apoplasmic pathway in direct proportion to the ratio of the symplasm membrane surface area and the endodermal membrane surface area. On this criterion it is clear that more than $90\%$ of the ion flux to the stele will cross the root cortex in the symplasm.

For water movement, on the other hand, it is likely that the apoplasm is the major pathway across the cortex. The apoplasm has a much higher hydraulic conductivity than the symplasm, and the observed rates of water exudation from excised roots can easily be provided by water uptake at the endodermal membranes, even if these membranes have 'typical' hydraulic conductivity values. Uptake of water at the endodermis will be a straight-forward osmotic uptake.

In the stele, the most common proposal is that the xylem parenchyma either leak or secrete ions into the xylem vessels. The argument concerning this point was set out in detail in the previous section and will not be taken up again here. There may be one other possibility, namely that the symplasm continues through to the xylem vessels, with plasmodesmatal connections from the xylem parenchyma cells through the pit areas of the vessel secondary walls. It should be said immediately that there are only isolated reports of such connections (e.g. Scott, 1949), but such an anatomical feature would be an attractive method of releasing ions into the xylem, and could be one way of reconciling the different views on the transport role of the xylem vessels.

The water pathway in the stele is not well determined; lignification of the cell walls interior to and including the inner tangential wall of the endodermis occurs at quite an early stage of root development and is well developed in regions of the root which are still active in absorption. Presumably this causes the hydraulic conductivity of the stele apoplasm to be quite low, and the symplasm may well be the major pathway for water movement across the stele. Calculations by Clarkson et al. (1971) indicate that the plasmodesmatal numbers in the mature endodermis of barley roots could conduct the observed water fluxes under reasonable values of driving force.

Finally, the most common formalism to describe the osmotic water flow of root pressure exudation (e.g. House and Findlay, 1966) assumes there is only a single osmotic membrane system within the root. It is highly likely that this is an oversimplification and that root pressure exudation can be more satisfactorily described by a model comprising two membranes in series (e.g. Curran and McIntosh, 1962). Such a model has been used for water movement across isolated maize root cortex by Ginsburg (1970). If water movement across the cortex is chiefly apoplasmic, the first osmotic membrane on a Curran and McIntosh model would be the outer tangential membrane of the endodermal cells, having a 'normal' $L_p$ value and a 'normal' $\sigma$ value of approximately unity. The second barrier would be whatever is the final membrane prior to the water emptying into the lumina of the xylem vessels, most probably the membranes of the xylem parenchyma. Depending on what condition one assumes for these membranes, the $L_p$ and $\sigma$ values might vary from those 'typical' for cells (likely if the parenchyma secretes ions) to values of $L_p$ which are very large and $\sigma$ values close to zero (likely if the parenchyma leak ions passively).

## 8.10 The common longitudinal transport model

The radial situation just described will apply in general at any plane along the longitudinal axis of the root, but it is well known that most of the various transport parameters do vary as one progresses back from the root apex. For example, the older regions are generally less water permeable, and the most obvious correlation is with rather gross anatomical alterations, such as progressive tertiary thickening of the endodermis and stele and cuticularisation of the epidermis, which leads in many species to the formation of an identifiable exodermis.

Although these structural changes are important, there are other, more subtle, factors involved. In many cereals, for example, very little $Ca^{2+}$ reaches the conducting tissue for translocation in the older regions of the seminal root where the tertiary endodermis is well developed (Clarkson et al., 1968), while $K^+$ is still transported in the same regions. The reason for this difference is not completely understood, but may be partly explained by the apoplasm being the chief pathway for $Ca^{2+}$, with $K^+$ moving mainly in the symplasm. In the younger regions of the root, the transport of these two cations is more similar; phosphate, like $K^+$, can apparently be transported across older regions of the root, but the pattern of uptake again differs from that of $K^+$. Sodium uptake proceeds quite rapidly into the young zones of *Zea mays* primary roots, but discrimination against $Na^+$ relative to $K^+$ increases markedly with increasing distance from the root tip (Shone et al., 1969).

These various alterations in transport as one proceeds along the root from the tip give rise, in a steady state situation, to the development of a standing gradient osmotic flow along the xylem vessels (Anderson et al., 1970). Fig. 8.10 shows the basis for this model; in words, ions are accumulated into the xylem to give a net salt influx, $J_s$, which may vary with distance from the root tip. The result of this salt influx is to produce a salt concentration in the xylem sap, which is unknown at any point within the xylem but which may be monitored at the cut end of the excised root as the concentration of the emergent xylem fluid. The local xylem salt concentration at any point within the root will provide the osmotic gradient to drive water inward at that point; the rate of water entry depends on the local radial hydraulic conductivity, $L_p$, and the local reflection coefficient, $\sigma$. In the steady state, the conservation equations for salt and for water within the xylem are as given in Fig. 8.9, and can be solved simultaneously to predict the rate of water flow from an excised root for given profiles along the root of ion transport rate, $T(x)$, and hydraulic conductivity and reflection coefficient, $\sigma(x)L_p(x)$. Details of the application of

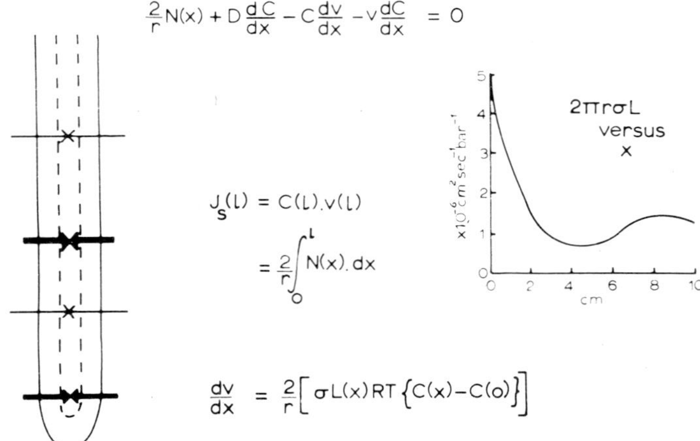



Fig. 8.10. A diagram for the most commonly accepted model for longitudinal transport along the root, taken from Anderson et al. (1970) where full details of the model are given.

this standing osmotic gradient model to exudation from excised primary roots of *Zea mays* can be found in Anderson et al. (1970).

## 8.11 *The transport of trace metals*

Compared to the number of studies which have been carried out on the uptake and translocation of the major plant nutrient ions, e.g. $K^+$, $NO_3^-$, $SO_4^{2-}$, $Cl^-$, $Mg^{2+}$, $Ca^{2+}$ and $Na^+$, our knowledge of the transport of trace metals is relatively poor. It has been known since the earliest days of plant mineral nutrition that deficiency symptoms develop if certain metal ions are not available to the plant, although for these trace elements, or micro-nutrients, the concentrations required to alleviate deficiency are small. This is probably the basic problem in studying trace element uptake; the rate of transport into the plant from solutions containing physiological levels of the trace element is very low, and in general cannot be artificially accelerated for study by increasing the external concentration because many of the trace elements are toxic at higher concentrations.

Iron has been by far the most widely studied of the trace metals and is in one sense atypical, because so many ferric salts with anions commonly found in plants are highly insoluble. It is generally supplied in the external solution as a highly stable chelate, usually with ethylenediamine-tetracetic acid (EDTA).

For a brief review of the chemistry of chelate formation, written for plant physiologists, see Beckett and Anderson (1973). There has been much discussion as to whether $Fe^{3+}$ must be displaced from the chelate prior to uptake by the root, or whether the $Fe^{3+}$-chelate complex is taken up as a single entity. One school of thought (e.g. Tiffin, 1970) believes that $Fe^{3+}$ is displaced by competition with other molecules in the cell membrane, presumably the carries molecules, while the second opinion (e.g. Hill-Cottingham and Lloyd-Jones, 1965) is that the $Fe^{3+}$–chelate complex is absorbed. Beckett and Anderson (1973) show in young maize roots that $Fe^{3+}$ is apparently displaced from EDTA prior to uptake from low external concentrations, while at higher concentrations the $Fe^{3+}$–EDTA complex is transported. However, it is known that EDTA can be metabolised to the eventual production of $CO_2$ in maize roots, which fact alone may account for the results.

Once into the root, $Fe^{3+}$ must be kept in solution during translocation through the xylem. Tiffin (1967, 1970) has shown that in mature root systems of tomato and soybean, $Fe^{3+}$ is translocated in the xylem as citrate with which it forms a sufficiently stable chelate to prevent precipitation with other anions. Note that $Fe^{3+}$ must remain chelated to something all the way from the EDTA in the external solution to its chief eventual fate in the porphyrin ring of the haem proteins.

Tiffin (1967) has also given data for the xylem translocation in mature tomato roots of $Mn^{2+}$, $Co^{2+}$ and $Zn^{2+}$; all are found to be translocated as simple hydrated cations, as are $Ca^{2+}$ and $Mg^{2+}$. It is assumed, though little is actually known, that all these are taken up into the symplasm as simple hydrated cations by carrier mechanisms similar to those which transport the major nutrient cations.


## 8.12 Effects of hormones on transport

This is an area of increasing interest and it is now known that various plant growth substances have marked effects on ion transport and on water permeability in root systems. Tal and Imber (1971), Glinka (1973) and Collins and Kerrigan (1973) found that ABA in the concentration range around $10^{-3}$ Mol m$^{-3}$ produced increased exudation rates in tomato, sunflower and maize roots respectively. Cram and Pitman (1972) found that similar concentrations of ABA caused the exudation rate to decrease in barley and maize roots, but this apparent conflict seems to have been resolved by the recent finding by Pitman et al. (1974) that the effect of ABA depends dramatically on temperature and on the culture solution conditions (see Ch. 9). It is generally noted by all workers that ABA causes tissue ion concentrations to rise and also increases the rate of ion transport to the xylem; water permeability is generally increased, the tissue loses turgor to some extent and the rate of exudation increases.

Kinetin has a similar effect on tissue concentration but causes xylem exudation fluxes of both ions and water to decrease (Tal and Imber, 1971; Collins and Kerrigan, 1973). At present the mode of action of hormones is not well understood, but it seems likely that they all act firstly at the cell membranes, causing alterations to transport rates and permeabilities, and possibly also on the symplasm, changing the rate of radial transport toward the xylem. Since symplasmic transport is chiefly diffusive in mechanism, the implication is that plasmodesmatal structure, or ionic mobility in the cytoplasm, is affected by the hormone.

The effect of hormones on water transport is also of much interest; ABA is found to cause the root hydraulic conductivity to increase in several species when it is applied to the bathing solution at about $10^{-3}$ Mol m$^{-3}$ (Tal and Imber, 1971; Glinka, 1972; Collins and Kerrigan, 1973; Pitman et al., 1974) and the effect again may be temperature dependent. Interestingly, Tal and Imber (1971) find the ABA effect if the hormone is applied as a spray to the plant leaves prior to decapitation. ABA is known to be translocated from the shoot to the root in cotton seedlings (Shindy et al., 1973). Kinetin causes root hydraulic conductivity to decrease (Tal and Imber, 1971; Collins and Kerrigan, 1973), but the effect takes much longer to be established than does the ABA alteration, 20 h as opposed to 60 min. IAA has also been found to cause increases in hydraulic conductivity in excised roots of pea and sunflower (Skoog et al., 1938), in tobacco (Joven, 1964) and in tomato (Tal and Imber, 1971). The IAA effects on ion transport are much less clear cut.


## 8.13 Transpirational effects on transport

Transpiration from the plant leaves exerts its primary effect on water transport through the root, by the simple addition of a hydrostatic pressure term to the total driving force on water. Under most natural environmental conditions, this 'transpiration tension' is by far the largest force for driving water through the intact plant. A good exposition of the whole subject can be found in Slatyer (1967). Very simply, water movement across the root to the xylem will be described by

$$J_v = L_p(\Delta P - \sigma RT \Delta C_s) + \phi_o \tag{8.3}$$

where $J_v$ is the volume (water) flux into the root, $L_p$ is the hydraulic conductivity of the root, $\Delta P$ is the transpiration tension, $\sigma$ the reflection coefficient of the effective osmotic membrane in the root, $R$ is the universal gas constant, $T$ the absolute temperature, $\Delta C_s$ the solute concentration difference between xylem sap and external solution and $\phi_o$ is a so-called non-osmotic water flux. This equation is an obvious modification of the one normally used to describe root pressure exudation,

$$J_v = L_p(-\sigma RT\Delta C_s) + \phi_o \qquad (8.4)$$

where the negative sign is simple sign convention indicating that the water flow is toward the higher salt concentration in the xylem. The non-osmotic term $\phi_o$ was introduced to take up an observed discrepancy, namely that water flow continues in the apparent absence of an osmotic gradient (Arisz et al., 1951; House and Findlay, 1966). This apparent anomaly is perhaps best resolved by describing root pressure exudation by a combination of the two-membrane radial model and the standing gradient osmotic model, as set out in earlier sections of this chapter.

The chief complication in applying the first equation to account for transpiration effects on water transport is that $L_p$ apparently varies with $\Delta P$ (Brouwer, 1954; Mees and Weatherley, 1957). Root hydraulic conductivity apparently increases five-fold as $\Delta P$ increases by 2 bars in tomato root systems. Two explanations have been offered for this observation. Firstly, that the actual cell membrane water permeability increases and secondly, that there is an increase in the effective absorbing area of the root. Meiri and Anderson (1970) observed an apparently similar effect in excised primary maize roots but were able to correlate it with the relatively trivial artifact of water infiltration of the inter-cellular air spaces of the root cortex; application of suction pressure to the cut end of the root caused a progressive increase in root density due to infiltration and this provided an additional conduit for water movement, so accounting for the observed increase in $L_p$.

Although transpiration tension undoubtedly acts primarily on water transport, there are many reports that effects are also produced on solute transport. For example Petterson (1966) has shown that $SO_4^{2-}$ translocation is enhanced by transpiration in sunflower and Lazaroff and Pitman (1966) show that the translocation of several cations is affected by transpiration rate in barley. The effect is different for different ions; Pitman (1965b) has found that the selectivity of $K^+/Na^+$ uptake in barley is decreased at increased transpiration rates. Total $(K^+ + Na^+)$ uptake remains constant, with $Na^+$ uptake increasing and $K^+$ decreasing (see Ch. 9). Herbicides, such as the triazines, are extremely lipid soluble and their rate of uptake into the roots is markedly increased as transpiration increases (e.g. Shone and Wood, 1972).

The simplest explanation of solute transport enhancement by transpiration is that the xylem compartment of the root is flushed through more effectively by the increased water flow and therefore is a more effective sink for both passive and active transport to the xylem. However, this is probably not the whole explanation and more work on the mechanisms of coupling solute and water flow in the plant root is needed.

## *References*

W.P. ANDERSON, Annu. Rev. Plant Physiol., 23 (1972) 51.

W.P. ANDERSON, D.P. AIKMAN and A. MEIRI, Proc. R. Soc. B, 174 (1970) 445.

W.P. ANDERSON, L. GOODWIN and R.K.M. HAY, in J. Kolek (Ed.) Structure and Function of Primary Root Tissues (1971) Slovak Acad. Sci., p. 379.

W.P. ANDERSON, D.L. HENDRIX and N. HIGINBOTHAM, Plant Physiol., 54 (1974) 712.

W.P. ANDERSON and C.R. HOUSE, J. Exp. Bot., 18 (1967) 544.

W.H. ARISZ, Protoplasma, 46 (1956) 5.

W.H. ARISZ, R.J. HELDER and R. VAN NIE, J. Exp. Bot., 2 (1951) 257.

D.A. BAKER, in W.P. Anderson (Ed.) Ion Transport in Plants (1973a) Academic Press, London, p. 511.

D.A. BAKER, Planta, 112 (1973b) 293.

J.T. BECKETT and W.P. ANDERSON, in W.P. Anderson (Ed.) Ion Transport in Plants (1973a) Academic Press, London, p. 595.

D.J.F. BOWLING, Planta, 69 (1966) 377.

D.J.F. BOWLING, in W.P. Anderson (Ed.) Ion Transport in Plants (1973a) Academic Press, London, p. 483.

D.J.F. BOWLING, Planta, 111 (1973b) 323.

D.J.F. BOWLING and R.M. SPANSWICK, J. Exp. Bot., 15 (1964) 422.

G.E. BRIGGS, A.B. HOPE and R.N. ROBERTSON, Electrolytes and Plant Cells (1961) Blackwell, Oxford.

R. BROUWER, Acta Bot. Neerl., 3 (1954) 264.

J.W.A. BURLEY, F.I.O. NWOKE, G.L. LEISTER and R.A. POPHAM, Am. J. Bot., 57 (1970) 504.

D.T. CLARKSON, A.W. ROBARDS and J. SANDERSON, Planta, 96 (1971) 292.

D.T. CLARKSON, J. SANDERSON and R.S. RUSSELL, Nature, 220 (1968) 805.

J.C. COLLINS and C.R. HOUSE, J. Exp. Bot., 20 (1969) 497.

J.C. COLLINS and A.P. KERRIGAN, in W.P. Anderson (Ed.) Ion Transport in Plants (1973) Academic Press, London, p. 589.

A.S. CRAFTS and T.C. BROYER, Am. J. Bot., 24 (1938) 415.

W.J. CRAM and M.G. PITMAN, Aust. J. Biol. Sci., 25 (1972) 1125.

P.F. CURRAN and J.R. MCINTOSH, Nature, 193 (1962) 347.

J. DAINTY, Annu. Rev. Plant Physiol., 13 (1962) 379.

R.F. DAVIS and N. HIGINBOTHAM, Plant Physiol., 44 (1969) 1383.

J. DUNLOP and D.J.F. BOWLING, J. Exp. Bot., 22 (1971) 445.

B. ETHERTON, Plant Physiol., 38 (1963) 581.

B. ETHERTON and N. HIGINBOTHAM, Science, 131 (1960) 409.

H. GINSBURG, J. Theor. Biol., 32 (1971) 147.

z. GLINKA, Plant. Physiol., 51 (1973) 217.

B.E.S. GUNNING, J.S. PATE, F.R. MINCHIN and I. MARKS, Soc. Exp. Biol. Symp., 28 (1974) 87.

J.L. HALL, R. SEXTON and D.A. BAKER, Planta, 96 (1971) 54.

J.L. HALL, in W.P. Anderson (Ed.) Ion Transport in Plants (1973) Academic Press, London, p. 11.

R.K.M. HAY and W.P. ANDERSON, J. Exp. Bot., 23 (1972) 585.

R.J. HELDER and J. BOERSMA, Acta Bot. Neerl., 18 (1969) 99.

N. HIGINBOTHAM and W.P. ANDERSON, Can. J. Bot., 52 (1974) 1011.

N. HIGINBOTHAM, R.F. DAVIS, S.M. MERTZ and L.K. SHUMWAY, in W.P. Anderson (Ed.) Ion Transport in Plants (1973) Academic Press, London, p. 493.

N. HIGINBOTHAM, J.S. GRAVES and R.F. DAVIS, J. Membrane Biol., 3 (1970) 210.

D.G. HILL-COTTINGHAM and C.P. LLOYD-JONES, J. Exp. Bot., 16 (1965) 233.

T.K. HODGES, R.T. LEONARD, C.E. BRACKER and T.W. KEENAN, Proc. Natl. Acad. Sci. U.S., 69 (1972) 3307.

T.K. HODGES and Y. VAADIA, Plant Physiol., 39 (1964) 109.

C.R. HOUSE and N. FINDLAY, J. Exp. Bot., 17 (1966) 344.

B. HYLMÖ, Physiol. Plant., 6 (1953) 333.

P. JARVIS and C.R. HOUSE, J. Exp. Bot., 21 (1970) 83.

C.B. JOVEN, Plant Physiol. Diss. Abstr., 25 (1964) 5519.

H. KITASATO, J. Gen. Physiol., 53 (1968) 60.

A. KYLIN, in W.P. Anderson (Ed.) Ion Transport in Plants (1973) Academic Press, London, p. 369.

A. LÄUCHLI, Planta, 75 (1967) 185.

A. LÄUCHLI, in W.P. Anderson (Ed.) Ion Transport in Plants (1973) Academic Press, London, p. 1.

A. LÄUCHLI and E. EPSTEIN, Plant Physiol., 48 (1971) 111.

A. LÄUCHLI, D. KRAMER, M.G. PITMAN and U. LÜTTGE, Planta, 119 (1974a) 85.

A. LÄUCHLI, D. KRAMER and R. STELZER, in U. Zimmermann and J. Dainty (Eds.) Membrane Transport in Plants and Plant Organelles (1974b) Springer Verlag, Berlin, in press.

A. LÄUCHLI, U. LÜTTGE and M.G. PITMAN, Z. Naturforsch., 28c (1973) 431.

A. LÄUCHLI, A.R. SPURR and E. EPSTEIN, Plant Physiol., 48 (1971) 118.

N. LAZAROFF and M.G. PITMAN, Aust. J. Biol. Sci., 19 (1966) 991.

R.A. LEIGH, R.G. WYN JONES and F.A. WILLIAMSON, in W.P. Anderson (Ed.) Ion Transport in Plants (1973) Academic Press, London, p. 407.

H. LUNDEGÅRDH, Physiol. Plant., 3 (1950) 103.

G.C. MEES and P.E. WEATHERLEY, Proc. R. Soc. B, 147 (1957) 381.

A. MEIRI and W.P. ANDERSON, J. Exp. Bot., 21 (1970) 899.

F.R. MINCHIN and D.A. BAKER, Planta, 113 (1973) 97.

J.S. PATE and B.E.S. GUNNING, Annu. Rev. Plant Physiol., 23 (1972) 173.

S. PETTERSON, Physiol. Plant., 19 (1966) 581.

M.G. PITMAN, Aust. J. Biol. Sci., 18 (1965a) 541.

M.G. PITMAN, Aust. J. Biol. Sci., 18 (1965b) 987.

M.G. PITMAN, Aust. J. Biol. Sci., 24 (1971) 407.

M.G. PITMAN, Aust. J. Biol. Sci., 25 (1972) 243.

M.G. PITMAN, U. LÜTTGE, A. LÄUCHLI and E. BALL, J. Exp. Bot., 25 (1974) 147.

J.H. PRIESTLY, New Phytol., 19 (1920) 189.

A.W. ROBARDS, Protoplasma, 72 (1971) 315.

H.D.W. SADDLER, J. Gen. Physiol., 55 (1970) 802.

F.M. SCOTT, Bot. Gaz., 110 (1949) 492.

F.M. SCOTT, in K.F. Baker and W.C. Synder (Eds.) Ecology of Soil Borne Pathogens (1965) Univ. Calif. Press, p. 145.

W.W. SHINDY, C.M. ASMUNDSON, O.E. SMITH and J. KUMAMOTO, Plant Physiol., 52 (1973) 443.

M.G.T. SHONE, J. Exp. Bot., 20 (1969) 698.

M.G.T. SHONE, D.T. CLARKSON and J. SANDERSON, Planta, 86 (1969) 301.

M.G.T. SHONE and A.V. WOODS, J. Exp. Bot., 23 (1972) 141.

F. SKOOG, T.C. BROYER and K.A. GROSSENBACHER, Am. J. Bot., 26 (1938) 749.

R.O. SLATYER, Plant Water Relations (1967) Academic Press, London.

C.L. SLAYMAN, J. Gen. Physiol., 49 (1965) 93.

R.M. SPANSWICK, Planta, 102 (1972) 215.

R.M. SPANWICK, in W.P. Anderson (Ed.) Ion Transport in Plants (1973) Academic Press, London, p. 113.

M. TAL and D. IMBER, Plant Physiol., 47 (1971) 849.

C.J. TANDLER, C.M. LIBANATI and C.A. SANCHIS, J. Cell Biol., 45 (1970) 355.

L.O. TIFFIN, Plant Physiol., 42 (1967) 1427.

L.O. TIFFIN, Plant Physiol., 45 (1970) 280.

M.T. TYREE, J. Theor. Biol., 26 (1970) 181.

D.S. VAN FLEET, Bot. Rev., 27 (1961) 165.

*Ion transport in plant cells and tissues*
*edited by D.A. Baker and J.L. Hall*
© *North-Holland Publishing Company, 1975*



# CHAPTER 9

# Whole plants

M.G. PITMAN

## Contents



## 9.1  Introduction

Previous chapters have considered the activity of roots and the properties that make them so efficient at penetrating the soil and procuring nutrients from mineral particles or from solution. For many purposes it has been convenient to follow the abstraction made by plant physiologists and study roots in isolation from the rest of the plant. Excised roots continue to respire, will take up salt and excrete ions at rates similar to those of whole plants and will even transport ions into the xylem where a standing gradient of osmotic pressure leads to exudation of solution from the cut end of the root. However, in other ways an excised root reflects the activity of the whole to about the same degree as an arm cut from the body, in that nutrients can no longer enter the organ nor can messages pass between it and the rest of the body.

This chapter is concerned with the relationship between the demands of the growing plant and the uptake of ions by its roots, and with the processes involved within the plant, (see Läuchli (1972) as a general reference). Transport of ions from the root to the shoot takes place in the water flow through the xylem vessels. Within the stem and leaves the living cells around the xylem act as sinks to reabsorb ions from the xylem sap and perhaps transport these ions to other living cells not in contact with the xylem. The vascular bundle in the leaf appears to be a system for lateral ion transport similar to the stele in the root, but removing ions from instead of secreting ions into the xylem. The phloem too is a pathway for the redistribution of ions in the shoot. Mature leaves are supplied with ions in the xylem but export ions in the phloem to young leaves and fruits. Phloem transport also returns ions from the shoot to the root though the amount moving in this way may be small. The intact plant is ingeniously organised for the recycling of impressive amounts of nutrients from redundant to juvenile sites or to storage systems.

The movement of ions in the plant is related to the supply of carbohydrates from the shoot to the root and probably to some hormonal 'messages' that may regulate the output from the root to the shoot. Apparently the phloem is the pathway for supply of both energy and these controlling factors.

Finally, it is worth remembering that the major part of ions taken up by the plant is transported to the shoot. Only 3% of the $Ca^{2+}$, 16% of the $K^+$ and 25% of the P absorbed by the plant is retained in the roots.


## 9.2 Amounts of ions in plants

The amounts of nutrients in plants are expressed in various ways. The simplest measure is perhaps amount per plant, either on a weight or gram molecule basis, but as this quantity increases with plant size it is more common to express the content relative to the weight of the plant. Very often the amount is quoted relative to dry weight as 'ppm' or percentage. This is a convenient measure as dry weight is often simpler to determine than fresh weight in large scale field experiments. It has the disadvantage, however, that it is more difficult to relate measurements based on dry weight to the function of the nutrient than if a fresh weight basis is used. For example, on a dry weight basis, $K^+$ deficiency is noticeable in many crop plants when $K^+$ is less than 8–10 mg $gDW^{-1}$ in the leaves. The $K^+$ content of leaves of various plant species growing on nutrient-poor sandstone near Sydney was about 2–3 mg $gDW^{-1}$. Apparently the sandstone plants are very efficient in using $K^+$, but in fact the low $K^+$ level was due to a very high ratio of dry to fresh weight. When expressed relative to leaf water, the concentration of $K^+$ in crop-plants and sand-stone plants is about the same (50 Mol $m^{-3}$) and this value is close to the concentration needed for maximum activity of many of the $K^+$-requiring enzymes (Evans and Sorger, 1966).

As far as possible, nutrient levels will be quoted here either as concentrations (Mol $m^{-3}$, relative to water content of the tissue) or as 'relative content' in units of $\mu$Mol $gFW^{-1}$ (relative to fresh weight) or $\mu$Mol $gDW^{-1}$ (relative to dry weight).

The orders of magnitude of relative content of various ions are given in Table 9.1 as a general guide. Of course, the content of ions relative to each

Table 9.1.
Approximate levels of various ions or
elements in plant shoots.

| Substance | Relative content ($\mu$mol $gFW^{-1}$) |
| --- | --- |
| N | 150–300 |
| K | 50–200 |
| Ca | 10–50 |
| Mg | 10–50 |
| Na | 5–50(150) |
| P | 25 |
| S | 10–20 |
| Fe | 0.5–1 |

other and in absolute terms will vary from one species to another or according to the availability in the soil or solution, or ontogenetically as the plant passes through different stages of growth.

## 9.2.1 *Effect of external concentration*

Availability of ions in the soil is difficult to express with precision. The ion may be part of soil minerals; it may be adsorbed on to soil particles or associated with charged exchange sites, or small amounts may be free in the soil solution. Each of these sources may interact in supplying the plant root with the nutrient it requires, but in general the solid phase of the soil is a reservoir of nutrients that can be exploited by the root. (Concentrations in various soils and problems of measuring nutrient availability are discussed by Epstein, 1972, and by Milthorpe and Moorby, 1974.)

To avoid the interactions between components of nutrient supply many studies of uptake have been made using culture solutions. To be comparable with soil solutions concentrations need to be as low as 10 or 100 mMol m$^{-3}$ for some ions and then there is a problem of supplying adequate nutrient to the roots since uptake depletes the solution. Asher et al. (1965) overcame this problem by growing plants in large volumes of solution that could be circulated past the roots. Clement et al. (1974) describe an adaptation to flowing culture solution systems using ion-sensitive electrodes to monitor concentrations.

Results from a series of papers by Asher et al. using such culture systems are given here to show how plant growth and nutrient content may be affected by external concentration (Asher and Loneragan, 1967; Asher and Ozanne, 1967; Loneragan, 1968; Loneragan and Snowball, 1969a, b). A large number of plant species were grown in this series of experiments but data have been averaged here to emphasise general trends. The concentrations were intended to be representative of concentrations in the soil.

Table 9.2.

Relative K$^+$ content and relative yield of 14 species. (Data from Asher and Ozanne, 1967.)

| K$^+$ in solution (mMol m$^{-3}$) | Relative yield (%) | Relative contents ($\mu$Mol gFW$^{-1}$) | |
|---|---|---|---|
| | | Tops | Roots |
| 1 | 7 | 22 | 12 |
| 8 | 64 | 110 | 42 |
| 24 | 86 | 144 | 66 |
| 95 | 91 | 162 | 83 |
| 1000 | 97 | 190 | 112 |

Table 9.2 shows the relative content of K$^+$ in tops and roots of 14 species of crop and pasture plants grown in culture solutions containing K$^+$ levels between 8 and 1000 mMol m$^{-3}$. Growth is shown relative to the largest yield for each species. One general point is that growth and relative K$^+$ content in the shoot were near the maximum values at 24 mMol m$^{-3}$ and further increase in K$^+$ concentration to 1000 mMol m$^{-3}$ had only a small further effect. Another general point is that the relative content of the shoot was nearly

twice that of the roots. Other experiments at much higher external concentration of $K^+$ show that these trends continue up to at least 60 Mol m$^{-3}$ $K^+$ (Pitman, 1965a). From what is known about $K^+$ in plant cells (Chs 5–8) it is most likely that most of the $K^+$ is in solution in the vacuoles of the plant cells.

Table 9.3.
Relative content of P and relative yield of 7 species. (Data from Asher and Loneragan, 1967.)

| Concentration of P in solution (mMol m$^{-3}$) | Relative yield (%) | Relative content ($\mu$Mol gFW$^{-1}$) | |
|---|---|---|---|
| | | Tops | Roots |
| 0.04 | 50 | 3.8 | 3.9 |
| 0.2 | 76 | 10.5 | 7.4 |
| 1 | 89 | 23 | 15 |
| 5 | 97 | 25 | 20 |
| 25 | 98 | 27 | 22 |

A somewhat similar pattern of response was found for P (Table 9.3), except that there was less difference between root and shoot; the relative content of P was only 10–20% that of $K^+$; and the concentration needed for near maximum growth was lower (0.2–1 mMol m$^{-3}$). In this table the results are the means of 7 species. One (Lupin) has been omitted from the original data as its pattern of response was different in that the relative content of P in the shoot was much higher at 24 mMol m$^{-3}$ than at 5 mMol m$^{-3}$.

Phosphate absorbed by the plant may be converted into a variety of compounds, such as sugar phosphates or nucleotides. Loughman and Russell (1957) showed that at very low concentrations 90–95% of the phosphate was incorporated into organic compounds falling to about 40% at higher concentrations when there was more total P in the plant.

The effect of $Ca^{2+}$ concentration on the amount in the plant is shown in Table 9.4. Two separate groups of plants are shown since the amount in legumes seems to be substantially higher than in cereals and grasses. At lower concentrations the relative content of $Ca^{2+}$ was nearly constant, but then rose almost proportionately to concentration, and the level in the shoot was then much larger than in the root. For barley at least, this trend continues to much higher $Ca^{2+}$ concentrations than the 1 Mol m$^{-3}$ maximum shown in Table 9.4, and appears to be due to an effect of transpiration on uptake to the plant (Fig. 9.6). Magnesium behaves to some extent like $Ca^{2+}$.

At the lower range, growth was found to be impaired below about 2.5 mMol m$^{-3}$ for legumes and below 10 mMol m$^{-3}$ $Ca^{2+}$ for cereals and grasses. More

                    *M.G. Pitman*

Table 9.4.

Relative content of $Ca^{2+}$ in 16 legume species and 11 grass and cereal species. (Data calculated from Loneragan and Snowball (1969a) using percentage dry weights from Asher and Ozanne, 1967.)

| Concentration of $Ca^{2+}$ in solution (mMol m$^{-3}$) | Relative content of $Ca^{2+}$ ($\mu$Mol gFW$^{-1}$) | | | |
| --- | --- | --- | --- | --- |
| | Legumes | | Gramineae | |
| | Tops | Roots | Tops | Roots |
| 0.3 | 4.5 | 1.1 | 1.7 | 0.9 |
| 0.8 | 4.4 | 1.4 | 1.5 | 0.8 |
| 2.5 | 4.7 | 1.6 | 1.7 | 0.7 |
| 10 | 12 | 1.7 | 3.3 | 0.9 |
| 100 | 39 | 2.2 | 10.5 | 1.5 |
| 1000 | 54 | 3.3 | 18 | 2.3 |

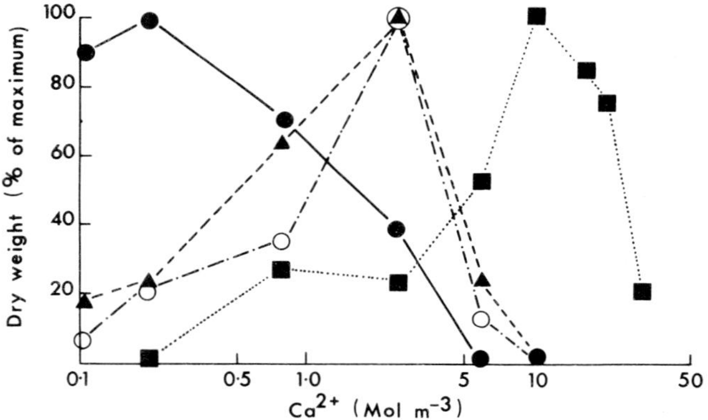


9.1. The growth of plants in cultures containing different levels of $Ca^{2+}$ supplied as $NO_3^-$. For each plant the weight of the shoot is expressed as a percentage of the maximum weight attained. Note that *Juncus squarrosus* (●——●) grew best at very low $Ca^{2+}$ concentrations and *Origanum vulgare* (■·····■) required higher concentrations. *Nardus stricta* (▲---▲) and *Sieglingia decumbens* (○-·-·-○) both showed inhibition of growth at higher $Ca^{2+}$ concentrations. (Redrawn from Jefferies and Willis, 1964.)

complicated responses of growth to $Ca^{2+}$ level are found (Fig. 9.1) in which high concentrations of $Ca^{2+}$ may be inhibitory to certain species but not to others. Such differences in response have been suggested to produce the well known calcicole/calcifuge distributions of plants, but it is likely that other factors than $Ca^{2+}$ are involved, such as pH of the soil.

With the exception of N, other nutrients seem to have parallel behaviour to either $K^+$, $Ca^{2+}$ or P. N uptake is more complex since both $NH_4^+$ and $NO_3^-$ can be taken up. In relation to external concentration Clement et al. (1974) found that in 7 mMol $m^{-3}$, $NO_3^-$ uptake was 70% maximum and there was only a small effect on growth between 7 and 710 mMol $m^{-3}$. Lycklama (1963) compared the effect of both $NO_3^-$ and $NH_4^+$ at varied concentrations and suggested $K_m$ for $NO_3^-$ uptake was about 40 mMol $m^{-3}$ which is higher than expected from Clement's data and may be due to differences in flowing-culture technique.

The amount of N in the plant is related to the dry weight. Fig. 9.2 shows the amounts of N in the shoots of wheat grown on soils with different fertiliser

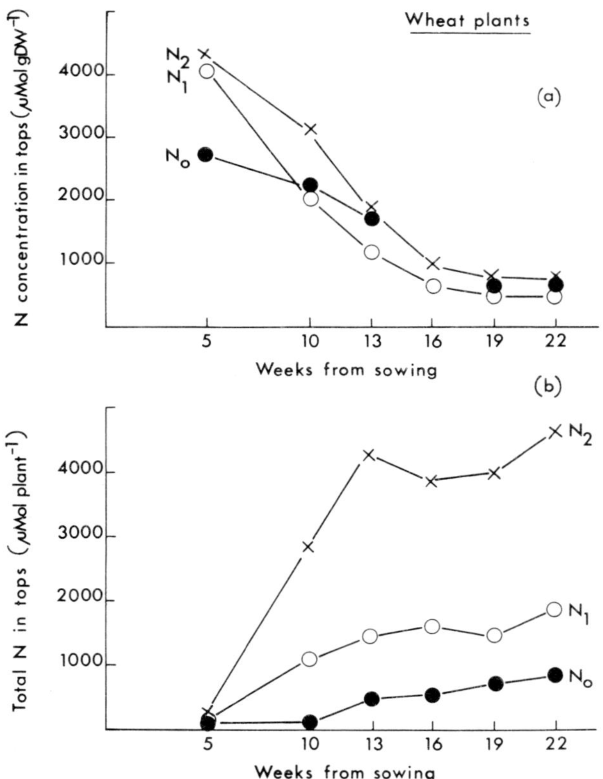


Fig. 9.2. Data for wheat plants grown at 3 different N-levels ($N_2 > N_1 > N_0$). There were about 180 plants $m^{-2}$. (a) Average concentration of N in above ground parts of the plant. (b) Uptake of N above ground plotted against time from sowing. (From Halse et al., 1969, with scales changed.)

application, and the same data expressed as a relative content. The plants grew better and contained more N per plant at the highest N-level but there was much less difference in relative content.

Fig. 9.2 shows another problem of expressing data simply as relative content; both parts of the figure are needed to show that less N was taken up at low N-levels but was used at about the same efficiency. In Tables 9.2–9.4 the content of the plant does not show the real extent to which plant growth was limited by availability of nutrients. Possibly a more descriptive measure of interaction of growth and internal content is the rate of absorption of the ion or nutrient from the soil.

## 9.3 *Supply of nutrients to the plant*

### 9.3.1 *Measurement of the rate of absorption*

Though isotopes are a convenient way of measuring ion uptake over short periods of, say, a few hours, estimates of uptake to the shoot by this method can be unreliable. The isotope becomes 'diluted' by exchange with ions in cell vacuoles in the root so that the specific activity at the site of transfer from the root to the shoot is indeterminate. Consequently, net uptake to the shoot may be underestimated. A more reliable method is to use the change in the content of plants between two harvests. One disadvantage of this method is that changes in content can only be measured accurately over periods of several days; another is that only net transfer from root to shoot is measured.

The rate of net uptake ($J$) is equal to $\dfrac{1}{W_{R}} \cdot \dfrac{dM}{dt}$ where $W_{R}$ is root weight and $M$ is content of the plant (moles per plant.) An average value can be written as

$$J = \frac{(M_2 - M_1)}{\overline{W}_{R}(T_2 - T_1)} \tag{9.1}$$

where $M_2$ and $M_1$ are contents of the plant at times $T_2$ and $T_1$, and $\overline{W}_{R}$ is the mean weight of root between these times. For young plants growing exponentially $\overline{W}_{R}$ can be expressed as the logarithmic mean, as in calculating relative growth rate. In this case $\overline{W}_{R} = (W_{R2} - W_{R1})/\log_e (W_{R2}/W_{R1})$. (For discussion of the use of suitable mean values see Evans, 1972).

The content of the plant ($M$) is equal to the average relative content of the plant times the total weight, or $M = W \cdot X$; hence $J$ is given by

$$J = \frac{1}{W_{\mathrm{R}}}\left[W \cdot \frac{\mathrm{d}X}{\mathrm{d}t} + X \cdot \frac{\mathrm{d}W}{\mathrm{d}t}\right]$$

$$= \frac{W}{W_{\mathrm{R}}}\left[\frac{\mathrm{d}X}{\mathrm{d}t} + X \cdot \frac{1}{W} \cdot \frac{\mathrm{d}W}{\mathrm{d}t}\right]$$

and as $\dfrac{1}{W} \cdot \dfrac{\mathrm{d}W}{\mathrm{d}t}$ is equal to relative growth rate ($R_{\mathrm{W}}$)

$$J = \frac{W}{W_{\mathrm{R}}} \cdot \left[\frac{\mathrm{d}X}{\mathrm{d}t} + X \cdot R_{\mathrm{W}}\right] \tag{9.2}$$

In many cases $X$ tends to be nearly constant so that Eq. (9.2) reduces to the very simple form

$$J = \frac{W}{W_{\mathrm{R}}} \cdot X \cdot R_{\mathrm{W}} \tag{9.3}$$

provided that $W/W_{\mathrm{R}}$ is constant (as it is effectively over long periods of growth). The relationships given here for calculation of $J$ were described by Williams (1948) and have been used in many publications (e.g. Loneragan, 1968; see also Milthorpe and Moorby, 1974).

Note that $X$ is an average value for the whole plant and $M$ is more realistically described as a sum of contents for each part of the plant. For example $M = W_{\mathrm{S}} \cdot X_{\mathrm{S}} + W_{\mathrm{R}} \cdot X_{\mathrm{R}}$ where $X_{\mathrm{S}}$ and $X_{\mathrm{R}}$ are relative contents of the shoot and root. In this case $J$ can be partitioned into net uptake to shoot and root:

$$J = J_{\mathrm{S}} + J_{\mathrm{R}} \tag{9.4}$$

$$J_{\mathrm{S}} = \frac{W_{\mathrm{S}}}{W_{\mathrm{R}}} \cdot X_{\mathrm{s}} \cdot R_{\mathrm{W}} \tag{9.5}$$

$$J_{\mathrm{R}} = \frac{W_{\mathrm{R}}}{W_{\mathrm{R}}} \cdot X_{\mathrm{R}} \cdot R_{\mathrm{W}} = X_{\mathrm{R}} \cdot R_{\mathrm{W}} \tag{9.6}$$

Again $X_{\mathrm{S}}$, $X_{\mathrm{R}}$ and $W_{\mathrm{S}}/W_{\mathrm{R}}$ are assumed constant.

As mentioned earlier, $J_{\mathrm{S}}$ is usually larger than $J_{\mathrm{R}}$ (i.e. most of the ions absorbed are transported to the shoot) since $J_{\mathrm{S}}/J_{\mathrm{R}}$ is proportional both to $X_{\mathrm{S}}/X_{\mathrm{R}}$ and to $W_{\mathrm{S}}/W_{\mathrm{R}}$; usually $X_{\mathrm{S}} > X_{\mathrm{R}}$ and $W_{\mathrm{S}} > W_{\mathrm{R}}$.

The relationship of $J_{\mathrm{S}}$ to relative growth rate is shown in Fig. 9.3 for barley seedlings. For these plants growing in culture solution the concentration of $K^{+}$ in the shoots ($X_{\mathrm{S,K}}$) is very nearly constant over the first two weeks of growth. For plants with the same ratio $W_{\mathrm{S}}/W_{\mathrm{R}}$ (i.e. those shown as $\bigcirc$), $J_{\mathrm{S}}$ was proportional to relative growth rate ($R_{\mathrm{W}}$) as predicted by Eq. (9.5). Plants growing in sand and having much lower ratios of $W_{\mathrm{S}}/W_{\mathrm{R}}$ had correspondingly

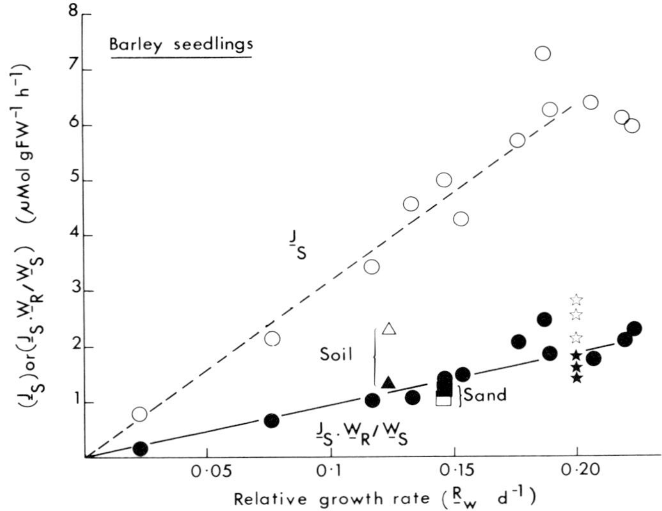



Fig. 9.3. Rate of ($K^+$ + $Na^+$) uptake to shoots of barley seedlings relative to root weight ($J_S$) (open symbols) and the product $J_S$ times $W_R/W_S$ plotted against relative growth rate. ($\bigcirc$, $\bullet$), plants grown on 10 Mol m$^{-3}$ K; ($\triangle$, $\blacktriangle$), plants grown in soil; ($\square$, $\blacksquare$), plants grown in sand; data from Pitman (1972); ($\star$, $\bigstar$), plants grown on 20, 200 and 2000 mMol m$^{-3}$ K$^+$; data from Johansen et al. (1968).

lower $J_S$. When allowance is made for differences in $W_S/W_R$ by calculating $J_S \cdot W_R/W_S$ this quantity is proportional to $R_W$ for all the seedlings as predicted from Eq. 9.5.

These data illustrate a problem of using an average value of $J$ based on total root weight. Differences in efficiency between young and old parts of the root are ignored, and if a flux relative to area of root is calculated from $J$ an 'average radius' must be used to related length of root to area. For example reduced $J_S$ for plants grown in sand compared with plants in solution (Fig. 9.3) could be due either to reduction in flux over the whole of the root, or restriction of uptake to only a proportion of the root surface.

Reduction in $W_S/W_R$ is commonly found as a response to low nutrient availability. For the plants studied in Table 9.3, $W_S/W_R$ was 4 to 6 at full nutrient level falling to about 1 in lower concentrations of P. Bradshaw et al. (1964) found a similar response of various grass species to N. For *Agrostis stolonifera* the response was:

| N-level (ppm) | 1 | 3 | 9 | 27 | 81 | 243 |
|---|---|---|---|---|---|---|
| $W_S/W_R$ | 1.3 | 1.5 | 1.9 | 2.3 | 3.4 | 5.2 |

### 9.3.2 *Effect of concentration on net rate of absorption*

Studies of the effect of concentration on ion influxes have been important as a way of characterising the uptake mechanism in terms of apparent 'enzyme constants' (Ch. 1). The effect of concentration on uptake to the whole plant is more complicated, since at high concentrations the demand of the shoot for growth is the major factor limiting uptake, while at low concentrations growth may be reduced to a level limited by the rate of nutrient absorption. A further complication is that the plant may have different morphology at low concentrations or may have higher concentrations of absorption sites (see later).

Net rates of absorption of various nutrients have been collected in Fig. 9.4 from the sources shown. The dashed lines show where growth was limited by the low concentration of the nutrient. Over the range shown by the solid lines,

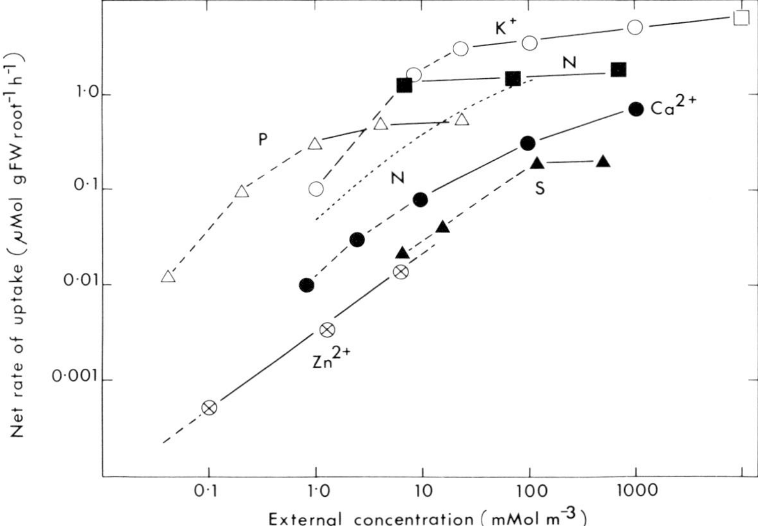


Fig. 9.4. Rates of net uptake of various elements calculated from change in external concentration or internal content. Dashed lines show where growth limitation was observed. Note this is a log/log plot to increase the range of data given. ($\bigcirc$) $K^+$, data from Asher and Ozanne (1967), assuming $R_w = 0.13$ $d^{-1}$, mean of 14 spp; ($\square$) from Pitman (1972) for same $R_w$ for barley; ($\triangle$) P, from Asher and Loneragan (1967), Loneragan and Asher (1967), mean of 8 spp; ($\blacksquare$) N, from Clement et al. (1974) and dotted line, N, from Lycklama (1963), both for *Lolium perenne*; ($\bullet$) $Ca^{2+}$, from Loneragan and Snowball (1969), mean of 30 spp; ($\blacktriangle$) S, from Bouma (1967a, b, c), for *Trifolium subterraneum*; ($\otimes$) $Zn^{2+}$, from Caroll and Loneragan (1968, 1969), mean of 8 spp.

relative growth rates were about 0.1–0.15 d$^{-1}$. As discussed above rates of absorption are affected both by relative growth rate and $W_R/W_S$.

At the lower concentrations where uptake (and growth) appears to be limited by availability of nutrient there seems greater efficiency of the roots at taking up P (and perhaps K$^+$) than the other ions. Measurements of P uptake by a variety of organisms shows that P can be absorbed at a rate to meet the demand of the organism from extremely low concentrations (less than 10 mMol m$^{-3}$). Since those concentrations are common in soils it seems that roots are adapted for more efficient uptake of P than other nutrients. The ability of roots to absorb P at low concentrations has already been discussed (Ch. 7).

Fig. 9.4 showed that net rate of uptake was virtually independent of concentration in the upper range of concentrations, that is, to about 1 Mol m$^{-3}$ for P and 10 Mol m$^{-3}$ for K$^+$. Other experiments showed that this range extends up to at least 60 Mol m$^{-3}$ both for barley and mustard for univalent cations i.e. (K$^+$ + Na$^+$) (Fig. 9.5). The response of net uptake to external concentrations differs from that found for short term influx to excised roots (Fig. 7.3) where influx continues to rise with concentration to about 50 Mol m$^{-3}$ (Mechanism II).

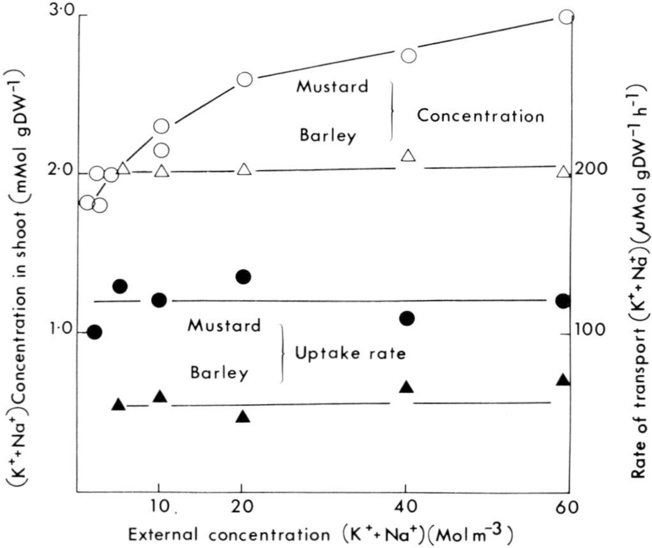


Fig. 9.5. Effect of external concentration of (K$^+$ + Na$^+$) on the concentration of (K$^+$ + Na$^+$) in the shoots of mustard seedlings (○) and barley seedlings (△), and on the rate of transport from the roots (●, ▲). Growth of barley was much less affected by external concentration than that of mustard. (Data from Pitman 1965a, 1966.)

Increasing concentration outside the plant also subjects the plant to decreased water potentials which may in turn reduce growth. Growth of mustard is particularly sensitive to water stress, and in the data of Fig. 9.5, growth was reduced by about 25% between 1 and 60 Mol m$^{-3}$ culture solutions. However the rate of uptake was constant and there was an increased concentration of univalent cations in the shoot (Fig. 9.5). In barley shoots the average concentration in the shoot changed less and growth was less affected, though rates of uptake were again independent of external concentration. These results emphasise the need to consider various components of ion uptake. Yield (i.e. growth), internal concentration and rates of uptake each show different aspects of the overall relationship between the plant and external concentration. It will be seen later that redistribution of ions within the plant further frustrates the drawing of simple conclusions.

Fig. 9.4 showed rates of Ca$^{2+}$ uptake increasing up to 1 Mol m$^{-3}$ and separate experiments show this trend continues at least up to 100 Mol m$^{-3}$ for barley; Mg$^{2+}$ uptake follows a similar pattern. This increased uptake appears partly due to transpiration (Fig. 9.6).

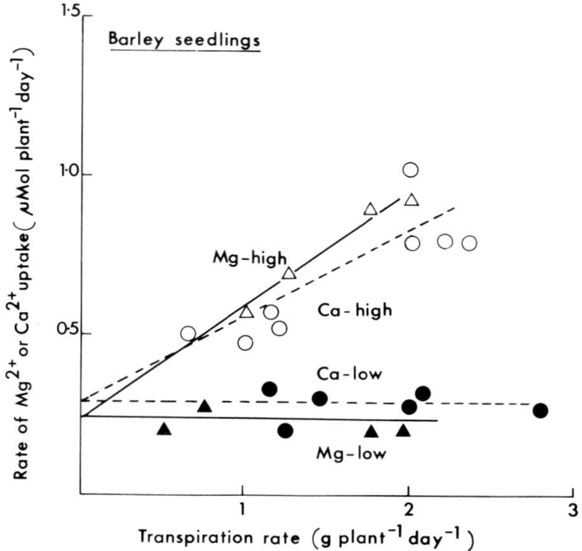


Fig. 9.6. Effect of transpiration rate on uptake of Ca$^{2+}$ (○, ●) and Mg$^{2+}$ (△, ▲) by barley seedlings from solutions containing either 15 Mol m$^{-3}$ (open symbols, high) or 0.5 Mol m$^{-3}$ (closed symbols, low). For both elements uptake appears to have two components, one dependent on both concentration and transpiration and the other not. (Data from Lazaroff and Pitman, 1966.)

For most ions there is a range of concentration above which plant growth is depressed. In Fig. 9.4, growth was reduced by $Zn^{2+}$ above 8 mMol m$^{-3}$. Loneragan and Asher (1967) suggested growth is reduced by phosphate when the level in the leaves increases above about 50 $\mu$Mol gFW$^{-1}$ and therefore if absorption by the roots was greater than about 0.4 $\mu$Mol gFW$^{-1}$ h$^{-1}$. For some species they found phosphate toxicity in concentrations as low as 24 mMol m$^{-3}$, though other species (and varieties) were unaffected.

Plant growth can also be reduced by high NaCl levels in the soil, though the reasons for salinity damage are not clear. Plants show very wide variability in their tolerance for salt and in the uptake of NaCl. This problem will be discussed again later (and see Ch. 10).

### 9.3.3 *Comparison of culture solutions and natural conditions*

Experiments using culture solutions show how plants can respond to particular conditions but plants living naturally face variations in distribution of nutrients in the soil, or seasonal fluctuations in abundance. Rorison (1969) points out that plants have developed different strategies for growing that may eventually achieve the same yield. He gives as an example *Scabiosa columbaria* which grows in calcareous soils where P levels are normally very low, but periodically may be locally abundant. This plant absorbs P in the root at concentrations well above those normally found in the shoot and in roots of other species. When P is available it can therefore be absorbed and stored as an internal source of P for subsequent growth. This behaviour is parallel to the storage of P as polyphosphate in Australian scrub plants. In low-nutrient heath communities of south eastern Australia a number of species (e.g. *Banksia ornata*) store phosphate as polyphosphate. Phosphate is present mainly as orthophosphate during the growing season (Dec–Mar) but as polyphosphate during the rest of the year (Specht and Groves, 1966; Jeffrey, 1968).

Differences in response of plants to available nutrients or to toxic levels of heavy metals or $Al^{3+}$ may not show up simply as effects on total yield but competition between species may amplify the different efficiencies into disparate distributions.

Plants rarely grow on soils in which $K^+$ is the most abundant univalent cation, and more commonly the shoots contain large amounts of $Na^+$. For example, the sclerophyllous plants of the Sydney Sandstone (p. 269) contained 90 Mol m$^{-3}$ $Na^+$ in addition to 50 Mol m$^{-3}$ $K^+$. Where $K^+$ is particularly low this $Na^+$ uptake may partially fill the plants need for high osmolarity of salt in the leaves, while selective transport processes retain $K^+$ preferentially in the cytoplasm (Ch. 10). This 'sparing' effect of $Na^+$ on $K^+$ has been seen often and may lead to doubling of yields at low $K^+$ levels (e.g. Smith, 1974; see also Fig. 9.7). Depending on the availability of $Na^+$, the concentration of $K^+$ in the leaves may have any value between that needed for growth and that needed for the osmotic potential of the vacuole, so the average $K^+$ concentration in the leaves is not a reliable guide to the nutritional status of the plants. Interactions in uptake between $K^+$ and $Na^+$ are considered again below.

At the other end of the concentration range the concentration of ions in the leaves may be determined by the water stress experienced by the plant during growth and this too may be different between culture solutions and natural conditions.

## 9.4 Transpiration and ion uptake

The results in the previous section show that the plant has some measure of control over uptake of most nutrients. The uptake of ions is not due simply to transpiration sweeping in nutrients from the soil and depositing them in the leaves, like a living wick. Some unionised compounds may be taken up in this way, showing the difference between transpiration-dependent and root-controlled uptake. For example, Jones and Handreck (1965) showed that the silica content of oat plants was equal to the volume of water transpired times the external concentration (silica is present in soils mainly as $SiO_2$, an uncharged molecule) and suggested measurements of silica in the plants could be used to determine water loss from an oat crop. Other species of plants, though, show restriction of uptake of silica by the roots in ways analagous to nutrient ions (Barber and Shone, 1966; Handreck and Jones, 1967).

Studies of concentrations in the xylem sap show that it may contain high concentrations of nitrogen compounds usually in the form of amide or amino nitrogen. The concentrations normally may fluctuate depending on the amount of transpiration and particularly on the season. Table 9.5 shows concentrations in xylem sap of apple trees showing peaks of P, N and $K^+$ during the spring–summer growing season.

Concentrations of ions in transpiring plants can also be estimated as the ratios of rates of transport to the shoots and rates of transpiration. For example, using the data of Fig. 9.3, we can take a rate of transpiration of $0.5$–$1.0$ g h$^{-1}$ gram of shoot$^{-1}$ as a 'normal' value and estimate concentrations in the xylem sap of about $1$–$2$ Mol m$^{-3}$ $K^+$, $0.1$–$0.2$ Mol m$^{-3}$ P and $0.1$–$0.2$ Mol m$^{-3}$ $Ca^{2+}$ (taking $W_S/W_R \simeq 3$). The concentrations would be lower, of course, if the rate of water flow were raised.

Table 9.5.
Concentrations (Mol m$^{-3}$) of nutrients in xylem sap of an apple tree, data calculated from Bollard (1953) as monthly averages.

| Month | N | K | P |
|---|---|---|---|
| August | 1.1 | 0.85 | 0.14 |
| September | 1.0 | 0.92 | 0.15 |
| October | 9.8 | 1.8 | 0.7 |
| November | 6.2 | 3.1 | 0.4 |
| December | 5.4 | 3.6 | 0.5 |
| January | 2.7 | 2.4 | 0.3 |
| February | 1.3 | 1.8 | 0.4 |
| March | 1.3 | 1.8 | 0.4 |
| April | 0.4 | 1.6 | 0.4 |
| May | 0.4 | 0.7 | 0.3 |
| June | 0.4 | 0.7 | 0.3 |
| July | 0.4 | 0.7 | 0.3 |
| August | 0.4 | 0.5 | 0.3 |

Earlier chapters (7 and 8) have shown that excised roots exude solution from the cut end due to secretion of ions into the xylem vessels. Concentrations of 30–50 Mol m$^{-3}$ in the exudate are common mainly due to salts of univalent cations. The secretion of ions seems to be an active process drawing on energy metabolism in the root and is sufficiently large to account for rates of transport of K$^+$ and Na$^+$ from root to shoot measured with intact plants (see Table 9.7).

Measurements of rates of transpiration by the barley seedlings used for Fig. 9.5 showed that it was independent of external concentration up to 40 Mol m$^{-3}$ and reduced by about 25% in 60 Mol m$^{-3}$ solution. Hence concentrations of univalent cations (K$^+$ + Na$^+$) in the xylem can be expected to be virtually the same over the whole range of external concentration. In solutions of low K$^+$ concentration (0.1 Mol m$^{-3}$) the xylem will be at a higher concentration than the solution but in 10–60 Mol m$^{-3}$ solutions the xylem concentration will be very much less. The root is acting as a barrier to diffusion, allowing build up of concentration in the xylem of transpiring plants growing in very dilute solutions (or in excised roots) but preventing inflow of ions to the xylem from solutions of high concentration. An extreme example of resistance to influx is shown by mangroves (Chs 10 and 11) in which the xylem sap of non-secreting species has been shown to contain about 5–20 Mol m$^{-3}$ Cl$^-$ when the roots were in sea water of about 400 Mol m$^{-3}$ Cl$^-$ (Scholander et al., 1962). The location of this barrier need not concern us here (see Ch. 8), but the interaction of ion and water fluxes across it and its permeability to ions has been extensively studied.

Water flow and ion transport across the root can be thought of analogously to the 'pump and leak' situation in the cell (p. 179). At low external concentrations the pump maintains a concentration in the xylem against (perhaps) a leak back from the xylem. At higher external concentrations the pump restricts and controls transport across the barrier and the 'leak' now acts potentially in the same direction as the pump. Transpiration can act in this system on the supply of ions to the pump and also on the leak itself. Experimentally, metabolic inhibitors have been used in an attempt to separate the pump from the leak (or active from passive components) or it has been assumed that 'leak' or passive components should be proportional to both water flux and external concentration.

Sutcliffe (1962) summarises experiments relating ion uptake and transpiration. In general it appears that uptakes of $NO_3^-$, phosphate, $NH_4^+$ and $K^+$ are strongly inhibited by factors affecting active transport (temperature of the roots, inhibitors of energy metabolism) but little affected by the rate of transpiration. Brouwer (1954, 1956) also found that uptake of $Cl^-$ to *Zea mays* and *Vicia faba* plants was strongly inhibited by DNP. However, Hylmö (1953) found $Cl^-$ (and $Ca^{2+}$) uptake to pea plants was increased at higher transpiration rates. Hylmö considered this to be evidence that there was a large passive component in uptake of $Cl^-$ and $Ca^{2+}$, i.e. a large leak.

Greenway (1965) measured $Cl^-$ uptake to 10-day old barley plants in 50 Mol m$^{-3}$ NaCl at varied rates of transpiration. He found that transport to the shoot only doubled as the rate of transpiration increased about sixteen times from 40 to 700 mg h$^{-1}$ gram of shoot$^{-1}$. Addition of DNP reduced uptake by about 50% at low and 30% at high rates of transpiration. (The concentration in the xylem was 15 Mol m$^{-3}$ at the lower and 2 Mol m$^{-3}$ at the higher rate of water flow, again showing the ability of the root to restrict ion entry to the plant.) Lüttge and Laties (1967) also found only 30% inhibition of transport of $Cl^-$ to the shoot from 20 to 40 Mol m$^{-3}$ KCl solution by CCCP but almost complete inhibition in 0.2 Mol m$^{-3}$ KCl. Since $Cl^-$ uptake was increased by raising the external concentration it may seem that this pattern of response is due to a constant active transport and an inward leak increasing with both external concentration and water flow.

A similar pattern was found for $Ca^{2+}$ and for $Mg^{2+}$ in barley (Fig. 9.6). Part of the $Ca^{2+}$ uptake appears to be active since Barber and Koontz (1963) found that DNP inhibited $Ca^{2+}$ transport to the shoot very strongly both from 0.5 and 5.0 Mol m$^{-3}$, though over longer periods in DNP (24 h) there was increased $Ca^{2+}$ uptake as if the barrier to diffusion in the root had been broken down. This effect of DNP makes it difficult to analyse uptake into 'pump' and 'leak' by use of inhibitors for there may be no assurance that the inhibitor has not changed the rate of leaking (e.g. as in $Cl^-$ experiments above).

It is doubtful too whether proportionality between uptake and external concentration is an exclusive test of a passive 'leak' process. Baker and Weatherley (1969) measured $K^+$ exudation from de-topped 4-week old *Ricinus* plants and found that the amount of $K^+$ exuded increased proportionately with external $K^+$ concentration. This increase was almost entirely due to an increased active component.

This response to external concentration is clearly different from that shown in Fig. 9.4 (and Fig. 9.5) but it may be relevant that the plants had been on low-$K^+$ nutrient solution for 3 days prior to the experiment. Using low-salt maize plants Lüttge and Laties (1966) also found that transport of $Cl^-$ and $K^+$ to the shoot increased with external concentration. Edwards (1970) measured uptake of P to the shoots of barley seedlings and found increased uptake and higher concentrations, but again the plants were transferred from one nutrient solution (5 mMol m$^{-3}$ for P) to a variety of others and we are likely to be seeing a transient rate of uptake rather than the steady-state rate given in Fig. 9.4.

Measurements of $K^+$ and $Na^+$ uptake to plants growing in culture solutions were made at two different rates of transpiration using barley and mustard seedlings (Pitman, 1965b, 1966). Table 9.6 shows that total net ($K^+ + Na^+$)

Table 9.6.
Effect of transpiration on uptake of $K^+$ and $Na^+$ to shoots of barley and mustard seedlings. Solution contained 15 Mol m$^{-3}$ $K^+$; 45 Mol m$^{-3}$ $Na^+$.

| | Transpiration relative to high level (%) | $[K^+ + Na^+]$ ($\mu$Mol/plant) | $[K^+ + Na^+]$ (mMol gDW$^{-1}$) | $K^+/Na^+$ |
|---|---|---|---|---|
| Barley | | | | |
| high | 100 | $108 \pm 2$ | – | $2.5 \pm 0.1$ |
| low | 50 | $106 \pm 4$ | – | $3.56 \pm 0.25$ |
| Mustard | | | | |
| high | 100 | – | $3.20 \pm 0.1$ | $0.73 \pm 0.07$ |
| low | 7 | – | $0.73 \pm 0.07$ | $2.6 \pm 0.1$ |

uptake was the same at high and low transpiration rates but that the ratio of $K^+/Na^+$ was decreased when transpiration was high. This decrease was more apparent at high than at low external concentrations. The selectivity of $K^+$

and Na$^+$ in the roots was unaffected by transpiration so this change in select-ivity in uptake to the shoot seems to be specifically related to the transport into the xylem. It was suggested that total (K$^+$ + Na$^+$) uptake was regulated by the 'pump' and that water flow affected the supply of ions across the cortex to the pump, wherever it was situated (Pitman, 1966).

This effect of water flow on selectivity of the pump may explain increased Cl$^-$ uptake at higher transpiration rates, only in this case water flow would be changing the ratio of Cl$^-$ to organic acids or amino acids in the xylem, balancing K$^+$, Na$^+$ and Ca$^{2+}$.

Concentrations of ions reaching the leaves may be lower than the estimates made above since cells along the path of the xylem extract ions from the solution. As the xylem solution passes through the stem and the leaf petiole it becomes depleted (Klepper and Kaufmann, 1966). Removal of Ca$^{2+}$ and Na$^+$ in the stem may also reduce the amounts of these ions transported to the leaves and reduce the 'load' of non-essential ions in the leaf.

## 9.5 *Selective uptake of K$^+$ relative to Na$^+$*

Selective uptake of K$^+$ in preference to Na$^+$ is found in many plant cells due to operation of an active K$^+$ influx and active Na$^+$ efflux (Ch. 5). Cells of plant roots show similar selectivity for K$^+$ (Pitman and Saddler, 1967; Jeschke, 1970, 1973). Within the whole plant the overall proportions of K$^+$/Na$^+$ are determined by a number of other factors relating to transport from root to shoot and translocation of ions from older to younger leaves. It is generally found that the ratio of K$^+$/Na$^+$ in the shoot is higher than in the roots and higher again than in the solution. Plants of different species are found to vary in selectivity of K$^+$ relative to Na$^+$ (Collander, 1941), with grasses and certain legumes showing high selectivity for K$^+$ and plants such as *Atriplex* at the other extreme.

A convenient way to study selective uptake to whole plants is to grow them in culture solution. The average content of K$^+$ and Na$^+$ in the shoot maintains more or less constant proportions over long periods of growth though total content increases experimentally (e.g. barley; Greenway et al., 1965). It appears that the root is in a 'steady-state' as far as transport of K$^+$ and Na$^+$ to the shoot is concerned. In barley, the proportions of K$^+$ to Na$^+$ in the xylem exudate from de-topped seedlings is very nearly the same as in the shoot as a whole (Table 9.7) so it appears that the xylem is the main pathway of transport into the shoot. However, within the shoot K$^+$ and Na$^+$ are distributed un-evenly due to K$^+$ retranslocation in the phloem out of older leaves (p. 293).

Table 9.7.

Comparison of $K^+/Na^+$ in shoots of whole barley seedlings and exudate from de-topped seedlings (Pitman, 1965a).

| Concentration (Mol m$^{-3}$) | | Ratio of $K^+/Na^+$ | | | |
| --- | --- | --- | --- | --- | --- |
| $K^+$ | $Na^+$ | Solution | Roots | Shoots | Exudate |
| 0.5 | 9.5 | 0.052 | 0.8 | 3.4 | 2.1 |
| 1.0 | 9.0 | 0.11 | 1.1 | 4.0 | 3.1 |
| 2.5 | 7.5 | 0.33 | 1.9 | 8.1 | 7.1 |
| 6.0 | 4.0 | 1.5 | 6.7 | 18 | 22 |
| 8.0 | 2.0 | 4.0 | 8.9 | 30 | 36 |

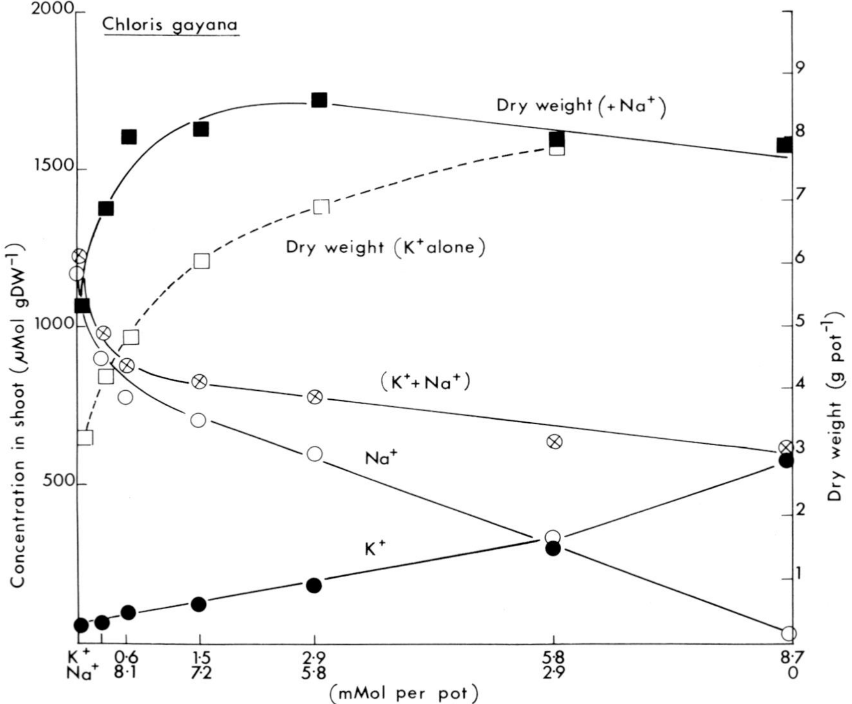


Fig. 9.7. Growth response of plants of *Chloris gayana* grown in pots with varied $K^+$ content (□), or ($K^+ + Na^+$) content (■) shows 'sparing' effect of $Na^+$ on $K^+$. The content of the plants grown on ($K^+ + Na^+$) shows interchange of $Na^+$ for $K^+$, but the total content is less affected: (○), $Na^+$; (●), $K^+$; (⊗), $K^+ + Na^+$. Redrawn from Smith (1974).

Fig. 9.7 shows that the total cation and the total ($K^+ + Na^+$) contents of the tropical grass *Chloris gayana* were virtually independent of the ratio of

$K^+/Na^+$ in the soil (Smith, 1974). Similar results have been found for barley (Pitman, 1965a) and mustard (Pitman, 1966).

Table 9.7 (and the data of Fig. 9.7) also show that the ratio of $K^+/Na^+$ in the plant depends very much on the proportions of $K^+/Na^+$ in the soil or solution. However, the ratio of $K^+/Na^+$ in the plant may also be affected by the total concentration of $(K^+ + Na^+)$ around the roots. Thus, when the external ratio of $K^+/Na^+$ was kept constant but total concentration of $(K^+ + Na^+)$ increased, there was a marked decrease in $K^+/Na^+$ in the shoots of both barley and mustard (Fig. 9.8). The patterns of response of the two

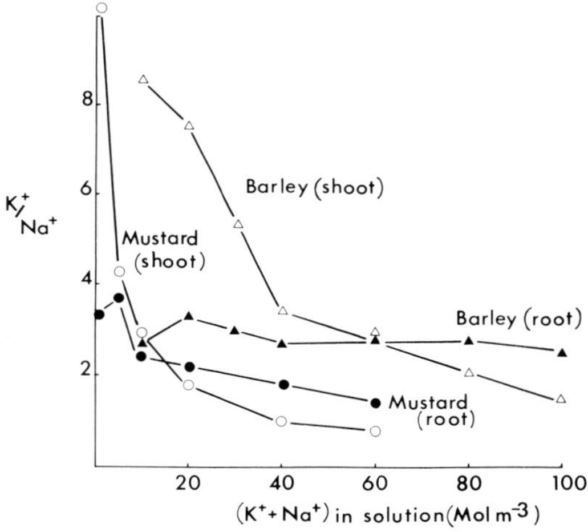


Fig. 9.8. Ratios of $K^+/Na^+$ in root (closed symbols) and shoot (open symbols) when external concentration of $(K^+ + Na^+)$ was varied but $K^+/Na^+$ constant and equal to 1/3. Note that $K^+/Na^+$ in the shoot is more affected by external concentration than in the roots. (Pitman 1965b, 1966.)

species were remarkably alike except that the reduction in selectivity occurred at lower external concentration for mustard than for barley. At low concentrations (in the range found in many soils) mustard can be as selective for $K^+$ as barley, but it becomes less effective in more saline conditions.

The ratio of $K^+/Na^+$ in the roots was much less affected by external concentration than that in the shoots, particularly for barley. Presumably the content of the roots is determined by fluxes into the cells but the content of the shoot is also affected by the rate of transpiration (Table 9.6). At higher concentrations $Na^+$ uptake was increased at the expense of $K^+$ when transpiration was occurring, but at low concentrations transpiration had no effect on $K^+$ and $Na^+$ uptakes. As will be shown below, there are other situations where changes in selectivity in the shoot appear to be due to changed select-ivity at the cellular level (Fig. 9.9).

There are many other published data for uptake of $K^+$ and $Na^+$ that show similar trends in selectivity. Often, though, comparison has been made by changing $Na^+$ concentration when $K^+$ was kept constant so that both $(K^+ + Na^+)$ concentration and the ratio of $K^+/Na^+$ varied. In this case it is less easy to separate effects due to concentration from those due to the $K^+/Na^+$ ratio.

### 9.5.1  *Divalent cations and selectivity of $K^+/Na^+$ uptake*

The ability of plants to take up $K^+$ selectively appears to depend on availab-ility of divalent cations. Fig. 9.9 shows the ratio of $K^+/Na^+$ in roots and

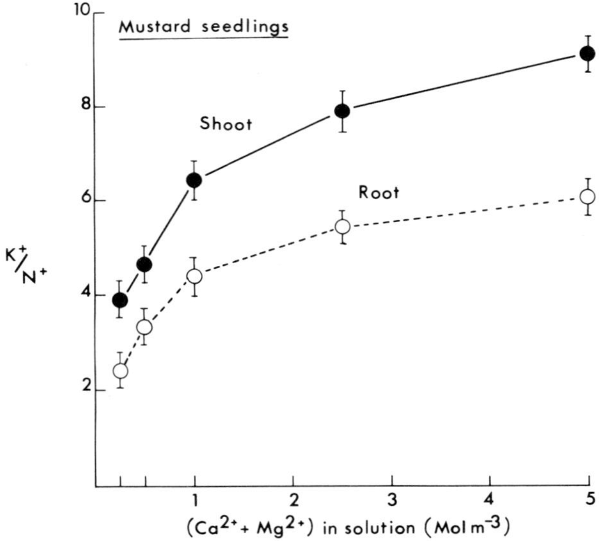


Fig. 9.9. Effect of varying divalent cation concentration on the ratio of $K^+/Na^+$ in root and shoot of mustard seedlings, when $K^+ = 0.5$ and $Na = 1.5$ Mol m$^{-3}$ in the solution. The ratio of $Ca^{2+}/Mg^{2+}$ was constant at $3:2$. (Data from Pitman, 1966.)

shoots of mustard seedlings when the concentration of $Ca^{2+}$ and $Mg^{2+}$ was varied. Note that uptake to both root and shoot was affected. The concen-tration of divalent cations needed for maximum selectivity varied with external concentration of $(K^+ + Na^+)$ and with the kind of plant. Barley was much less sensitive to $Ca^{2+}$ level than mustard, but in both cases maximum selectivity was only found when $Ca^{2+}$ was greater than about 2.5 Mol m$^{-3}$. Hyder and Greenway (1965) have suggested this role of divalent cations may be important in the resistance of plants to high salinity.

### 9.5.2 Conclusions

The selectivity of a plant for $K^+$ appears to be determined at two levels. Firstly, the selective transport of $K^+$ into the root cells and secondly, the transport of $K^+$ and $Na^+$ across the root to the xylem. The selectivity of the uptake process to the root cells is determined by $K^+/Na^+$ in the solution and by availability of $Ca^{2+}$. It thus has properties like selective $K^+$ and $Na^+$ transport in algal cells (Ch. 5) and, like ($Na^+$, $K^+$)-ATPase systems, requires divalent cations for activation (Ch. 2). Transport across the root operates proportionally to the selectivity in the root cells (Pitman, 1966) but can be affected too by the water flow across the root to the shoot and the concentration of ions external to the root. A further component of $K^+$ and $Na^+$ uptake is translocation in the phloem and this too will affect the selectivity of $K^+$ in various parts of the shoot as discussed below.

## 9.6 Translocation in the phloem

The xylem is the primary pathway for movement of nutrients to the shoot, but the phloem is involved in redistribution of ions within the shoot. Ions are retranslocated from older to younger leaves, to the roots and to developing fruits and seeds, which in many cases are supplied entirely via the phloem.

Cereal plants show redistribution of nutrients during the growth of the plant. As the stem elongates the oldest leaves become senescent and eventually, when the grain is being formed, photosynthesis is carried out almost entirely by the flag leaf. Eventually, the grain weighs about 45% of the total plant and contains 70–80% of the total N, 35% of the total $K^+$ and 85% of the total P in the plant. As the grain develops, N and other nutrients are translocated out of the plant as a whole into the grain, so that at this stage the plant is acting as its own resource for N and uptake is negligible (Fig. 9.10).

Greenway (Table 9.8) has measured translocation of $K^+$, $Na^+$ and $Cl^-$ into developing grain of barley plants grown under different salinities, and shown how efficient the transport system is in discriminating against $Na^+$. Despite large $Na^+$ and $Cl^-$ concentrations in the leaves, only small amounts of $Na^+$ and $Cl^-$ were transported to the grain of the saline-treated plants, whereas $K^+$ translocation was the same from each treatment. The amount of $K^+$ translocated to the grain was about 1% of the $K^+$ in the plant as a whole.

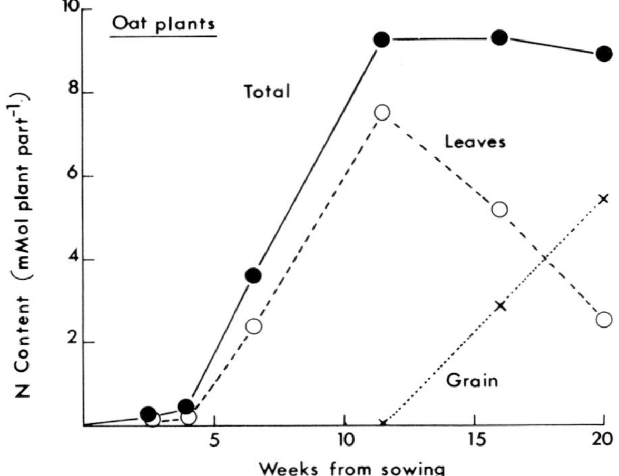



Fig. 9.10. Amounts of N in oats plants growing at high N and P levels. Note the redistribution from leaves to grain from about 12 weeks onwards. (Data from Williams, 1936.)

Table 9.8.

Translocation of $K^+$, $Na^+$ and $Cl^-$ to grain of barley plants and levels in the leaves. (Data from Greenway et al., 1965.)

| Ion | Concentration in solution $(Mol\ m^{-3})$ | Concentration in leaves $(mMol\ gDW^{-1})$ | Concentration in grain $(mMol\ gDW^{-1})$ | Amount translocated $(mMol)$ |
|---|---|---|---|---|
| (a) $K^+$ | 6 | 1.3 | 0.29 | 2.36 |
| $Na^+$ | 1 | 0.1 | 0.02 | 0.16 |
| $Cl^-$ | 1 | 0.2 | 0.05 | 0.40 |
| (b) $K^+$ | 6 | 0.5 | 0.25 | 2.00 |
| $Na^+$ | 50 | 1.0 | 0.04 | 0.32 |
| $Cl^-$ | 50 | 1.0 | 0.09 | 0.72 |

Mature barley grains weigh about 44 mg and contain $0.18\ mMol\ gDW^{-1}\ K^+$, $0.08\ mMol\ gDW^{-1}\ Mg^{2+}$ and $0.01\ mMol\ gDW^{-1}\ Ca^{2+}$, i.e. $Mg^{2+}$ is translocated to the grain almost as fast as $K^+$, and very much more rapidly than $Ca^{2+}$. Translocation of $Ca^{2+}$ is generally slow both into fruits and out of leaves. Use of tracers shows small amounts of $^{45}Ca$ move in the opposite direction to flow in the xylem (and so presumably in the phloem) but only when levels of $Ca^{2+}$ in the leaves are high (e.g. Millikan and Hanger, 1966; Ringoet et al., 1968). Often the supply of $Ca^{2+}$ in the phloem is inadequate to meet the demands of growing fruits and a number of diseases of cultivated fruits are due to lack of $Ca^{2+}$ in the fruit (for example 'blossom end rot' of tomato).

The mobility of ions in the phloem has been studied extensively using the technique of applying a radioactive nuclide to the surface of a leaf and observing its subsequent redistribution either by autoradiography, or by measuring radioactivity. Basal movement from the site of application has been used as a measure of relative efficiency of translocation in the phloem. The method has limitations, particularly as there is some uncertainty about the specific activity of the ion at the site of loading to the phloem. Another problem is that a rapidly translocated nutrient, such as P, may pass from the leaves to the roots and then be returned to the leaves in the xylem so that measurements made over a long period may underestimate translocation (Greenway and Gunn, 1966). There is the basic assumption too that water always flows acropetally in the xylem.

Steucek and Koontz (1970) collected data for translocation of foliar – applied nutrients in bean plants (*Phaseolus*) (Table 9.9). It can be seen that $Mg^{2+}$ is very much more labile than $Ca^{2+}$, as already discussed for seeds. $Na^+$ translocation was very rapid, confirming other measurements showing large net export of $Na^+$ from *Phaseolus* leaves (see Fig. 9.11). *Phaseolus* seems to be different in this respect from many other plants where $Na^+$ translocation

Table 9.9.

Export of radioactive tracers from leaves of bean plants (*Phaseolus*) expressed relative to the amount absorbed in the leaf. (a) From Bukovac and Wittwer, 1957; (b) from Steucek and Koontz, 1970; (c) from Biddulph et al., 1959).

| Tracer | Tracer exported in 24 h (%) |
|---|---|
| $^{22}$Na | 39 (a) |
| $^{86}$Rb | 34 (a) |
| $^{42}$K | 25 (a) |
| $^{32}$P | 25 (a) |
| $^{35}$S | 12 (a) |
| $^{36}$Cl | 7 (a) |
| $^{28}$Mg | 7 (b) |
| $^{45}$Ca | 0.05 (c) |

         *M.G. Pitman*

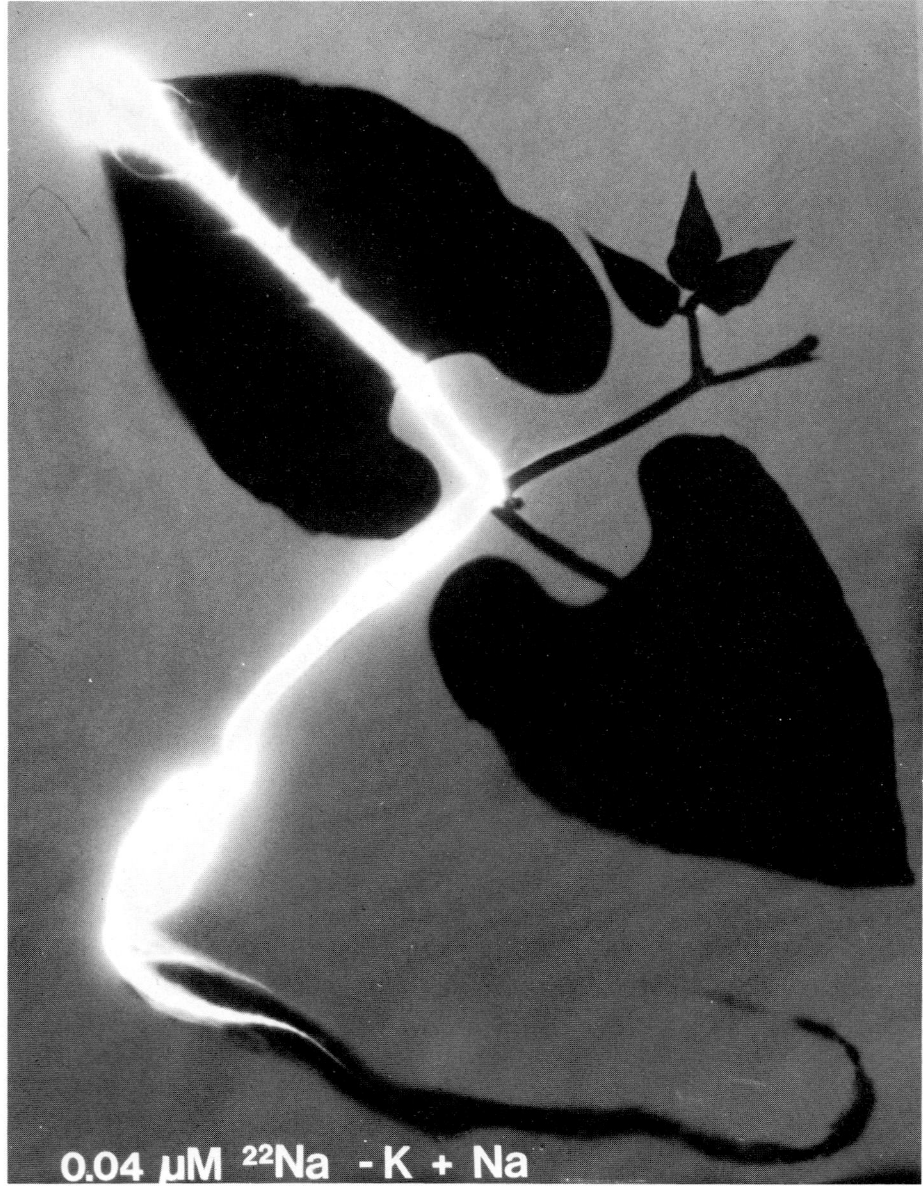



Fig. 9.11. Retranslocation of $^{22}$Na 48 h after application to the tip of a primary leaf of *Phaseolus vulgaris*. The autoradiograph has been superimposed on a photograph on the freeze dried plant. Provided by Prof. Dr. H. Marschner, Technische Universität Berlin.

is usually very slow. In the leaves of barley plants for example, the ratio of $K^+/Na^+$ falls as the leaves become older because there is greater export of $K^+$ than $Na^+$ from older to younger leaves. Plants grown on a solution containing 2.5 Mol m$^{-3}$ $K^+$ + 7.5 Mol m$^{-3}$ $Na^+$ maintained a steady proportion of $K^+/Na^+$ in the shoot as a whole over 10 days of seedling growth, but the ratios of $K^+/Na^+$ in 1st, 2nd and 3rd leaves were 7.5, 15.4 and 23, respectively (Pitman, 1965a).

Export of $K^+$, $Na^+$, $Cl^-$ and P from leaves has been estimated from measurements of total content and tracer uptake to the leaf from the roots. For example, the total content of $K^+$ in the oldest leaf of barley was found to change by only 0.34 $\pm$ 0.4 $\mu$Mol d$^{-1}$ plant$^{-1}$ and the amount of tracer supplied to the leaf from the roots increased by 1.9 $\mu$Mol d$^{-1}$ plant$^{-1}$. Apparently, there was an export of $K^+$ from this leaf at a rate of 1.6 $\mu$Mol d$^{-1}$. In fact, the rate of import was higher due to dilution of tracer with $K^+$ from the vacuoles of root cells (Greenway and Pitman, 1965). Rates of export from the first leaf of barley expressed as an amount per day as a percentage of content of the leaf were:

| | | | |
|---|---|---|---|
| $K^+$ | 3.1 $\mu$Mol d$^{-1}$ | (7%) | (Greenway and Pitman, 1965) |
| $Na^+$ | 0.7 $\mu$Mol d$^{-1}$ | (0.8%) | (Greenway et al., 1965) |
| $Cl^-$ | 1.6 $\mu$Mol d$^{-1}$ | (2.7%) | (Greenway et al., 1965) |
| P | 0.8 $\mu$Mol d$^{-1}$ | (14%) | (Greenway and Gunn, 1966) |
| Sugars 7 | $\mu$Mol d$^{-1}$ | (50–100%) | (Greenway and Pitman, 1965) |

Measurements of P and $K^+$ retranslocation in leaves of different ages show that three stages can be distinguished, which roughly parallel patterns of sugar translocation. During the early stages of development of the leaf (to about 20% of its final weight) nutrients are supplied predominantly in the phloem. There then follows an intermediate stage when the leaves are increasing their content of the ion, possibly now via the xylem, but are exporting the ion in the phloem. At the final stage when the leaf is mature there is concurrent import and export, but little increase in net content. (At even later stages when the leaf becomes senescent the export may be larger than the import; see Thrower, 1962; Hopkinson, 1964; Greenway and Gunn, 1966.)

In general, the mobility of ions in the phloem is related to the content of the phloem sap. MacRobbie (1971) quotes the following values she collected from various sources:

| | | | | | |
|---|---|---|---|---|---|
| $K^+$ | = 20–85 | Mol m$^{-3}$ | $Ca^{2+}$ | = 0.25–0.5 | Mol m$^{-3}$ |
| $Na^+$ | = 0.06–0.3 | Mol m$^{-3}$ | $Mg^{2+}$ | = 2.3–23 | Mol m$^{-3}$ |
| P | = 3–10 | Mol m$^{-3}$ | | | |

$NO_3^-$ and $SO_4^{2-}$ were both absent and these elements are presumably translocated as organic compounds. Further information is given by Ziegler (1975). It appears that the efficiency of translocation of elements is largely due to the efficiency of loading into the phloem; once in the phloem redistribution can take place. Differences in $Na^+$ mobility between *Phaseolus* and barley (for example) are then presumably due to greater efficiency of the phloem in *Phaseolus* at absorbing $Na^+$. The low mobility of $Ca^{2+}$ in the plant also reflects its low rate of entry into the phloem.

## 9.7 *Regulation of nutrient content in leaves*

The leaf is a closed system receiving nutrients via the xylem and at early stages of development via the phloem. Ions may be exported again in the phloem. In most physiological experiments these are the only mechanisms contributing to the amount in the leaf, though other components of loss from the leaf need to be recognised. Under very humid conditions guttation may occur. In field conditions water will be present occasionally on the leaves due to rain or mist, and may leach ions from the leaves (p. 300).

The amount of an ion in the leaf is perhaps less useful as a measure of physiological activity than its concentration (expressed for convenience in any of the ways given earlier). The question discussed here is the extent to which the concentration in the leaf is regulated and by what mechanisms. Results given earlier in this chapter seem to show some regulation in the shoot as a whole. The average concentrations of P, N, $K^+$ (or $K^+ + Na^+$ when both ions are present) were relatively steady with age, or when external concentrations or relative growth rate were varied. In contrast, other ions such as $Ca^{2+}$ or $Mg^{2+}$ increased with age and external concentration (Figs 9.5, 9.6, 9.12 and Tables 9.2, 9.3, 9.4). However, the average concentration of the shoot masks variations that occur during development of leaves due to changing rates of import and export (p. 292).

The amounts of ions in a leaf are determined by the balance between uptake and export, but the concentration of the ions is affected by growth of the leaf too. (Concentration in this sense can be expressed as amount relative to cell water or to fresh-weight of the leaf, or to dry weight.) Measurements show that for many ions the concentration in developing leaves is often low at first as growth of the leaf outstrips import, but then concentration may be steady due either to balance between expansion of the leaf and net import of ions, or to reduction in net import of ions as export increases (Greenway and Thomas, 1965). As the leaf passes through maturity and senescence, growth

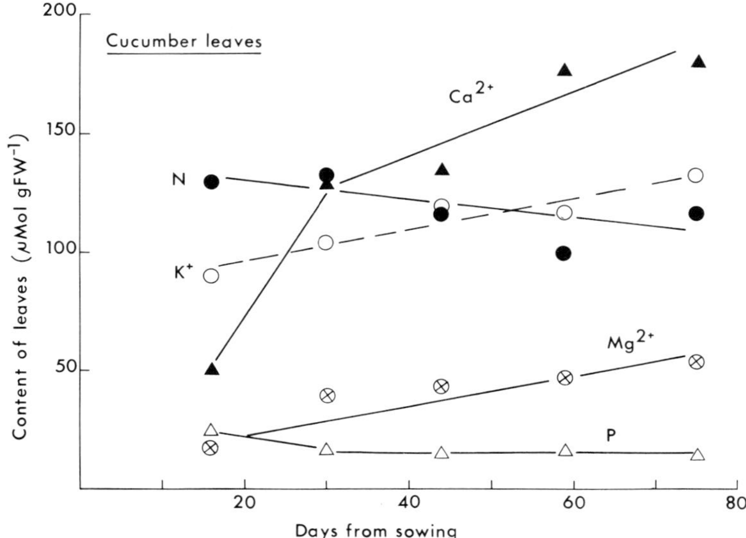



Fig. 9.12. Concentrations of N, P, K$^+$, Ca$^{2+}$ and Mg$^{2+}$ in leaves of cucumber plants. Note that the relative increase in Ca$^{2+}$ and Mg$^{2+}$ concentration (2–3 fold) was very much larger than changes in N, P or K$^+$. The plants stopped production of leaves from 44 days. (Data from Vogel and Weber, 1932.)

is negligible and changes in concentration depend on the difference between import and export.

Concentrations of labile elements such as P or K$^+$ in individual leaves may fall, particularly when the supply from the roots is low (Fig. 9.13). There is evidence for regulation of univalent cation concentration in the leaf since this may be constant over long periods in mature leaves, except when the leaf is subject to excess salinity (Cl$^-$, Fig. 9.13) or water stress (p. 296). In contrast, concentrations of Ca$^{2+}$ and Mg$^{2+}$ often increase with age (Fig. 9.12) due to the slow export of these ions from the leaves and there seems to be little or no regulation of their concentration in the leaf.

Processes affecting the concentrations of the non-metabolised ions such as K$^+$, Na$^+$ or Cl$^-$ in the leaf are shown in Fig. 9.14. The difference between import to the leaf and export in the phloem determines net changes in the amount of each ion in the leaf, but the concentration in the leaf cells is determined by active transport acting against the efflux from the vacuole. (As for example in algal cells (Ch. 5) barley root cells (Ch. 7) and storage tissues (Ch. 6.) Export from the leaf and net uptake to cells of the leaf compete for ions in the free space and a critical question is whether the rate of influx is regulated by vacuolar content or whether it 'floats', controlled only by the availability of ions in the free space.

Measurements of fluxes into cells of the leaf are difficult to make, due to the problem of supplying ions to the free space at a known concentration. One way of overcoming this problem has been to use narrow slices of the leaf (e.g. Smith and Epstein, 1964), which seem to have the same rates of absorption as the intact leaf when freshly cut (Pitman et al., 1974a). It was found that, when put into KCl solutions of increasing concentration, the slices took up $K^+$ and $Cl^-$ to higher levels than found in the leaf. It appears then that the constant concentrations in the leaf are due not to regulation at the cellular level, but balance of import and export from the leaf (Pitman et al., 1974a). However, it is clear that the efficiency of the transport mechanism may change during development of the leaf (Jacoby et al., 1973).

It has been generally found that $K^+$ or $Cl^-$ influx to freshly cut leaf slices is very nearly the same in light and dark and the transport process can be coupled to energy metabolism independently of photosynthesis. An obvious advantage of this arrangement is that it avoids large fluctuations in concentration in the vacuoles between day-time and night-time.

The concentration of $K^+$ and $Na^+$ in the leaves of many species increases when grown under water stress, usually due to decreased water potential around the roots. Often the increased concentration is accompanied by decreased leaf area and net assimilation. Fig. 9.5 for example, showed ($K^+$ + $Na^+$) increased in the shoot of mustard plants with increased external concentration; growth was reduced too. A more striking response has been shown in *Atriplex* (Gale and Poljakoff-Mayber, 1970; Gale et al., 1970). Fig. 9.15 shows that in normal air the growth was low in solutions containing little NaCl, increased to a maximum and then decreased with further increase in NaCl. In humid air the growth was higher in solutions of low NaCl concentration. This growth response was interpreted as showing a need for the plant to take up enough NaCl to maintain high concentrations in the leaves when under water stress. Measurements of water potential of the leaf sap (Fig. 9.15) showed that this changed with external water potential to about the same extent. No information is available about how the higher NaCl concentrations resulted and in terms of Fig. 9.14 it would be interesting to know if water stress acted on the influx–efflux balance of the cell or the import to the leaves.

While salt bush is adapted to make use of high levels of $Na^+$ and $Cl^-$ in the leaves and other plants can tolerate concentrations of 300–400 Mol m$^{-3}$ (Ch. 10), at the other end of the range there are many species that are very sensitive to increased salinity. Greenway, in various publications, has stressed the importance of regulation of $Cl^-$ and $Na^+$ concentration in the leaves as

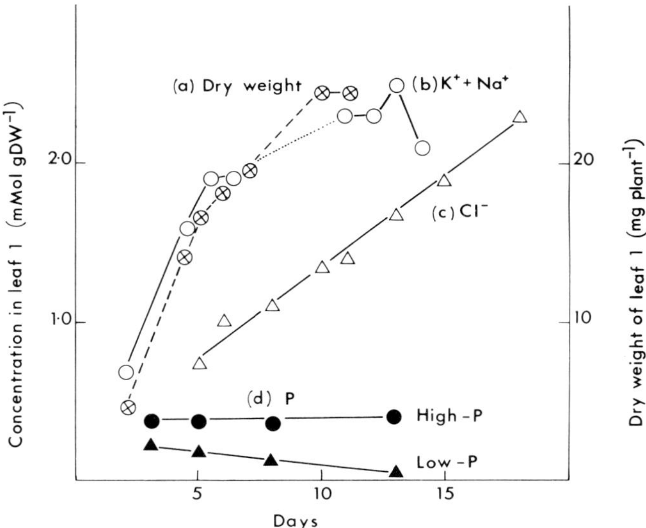



Fig. 9.13. Changes in dry weight of the first leaf of barley seedlings and in concentration of certain ions. (a) Dry weight; (b) $(K^+ + Na^+)$ concentration, plants grown in 10 Mol m$^{-3}$ K$^+$ or $(K^+ + Na^+)$ (data of Pitman, 1965a and Greenway and Pitman, 1965); (c) Cl$^-$ concentration in leaves of plants transferred to 100 Mol m$^{-3}$ NaCl (data of Greenway et al., 1966); (d) changes in P concentration for plants grown on high and low P (data of Greenway and Gunn, 1966).

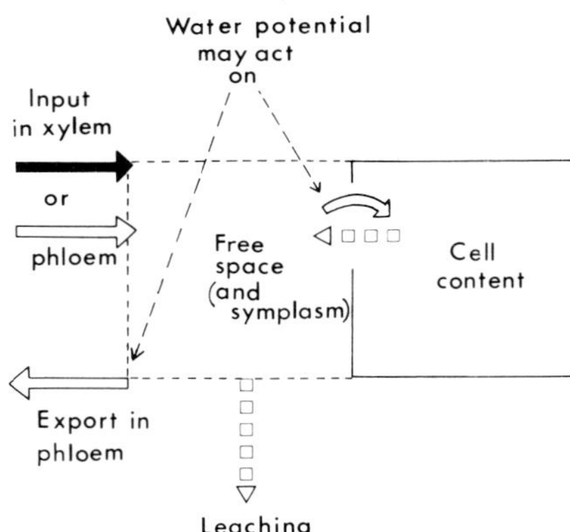



Fig. 9.14. Regulation of ion content in the leaf involves many processes. The observed concentration is that in the cells of the leaf, but this is in contact with the free space and the symplasm in the leaf and so can change in response to variation of input in xylem or phloem, export in the phloem or leaching across the cuticle. The cell content also appears to be affected by water potential in certain species.

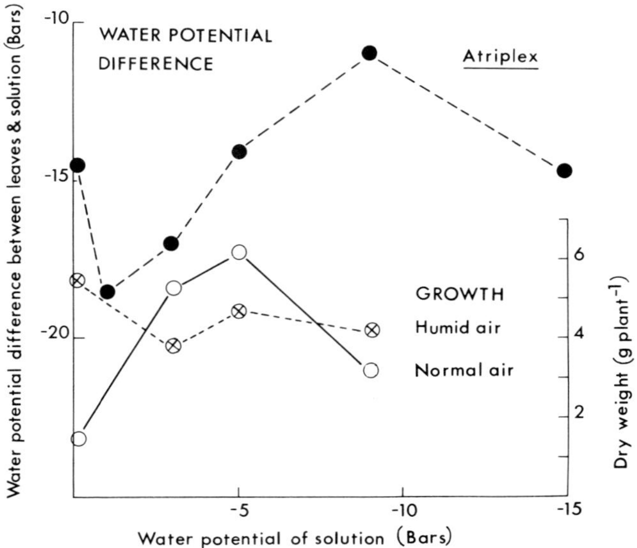



Fig. 9.15. Effect of water stress on growth and water potential difference of leaves of *Atriplex*. (●), water potential difference between leaves and solution of NaCl; (○), growth of plants in normal (dry) air; (⊗), growth of plants in humid air. (Data from Gale and Poljakoff-Mayber, 1970 and Gale et al., 1970.)

contributing to the ability of moderately resistant plant to tolerate saline conditions. Particularly in young developing leaves the continuing growth provides space for accumulation of $Na^+$ and $Cl^-$ taken into the leaf. In older leaves there seemed to be less ability to regulate $Na^+$ or $Cl^-$ concentrations, but the increased content of mature/senescent leaves provides a means for the plant to remove NaCl (see also Ch. 10). It is interesting that the saline-sensitive *Phaseolus* has a well-developed mechanism for removal of $Na^+$ and $Cl^-$ via the phloem (Fig. 9.11 and Greenway et al., 1966)

An interesting problem concerning salinity from leaves is shown by damage to Norfolk Island Pine trees growing along metropolitan beaches in Sydney. Since the late 1940s it has been noticed that foliage has died on the trees and new trees do not become established along the beachfront. This damage has been correlated with increased levels of $Na^+$ and $Cl^-$ in the leaves and only occurred in the metropolitan area. Elsewhere, trees exposed to excessive sea spray showed their normal resistance to salt. It was suspected that the NaCl entered the leaves due to detergent carried from nearby sewage outlets in the sea spray and this theory was supported by experiments in which foliage was sprayed with detergent alone, NaCl alone or NaCl plus detergent. Damage was only found when salt and detergent were present together. The resistance of the Norfolk Island Pine to salinity seems to be due to the ability to exclude NaCl from the roots and from penetration into the leaves across the cuticle or as droplets of spray through the stomata (see Fig. 9.16). The presence of detergent allows NaCl to pass across the cuticle or in through the stomata and once in the leaf there is no mechanism for its effective removal.

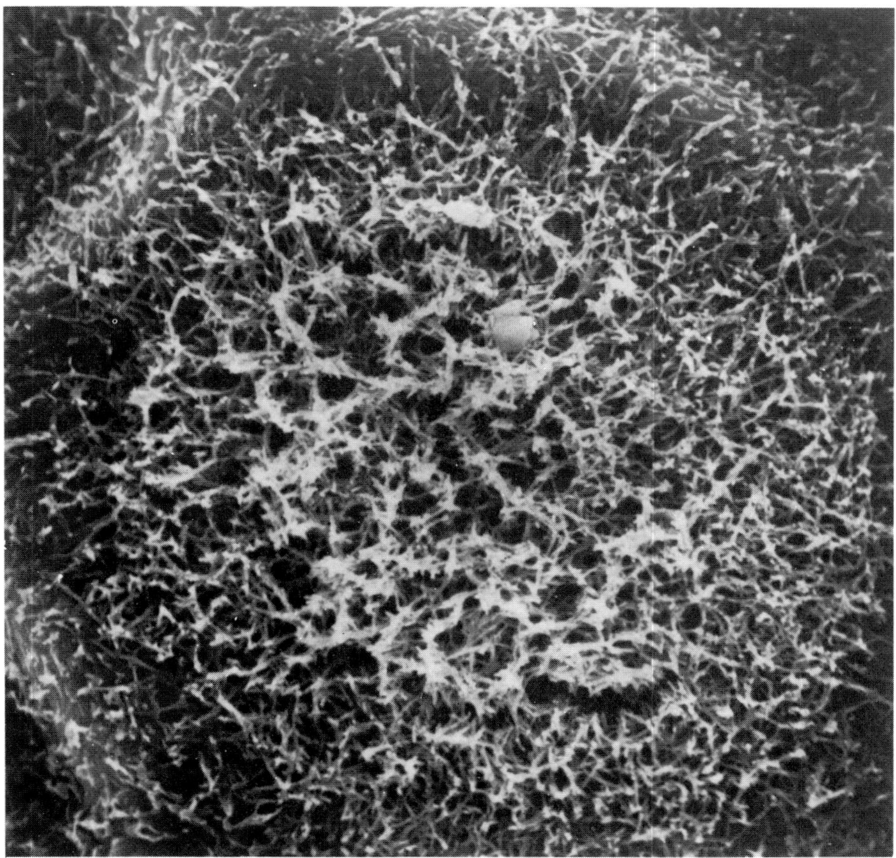

Fig. 9.16. Scanning electron micrograph of waxy hairs over the stoma of Norfolk Island Pine leaf. Droplets of spray are prevented from penetrating the stomatal cavity (photo by C. Nockolds, A. Grieve and M.G. Pitman).

### 9.7.1 *Leaching of ions from leaves and foliar absorption of nutrients*

In most physiological experiments using culture solutions, leaching of nutrients from the leaves is not likely to affect the results given in Tables 9.2–9.4 and Fig. 9.4, nor the conclusions drawn about regulation of nutrient uptake. However, large losses of nutrients may take place under natural conditions where water from rain or mist can wash over the leaves. Table 9.10 gives

Table 9.10.

Effect of rain in leaching nutrients from leaves of deciduous forest trees.

| Element | *Quercus petraea* | | Maple/birch/beech | |
| --- | --- | --- | --- | --- |
| | Content of leaves[a] ($\mu$Mol gDW$^{-1}$) | % removed by rain during season | Content of leaves[b] ($\mu$Mol gDW$^{-1}$) | % removed by rain during season |
| N | 760 | −1.8 | 1700 | 13 |
| P | 19 | 40 | 19 | 12 |
| K | 69 | 240 | 270 | 91 |
| Na | 19 | 1200 | 0.8 | 500 |
| Ca | 155 | 42 | 160 | 30 |
| Mg | 38 | 120 | 46 | 50 |
| S | – | – | 15 | 370 |

[a] At leaf fall, Carlisle et al. (1966).
[b] Standing crop, Eaton et al. (1973).

examples of content of litter or foliage from forests in N. America and N.W. England and the relative amounts removed each year in rainfall. (These figures allow for nutrients entering the system in the rainfall above the canopy.) Large proportions of $Na^+$ were removed from the leaves even though the content of the leaves differed. $K^+$, $Ca^{2+}$, $Mg^{2+}$ and S were also removed in relatively large amounts from the leaves, but there was some difference in loss of N and P.

Leaching of nutrients from leaves (mainly of horticultural crops) has been reviewed by Tukey (1970) and the general pattern follows the results of the above table. Cations tend to be leached preferentially to anions and the solution from the leaves is usually slightly alkaline. The rate of leaching is often lower for young than for older leaves and, of course, leaves differ from one species to another in the development of cuticle. Loss of ions was suggested to depend on the content of ions in the free space and so be related to supply and retranslocation of ions within the leaf. Under natural conditions it is clearly a process contributing to the overall balance of ions in the leaf, though care should be taken to put the losses into perspective by expressing them in the same units as entry of ions to the shoot. Leaching of ions from leaves by rain has its converse in supplying deficient plants with nutrients as a spray on the leaves.

Using isolated cuticles, measurements have been made of the rate of penetration of ions (McFarlane and Berry, 1974). The results suggest that the cuticle contains negatively charged sites and show $K^+$ penetrating slightly more rapidly than $Na^+$ and univalent cations diffusing faster than divalent cations and faster again than trivalent ions. The ratio of $K^+/Ca^{2+}$ was 4.5 and of $K^+/Fe^{3+}$ was 10. No information was given on anion permeability. This is an area of study that could well be integrated with studies of turnover in natural communities or crops.

## 9.8 Regulation of ion content of the whole plant

Discussion of regulation of ion content of leaves and observations made earlier in this chapter about net rates of transport from root to shoot emphasises the need to look at this latter process in considering regulation of transport in the whole plant. Leaves have varied import and export as they age and redistribution of nutrients is an important and continuing process in the shoot, yet the supply of ions to the shoot appears to be in balance with the overall demand of the shoot and related to the growth of the plant (Fig. 9.3, Table 9.7). The process most closely concerned with supply of ions to the shoot is the mechanism of transport from the root into the xylem. Weatherley (1969) gives an excellent account of the various processes involved in ion uptake to the shoot.

An open question is the extent to which export from the shoot to the root may remove surplus ions from the shoot, and so contribute to maintenance of nutrient concentrations in the shoot. Ben Zioni et al. (1971) suggested that $NO_3^-$ uptake in tobacco could be regulated by the rate at which potassium malate was translocated from the shoot to the root, providing $KHCO_3$ which then exchanged for $KNO_3$. An aspect of this system is that downward transport of $K^+$ in the phloem must be at least as fast as $KNO_3$ transport upward in the xylem. Translocation of tracers shows that $K^+$ (or $Rb^+$) can move more rapidly from shoot to root (this forms the basis for one method of estimating root density in the soil). It seems clear though that the large rates of translocation required to balance $NO_3^-$ uptake are not general in plants. In barley seedlings grown on full nutrient solution the export of $K^+$ from shoot to root was not greater than 0.5 $\mu$Mol gFW$^{-1}$ h$^{-1}$ compared with net transport of 4.0 $\mu$Mol gFW$^{-1}$ h$^{-1}$ (Pitman, 1972). Greenway and Thomas (1965) and Greenway et al. (1966) found translocation of Na$^+$ and Cl$^-$ from shoots to roots of barley, *Phaseolus* or *Atriplex* was much less important in regulation of NaCl content than limitation of uptake to the root.

One factor that affects the rate of import to the shoot is the availability of metabolites. Bowling (1968) cooled the stem of *Helianthus* plants to reduce translocation of sugars to the root, and found there was a 60% reduction in K$^+$ uptake by the roots, though transpiration was unaffected. Hatrick and Bowling (1973) have also shown that Rb$^+$ uptake (and respiration) was correlated with translocation of sugar from the shoot to the root of barley seedlings. In other experiments the uptake of K$^+$ by the roots and transport to the shoot increased by nearly 100% one hour after the start of the light period in barley seedlings grown in 2 h light/22 h dark, and this increase coincided with the appearance in the root of sugars produced in photosynthesis (Pitman, 1972). However no difference in the rate of K$^+$ uptake was found between light and dark using plants grown in 16 h light/8 h dark. It seems that while translocation of nutrients to the root may regulate uptake of ions, particularly when growth is restricted, other controls may also be involved when growth rates are high.

Differences in rates of uptake can be due to the amount of 'carrier' available in the roots. Clark et al. (1973) showed that a maize mutant that was inefficient at using Fe, lacked the ability to transport Fe from the root to the xylem. In other ways the mutant was like normal varieties; in particular, there was little difference in uptake of Fe to the root, production of citrate (which chelates with Fe) and transport of P to the shoot. It appears that the mutant lacks a specific enzyme concerned with Fe transport. A further example of changed levels of 'carrier' is that certain plants grown under nutrient deficiency have shown exaggerated rates of uptake when transferred to normal soils (P uptake: Bowen, 1970; P and S uptake: Bouma, 1967a).

A further interesting aspect of Fe uptake is that H$^+$ release by the roots is related to the Fe status of the plants. Fe-deficient sunflower plants have been shown to acidify the culture solution but then, when Fe is available, make the culture solution alkaline as the plants re-green. Cycles of pH change were induced in the culture solution as the plants became Fe-stressed and recovered (Raju and Marschner, 1972).

Much interest has been given in recent years to the possible involvement of phytohormones in regulation of ion transport from root to shoot. The close correlation between different growth rates and uptake has already been mentioned. It is almost a basic law of botany that changes in plant morphology are mediated by changes in levels of hormones, whether growth is affected by environment or ontogenetic control. For example, the form of plants can be altered when under nutrient stress (e.g. p. 296) or when grown at varied temperatures or salinities. In each case there is evidence for involvement of changed hormone content.

Plant roots produce cytokinins (and gibberellins) which have been suggested to affect growth of the plant as a whole (e.g. Vaadia and Itai, 1969). Specifically, changes in cytokinin levels in xylem exudate from roots have been shown to be associated with water stess (Itai and Vaadia, 1965) and other environmental conditions. Wagner and Michael (1971) have shown that cytokinin in exudate from sunflower seedlings was reduced when N was omitted from the culture solution but increased when N was replaced. Skene and Kerridge (1967) showed temperature affected the level of cytokinin in root exudate from *Vitis vinifera* and Atkin et al. (1973) have shown that cytokinin and gibberellic acid in exudate from maize plants increased strongly with temperature to a maximum at 28° C.

The shoot too produces hormones which may be translocated to the roots in the phloem. In particular, abscisic acid (ABA) seems to be produced mainly in leaves and its level responds rapidly to water stress (e.g. Hiron and Wright, 1973). Though there is no evidence for production of ABA in wilted roots it can be translocated from the shoot to the root (Hocking et al., 1972).

Both ABA and cytokinins have been shown to have a specific inhibitory effect on transport of ions from root to shoot. Transport into the xylem (i.e. exudate from the cut end of a root) can be inhibited by stopping energy metabolism but then uptake at the root surfaces is inhibited too (Ch. 8). Inhibition of transport to the xylem could therefore be due to inhibition of uptake. Cytokinins have been shown to inhibit transport to the xylem without inhibiting uptake to the root as a whole and so appear to be having a direct effect on movement of ions into the xylem and hence on transport from root to shoot. ABA can also inhibit transport without effect on uptake to the root, but this may be an indirect action due to interference with endogenous levels of hormone in the root since both stimulations and inhibitions have been reported depending on the nutrient status of the roots and the temperature (Cram and Pitman, 1972; Collins and Kerrigan, 1974; Pitman et al., 1974).

Another group of substances which act on transport to the xylem without inhibition of uptake are certain inhibitors of enzyme synthesis. Cycloheximide, for example, inhibits protein synthesis and also inhibits transport to the xylem without effect on uptake in short-term experiments (Läuchli et al., 1973). (In longer experiments cycloheximide disturbs nucleotide levels in the cell (Cocucci and Marré, 1973) which may lead to reduced uptake of ions.) The anologue of phenylalanine, *p*-fluorophenylalanine, also inhibits transport but not uptake to the cells (Schaefer et al., 1975). This compound does not stop protein production but renders the protein ineffective by incorporation of the analogue instead of phenylalanine. It seems that transport into the xylem involves a protein which turns over rapidly in the root. Both ABA and cytokinin have been implicated in regulation of RNA and protein metabolism and it is clearly an attractive hypothesis that plant hormones can act on ion transport through control of the number of active sites available.

This hypothesis relating transport to specific proteins and plant hormones provides a basis for interpreting regulation of transport from root to shoot. This system, plus the role of carbohydrate transport, is shown in Fig. 9.17. The xylem is the pathway for upward movement of ions and hormones (cytokinins and gibberellins?). The phloem is the pathway for downward movement of carbohydrates, hormones and possibly ions (K$^+$?) as well as possible upward movement of nutrients. The balance of supply and export of

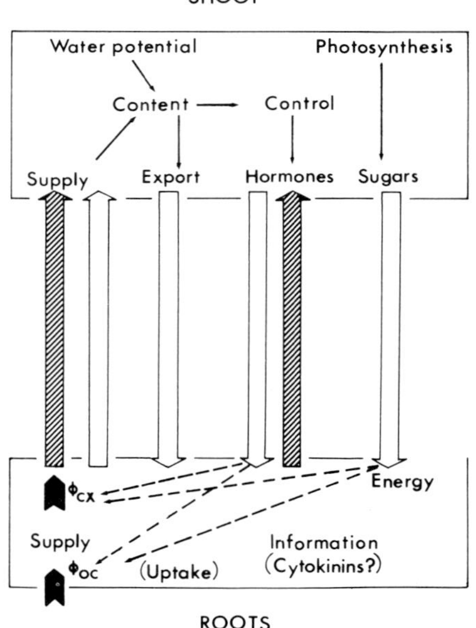


Fig. 9.17. Interaction of processes between root and shoot. In the roots ions enter by the active flux $\phi_{oc}$ and are transported to the xylem by $\phi_{cx}$. Energy for these processes is supplied by the phloem, (blank arrow). Control may also be affected by hormones in the phloem or produced in the roots. Hormone transport in the xylem (dashed arrow) may interact with processes in the leaves. The water potential of the leaves can act on both energy and control factors exported to the root and so affect the content of leaf cells. (See also Pitman, 1972 and Lüttge, 1973.)

nutrients and activity of transport into the cells determines the content in the leaves, and also the retranslocation of ions to younger organs. This is clearly a simplification, as messages carried in the xylem and phloem are likely to be based on combinations of phytohormones rather than absolute levels.

## 9.9 Future research

Work on ion uptake in whole plants spreads in many different directions of interest. Studies with plants in culture solutions gives some valuable information about how certain plants control nutrient uptake, but the number of plants studied is limited. Investigation of responses of plants in the field, or comparisons between ecotypes can be expected to fill in the range of experience needed to build a full picture of the plants response to nutrients. Variations in the solutions that plants have found to the problems of living successfully in an environment are infinite, and we can only experience the virtuosity of these responses by exploring new species and new environments apart from the standard agricultural plants and well-worked ecotypes.

Genetic differences between plants will provide useful comparisons which may help our understanding of the mechanisms involved. The work by Clark et al. (1973) on Fe uptake is one excellent example. Comparisons of plants differing in response to salinity or heavy metal requirements have been initiated in several laboratories and can be expected to yield valuable results.

A basic need at present is the integration of information about whole plant behaviour and properties of individual cells within the plant. The whole area of regulation of nutrient transport in the plant has received little attention in the past, largely because mechanisms for control of uptake were not obvious. There are two levels at which progress might be expected. One is the study of the properties and structure of individual cells within specialised regions such as the bundle sheath in leaves or the stele in the root. The other is the role of phytohormones in the plant. There seems good evidence that protein synthesis and ion transport can be related through the action of phytohormones. This is an area in which we have only the scantest knowledge of the interactions of transport and phytohormones.

However phytohormones operate, the most important development in whole plant physiology, in my view, is that attempts are being made to look for controls and regulatory systems. Such studies will put work with isolated roots or leaf slices into their proper context. In this way we can hope to see the plant treated not as a 'vegetable' but as an integrated organism.

# *References*


C.J. ASHER and J.F. LONERAGAN, Soil Sci., 103 (1967) 225.

C.J. ASHER and P.G. OZANNE, Soil Sci., 103 (1967) 155.

C.J. ASHER, P.G. OZANNE and J.F. LONERAGAN, Soil Sci., 100 (1965) 149.

R.K. ATKIN, G.E. BARTON and D.K. ROBINSON, J. Exp. Bot., 24 (1974) 475.

D.A. BAKER and P.E. WEATHERLEY, J. Exp. Bot., 20 (1969) 485.

D.A. BARBER and M.G.T. SHONE, J. Exp. Bot., 17 (1966) 569.

D.A. BARBER and H.V. KOONTZ, Plant Physiol., 38 (1963) 60.

A. BEN ZIONI, Y. VAADIA, S. HERMAN LIPS, Physiol. Plant., 24 (1971) 288.

O. BIDDULPH, R. CORY and S. BIDDULPH, Plant Physiol., 34 (1959) 512.

E.G. BOLLARD, J. Exp. Bot., 4 (1953) 363.

D. BOUMA, Aust. J. Biol. Sci., 20 (1967) 51.

D. BOUMA, Aust. Biol. Sci., 20 (1967) 601.

D. BOUMA, Aust. J. Biol. Sci., 20 (1967) 613.

G.D. BOWEN, Soil Sci. Plant Anal., 1 (1970) 293.

D.J.F. BOWLING, J. Exp. Bot., 19 (1968) 381.

A.D. BRADSHAW, M.J. CHADWICK, D. JOWETT and R.W. SNAYDON, J. Ecol., 52 (1964) 665.

R. BROUWER, Acta. Bot. Neerl., 3 (1954) 264.

R. BROUWER, Acta. Bot. Neerl., 5 (1956) 287.

M.J. BUKOVAC and S.H. WITTWER, Plant Physiol., 32 (1957) 428.

A. CARLISLE, A.H.F. BROWN and E.J. WHITE, J. Ecol., 54 (1966) 87.

M.D. CARROLL and J.F. LONERAGAN, Aust. J. Agric. Res., 19 (1968) 859.

M.D. CARROLL and J.F. LONERAGAN, Aust. J. Agric. Res., 20 (1969) 457.

R.B. CLARK, L.O. TIFFIN and J.C. BROWN, Plant Physiol., 52 (1973) 147.

C.R. CLEMENT, M.J. HOPPER, R.J. CANAWAY and L.H.P. JONES, J. Exp. Bot., 25 (1974) 81.

M.C. COCCUCCI and E. MARRE, Plant Sci. Lett., 1 (1973) 293.

J.C. COLLINS and A.P. KERRIGAN, New Phytol., 73 (1974) 309.

R. COLLANDER, Plant Physiol., 16 (1941) 691.

W.J. CRAM and M.G. PITMAN, Aust. J. Biol. Sci., 25 (1972) 1125.

J.S. EATON, G.E. LIKENS and F.H. BORMANN, J. Ecol., 61 (1973) 495.

D.G. EDWARDS, Aust. J. Biol. Sci., 23 (1970) 255.

E. EPSTEIN, Mineral Nutrition of Plants : Principles and Perspectives (1972) Wiley, New York.

G.C. EVANS, The Quantitative Analysis of Plant Growth (1972) Blackwell, Oxford.

H.J. EVANS and G.J. SORGER, Annu. Rev. Plant Physiol., 17 (1966) 47.

J. GALE and A. POLJAKOFF-MAYBER, Aust. J. Biol. Sci., 23 (1970) 937.

J. GALE, R. NAAMAN and A. POLJAKOFF-MAYBER, Aust. J. Biol. Sci., 23 (1970) 947.

H. GREENWAY, Aust. J. Biol. Sci., 18 (1965) 249.

H. GREENWAY and A. GUNN, Planta, 71 (1966) 43.

H. GREENWAY, A. GUNN and D.A. THOMAS, Aust. J. Biol. Sci., 19 (1966) 741.

H. GREENWAY, A. GUNN, M.G. PITMAN and D.A. THOMAS, Aust. J. Biol. Sci., 18 (1965) 525.

H. GREENWAY and M.G. PITMAN, Aust. J. Biol. Sci., 18 (1965) 135.

H. GREENWAY and D.A. THOMAS, Aust. J. Biol. Sci., 18 (1965) 505.

K.A. HANDREK and L.H.P. JONES, Aust. J. Biol. Sci., 20 (1967) 483.

N.J. HALSE, E.A.N. GREENWOOD, R. LAPINS and C.A.P. BOUNDY, Aust. J. Agric. Res., 20 (1969) 987.

A.A. HATRICK and D.J.F. BOWLING, J. Exp. Bot., 24 (1973) 607.

R.W.P. HIRON and S.T.C. WRIGHT, J. Exp. Bot., 24 (1973) 769.

T.J. HOCKING, J.R. HILLMAN and M.B. WILKINS, Nature New Biol., 235 (1972) 124.

J.M. HOPKINSON, J. Exp. Bot., 16 (1964) 125.

S.Z. HYDER and H. GREENWAY, Plant Soil, 23 (1965) 258.

B. HYLMÖ, Physiol. Plant., 6 (1953) 333.

C. ITAI and Y. VAADIA, Physiol. Plant., 18 (1965) 941.

B. JACOBY, S. ABAS and B. STEINITZ, Physiol. Plant., 28 (1973) 209.

R.L. JEFFERIES and A.J. WILLIS, J. Ecol., 52 (1964) 691.

D.W. JEFFREY, Aust. J. Bot., 16 (1968) 603.

W.D. JESCHKE, Planta, 94 (1970) 240.

W.D. JESCHKE, in W.P. Anderson (Ed.) Ion Transport in Plants (1973) Academic Press, London, pp. 285–296.

C. JOHANSEN, D.G. EDWARDS and J.F. LONERAGAN, Plant Physiol., 43 (1968) 1722.

L.H.P. JONES and K.A. HANDREK, Plant Soil, 23 (1965) 79.

B. KLEPPER and M.R. KAUFMANN, Plant Physiol., 41 (1966) 1743.

A. LÄUCHLI, Annu. Rev. Plant Physiol., 23 (1972) 197.

A. LÄUCHLI, U. LÜTTGE and M.G. PITMAN, Z. Naturforsch., 28c (1973) 431.

N. LAZAROFF and M.G. PITMAN, Aust. J. Biol. Sci., 19 (1966) 991.

J.F. LONERAGAN, Trans. 9th Internatl. Congr. Soil Sci., Vol. 2 (1968) 173.

J.F. LONERAGAN and C.J. ASHER, Soil. Sci., 103 (1967) 311.

J.F. LONERAGAN and K. SNOWBALL, Aust. J. Agric. Res., 20 (1969) 465.

J.F. LONERAGAN and K. SNOWBALL, Aust. J. Agric. Res., 20 (1969) 479.

B.C. LOUGHMAN and R.S. RUSSELL, J. Exp. Bot., 8 (1957) 280.

U. LÜTTGE, Stofftransport der Pflanzen (1973) Springer Verlag, Berlin.

U. LÜTTGE and G.G. LATIES, Plant Physiol., 41 (1966) 1531.

U. LÜTTGE and G.G. LATIES, Plant Physiol., 42 (1967) 181.

J.C. LYCKLAMA, Acta. Bot. Neerl., 12 (1963) 361.

J.C. MCFARLANE and W.L. BERRY, Plant Physiol., 53 (1974) 723.

E.A.C. MACROBBIE, Biol. Rev., 46 (1971) 429.

C.R. MILLIKAN and B.C. HANGER, Aust. J. Biol. Sci., 19 (1966) 1.

F.L. MILTHORPE and J. MOORBY, An Introduction to Crop Physiology (1974) Cambridge University Press.

M.G. PITMAN, Aust. J. Biol. Sci., 18 (1965) 10.

M.G. PITMAN, Aust. J. Biol. Sci., 18 (1965) 987.

M.G. PITMAN, Aust. J. Biol. Sci., 19 (1966) 257.

M.G. PITMAN, Aust. J. Biol. Sci., 25 (1972) 905.

M.G. PITMAN, U. LÜTTGE, A. LÄUCHLI and E. BALL, Z. Pflanzenphysiol., 72 (1974) 75.

M.G. PITMAN, U. LÜTTGE, A. LÄUCHLI and E. BALL, J. Exp. Bot., 25 (1974) 147.

M.G. PITMAN and H.D.W. SADDLER, Proc. Natl. Acad. Sci. U.S., 57 (1967) 44.

V.K. RAJU and H. MARSCHNER, Z. Pflanzenernähr. Bodenkunde, 133 (1972) 227.

A. RINGOET, G. SAUER and A.J. GIELINK, Planta, 80 (1968) 15.

I.H. RORISON, Ecological Aspects of the Mineral Nutrition of Plants (1969) Blackwell, Oxford, p. 155.

N. SCHAEFER, R. WILDES and M.G. PITMAN, Aust. J. Plant Physiol., 2 (1975) 61.

P.F. SCHOLANDER, H.T. HAMMEL, E. HEMMINGSEN and W. GAREY, Plant Physiol., 37 (1962) 722.

K.G.M. SKENE and G.H. KERRIDGE, Plant Physiol., 42 (1967) 1131.

F.W. SMITH, Aust. J. Agric. Res., 25 (1974) 407.

R.C. SMITH and E. EPSTEIN, Plant Physiol., 39 (1964) 338.

R.L. SPECHT and R.H. GROVES, Aust. J. Bot., 14 (1966) 201.

G.L. STEUCEK and H.V. KOONTZ, Plant Physiol., 46 (1970) 50.

J.F. SUTCLIFFE, Mineral Salts Absorption in Plants (1962) Pergamon Press, London.

S.L. THROWER, Aust. J. Biol. Sci., 15 (1962) 629.

H.B. TUKEY, Jr, Annu. Rev. Plant Physiol., 21 (1970) 305.

F. VOGEL and E. WEBER, Gartenbauwiss., 6 (1932) 478.

H. WAGNER and G. MICHAEL, Biochem. Physiol. Pflanz., 162 (1971) 147.

P.E. WEATHERLEY, in I.H. Rorison (Ed.) Ecological Aspects of the Mineral Nutrition of Plants (1969) Blackwell, Oxford, p. 323.

R.F. WILLIAMS, Aust. J. Exp. Biol. Med., 14 (1936) 165.

R.F. WILLIAMS, Aust. J. Sci. Res. B, 1 (1948) 333.

H. ZIEGLER, Encyclopedia of Plant Physiology (New Series) (1975) Springer Verlag, Berlin, in press.

*Ion transport in plant cells and tissues*
*edited by D.A. Baker and J.L. Hall*
© *North-Holland Publishing Company, 1975*



# CHAPTER 10

# Halophytes

T.J. Flowers

## Contents



## *10.1 Introduction*

Halophytes are plants which complete their life cycle in an environment which has a high salt content and may be contrasted with glycophytes which are plants that are intolerant of such high salt levels. This definition excludes the so-called pseudohalophytes which sometimes grow in salty habitats when conditions are not saline: for example, in the presence of fresh water during a rainy season. The definition is vague, however, in two important aspects, namely the nature of the salts involved and their concentrations.

Perhaps the best known and most easily characterised of the saline habitats, at least as far as the nature of salts is concerned, are the maritime salt marshes, which are inundated at regular intervals by sea water. These areas are dominated by $Na^+$ and $Cl^-$, which together make up some 86% of the weight of ions present in sea water (Table 10.1). Although maritime salt marshes often

Table 10.1.
The major constituents of an ocean water after
Harvey (1966).

| Element | Concentration | |
| --- | --- | --- |
| | $g\ kg^{-1}$ | equiv. $m^{-3}$ |
| Total | 35.1 | – |
| Sodium | 10.8 | 483 |
| Magnesium | 1.3 | 109 |
| Calcium | 0.41 | 21 |
| Potassium | 0.39 | 10 |
| Strontium | 0.01 | 0.2 |
| Chloride | 19.4 | 558 |
| Sulphate | 2.7 | 58 |
| Bromide | 0.07 | 0.8 |
| Borate | 0.03 | 0.4 |

form extensive areas, they are only one of a number of types of saline habitat. There are also large tracts of land, particularly in arid or semi-arid regions of the world, which have saline soils or are influenced by saline ground water (see Craig, 1970). Although $Na^+$ and $Cl^-$ are again often the dominant ions, many other salts may be present and in a wide variety of concentrations. Predominant amongst these salts are $Na_2CO_3$, the sulphates of $Mg^{2+}$, $Na^+$ and $Ca^{2+}$, together with $MgCl_2$. Other salts do occur in high concentrations, for example, KCl and potassium borates (see Waisel, 1972; Kovda et al., 1973) but less commonly.

The salt concentration in which halophytes normally grow is not easily defined, due to changes in soil water content. For maritime species the concentration is approximately set by the composition of sea water (Table 10.1): higher concentrations will naturally occur during periods of high insolation between tides and lower concentrations during periods of rainfall. Salt concentrations in the soil solution of inland saline soils are likely to be much more variable, however, than those of maritime habitats since they are not continually inundated with a solution of fairly constant composition. The salt concentration which plants in these habitats experience is therefore subject to wide fluctuation.

Plants which grow in sea-water concentrations of salt (about 3.3%, w/v) are clearly halophytes. A problem of definition may arise, however, for plants which will not tolerate such high salt concentrations: i.e. in drawing a dividing line in a continuous distribution. For NaCl, which, as we have seen, is the dominant salt of most saline soils, the line is normally drawn at a concentration of 0.5% in the soil solution (about 85 Mol m$^{-3}$) and many classifications of halophytes have been drawn up on this basis (Chapman, 1966; Waisel, 1972) some of these classifications take account of other salts while others do not. In summary, we might then slightly elaborate our definition of a halophyte to include plants which complete their life cycle in an environment which has a high (0.5% to perhaps 6%) salt (normally NaCl) content.

In this chapter we shall deal mainly with the so-called euhalophytes (true halophytes) which are plants that tolerate at least 1% NaCl and which show stimulated growth in such salt concentrations (Fig. 10.3). Marine euhalophytes are restricted to the algae (see Ch. 5), fungi and a few angiosperms. Those species found in intertidal zones, salt marshes and salt deserts are predominantly angiosperms although there are just a few salt-tolerant bryophytes and pteridophytes (Uphof, 1941; Bates and Brown, 1974). All such plants maintain water within their symplast against a considerable osmotic gradient, since the cellular water potential must be more negative than that of the surroundings if the plant is to maintain its water content. The nature of the osmotic adjustment necessary to achieve this end forms a substantial part of this chapter. The balance of the chapter is concerned with the mechanism of this osmotic adjustment, its relation to ion uptake and with the ability of halophytes to accumulate potassium from an environment dominated by sodium ions. Consideration of the wider aspects of the uptake of other essential ions which are present only in very low concentrations is precluded by a lack of data.

## 10.2 Osmotic adjustment

### 10.2.1 Osmotic potentials

The osmotic potential of an ocean water such as that described in Table 10.1 is some $-23$ bars (Harvey, 1966). This value might therefore be expected to set the upper limit of the water potential of maritime halophytes. Such a water potential, which is low as far as the majority of plants are concerned, results

312                              *T.J. Flowers*

from an even lower (more negative) osmotic potential (Table 10.2). This allows the plant to maintain the positive turgor pressure necessary for growth while having a low water potential; the matric potential is assumed to be a relatively small component of the total water potential as is the case in glycophytes (Wiebe, 1966). As might be expected the osmotic potentials of halophytes are significantly lower than those of glycophytes. The so-called osmotic spectra of Fig. 10.1 illustrate the range of values in plants from differing habitats and

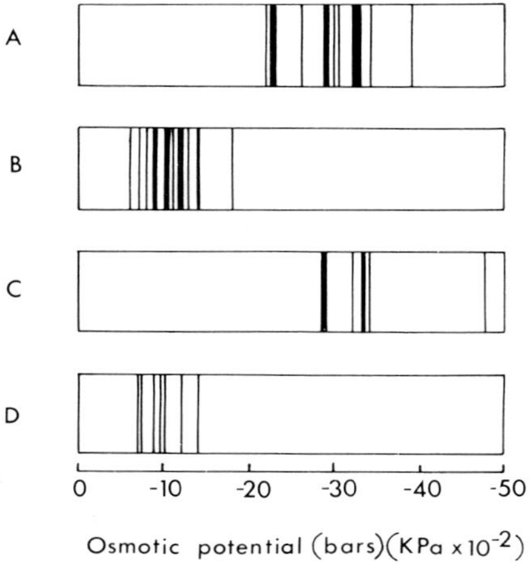


Fig. 10.1. 'Osmotic spectra' of halophytes and glycophytes. The osmotic values of species growing within a given habitat are represented by vertical lines. (A) Saltmarsh species, as Table 10.2. (B) Moorland species from Hungary. (C) Mangroves of East Africa. (D) Woody species of a tropical rainforest. (From Steiner, 1939.)

show that those species growing in saline conditions (A and C) do clearly have the more negative osmotic potentials. In fact values as low as $-150$ bars and less have been recorded for some species (Chapman, 1960; Waisel, 1972).

### 10.2.2 Ion contents

The importance of this osmotic adaptation in terms of ion uptake is that in most halophytes the free energy of water is lowered by a massive increase in the ion content of the plant, which can thus reach very high values. For example, in species of the genera *Suaeda*, *Salicornia* and *Atriplex*, 20–40% of the dry weight may be composed of inorganic salts.

The major components of the salt content are $Na^+$ and $Cl^-$ ions (Table 10.2): for these data Steiner calculated that chlorides (i.e. chloride ion + cation)

Table 10.2.
The solute potentials of some halophytes of a
saltmarsh on the eastern seabord of the U.S.A.
(from Steiner, 1934).

| Species | Solute potential (bars) |
|---|---|
| *Spartina glabra* | $-34.6$ |
| *Distichlis spicata* | $-32.0$ |
| *Juncus gerardii* | $-39.7$ |
| *Salicornia bigelowii* | $-49.0$ |
| *Atriplex hastata* | $-31.6$ |
| *Limonium carolianum* | $-33.3$ |
| *Suaeda linearis* | $-29.0$ |

could account for between 40 and 90% of the osmotic potential of the halophytes he analysed. Steiner and Eschrich (1958) in a review of data from glycophytes found that the $Cl^-$ content only accounted for between 0 and 40% of the osmotic potential, with 11 out of the 16 results having $Cl^-$ contents of less than 10%. In an extensive survey of the literature Arnold (1955), deduced that chlorides accounted for between 35 and 95% (average of 65%) of the osmotic potential of halophytes, between 15 and 65% (average of 35%) of marsh glycophytes and between 0 and 65% (average of 17%) of other glycophytes.

There is a paucity of more complete analyses of the ion content of halophytes since the earlier literature concentrated on $Cl^-$ and solute potential. Where more complete determinations have been made it is clear that while $Na^+$ and $Cl^-$ dominate the ion content (Table 10.3), $K^+$ levels may also be somewhat higher than in glycophytes (compare Table 10.3 and Table 9.1). Where analyses of roots and shoots have been made, it seems that the majority of the $Na^+$ and $Cl^-$ present in the plant is in the shoots. For those examples shown in Table 10.3, the root $Na^+$ contents (mequiv. g $DW^{-1}$) for *Suaeda maritima* and the two species of *Atriplex, nummularia* and *vesicaria* were 38%, 46% and 39% of that in the shoot respectively. Furthermore, since the shoot/root dry weight ratios in the three species were, *S. maritima* 9.4, *A. nummularia* 7.5 and *A. vesicaria* 6.7, between 94 and 96% of the $Na^+$ content of the plant was in the tops (the absolute value of any of these ion contents can of course vary enormously depending on the exact medium on which the plants are grown, see Fig. 10.2).

$Na^+/K^+$ ratios are generally high in halophytes in contrast to glycophytes, although there is a relative enrichment of $K^+$ in the plant, in comparison to the medium. $Na^+/K^+$ ratios in the growth media listed in Table 10.3 range

Table 10.3.
The ion contents of a number of halophytes.

| Species | | Ion content (mequiv. gDW$^{-1}$) | | | | | | Medium (equiv. m$^{-3}$ solution) | | | | | | |
|---|---|---|---|---|---|---|---|---|---|---|---|---|---|---|
| | | Ca$^{2+}$ | Mg$^{2+}$ | K$^+$ | Na$^+$ | Cl$^-$ | Na$^+$/K$^+$ | Ca$^{2+}$ | Mg$^{2+}$ | K$^+$ | Na$^+$ | Cl$^-$ | Na$^+$/K$^+$ | |
| **Plant leaves** | | | | | | | | | | | | | | |
| *Rhizophora mucronata* | a1 | – | – | 0.11 | 0.84 | 1.02 | 7.6 | Plants growing in the intertidal zone | | | | | | |
| *Aegitalis annulata* | a2 | – | 0.99 | 0.20 | 0.62 | 0.63 | 3.1 | Queensland, Australia | | | | | | |
| *Plantago maritima* | b | 0.31 | 0.24 | 0.43 | 1.4 | – | 3.3 | 14.4 | 63.2 | 6.5 | 295 | – | 41 | (soil solution) |
| *Triglochin maritima* | b | 0.12 | 0.20 | 0.46 | 1.1 | – | 2.4 | 16.8 | 60.2 | 8.2 | 353 | – | 43 | (soil solution) |
| *Suaeda monoica* | c | 0.40 | – | 0.87 | 6.8 | 1.91 | 7.8 | Plants growing in saline soil in Sedom, Israel | | | | | | |
| *Suaeda maritima* | d | 0.28 | 0.79 | 0.44 | 5.1 | – | 11.6 | Plants growing in a saltmarsh Sussex, England | | | | | | |
| *Suaeda maritima* | d | 0.11 | 0.29 | 0.37 | 6.7 | – | 18.0 | 8.0 | 4.0 | 7.0 | 340 | 340 | 48.6 | (culture solution) |
| *Atriplex nummularia* | e | – | – | 0.60 | 5.0 | 3.6 | 8.3 | 5.0 | 5.0 | 5.0 | 300 | 300 | 60 | (culture solution) |
| *Atriplex hastata* | f | – | – | 0.85 | 3.7 | 1.4 | 4.3 | 8.0 | 4.0 | 7.0 | 300 | – | 48.6 | (culture solution) |
| *Atriplex vesicaria* | g | – | – | 0.26 | 5.8 | 4.3 | 22.1 | 8.0 | 4.0 | 7.0 | 300 | – | 48.6 | (culture solution) |
| *Salicornia herbacea* | h | – | – | – | 5.6 | 4.7 | – | | | | | | | (culture solution) |
| **Plant roots** | | | | | | | | | | | | | | |
| *Suaeda maritima* | d | 0.11 | – | 0.60 | 2.5 | – | 4.2 | 8.0 | 4.0 | 7.0 | 340 | 340 | 48.6 | (culture solution) |
| *Atriplex nummularia* | e | – | – | 0.60 | 2.3 | 1.7 | 3.8 | 5.0 | 5.0 | 5.0 | 300 | 300 | 60.0 | (culture solution) |
| *Atriplex vesicaria* | g | – | – | 0.56 | 2.2 | 1.2 | 3.9 | 8.0 | 4.0 | 7.0 | 300 | 300 | 48.6 | (culture solution) |

a1, Atkinson et al. (1967) salt-accumulating mangrove: data for leaf 3; a2, Atkinson et al. (1967) salt-secreting mangrove: data for leaf 3; b, Parham (1970), perennial herb from saltmarsh in eastern England; c, Waisel (1972), perennial shrub of the eastern Miditerranean region; d, Yeo (1974); e, Greenway (1968); f, Black (1956); g, Black (1960); h, van Eijk (1934).

from 40 to 60 while the same ratio in the plants listed only ranges from 3 to 22. In a number of algae, $K^+$ can be accumulated to high values in place of $Na^+$ (Steiner and Eschrich, 1958; Gutknecht and Dainty, 1968; Table 5.1) although this total replacement rarely appears to occur in the higher halophytes (Table 10.3). It has, however, been reported that the amount of $K^+$ accumulated in *Allenrolfea* and *Suaeda divaricata* can exceed the amount of $Na^+$ (Chapman, 1968). Such a situation could reflect a low $Na^+$ content in the growth medium (see below).

The anion/cation balance in halophytes has not been investigated in detail. The massive uptake of $Na^+$ is clearly accompanied by the uptake of $Cl^-$ (Fig. 10.2): in some species $Cl^-$ approximately balances $(Na^+ + K^+)$ (Fig.

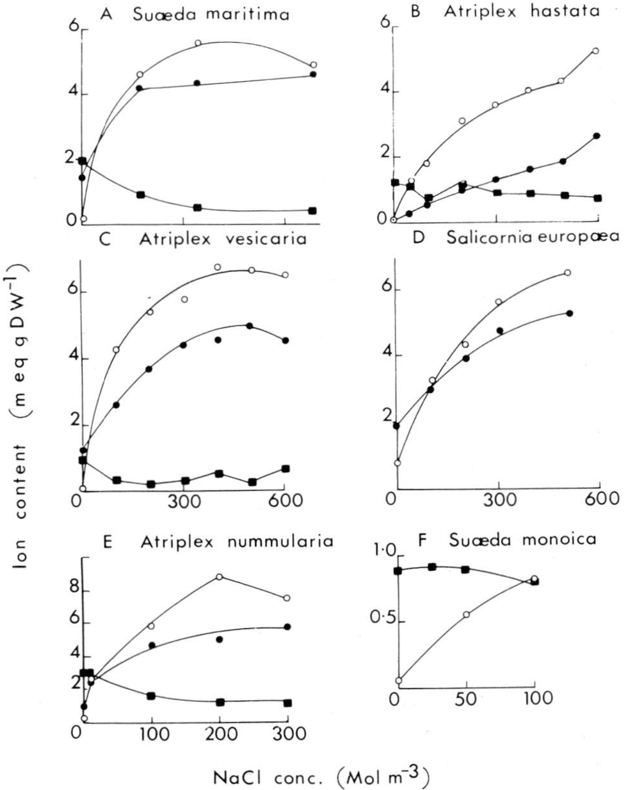


Fig. 10.2. The effect of the NaCl content of the culture solution on the content of $Na^+$ (○), $Cl^-$ (●) and $K^+$ (■) in various halophytes. (A) *Suaeda maritima* (Flowers, unpublished); (B) *Atriplex hastata* (Black, 1956); (C) *Atriplex vesicaria* (Black, 1960); (D) *Salicornia europoea* (van Eijk, 1934); (E) *Atriplex nummularia* (Greenway, 1968); (F) *Suaeda monoica* (Waisel, 1972).

10.2A) although in others there is clearly a large imbalance (Fig. 10.2B). The presence of organic anions in plants has been known for many years and it is now clear that these, and specifically oxalate, play an important role as counter-anions. Thus, high levels of oxalates occur in many species of the Chenopodiaceae found in semi-arid conditions (Osmond, 1967). Oxalate levels increase with increasing $Na^+$ concentration in the medium (Williams, 1960; Osmond, 1967) and in *Atriplex spongiosa* oxalate balances some 75% of the cation excess – that is $\Sigma(Na^+ + K^+ - Cl^- - NO_3^-)$ – (Osmond, 1967). The majority of the oxalate is assumed to be present in the vacuoles, since Osmond calculated that if it were confined to the cytoplasm, the concentration would approach $3 \times 10^3$ Mol $m^{-3}$ (a highly toxic level); if the oxalate were present in the vacuoles, then the concentration would be approximately 200 Mol $m^{-3}$ (see later sections for further evidence for the compartmentation of ions in halophyte cells).

Where halophytes are grown in normal culture solutions in the absence of high NaCl concentrations, then it is apparent that they accumulate ions other than $Na^+$ and $Cl^-$ to high levels: i.e. they have a high constitutive capacity for ion uptake. In those halophytes that have been investigated this results in high $K^+$ (Fig. 10.2) and probably $Ca^{2+}$ and $Mg^{2+}$ levels in plants grown in low $Na^+$. Analyses of *Suaeda maritima* plants grown under these conditions show $K^+$ contents of nearly 2 mequiv. $gDW^{-1}$ (about twice the level in peas growing in an identical culture solution, Flowers, 1972a), and indicate that $NO_3^-$ may act as a counter-anion. The high capacity for ion uptake is reflected in the fact that halophytes growing in non-saline habitats have much lower (larger negative values) solute potentials than glycophytes alongside which they are growing. Walter (1962) found that under such conditions the osmotic potential of the halophytes was about double that of the glycophytes.

The high capacity of halophytes for ion uptake was noted by Collander in his experiments undertaken in the early 1940s (Collander, 1941). Twenty different species of phanerogam were grown in various culture solutions and the ion content of the plant then analysed. As far as $K^+$ was concerned, differences between the maximum and minimum values found were only some 2–3 times, but for $Na^+$, the maximum values ranged between some 30 and 90 times the minimum values depending upon the $Na^+$ concentration of the culture solution. The plants with the ability to take up large amounts of $Na^+$ from the low $Na^+$ concentrations used (the maximum was 4 equiv. $m^{-3}$) were on the whole the known halophilic species.

Thus, in the presence of high NaCl levels in the growth medium, the ion content of halophytes tends to be dominated by these ions, whereas in their absence $K^+$ is the dominant cation. As the NaCl concentration in the medium

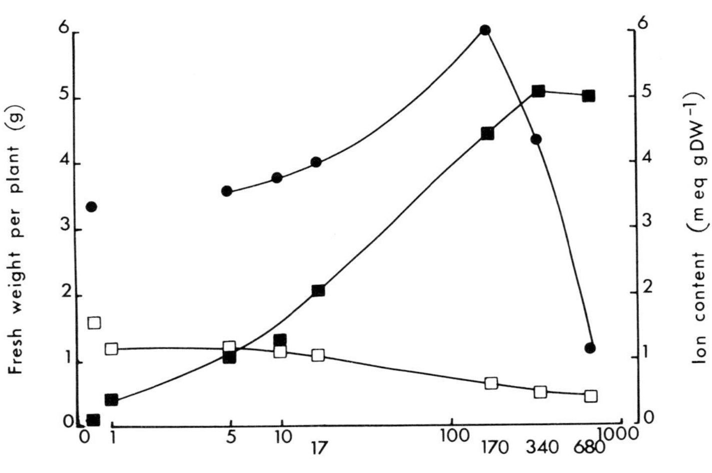

Fig. 10.3. The effect of NaCl in the culture solution on the growth expressed as fresh weight per plant (●), and content of $Na^+$ (mequiv. $gDW^{-1}$, ■) and $K^+$ (mequiv. $gDW^{-1}$, □) of *Suaeda maritima* plants. (Data of Yeo, 1974.)

is varied, so therefore, does the level in the plant (Figs 10.2, 10.3) indicating an ability to regulate the ion content. Two types of regulation can be readily distinguished. Firstly, those plants which increase their salt content but maintain a constant solute potential by increasing their water content; i.e. the succulence of the plants increases. Secondly, those plants with salt glands which are able to regulate their ion content by secreting excess ions (see Ch. 11). There are also species such as *Juncus gerardii*, however, where the salt content as well as the osmotic potential increases throughout the growing season (see Steiner, 1939).

Thus, in halophytes, massive ion accumulation appears to occur as a part of the salt-tolerance mechanism. This accumulation of ions is associated with growth stimulation and there is an increase in both fresh and dry weight with increasing salt concentration in the medium up to an optimal value (see Jennings, 1968 and Fig. 10.3). In the relatively salt-tolerant glycophytes (i.e. those species and varieties which are perhaps borderline halophytes) a rather different mechanism seems to apply. With these plants, salt tolerance seems to be associated with ion.exclusion rather than ion accumulation (see Greenway, 1973); for example, the soybean varieties or species showing marked growth inhibition by NaCl are those with high $Cl^-$ contents, while those showing greater levels of tolerance are those which are able to exclude $Cl^-$ (Fig. 10.4).

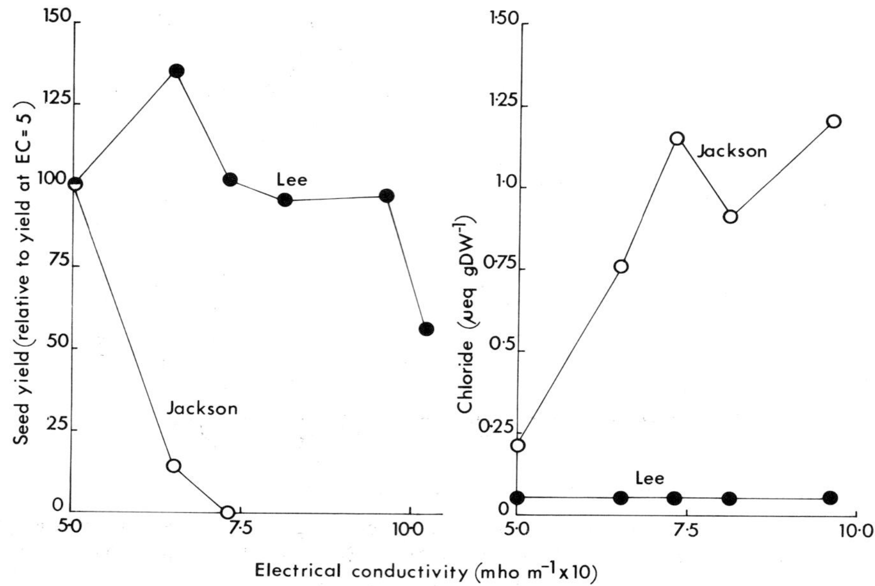



Fig. 10.4. The yield and Cl⁻ contents of two varieties (Jackson, ○, and Lee, ●) of *Glycine max* (soybean) in relation to the ion content of the irrigation water (expressed as conductivity). The Cl⁻ contents are for leaves on young plants. (Data of Abel and McKenzie, 1964 and reproduced from Greenway, 1973.)

## 10.3 Compartmentation of ions in halophytes

### 10.3.1 Inferences from metabolic studies

When the ion content of halophyte leaves is expressed on the basis of their water content, concentrations of $Na^+$ and $Cl^-$ lying between $0.5 \times 10^3$ and $1.0 \times 10^3$ Mol m$^{-3}$ are commonly obtained. For example, in *Suaeda maritima*, $Na^+$ and $Cl^-$ concentrations of 860 equiv. m$^{-3}$ and 730 equiv. m$^{-3}$ respectively were found for plants grown in sand and irrigated with a culture solution containing 680 Mol m$^{-3}$ NaCl (Flowers, 1972a). Since many metabolic processes in both plants and animals are sensitive, at least in vitro, to high ionic strength, the high salt concentrations found in the tops of halophytes presents two important alternatives: either the ions are mostly excluded from the cytoplasm or the metabolism has evolved in such a way that it functions in the presence of high ion concentrations. Such adaptations are known in lower organisms where the enzymes of the halophilic bacteria are adapted to

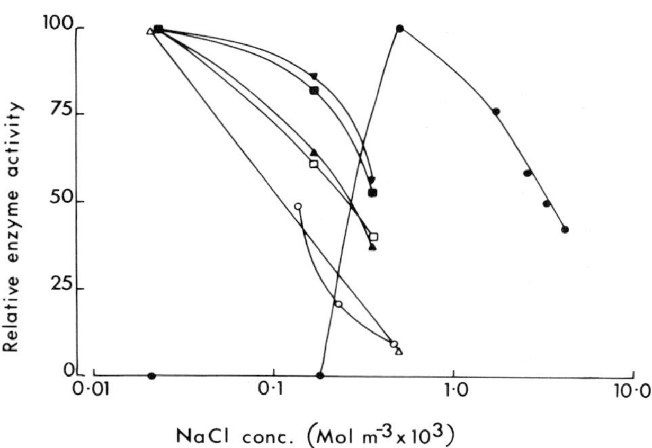

Fig. 10.5. The effect of NaCl concentration on the activity of malate dehydrogenase measured in vitro. Data are presented for both halophilic (solid symbols) and salt-sensitive (open symbols) organisms. *Halobacterium salinarium* (●) and pig liver ( △), Holmes and Halvorsen (1965); *Suaeda maritima*, Flowers (1972a) and *Beta maritima*, Flowers (1972b) ( ▲); *Pisum sativum* (□), Flowers (1972a); *Salicornia ramosissima* (▼) and *Halimione portulacoides* (■), Flowers (1972b); *Glycine max* (○), Flowers and Hanson (1969).

function in extremely high salt concentrations (Fig. 10.5) which match the estimated internal ion concentrations in these bacteria (Larsen, 1967). Recently, the in vitro response of a number of enzymes from halophytes and glycophytes have been compared (Flowers, 1972a, b; Greenway and Osmond, 1972; Heimer, 1973; Austenfeld, 1974). In contrast to the bacteria, the plant enzymes are quite sensitive to NaCl in vitro (Fig. 10.5) and this situation also seems to hold for marine algae (Johnson et al., 1968; Ben-Amotz and Avron, 1972; Heimer, 1973) and to some extent for the fungi (Troke, 1975), although Greenway and Sims (1974) point out that the degree of inhibition is also a function of the substrate concentration. For example, 200 Mol m$^{-3}$ KCl causes an 80% inhibition of malate dehydrogenase (decarboxylating) in the presence of 0.1 Mol m$^{-3}$ malate, but only a 40% inhibition in the presence of 4 Mol m$^{-3}$ substrate. The interpretation of all such enzyme data is, of course, open to question, since it is possible that changes in enzyme response to NaCl result from the isolation of the enzyme from its cellular environment. Furthermore, it is not always clear what percentage of the total activity of any enzyme is required in vivo for the normal metabolism of any given species. However, not only are individual enzymic reactions sensitive to added NaCl, but so are the multiple reactions catalysed by very different types of organelle, namely ribosomes and mitochondria. The incorporation of [$^{14}$C]leucine into

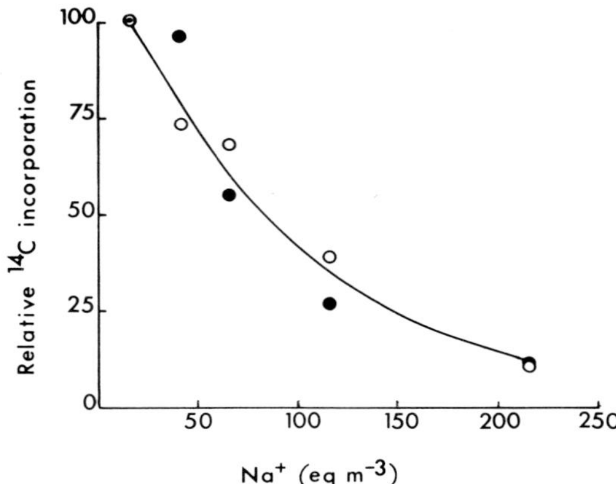



Fig. 10.6. The effect of NaCl on the incorporation of [$^{14}$C]leucine into protein in the isolated ribosomes of *Suaeda maritima*. Incorporation (ordinate) is expressed relative to that in the assay medium without added NaCl, but which contained 15 equiv. m$^{-3}$ Na$^+$ from the salts of ATP, GTP, etc. The plant material from which the ribosomes were isolated was grown both in the presence of 340 Mol m$^{-3}$ NaCl (●) and in the absence of NaCl (○). (From Hall and Flowers, 1973.)

protein by ribosomes from *Suaeda maritima* is extremely sensitive to the concentration of monovalent chlorides present in the assay medium, above those required for enzyme activation (Fig. 10.6). The inhibition is precisely that which is found in non-halophilic organisms. As far as mitochondria are concerned, the in vitro response of these organelles from *Pisum* (a very salt-sensitive glycophyte) and *Suaeda maritima* to NaCl is identical. In both species oxidative phosphorylation is inhibited by increasing NaCl concentrations as the respiration rate is inhibited and acceptor control ratios fall (Flowers, 1974). The inference drawn from these biochemical studies (showing that the cytoplasmic machinery of halophytes has not evolved to become salt tolerant) is that the cytoplasm must be maintained at a relatively low ionic strength and that the bulk of the salt is sequestered in the vacuole, as was argued for oxalate by Osmond (see p. 316).

## 10.3.2 Physical measurements

The conclusions drawn from the biochemical data contrast with results reported by Waisel and Eshel (1971) who claimed, as a result of electron microprobe analyses, that cytoplasmic ion concentrations were high in the halophyte *Suaeda monoica*. However, their photographs are difficult to interpret unequivocally and, furthermore, it is not possible to estimate the amounts of the ions analysed. Direct confirmation of the biochemical data must await further and more sophisticated electron microprobe analyses.

There is, however, a variety of other evidence for ionic compartmentation in halophytes. Work on marine algal cells, which are large enough to allow analysis of the ion contents of sap and (although less reliably) of the cytoplasm has shown that the $Na^+$ content in the cytoplasm is in general low, although $K^+$ contents may be high (Gutknecht and Dainty, 1968; Table 5.1). Compartmental analysis of the halophyte *Triglochin maritima* indicated that the cytoplasmic concentration of $Na^+$ in the root cells ranged between 70 and 150 equiv. $m^{-3}$, depending upon the external NaCl concentration (Jefferies, 1973), while in *Suaeda maritima* the $Na^+$ concentration in the leaf cytoplasm of plants growing in 340 Mol $m^{-3}$ NaCl was estimated to be some 165 equiv. $m^{-3}$ (Yeo, 1974). Plants growing in culture solution in the absence of NaCl contained approximately 140 equiv. $m^{-3}$ $K^+$ in the cytoplasm (all such values are approximate, however, as the analysis involves many assumptions; see Epstein, 1972). Efflux analysis of *Atriplex spongiosa* leaf slices suggested that $Na^+$ was restricted to the vacuoles, whereas $K^+$ had a more uniform distribution throughout the cell (Osmond, 1966). In *Suaeda maritima*, $Rb^+$ which can be seen under the electron microscope and which is absorbed in quite large quantities appears to be restricted to the vacuoles, although again there are limitations to the technique (Hall et al., 1974).

### 10.3.3 Compartmentation within organelles

Although the major discontinuity in the distribution of $Na^+$ and $Cl^-$ probably lies between the cytoplasm and the vacuole, there is some evidence to suggest that there may be discontinuities within the cytoplasm itself. The metabolic evidence described above would suggest that the continuous phase of the cytoplasm contains only low concentrations of $Na^+$ and $Cl^-$. There is, however, a considerable amount of work to suggest that chloroplasts in particular are a major site of ion accumulation in green cells (Ch. 4). In the halophyte *Limonium vulgare* the chloroplasts accumulate large quantities of $Cl^-$ (Ziegler and Lüttge, 1967; Larkum, 1968) and $Na^+$ (Larkum, 1968) and the $Na^+$ levels in particular are higher than those in glycophytes (Table 4.3); concentrations of $Na^+$ and $Cl^-$ in *Limonium* can thus reach values of 500 and 700 Mol $m^{-3}$, respectively (Larkum and Hill, 1970). In glycophytes $Na^+$ concentrations are normally less than 50 Mol $m^{-3}$ although $Cl^-$ concentrations approach 350 Mol $m^{-3}$ (Table 4.3). However, not all plants appear to contain high $Cl^-$ concentrations within their chloroplasts and van Steveninck et al. (1973)

reported that the halophyte *Aegiceras corniculatum* and the glycophyte *Nymphoides indica* contain very little $Cl^-$ in this organelle.

The distribution of ions in other subcellular organelles of halophytes has not been investigated in detail. Ziegler and Lüttge (1967) reported the presence of $Cl^-$ in the nuclei of the mesophyll cells of *Limonium vulgare* and in the nuclei and mitochondria of the cells of the salt glands (see Ch. 11).

### 10.3.4 Osmotic adjustment in the cytoplasm

The probability of a discontinuous distribution of ions within halophyte cells raises the question of osmotic adjustment within the cytoplasm: viz. if the vacuole has a very low water potential due to its high concentration of ions, how is the water potential of the cytoplasm adjusted to equal or nearly equal that of the vacuole? In the cytoplasm both the matric potential and Donnan equilibria could be important in osmotic adjustment, although neither have been evaluated. There is indirect evidence, however, of the presence of a solute component of the total potential in the cytoplasm. The halophilic alga *Dunaliella parva*, for example, produces glycerol in amounts which reflect the osmotic potential of the medium (Ben-Amotz and Avron, 1973). Furthermore, nitrate reductase from *Dunaliella* is unaffected by high glycerol concentrations, although this enzyme from other sources is inhibited (Heimer, 1973). It is clear, therefore, that glycerol could act to balance the osmotic potential of the cytoplasm as well as the vacuole in this species. A number of other species, particularly algal, have been shown to regulate the concentration of other organic molecules in relation to the concentration of ions in their growth medium. The rate of production of mannisidomannitol by the lichen *Lichina pygmaea* increases as the concentration of ions in the medium in which it is growing increases and decreases as the latter decreases (Feige, 1972). In the flagellate *Ochromonas malhamensis* α-galactosylglycerol seems to play a similar osmoregulatory role (Kauss, 1973) and in *Monochrysis lutheri* this function is fulfilled by 1,4/2,5-cyclohexanetrol (Craigie, 1969) while in the fungus *Dendryphiella salina* the synthesis of arabitol and mannitol is clearly linked to the ionic status of the medium (see Jennings, 1973). Since these solutes must presumably be produced in the cytoplasm they could regulate the cytoplasmic water potential, although they may be produced or degraded in sufficient quantities to regulate the overall water potential.

As far as higher plants are concerned there have been few investigations of the type described for fungi and algae. The concentration of sugars could clearly be important in this respect. Work on *Suaeda maritima* in this laboratory has shown that sucrose concentrations, based on the overall water content, are maintained constant at about 1 Mol m$^{-3}$ at varying salinities. Since the water content increases much as the fresh weight (Fig. 10.3) there is an increase in sucrose from about 1.00 $\mu$Mol plant$^{-1}$ grown in culture solution alone to 1.75 $\mu$Mol plant$^{-1}$ for plants grown in 340 Mol m$^{-3}$ NaCl. Sucrose contents expressed as gFW$^{-1}$ are 0.84 and 0.66 $\mu$Mol for plants grown in the absence and presence of salt, respectively, and although these levels are low they are approximately one third the level of proline reported to be present in *Suaeda maritima* (Stewart and Lee, 1974).

The importance of proline in halophytes has recently been highlighted by Stewart and Lee who found that in most halophytes proline accounts for more than 30% of the amino acid pool (the only exception in eleven species tested being *Plantago maritima*). Such a high proportion of proline is not found in glycophytes although the levels in some species can increase under salt or water stress. The proline content of halophytes (excepting *Plantago maritima*) analysed by Stewart and Lee fell into 3 groups ranging between 2 and 4 (3 species), 15 and 30 (5 species) or 65 and 85 (2 species) $\mu$Mol gFW$^{-1}$. A level of 25 $\mu$Mol gFW$^{-1}$ would amount to a concentration of 28 Mol m$^{-3}$ if the solute were distributed throughout a cell with a water content of 90% of the fresh weight. However, if this proline were confined to 10% of the water content (the cytoplasm) the concentration of proline in that fraction would be some 280 Mol m$^{-3}$, a concentration high enough to have a significant effect on the osmotic potential of the cytoplasm. If the proline content were only 2.5 $\mu$Mol gFW$^{-1}$ it must make a proportionately smaller contribution to the osmotic potential, unless the cytoplasm constitutes a much smaller proportion of the total volume.

A further factor which may play a part in osmoregulation is the level of metabolic intermediates present in the cytoplasm. Greenway and Sims (1974) postulated that the control of enzyme activity by ions present in the cytoplasm leads to an increase in the concentration of metabolites (which in turn modifies the enzyme response to ions). The elevated metabolite concentration could then serve to adjust the cytoplasmic water potential.

## 10.4 Ion uptake

The discussion so far has centred upon a rather static picture of the ionic relations of halophytes derived from analyses of plants which it is perhaps important to summarise at this point. Firstly, halophytes clearly have the ability to accumulate massive quantities of ions during osmotic adjustment. These ions are most frequently Na$^+$ and Cl$^-$, but organic anions may be of considerable importance in some species. This ability to accumulate such large quantities of ions seems to be a constitutive property of the plants since osmotic potentials are normally more negative than those of glycophytes when both are growing under non-saline conditions. Furthermore, where $Na^+$ and $Cl^-$ are present in low levels in culture solutions, which do not have a marked excess of any other ions, $K^+$ (and $NO_3^-$) levels in the plants are high: i.e. the mechanisms involved in osmotic adjustment will operate with $K^+$, but $K^+$ uptake is suppressed or overtaken in the presence of $Na^+$. Secondly, the current evidence suggests that the distribution of ions within the cells is discontinuous – i.e. that the 'soluble phase' of the cytoplasm is relatively low in ions, while the vacuole contains the majority of those ions absorbed. Thirdly, $K^+$ levels in the tissue relative to the $Na^+$ levels are higher than those in the medium: i.e. the plants are enriched in $K^+$ relative to the medium (see Table 10.3). Such an enrichment in $K^+$ is indicative of the ability of halophytes to absorb essential elements in the presence of vast excesses of other elements. It can thus be inferred that these plants do have very specific ion-uptake mechanisms. In the remainder of this section the characteristics of the mechanisms involved in the accumulation of ions are discussed in relation to the uptake of $Na^+$ and $K^+$, although the information available on ion uptake in halophytes is extremely scant.

### 10.4.1 $Na^+$ uptake into whole plants

The addition of 100 Mol m$^{-3}$ NaCl to *Atriplex vesicaria* plants which had previously been in 6 equiv. m$^{-3}$ $Na^+$ resulted in an increase in the $Na^+$ content of the young leaves of about 0.24 mequiv. gDW$^{-1}$ day$^{-1}$, while the ion contents of the roots and of the mature leaves did not change significantly over the 10 days of the experiment (Black, 1960). Similar rates of uptake of $Na^+$ (determined by analysis of the plant tops) can be calculated for young *Suaeda maritima* plants; the values range from 0.12 mequiv. gDW$^{-1}$ day$^{-1}$ from a concentration of 140 Mol m$^{-3}$ NaCl to 0.32 mequiv. gDW$^{-1}$ day$^{-1}$ where the concentration of NaCl was increased from 140 to 300 Mol m$^{-3}$ over the uptake period, indicating that the concentration of $Na^+$ affects its rate of accumulation. More rigorous investigation of the effect of concentration on the rate of uptake of $Na^+$ has only been undertaken for a few halophyte species. For *Suaeda maritima* seedlings the rate of uptake as measured over a 24 h period with $^{22}$Na, ranges from 0.32 $\mu$equiv. gDW of roots $^{-1}$ h$^{-1}$ in a concentration of 0.01 Mol m$^{-3}$ NaCl to 399 $\mu$equiv. gDW of roots$^{-1}$ h$^{-1}$ in 300 Mol m$^{-3}$ NaCl (Table 10.4). The relationship between the logarithm of the concentration and the logarithm of the rate of uptake is linear and

Table 10.4.

The effect of concentration and anion on the uptake of $Na^+$ (measured as $^{22}Na$ accumulated over a 24 h period) into *Suaeda maritima* seedlings and expressed per unit weight of the roots. Data from Yeo (1974).

| $Na^+$ in solution (equiv. $m^{-3}$) | Rate of Na uptake ($\mu$equiv. $gDW^{-1}$ $h^{-1}$) | | |
|---|---|---|---|
| | $Cl^-$ | $NO_3^-$ | $SO_4^{2-}$ |
| 0.01 | 0.32 | 0.39 | 0.35 |
| 0.1 | 1.94 | 2.55 | 1.10 |
| 1.0 | 5.6 | 7.0 | 3.0 |
| 10.0 | 48.4 | 51.5 | 15.4 |
| 100.0 | 182.0 | 154.0 | 60.3 |
| 300.0 | 399.0 | 382.0 | 228.0 |

maximum rates of uptake measured were some 400 $\mu$equiv. gDW of roots$^{-1}$ $h^{-1}$ which is about twice the rate of $Na^+$ uptake measured for beetroot disks in 40 Mol $m^{-3}$ NaCl (Poole, 1971): barley roots in a similar concentration of NaCl (50 Mol $m^{-3}$) have an uptake rate of some 30 $\mu$equiv. $gFW^{-1}$ $h^{-1}$ (Rains and Epstein, 1967) which is equivalent to approximately 300 $\mu$equiv. $gDW^{-1}$ $h^{-1}$ – i.e. the rate of $Na^+$ uptake in halophytes is of the same order of magnitude, when expressed per unit weight of tissue, as that in glycophytes. The effect of $Na^+$ concentration on $Na^+$ uptake is influenced to some extent by the nature of the anion present in the external solution. Thus, $Na^+$ uptake was inhibited from a solution of $Na_2SO_4$ as compared with NaCl, but uptake from $NO_3^-$ and $Cl^-$ were essentially the same. A similar inhibitory effect of $SO_4^{2-}$ has been reported for *Suaeda monoica* (see Waisel, 1972) and for barley (Rains and Epstein, 1967a).

In *Plantago maritima* the rate of uptake of $Na^+$ by isolated roots increases linearly with increasing $Na^+$ concentration, while for *Triglochin maritima* $Na^+$ uptake increased continuously with time (Parham, 1971).

There is some evidence to suggest that the mechanism of $Na^+$ uptake varies with the external concentration. Firstly, at low concentrations of $Na^+$ (0.05 Mol $m^{-3}$) $Na^+$ uptake into *Suaeda maritima* is inhibited in direct proportion to the logarithm of the KCl concentration (Fig. 10.7A), while the uptake of $Na^+$ from a 5 Mol $m^{-3}$ $Na^+$ concentration is much less dependent on the $K^+$ concentration (Fig. 10.7B). Secondly, with increasing concentrations of $Na^+$, the inhibitory effect of low temperature on this uptake of $Na^+$ into seedlings decreases. This result correlates with a requirement for light for $Na^+$ uptake from low concentration which is absent at higher concentrations (Yeo, 1974).

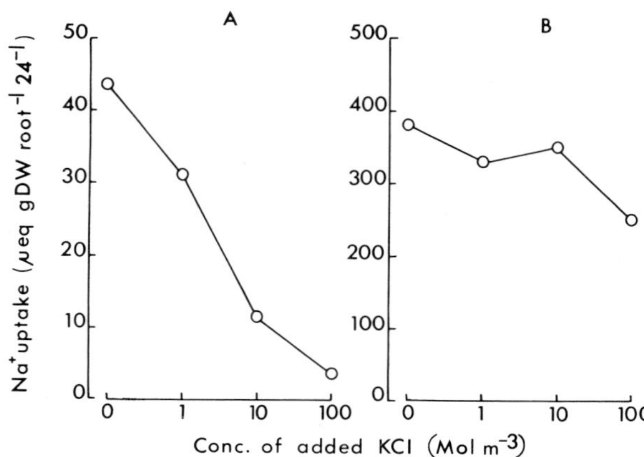

Fig. 10.7. The effect of the concentration of KCl on the uptake of $^{22}Na^+$ into *Suaeda maritima* seedlings. (A) NaCl at 0.05 Mol m$^{-3}$; (B) NaCl at 5.0 Mol m$^{-3}$. Data from Yeo (1974).

In essence these differences indicate that uptake from a low concentration is more dependent upon metabolic energy than uptake from a high concentration.

### 10.4.2 Na$^+$ accumulation by leaves

As we have already seen, the majority (95%) of Na$^+$ accumulated by halophytes reflects accumulation in the leaves. Consequently, similar conclusions can be drawn for ion accumulation in the leaves of *Suaeda maritima* to those drawn for uptake into the whole seedling. Again Na$^+$ uptake from low Na$^+$ concentrations is almost unaffected by the K$^+$ concentration. It is a notable feature of the leaves of *Suaeda maritima*, that Na$^+$ is strongly retained within the vacuoles, the half-time for efflux from the vacuolar compartment of plants grown in the presence of 340 Mol m$^{-3}$ NaCl being some 148 h (Yeo, 1974). Na$^+$ exchange from *Atriplex spongiosa* leaf slices is also slow, suggesting that it is confined to the vacuoles, whereas K$^+$ loss occurred as if it were distributed in the free space, cytoplasm and vacuoles (Osmond, 1968). These data indicate that once absorbed, the Na$^+$ is strongly retained in the cells. This correlates well with the slow rate of movement of Na$^+$ out of leaves of *Atriplex halimus* labelled with $^{22}Na$ to other parts of the plant (Waisel, 1972) and its lack of retranslocation in *Suaeda maritima* (Yeo, 1974).

### 10.4.3 Na$^+$ transport from root to shoot

The transport of ions from the root to the shoot in halophytes has been the subject of a few investigations, mostly using mangroves. Early experiments of

Scholander et al. (1962) indicated that those mangrove species which do not secrete salt from glands on the leaves are relatively lower in Na$^+$ in the xylem than those species which do have salt glands. Even the species without glands, however, had salt concentrations some 10–50 times those of the common glycophytes. Recently, Atkinson et al. (1967) found that in the mangrove, *Aegialitis*, the xylem sap contained between 85 and 122 equiv. m$^{-3}$ Cl$^-$ and transpiration delivered some 100 $\mu$equiv. leaf$^{-1}$ day$^{-1}$. The ion concentration in the leaves was maintained at about 400 equiv. m$^{-3}$ by the operation of the salt glands. In *Rhizophora*, however, a mangrove without salt glands, the xylem sap contained only some 17 equiv. m$^{-3}$ Cl$^-$. Transpiration supplied 17 $\mu$equiv. Cl$^-$ leaf$^{-1}$ day$^{-1}$ and the Cl$^-$ concentration in the leaf could be maintained at about 500–600 equiv. m$^{-3}$ by an increase in dry weight of 3% day$^{-1}$. These results support the conclusion of Scholander et al. (1966) that xylem sap concentrations in halophytes are generally low and it is clear that the roots exert a considerable control over the transport of ions to the shoots (see also Ch. 11).

Recent work on *Suaeda maritima* has identified a discrimination between the transport of Na$^+$ and K$^+$ to the shoot (Yeo, 1974). During the uptake of Na$^+$ from a 5 Mol m$^{-3}$ solution, if the KCl concentration around the roots is increased from zero to 100 Mol m$^{-3}$, then the percentage of Na$^+$ which is transported to the shoots increases (Fig. 10.8): i.e. as K$^+$ becomes available, so the amount of Na$^+$ in the shoot increases. The converse is true with K$^+$;

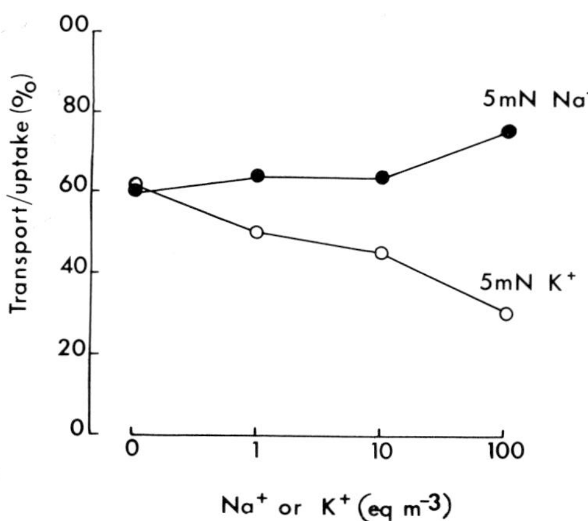


Fig. 10.8. The effect of Na$^+$ and K$^+$ concentration on the transport of K$^+$ and Na$^+$, respectively, to the shoot of *Suaeda maritima* seedlings. The uptake of $^{86}$Rb$^+$ from a 5 Mol m$^{-3}$ KCl solution and of $^{22}$Na$^+$ from 5 Mol m$^{-3}$ NaCl was measured over 24 h and roots and shoots counted separately. $^{86}$Rb$^+$ (○); $^{22}$Na$^+$ (●). (Data from Yeo, 1974.)

as $Na^+$ is increased from zero to 100 Mol $m^{-3}$, then $K^+$ (measured as [86]Rb uptake from a 5 Mol $m^{-3}$ KCl solution) transported to the shoot decreases. Thus, in this species at least, there appears to be a $K^+$-stimulated $Na^+$ transport to the shoot. This could result either from a $K^+$ stimulation of $Na^+$ transport to the xylem or from a reduction in the retention of $Na^+$ by the roots in the presence of $K^+$. The situation in which there is a preferential transport of $Na^+$ to the shoot contrasts with that in most glycophytes, where $Na^+$ tends to be retained in the roots.

### 10.4.4 $Na^+$ uptake by the roots of halophytes

In *Suaeda maritima* $K^+$ has a similar effect on the retention of $Na^+$ by the roots to that already described for the shoots. Thus, at low concentrations of $Na^+$ (0.05 Mol $m^{-3}$) $K^+$ concentrations greater than 1 Mol $m^{-3}$ have a markedly inhibitory effect on the retention of ions by the roots. In 5 Mol $m^{-3}$ $Na^+$, on the other hand, $K^+$ is almost without effect on the retention of $Na^+$ by the roots themselves (Yeo, 1974). This does suggest that under the conditions of Fig. 10.8, $K^+$ stimulates $Na^+$ transport to the shoot.

Interestingly, $Na^+$ efflux from the root cell vacuoles of *Triglochin* has a relatively short half-time (a maximum of 18.6 h) in contrast to the situation described for the shoots of *Suaeda maritima*. This half-time is shorter than for many glycophytes (Jefferies, 1973) and could be related to the transport of $Na^+$ to the shoot and its subsequent retention or export to salt glands (depending upon species). In fact, in *Suaeda maritima* the $t_{\frac{1}{2}}$ for the efflux of $Na^+$ from the root cell vacuoles is noticeably less than that from the leaf cell vacuoles (27 h as opposed to 145 h for the leaves, Yeo, 1974) and is of the same order as the value for *Triglochin*. It does, therefore, appear that leaf and root cells do differ markedly in the permeability of their tonoplast membranes towards $Na^+$. A further feature of *Triglochin* roots is that Jefferies concluded, as a result of compartmental analysis and potential measurements, that $Na^+$ appears to be in equilibrium between the cytoplasm and the vacuole at all salinities.

As far as $Na^+$ uptake is concerned, halophytes appear therefore to be essentially similar to glycophytes in that the rate of $Na^+$ uptake increases with increasing $Na^+$ concentration. There appears to be a difference in the mechanism of uptake at low and high concentrations as judged by the influence of temperature and light on uptake as well as by the effects of $K^+$. However, there does appear to be a marked contrast to glycophytes in that $Na^+$ seems to be preferentially transferred to the shoot. This may be related to the need for osmotic adjustment in the shoot and its ability to retain $Na^+$ in a vacuolar compartment. This appears to contrast with the roots for halophytes, where $Na^+$ efflux may be rapid. We must not, of course, lose sight of the fact that $Na^+$ uptake into halophytes normally takes place from a concentration of ions which is toxic to glycophytes.

### 10.4.5 $K^+$ uptake

In contrast to the enormous concentration of $Na^+$ which is available to most halophytes, $K^+$ uptake occurs in an environment where $K^+$ is out-numbered by about 50:1 by $Na^+$ (in sea water) – although the $K^+$ concentration of 10 equiv. $m^{-3}$ in sea water is high in comparison with most environments. As we have already seen $K^+$ appears to be able to substitute for $Na^+$ in that halophytes growing in low $Na^+$ environments contain high $K^+$ levels; $K^+$ uptake must then be suppressed by the addition of $Na^+$, since plants growing in increasing $Na^+$ concentrations contain decreasing amounts of $K^+$ (Fig. 10.2). Suppression of $K^+$ uptake, over a ten day period, was reported for *Atriplex vesicaria* after the addition of $Na^+$ to the culture solution (Black, 1960). This conclusion is borne out by experiments on *Suaeda maritima*, where $K^+$ accumulation from a 5 equiv. $m^{-3}$ solution into the shoots of seedlings is suppressed as the $Na^+$ concentration increases (Fig. 10.9A). In contrast, $K^+$ uptake from a 0.05 equiv. $m^{-3}$ solution is almost unaffected until the concentration of $Na^+$ is greater than 10 equiv. $m^{-3}$ (Fig. 10.9B). Interestingly, in *Avicennia marina* leaves, $Na^+$ stimulates $K^+$ uptake from both 1 and 10 equiv. $m^{-3}$ $K^+$ (Rains and Epstein, 1967b); $Na^+$ concentrations of 500 equiv. $m^{-3}$ eventually suppressed $K^+$ uptake from the 1 equiv. $m^{-3}$ $K^+$ (by 73%), but in 10 equiv. $m^{-3}$ $K^+$ the rate of uptake was always greater in the presence of $Na^+$ than in its absence. Rains and Epstein also reported that the uptake at low concentrations resembled that from barley in its selective preference for $K^+$.

Accumulation of $K^+$ into the roots of *Suaeda maritima* from a 5 equiv. $m^{-3}$ KCl solution over a 24-h uptake period was suppressed by $Na^+$ (Fig. 10.9A). The uptake of $K^+$ into the roots of *Triglochin maritima* in a 0.1 equiv. $m^{-3}$ solution of $K^+$ is, however, markedly stimulated by $Na^+$ (about 400% by 10 equiv. $m^{-3}$ $Na^+$; Fig. 10.9D) and is only suppressed, by about 14%, at concentrations of $Na^+$ greater than that in seawater (Parham, 1970; Jefferies, 1973). The stimulation is not apparent in $Na_2SO_4$, LiCl, $MgCl_2$ or $NH_4Cl$ and is inhibited by CCCP. $K^+$ uptake from a 10 equiv. $m^{-3}$ concentration of $K^+$ is, on the other hand, almost unaffected by $Na^+$ concentration (Fig. 10.9B) and by CCCP. Jefferies concluded that at high salinities such as those of sea water, $K^+$ is entering by passive diffusion and that the influx pump only operates

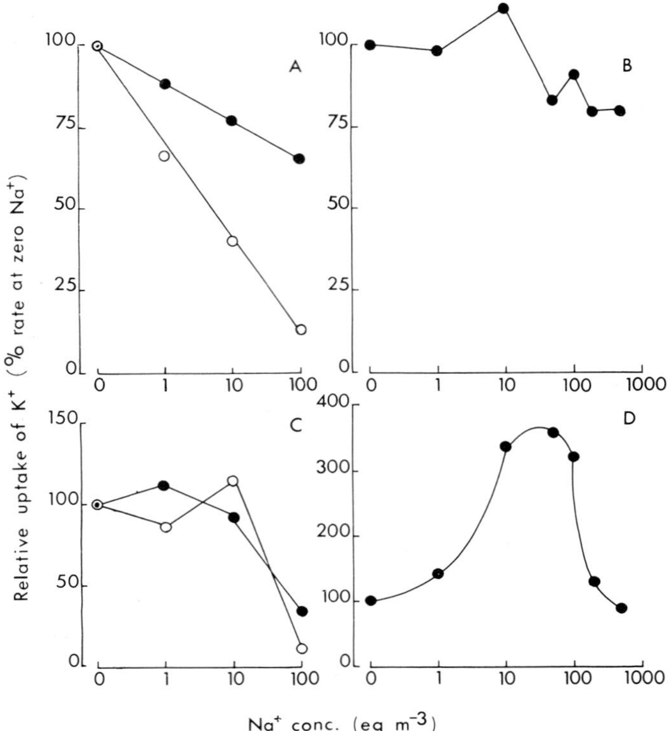

Fig. 10.9. The effect of $Na^+$ concentration on the uptake of $K^+$ into roots (●) and shoots (○) and measured as $^{86}Rb^+$, in *Suaeda maritima* and as $^{42}K^+$ in *Triglochin maritima*. (A) *S. maritima*, $K^+$ 5.0 equiv. m$^{-3}$; (B) *T. maritima*, $K^+$ 10.0 equiv. m$^{-3}$; (C) *S. maritima*, $K^+$ 0.05 equiv. m$^{-3}$; (D) *T. maritima*, $K^+$ 0.1 equiv. m$^{-3}$; (A) and (C) data of Yeo (1974); (B) and (D) data of Jefferies (1973).

at low salinities. He also concluded that $K^+$, as $Na^+$, was in electrochemical equilibrium between cytoplasm and the outside solution. It is, however, difficult to understand the importance of a specific mechanism operating at low $K^+$ concentrations to halophytes, unless it is a relic and indicative of the recent evolution of the halophilic habit in angiosperms.

## 10.5 Conclusions

The rate of uptake of $Na^+$ increases as its concentration in the external solution increases. At all concentrations, however, there is evidence in *Triglochin maritima* for a passive equilibrium between solution and roots. Furthermore, as far as uptake into the shoot of *Suaeda maritima* is concerned, metabolism appears to be of decreasing importance as the $Na^+$ concentration rises. The shoots are all important in the accumulation of ions in halophytes, however, and halophytes appear to differ from glycophytes in that $Na^+$ is transported upwards and perhaps preferentially over storage in the roots. The ability to accumulate large concentrations of ions in the tops seems to be correlated with the ability to retain these ions within the vacuoles, a property of the relatively $Na^+$-impermeable tonoplast. That the ions are chiefly accumulated in the vacuoles is inferred from both biochemical and efflux data. The biochemical data would indicate that the maximum $Na^+$ concentration in the cytoplasm is perhaps 150 Mol m$^{-3}$: the vacuolar concentration must, in most halophytes, be in excess of 500 Mol m$^{-3}$. If the concentrations in the cytoplasm and vacuole were 100 and 500 Mol m$^{-3}$, respectively, a potential difference across the tonoplast of some 40 mV would be required for the ions to be in passive equilibrium. Such a value would be high for a trans-tonoplast potential, but this membrane may have unusual properties in its low permeability to $Na^+$. The alternative to such a high potential in the face of the same concentration difference would be a lower potential associated with an efflux pump operating from cytoplasm to vacuole. Jennings (1968) speculated over the role of a reversible ATPase in the accumulation of ions in halophytes. His argument was centred on an analogy drawn with red blood cells and he postulated that the transport of $Na^+$ from high to low concentration could be coupled to a reversible ATPase and therefore allow the synthesis of ATP during the $Na^+$ accumulation. Such a process, if it were to occur, must be confined to the roots of halophytes where large concentration differences might occur between the external solution and the cytoplasm of root cells: the available evidence suggests that the xylem sap is low in $Na^+$ and so, if anything, uphill transport will occur between xylem and the site of accumulation in leaf cell vacuoles.

In spite of the speculation as to the role of ATPase in salt transport in halophytes, however, there have been few investigations of the properties of those enzymes isolated from salt-tolerant plants. In *Avicennia* the activity of an ATPase associated with what appears to be a ribosomal fraction was stimulated by $Na^+$ and by $K^+$ (Kylin and Gee, 1970) and there were indications of synergism between the two ions as has been reported for animal ATPases (see Ch. 2).

In a halophilic isolate of the alga *Chlamydomonas*, the presence of a $K^+$- or $Na^+$-stimulated ATPase associated with the microsomal fraction has been reported (Yamamoto and Okamoto, 1968). Enzyme activity was stimulated to a greater extent by $K^+$ than by $Na^+$ and required $Ca^{2+}$ rather than $Mg^{2+}$

for this activity. There was, however, no evidence of the synergism between $Na^+$ and $K^+$ which is characteristic of the animal ATPase, although there was some (20%) inhibition by 0.1 Mol m$^{-3}$ ouabain. Ouabain sensitivity could not be detected by Batterton and van Baalen (1971) in ATPase activity in a number of blue-green algae, although they did find indications of higher levels of a ($Na^+$, $K^+$)-activated ATPase in isolates with greater salt-tolerance. In *Suaeda maritima* hydrolysis of ATP by 100,000 *g* supernatants is effected by an acid phopshatase, the specific activity of which appears to increase when the plants are grown in the presence of NaCl. The $K_m$ of the enzyme decreases and there is a small change in the response of the enzyme to NaCl and KCl, although the relation of this enzyme to salt transport is unknown (Flowers, unpublished). The inhibitory effects of NaCl on ATPase activity from *S. monoica* and *Atriplex halimus* and the contrasting stimulating effects on the glycophytes *Zea mays* and *Phaseolus vulgaris* reported by Horowitz and Waisel (1970) are hard to evaluate since activity was determined only at pH 8.8.

Little is known of the mechanism of transport of ions across the cytoplasm to the vacuole in any plants, yet alone halophytes. The evidence for pinocytotic transport has been recently reviewed (Baker and Hall, 1973) and such a mechanism would be consistent with maintaining a low ion concentration in the cytoplasm: certainly, many small vesicles are visible in the cytoplasm of salt glands and may be working in reverse to normal ion accumulation (see Ch. 11).

Although the majority of the work reviewed in this chapter has been concerned with cation transport, anions must clearly play an important role. The nature of the anion does influence the uptake of $Na^+$ in halophytes (Table 10.4) as in glycophytes: furthermore, the growth of halophytes is dependent upon the nature of both the anion and the cation (Williams, 1960; Yeo, 1974). In a number of halophytes there is a marked imbalance between the levels of inorganic anions and cations and organic anions play an important role. There is evidence for the presence of $Cl^-$ pumps in marine algae (Ch. 5) and of $Cl^-$ accumulation in chloroplasts of both glycophytes and halophytes and in *Limonium* a $Cl^-$-stimulated ATPase has been identified from the salt glands (see Ch. 11).

Halophytes thus provide an important system for ion uptake studies: they are plants with an extreme ability to accumulate ions and might therefore provide a source of material in which the mechanism of ion transport could be identified due either to its high specific activity or high concentration. There are also important agricultural applications to the study of halophytes. In arid regions of the world, salt levels invariably build up in soils where irrigation is practised with water of high mineral content and this eventually leads to the loss of yield from the crop. In extreme circumstances it can mean expensive land reclamation schemes. The breeding of crop plants for salt tolerance is an entirely empirical process at the present, due to our lack of specific knowledge of what makes plants salt tolerant. An understanding of the mechanism of salt tolerance would provide the basis for expediting plant breeding programmes to provide salt-tolerant crop plants. Salt tolerance in the crop then provides a much greater latitude in the management of irrigation schemes.

## *References*

G.H. ABEL and A.J. MCKENZIE, Crop. Sci., 4 (1964) 157.

A. ARNOLD, Die Bedeutung der Chlorionen fur die Pflanze (1955) Gustav Fischer, Frankfurt.

M.R. ATKINSON, G.P. FINDLAY, A.B. HOPE, M.G. PITMAN, H.D.W. SADDLER and K.R. WEST, Aust. J. Biol. Sci., 20 (1967) 589.

F.A. AUSTENFELD, Z. Pflanzenphysiol., 71 (1974) 288.

D.A. BAKER and J.L. HALL, New Phytol., 72 (1973) 1281.

J.W. BATES and D.H. BROWN, New Phytol., 73 (1974) 483.

J.C. BATTERTON and C. VAN BAALEN, Arch. Mikrobiol., 76 (1971) 151.

A. BEN-AMOTZ and M. AVRON, Plant Physiol., 51 (1973) 875.

R.F. BLACK, Aust. J. Biol. Sci., 9 (1956) 67.

R.F. BLACK, Aust. J. Biol. Sci., 13 (1960) 249.

V.J. CHAPMAN, Salt Marshes and Salt Deserts of the World (1960) Leonard Hill/Interscience, New York.

V.J. CHAPMAN, in H. Boyko (Ed.) New Approaches to Old Problems (1966) Junk, The Hague, p. 23.

V.J. CHAPMAN, in H. Boyko (Ed.) Saline Irrigation for Agriculture and Forestry (1968) Junk, The Hague, p. 201.

R. COLLANDER, Plant Physiol., 16 (1941) 691.

J.R. CRAIG, in R.B. Mattox (Ed.) Saline Water (1970) Contribution 13 of Committee on Desert and Arid Zones Research, Southwestern and Rocky Mountains Division, Am. Assoc. Adv. Sci., p. 3.

J.S. CRAIGIE, J. Fish. Res. Bd. Canada, 26 (1969) 2959.

E. EPSTEIN, Mineral Nutrition of Plants: Principles and Perspectives (1972) Wiley and Sons, New York.

G.B. FEIGE, Z. Pflanzenphysiol., 68 (1972) 121.

T.J. FLOWERS, J. Exp. Bot., 23 (1972a) 310.

T.J. FLOWERS, Phytochemistry, 11 (1972b) 1881.

T.J. FLOWERS, J. Exp. Bot., 25 (1974) 101.

T.J. FLOWERS and J.B. HANSON, Plant Physiol., 44 (1969) 939.

H. GREENWAY, Isr. J. Bot., 17 (1968) 169.

H. GREENWAY, J. Aust. Inst. Agric. Sci., 39 (1973) 24.

H. GREENWAY and C.B. OSMOND, Plant Physiol., 49 (1972) 256.

H. GREENWAY and A.P. SIMS, Aust. J. Plant Physiol., 1 (1974) 15.

                                    *T.J. Flowers*


J. GUTKNECHT and J. DAINTY, Oceanogr. Mar. Biol. Annu. Rev., 6 (1968) 163.

J.L. HALL and T.J. FLOWERS, Planta, 110 (1973) 361.

J.L. HALL, A.R. YEO and T.J. FLOWERS, Z. Pflanzenphysiol. 71 (1974) 200.

H.W. HARVEY, The Chemistry and Fertility of Sea Waters (1966) Cambridge University Press.

V.M. HEIMER, Planta, 113 (1973) 279.

P.K. HOLMES and H.O. HALVORSEN, J. Bacteriol., 90 (1965) 312.

C.T. HOROWITZ and Y. WAISEL, Proc. 18th Internatl. Hort. Congr. (1970) p. 89.

R.L. JEFFERIES, in W.P. Anderson (Ed.) Ion Transport in Plants (1973) Academic Press, London, p. 297.

D.H. JENNINGS, New Phytol., 67 (1968) 899.

D.H. JENNINGS, in W.P. Anderson (Ed.) Ion Transport in Plants (1973) Academic Press, London, p. 323.

M.K. JOHNSON, E.J. JOHNSON, R.P. MACELROY, H.L. SPEER and B.S. BRUFF, J. Bacteriol., 95 (1968) 1461.

H. KAUSS, Plant Physiol., 52 (1973) 613.

V.A. KOVDA, C. VAN DEN BERG and R.M. HAGAN (Eds.) Irrigation, Drainage and Salinity: An International Sourcebook (1973) Hutchinson/FAO/UNESCO.

A. KYLIN and R. GEE, Plant Physiol., 45 (1970) 169.

A.W.D. LARKUM, Nature, 218 (1968) 447.

A.W.D. LARKUM and A.E. HILL, Biochim. Biophys. Acta., 203 (1970) 133.

C.B. OSMOND, Aust. J. Biol. Sci., 20 (1967) 575.

C.B. OSMOND, Aust. J. Biol., 21 (1968) 1119.

M.R. PARHAM, A Comparative Study of the Mineral Nutrition of Selected Halophytes, PhD. Dissertation (1970) University of East Anglia.

R.J. POOLE, Plant Physiol., 47 (1971) 735.

D.W. RAINS and E. EPSTEIN, Plant Physiol., 42 (1967a) 314.

D.W. RAINS and E. EPSTEIN, Aust. J. Biol. Sci., 20 (1967b) 847.

P.F. SCHOLANDER, H.T. HAMMEL, E.A. HEMMINGSEN and W. GAREY, Plant Physiol., 37 (1962) 722.

P.F. SCHOLANDER, E.D. BRADSTREET, H.T. HAMMEL and E.A. HEMMINGSEN, Plant Physiol., 41 (1966) 529.

M. STEINER, Jahrb. Wiss. Botanik, 81 (1934) 94.

M. STEINER, Ergeb. Biol., 17 (1939) 151.

M. STEINER and W. ESCHRICH, in W. Ruhland (Ed.) Handbuch der Pflanzenphysiologie (1958) Vol. 4, Springer Verlag, p. 334.

G.R. STEWART and J.A. LEE, Planta, 120 (1974) 279.

P.F.T. TROKE, Aspects of Salt Tolerance in Fungi, D. Phil. Thesis (1975) University of Sussex.

J.C.TH. UPHOF, Bot. Rev., 7 (1941) 1.

M. VAN EIJK, Proc. K. Ned. Akad. Wet., 37 (1934) 556.

R.F.M. VAN STEVENINCK, A.R.F. CHENOWETH and M.E. VAN STEVENINCK, in W.P. Anderson (Ed.) Ion Transport in Plants (1973) Academic Press, London, p. 25.

Y. WAISEL, Biology of Halophytes (1972) Academic Press, London.

Y. WAISEL and A. ESHEL, Experimentia, 27 (1971) 230.

H. WALTER, Die Vegetation der Erde (1962) Vol. 1, Jena.

H.H. WIEBE, Plant Physiol., 41 (1966) 1439.

M.C. WILLIAMS, Plant Physiol., 35 (1960) 500.

M. YAMAMOTO and H. OKAMOTO, Z. Allg. Mikrobiol., 8 (1968) 71.

A.R. YEO, Salt Tolerance in *Suaeda maritima* (1974) D. Phil. Thesis. University of Sussex.

H. ZIEGLER and U. LÜTTGE, Planta., 74 (1967) 1.

CHAPTER 11

# Salt glands

U. Lüttge

# Contents



335

## *11.1 The phenomenon of salt excretion : an ecological adaptation*

### *11.1.1 The efficiency of salt excretion by salt glands*

The massive excretion of NaCl by salt glands is a very effective adaptation to saline habitats, though this is certainly not the only mechanism by which halophilic plants can withstand salinity of their medium (see Ch. 10). The efficiency of salt excretion by glands is best illustrated by a comparison of salt regulation in the mangrove species *Rhizophora mucronata*, which partially excludes salt from the transpiration stream, with the mangrove species *Aegialitis annulata*, which excretes salt by glands on its leaves (Atkinson et al., 1967). In *Rhizophora* the $Cl^-$ concentration of the xylem sap is 17 equiv. $m^{-3}$ delivering 17 $\mu$equiv. $leaf^{-1}$ $day^{-1}$. By contrast the xylem sap of *Aegialitis* contains 85–122 equiv. $m^{-3}$ $Cl^-$, supplying a leaf with about 100 $\mu$equiv. $Cl^-$ $day^{-1}$. Notwithstanding these large differences in the supply, NaCl contents and concentrations in the leaves of *Rhizophora* and *Aegialitis* are reasonably similar. In mature leaves the levels are even lower in *Aegialitis* with its much higher xylem sap concentrations, than in *Rhizophora* (Table 11.1).

Table 11.1.

$Na^+$ and $Cl^-$ amounts and concentrations in the leaves of a salt-excluding mangrove (*Rhizophora*) and of a salt-excreting mangrove (*Aegialitis*). Increasing sample numbers refer to leaves of increasing age. Data from Atkinson et al. (1967).

| Sample | $Na^+$ | | | | | | $Cl^-$ | | | | | |
|---|---|---|---|---|---|---|---|---|---|---|---|---|
| | 1 | 2 | 3 | 4 | 5 | 6 | 1 | 2 | 3 | 4 | 5 | 6 |
| *Rhizophora :* | | | | | | | | | | | | |
| amounts [$\mu$equiv. $leaf^{-1}$] | 61 | 290 | 420 | 480 | 520 | 645 | 74 | 520 | 510 | 585 | 580 | 730 |
| conc. [mequiv.] | 305 | 313 | 431 | 435 | 461 | 461 | 370 | 562 | 522 | 530 | 515 | 522 |
| *Aegialitis :* | | | | | | | | | | | | |
| amounts [$\mu$equiv. $leaf^{-1}$] | 420 | 325 | 275 | 280 | 235 | – | 415 | 290 | 270 | 260 | 255 | – |
| conc. [mequiv.] | 518 | 480 | 411 | 388 | 356 | – | 512 | 429 | 405 | 361 | 386 | – |

In *Rhizophora* a mean growth rate of 3% increase in dry weight per day keeps the chloride at the observed level. In *Aegialitis* elimination of NaCl by the salt glands is responsible for the balance. The rate of salt excretion per effective gland surface (i.e. cross-section of the sub-basal gland cell through which all ions excreted must move) is 250 $\mu$Mol $m^{-2}$ $s^{-1}$ (Atkinson et al., 1967). This rate is 50 times higher than the largest $Cl^-$ fluxes observed in plant cells (e.g. 5–7 $\mu$Mol $m^{-2}$ $s^{-1}$ in the marine alga *Acetabularia*; see

MacRobbie, 1971b). The density of salt glands in the epidermis of halophyte leaves ranges between $650 \times 10^4$ and $4800 \times 10^4$ glands $m^{-2}$ (Waisel, 1972). In the upper surface of *Aegialitis* leaves the gland density is $900 \times 10^4$ $m^{-2}$ (Atkinson et al., 1967). For comparison the density of stomata on leaves is between $10^8$ and $10^9$ $m^{-2}$ (Nultsch, 1971). A typical leaf of *Aegialitis* can eliminate by its glands 100 $\mu$equiv. salt $day^{-1}$ (Atkinson et al., 1967). When the atmosphere is dry the large amounts of salt excreted can often be seen as salt crystals on the surfaces of gland-bearing leaves of halophytes (Fig. 11.1).

The excretory process in *Aegialitis annulata* is very specific in respect to the ionic species eliminated. The excreted fluid contains 450 equiv. $m^{-3}$ $Cl^-$, 355 equiv. $m^{-3}$ $Na^+$ and only 27 equiv. $m^{-3}$ $K^+$ (Atkinson et al., 1967). As shown by the following tabulation of $Na^+/K^+$ ratios, excretion effectively eliminates $Na^+$ as compared with $K^+$:

| | |
|---|---|
| excreted solution | 13 : 1 |
| leaf tissue | 3 : 1 |
| xylem exudate | 8 : 1 |

Similarly, salt glands of *Aeluropus littoralis* (Pollak and Waisel, 1970) and *Tamarix aphylla* (Table 11.5) preferentially exude $Na^+$.

## 11.1.2 Evolutionary aspects and taxonomy

Glands specialised in transport of salts are not merely an evolutionary invention selected as an adaptation to salinity. Although in halophytes we find the most efficient salt glands, there are other examples.

The lower epidermis of water lily (*Nymphaea*) leaves floating on the surface of fresh water lakes, carries hydropote glands which move ions from the outer medium into the leaf tissue (Lüttge, 1964a; Lüttge and Krapf, 1969; Lüttge et al., 1971; see Fig. 11.2g, h). Digestive glands of the pitcher walls of the carnivorous plant *Nepenthes* (see Fig. 11.2i, j) transport various ions between the pitcher wall tissue and the pitcher fluid, i.e. $K^+$, $Na^+$ and $Cl^-$, and other ions when available from the digested prey. The $Cl^-$ concentration in the secreted pitcher fluid is 17–27 Mol $m^{-3}$ (Morrissey, 1955; Lüttge, 1966a; Nemček et al., 1966). Hydathodes are commonly considered as structures excreting water. However, from thermodynamical considerations it is almost certain that water molecules per se are never transported actively. Thus, the elimination of water by hydathodes is always an osmotic consequence of the transport of some inorganic or organic solutes. Active steps in this transport occur at the level of the roots, where root pressure drives exudation and the hydathodes are merely openings in the epidermis. Alternatively, or in addition, active processes may participate at the level of the hydathodes, if the hydathodes are

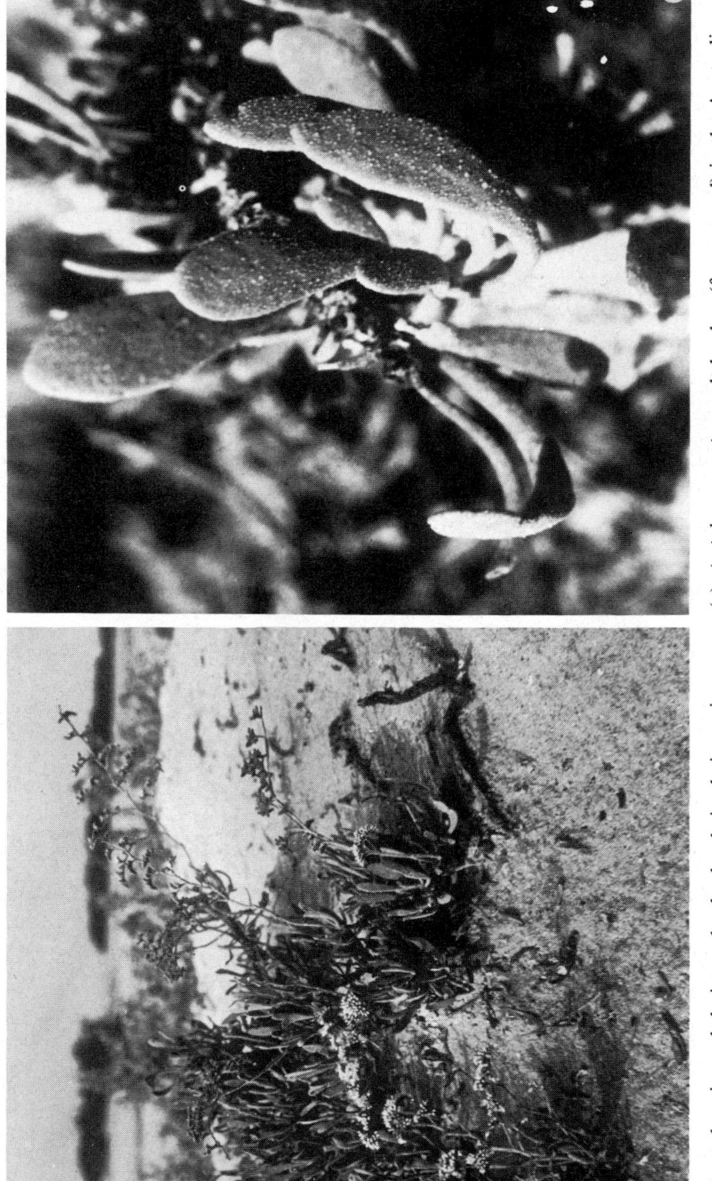

Fig. 11.1. Plants bearing salt hairs and salt glands in their environment. (a) *Atriplex spongiosa* salt bushes (foreground) in the Australian salt bush steppe (near the Koonamoore field station of the University of Adelaide, about 210 miles North of Adelaide and 130 miles West of Broken Hill; Flinders Ranges in the background). (b) Tip of an *A.spongiosa* branch with the spongy fruits. Salt hairs and a salt crust on the leaves. About 4 ×. (c) Leaves of an *A.spongiosa* plant grown on water culture in the greenhouse. Salt hairs on the leaves clearly discernible. About 2 ×. (d) *Limonium axillare* on the Sinai (Red Sea near Nabek). (e) *L.axillare* leaves with the salt excreted by the glands crystallised on their surface. About 3 ×. (a–c original photos of the author; d–e by courtesy of Klaus Winter, Darmstadt.)

          *U. Lüttge*

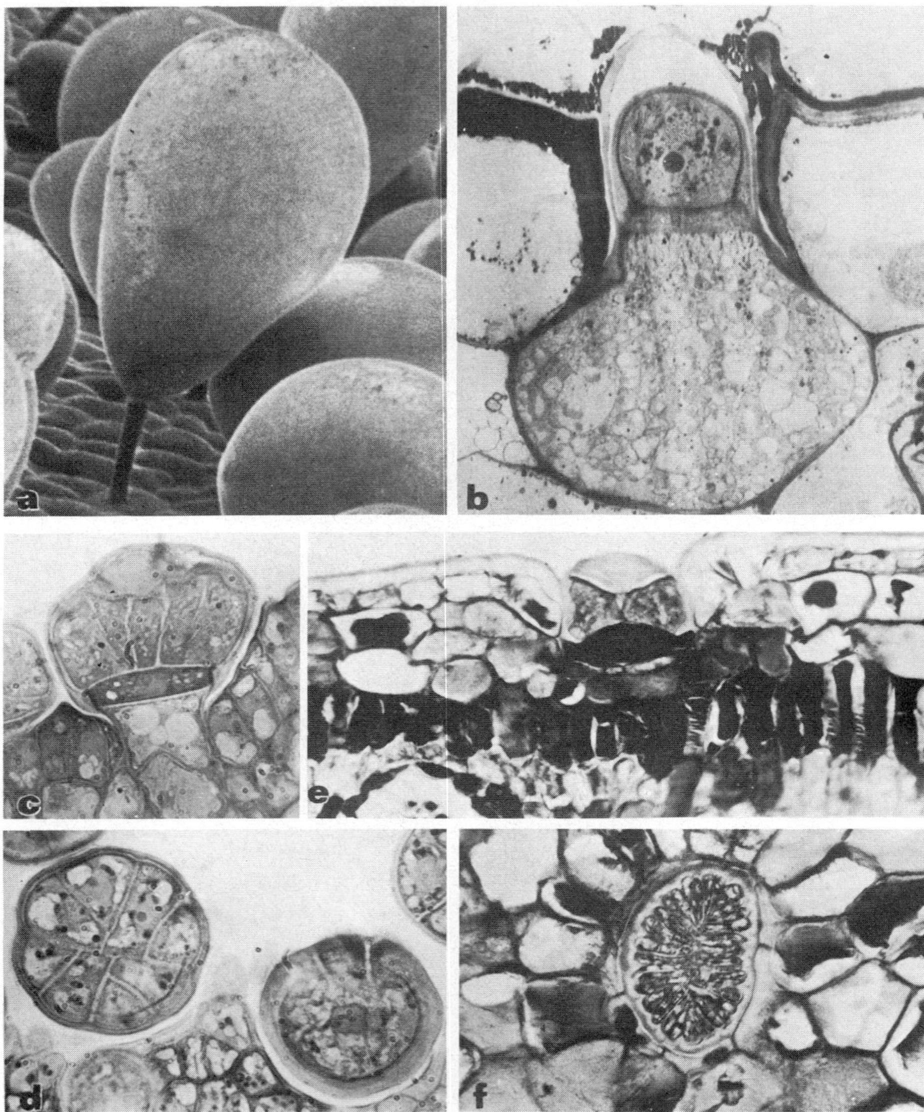

Fig. 11.2. Salt transporting glands. (a) Scanning electron micrograph of an *A.spongiosa* salt hair (stalk and bladder cells), magnification × 250. (With kind permission from Troughton and Donaldson, 1972.) (b) Salt gland of *Spartina foliosa* in a cross-section of a leaf, magnification × 2500. (With kind permission from Levering and Thomson, 1971.) (c) and (d) Salt glands of the mangrove species *Avicennia marina*, (c) longitudinal section, magnification × 750, (d) tangential section of a leaf, magnification × 750. (With kind permission from Shimony et al., 1973.) (e) and

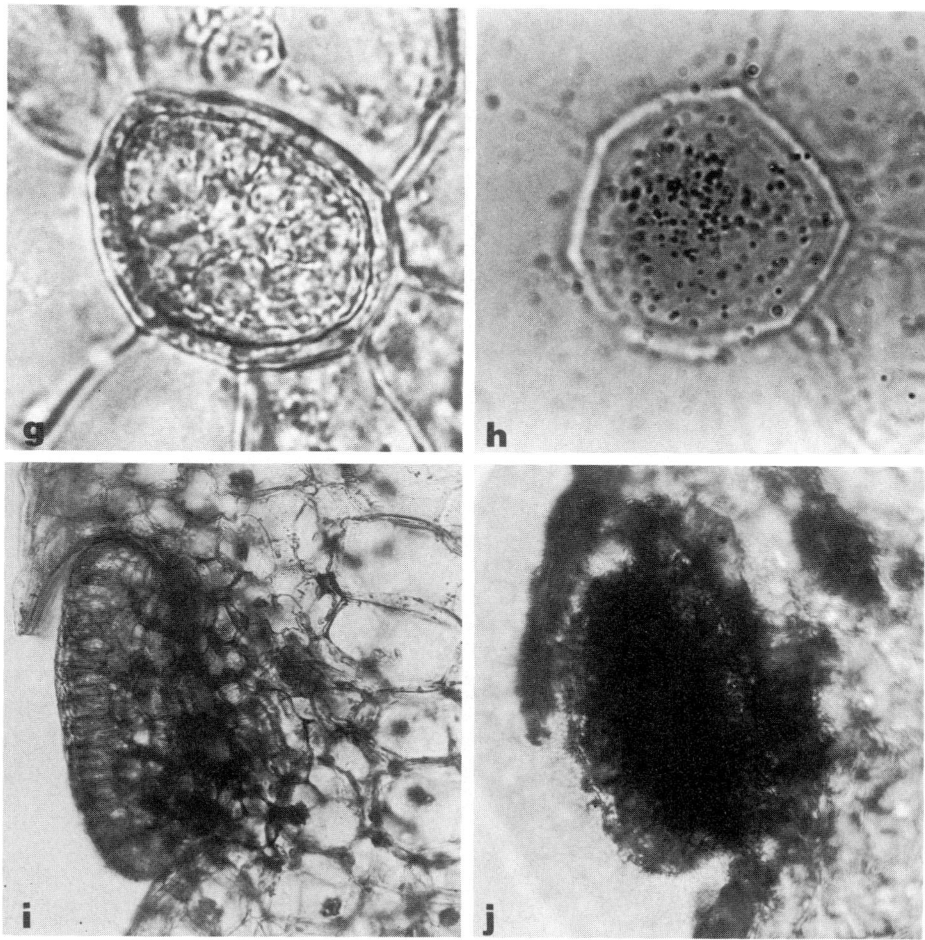

(f) Salt glands of the mangrove species *Aegiceras corniculatum*, (e) transverse section, magnification × 220, (f) tangential section of a leaf, magnification × 220. (With kind permission from Cardale and Field, 1971.) (g) and (h) Hydropote glands of *Nymphaea*, tangential sections of the lower surface of floating leaves, (h) microautoradiograph showing preferential accumulation of label in the hydropote gland after uptake of $^{35}$S-labelled sulphate, magnification × 1300. (From Lüttge, 1964a and 1966c.) (i) and (j) Digestive glands of *Nepenthes* pitchers in longitudinal sections of the inner pitcher wall, (j) microautoradiograph of a heavily labelled gland after uptake of $^{35}$S-labelled sulphate from the fluid in the pitcher lumen; magnification × 475. (From Lüttge, 1965, 1966c.)

gland-like structures. With increasing concentration of inorganic (salts) or organic solutes (e.g. sugars) in the eliminated fluid, one observes a gradual change from hydathodes to salt glands and nectary glands respectively (Frey-Wyssling, 1935; Frey-Wyssling and Häusermann, 1960; see also Lüttge, 1975).

Hence, we can deduce from the present diversity of phenomena, that evolution had considerable material to play with, eventually developing salt glands as an effective ecological adaptation. This is also reflected in the wide taxonomical distribution of salt-excreting genera in various sub-classes of angiosperms. Not all of these sub-classes are closely related to each other, which suggests a poly-phyletic evolution of salt excretion (Table 11.2).

### 11.1.3 Ecological distribution

We have used above the example of the mangrove *Aegialitis* to give a quantitative idea of the efficiency of salt excretion. Mangroves are hygrohalophytes growing in coastal swamps on very wet soil or in an aqueous medium of high salt concentration. Other hygrohalophytes are growing in coastal salines, e.g. salt marshes in the temperate zone, or near inland salines in the arid zone (see Ch. 10 and Waisel, 1972).

Salt excretion is also performed by xerohalophytes in the arid zone, where salinity is a consequence of the shortage of water. As an example, in the next section we will consider species of the genus *Atriplex* where salt elimination is mediated by salt hairs.

### 11.1.4 Salt elimination by salt hairs

Salt glands normally excrete salt to the outer surface of the plant. However, one also observes so-called salt hairs, where gland-like cells excrete salt into large vacuoles of bladder-shaped cells (Fig. 11.2a). As we will see below the physiological mechanism of excretion is not principally different in the two cases (see Fig. 11.15).

Salt hairs are particularly found on the leaves of chenopod species (Table 11.2) and in the genus *Atriplex* they serve as an adaptation to salinity. Fig. 11.3 shows that elimination of salt from these leaves into the vacuoles of epidermal bladder cells may very effectively reduce the salt load in the metabolically active photosynthesising tissue of *Atriplex* plants growing in highly saline media. In desert annuals, e.g. *Atriplex spongiosa*, which thrive for only a few weeks after seed germination becomes possible due to a strong rainfall, this elimination of salt into the bladder vacuoles may provide sufficient regulation

Table 11.2.

Phylogenetic relations between taxa comprising salt-excreting genera of terrestrial halophytes. An enumeration of salt-excreting genera was obtained from Waisel (1972, p. 142) and from Hill and Hill (1975). Taxa were arranged according to the phylogenetical scheme given by Ehrendorfer (1971, p. 741, Fig. 742); and the angiosperm taxonomy of Ehrendorfer (1971) was followed largely, but Engler and Diels (1936) had to be consulted for *Frankenia* and *Laguncularia*, because in Ehrendorfer (1971) these genera are not mentioned. Taxa are : C., class; S.C., sub-class; G.O., group of orders; O., order; F., family; G., genus. Arrows indicate probable direction of evolution of genera. Asterix : genus with salt hairs.

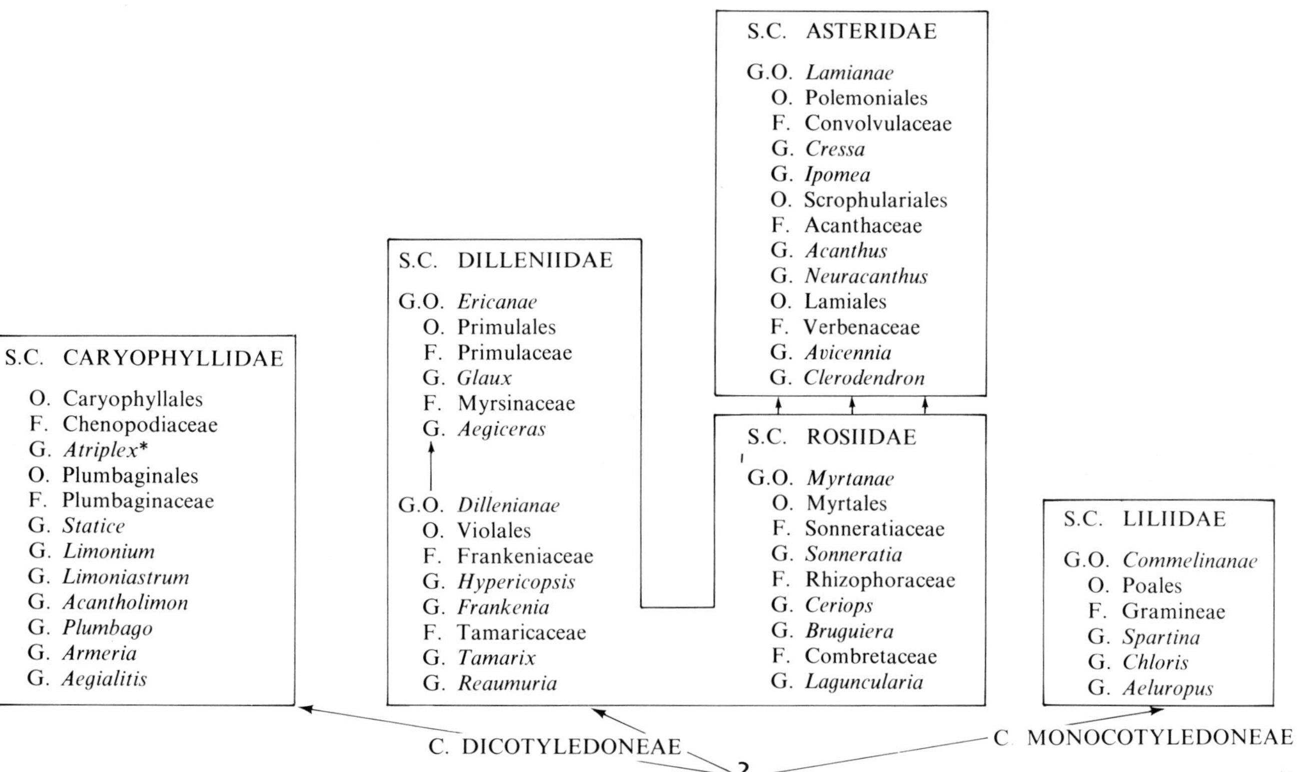

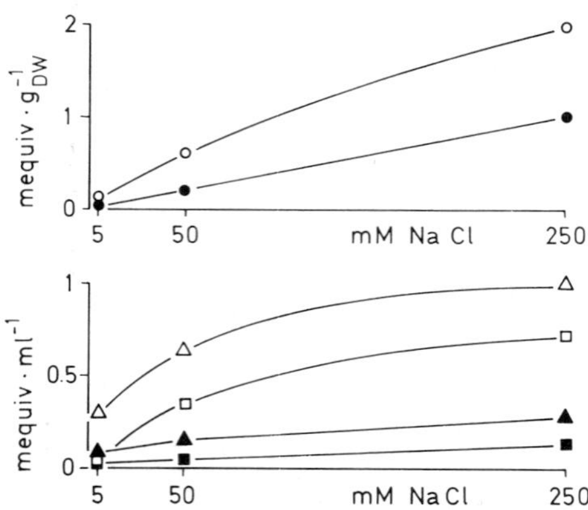

Fig. 11.3. Distribution of Na$^+$ ($\triangle$, $\blacktriangle$) and Cl$^-$ ($\bigcirc$, $\bullet$, $\square$, $\blacksquare$) between salt hairs (open symbols) and lamina (closed symbols) of *A.spongiosa* plants grown in culture solutions containing increasing amounts of NaCl. The discrepancy between salt levels in salt hairs and lamina is much more pronounced, when the values are expressed as concentrations ($\triangle$, $\blacktriangle$, $\square$, $\blacksquare$) mequiv. ml$^{-1}$ H$_2$O) as compared to a dry weight basis (mequiv. gDW$^{-1}$) ($\bigcirc$, $\bullet$). Data from Osmond et al. (1969).

for the entire lifespan of a leaf. In perennials, such as *Atriplex nummularia*, collapsing and drying salt hairs eventually form a highly reflective silvery salt crust on the leaves, which at times may be washed away by rain; new salt hairs are formed continuously.

*11.1.5  Fractionation of Atriplex leaves by a small desert rodent*

By contrast to *A. spongiosa* (Fig. 11.3), there is a high salt concentration in the leaf lamina of the perennial salt-bush shrub *Atriplex nummularia*, already at low external NaCl concentrations. However, salt levels in the lamina of *A. nummularia* do not increase much further as external concentrations rise (Fig. 11.4). On the other hand, similarly to *A. spongiosa*, the salt concentration in the bladders of *A. nummularia* is always very much higher than in the leaf mesophyll and increases considerably with increasing NaCl concentrations of the medium (Fig. 11.4).

To perform analyses like those shown in Figs. 11.3 and 11.4, the development of techniques allowing separation of the bladder and lamina tissues was required (Osmond et al., 1969). In the Californian desert, Kenagy (1972, 1973) has observed that a small rodent (*Diplodomys microps*; family *Heteromyidae*) feeds on the leaves of the salt bush *Atriplex confertifolia*, which are similar to those of *A. nummularia*. This kangaroo rat species may eat the whole leaves of

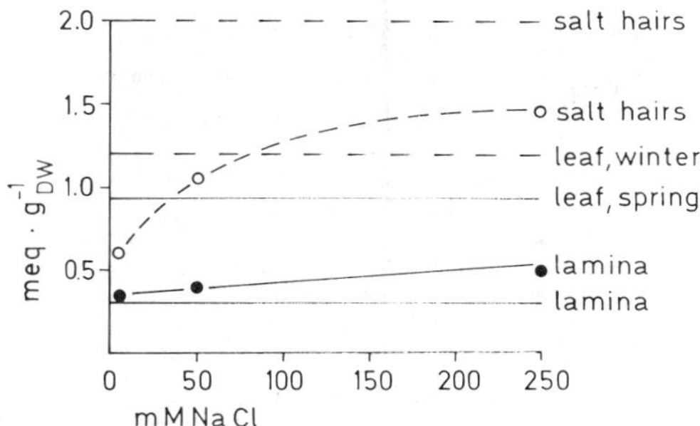



Fig. 11.4. Distribution of Cl$^-$ between salt hair ($\bigcirc$) and lamina ($\bullet$) of *A.nummularia* plants grown in culture solutions containing increasing amounts of NaCl. (From Osmond et al., 1969.) Horizontal lines give, for comparison, Na$^+$ concentrations of leaf material of *A.confertifolia* eaten (solid lines) and not eaten (broken lines) respectively by *D.microps*. (From Kenagy, 1973.) 'Leaf' refers to whole leaf, 'lamina' and 'salt hairs' refer to the separated tissues. Spring-type leaves of *A.confertifolia* are low enough in salt to be eaten entirely. By fractionation of high-salt leaves into lamina and salt hairs, edible leaf lamina material is obtained by the animal (see also Fig. 11.5.).

*A. confertifolia* when the leaves contain low salt levels, as during the springtime when new leaves are developing. However, during most of the year these animals remove the salt-loaded surface tissues and eat only the inner tissue. Achieving a similar result to that of the scientific investigators, but employing a natural, mechanical procedure, *D. microps* is therefore also capable of fractionating the *Atriplex* leaves. Using its highly specialized, broadened and chisel-shaped lower incisors *D. microps* shaves off the salt hair tissue from the leaf lamina; it discards the salt hairs and eats the lamina (Fig. 11.5). A comparison of the salt levels recorded in salt hairs and lamina respectively of *A. nummularia* with an analysis of the *A. confertifolia* tissue discarded or eaten by *D. microps* is made in Fig. 11.4. At external NaCl concentrations greater than about 50 Mol m$^{-3}$ bladders of *A. nummularia* contain NaCl levels which are above those acceptable to the animal. By contrast, even at 250 Mol m$^{-3}$ NaCl, salt levels in the lamina of *A. nummularia* are well below those which are still eaten by *D. microps*.

## 11.2 Physiological anatomy and cytology of salt glands

### 11.2.1 Morphological distribution of salt glands

All aerial parts of halophytes may bear salt glands in their epidermis. However, the typical situation of salt glands is on the leaves. Glands are inserted in or may emerge from the epidermis (Fig. 11.2).

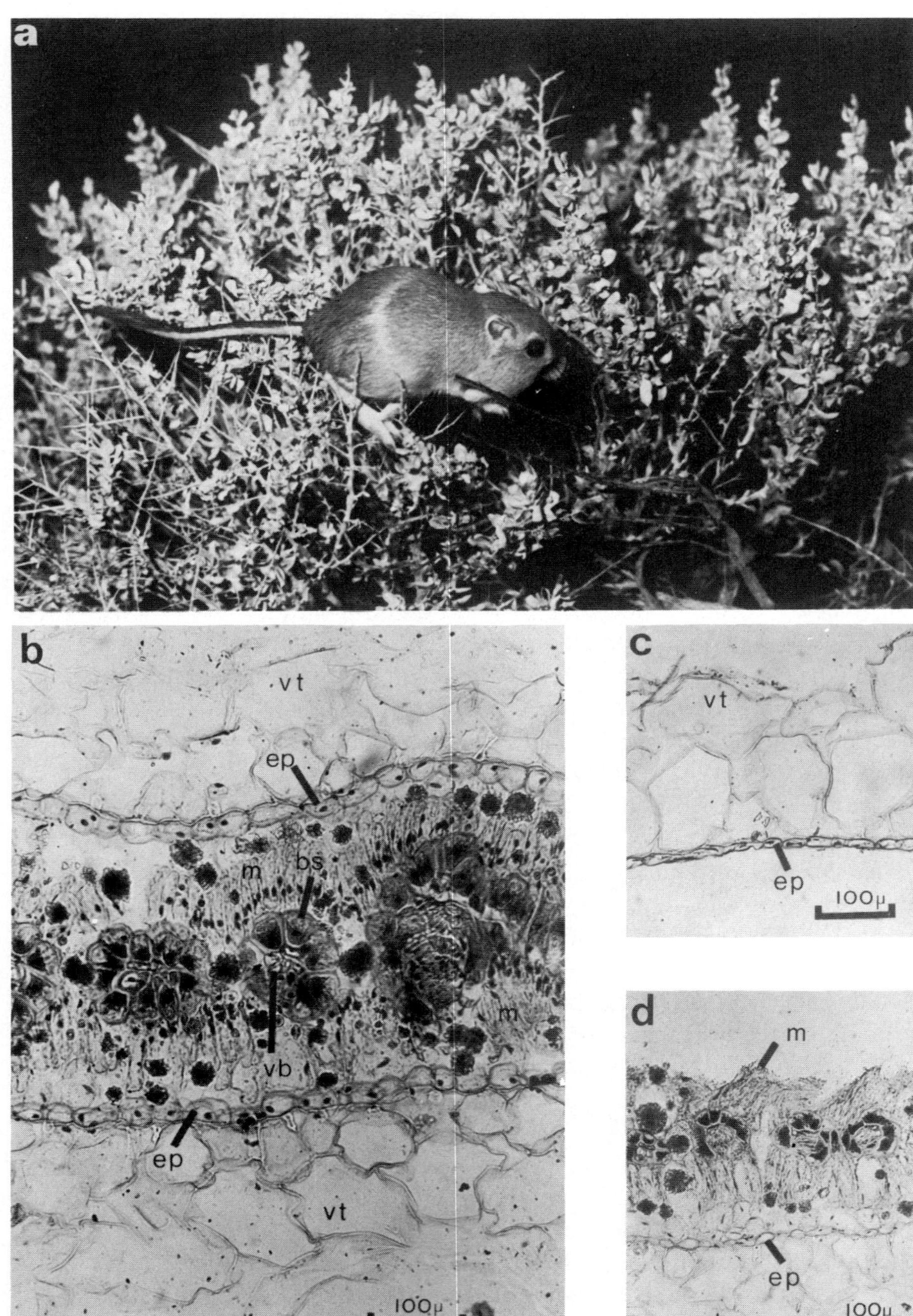
a
b
vt
ep
m bs
vb
ep
vt
100µ
c
vt
ep
100µ
d
m
ep
vt
100µ

### 11.2.2 Gland anatomy

Salt hairs are usually differentiated into excretory stalk cells and salt-accumulating bladder cells (Figs 11.2a and 11.15).

Salt glands are often more complex structures comprising between 2 to 40 cells (or even more). A few examples are shown in Fig. 11.2. Frequently there appear to be various types of cells distinguished by their shape and anatomical placement. Stalk cells, basal cells, or collecting cells may be distinguished from the salt-eliminating excretory cells. However, one generally knows very little about the real function of the various cell types differentiated anatomically. Therefore there is no immediate need in this chapter to describe fully the structures of the many glands which have been investigated anatomically. In a critical review, Hill and Hill (1973b) have written about this problem in the following way: 'Microscopists have an unfortunate tendency to propose physiological mechanisms on the basis of the interpretation of anatomical details, and salt glands have not been spared'. It therefore appears to be more profitable to depict only one particular example in detail. We chose the gland of *Limonium* for this purpose (Fig. 11.6). This gland has the shape of a small cylinder, each quarter section of which is made up of 4 differently shaped cells. The complex of 16 gland cells is in contact with the leaf mesophyll via 4 'collecting cells'. Intercellular spaces between the gland cells serve as channels in which the excreted salt solution can move to the 4 cuticular pores (Fig. 11.7) through which it is released to the outside.

We shall return to this example below during consideration of physiological cytology of glands and of gland function.

### 11.2.3 Physiological gland cytology

*General aspects of gland cytology*
Gland cells generally lack large central vacuoles and therefore their cytoplasm usually appears very dense (Figs 11.6 and 11.8). The nuclei are relatively large compared with the cell volume (Figs 11.6 and 11.8). There are also many and

---

Fig. 11.5. *Diplodomys microps* feeding on leaves of the perennial spiny salt bush shrub *Atriplex confertifolia*. (a) *D. microps* in the top of the saltbush. (b) Cross section of a leaf of *A. confertifolia*. (c) Surface tissue (salt hairs) removed and discarded by *D. microps*. (d) Leaf with the salt-accumulating salt hair tissue sheared off one side. bs = bundle sheath parenchyma; ep = epidermis, m = mesophyll, vb = vascular bundles, vt = vesicular trichomes (salt hairs). (With kind permission from Kenagy, 1972; copyright 1972 by the American Association for the Advancement of Science; 1973.)

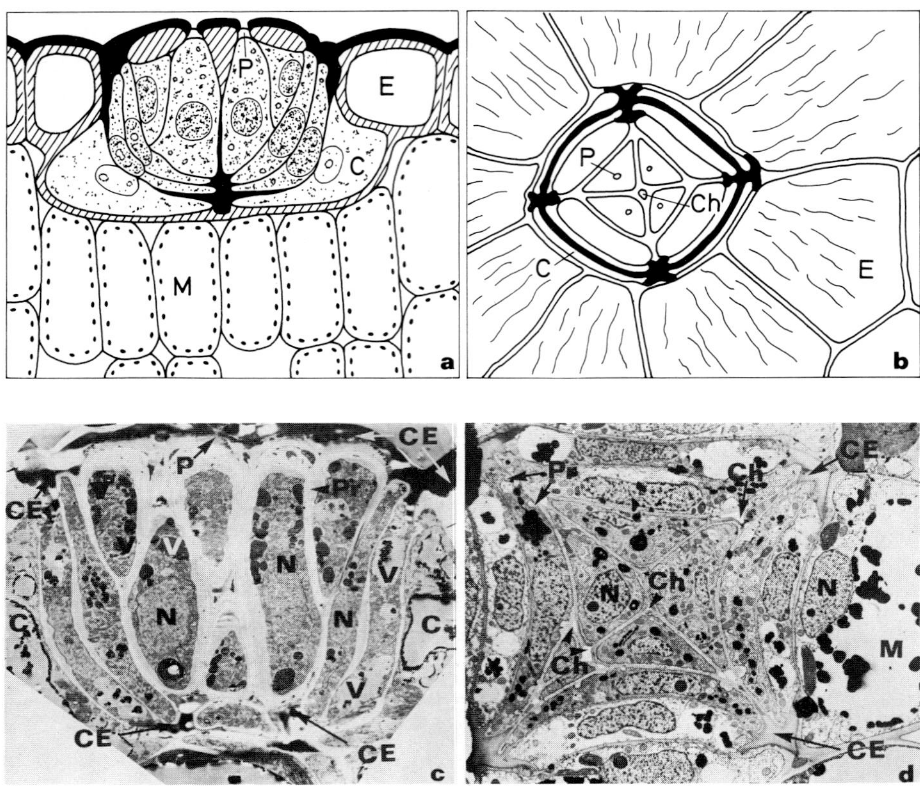



Fig. 11.6. Anatomical and cytological structure of *Limonium vulgare* salt glands. (a) Drawing of a gland as seen in a cross section of the leaf, showing two quarter sections of the gland complex with 8 of the 16 gland cells and 2 of the 4 'collecting cells' which provide contact with the mesophyll. The cuticular envelope of the gland (cutinisations are drawn with heavy black) is interrupted twice in 4 places: firstly on the outer surface forming 4 narrow excretory pores (see Fig. 11.7); secondly adjacent to the collecting cells, where the openings of the cuticular envelope provide contact with surrounding cells; the cell wall is passed here by numerous plasmodesmata. (b) Drawing of a surface view of the gland complex with cuticular envelope (heavy black), excretory pores and excretory channel between gland cells. (c) Longitudinal section of a gland in a cross section of a leaf corresponding to (a). Parts of the cuticular envelope with an excretory pore can be clearly seen. Other important cytological features are: rather large nuclei relative to cell volume, dense cytoplasm, small vacuoles with electron-dense material, cell wall protuberances. (d) Cross section of a gland, corresponding to (b). The excretory channels between the gland cells become clearly discernable. Other cytological details as in (c). (a) and (b) after Ruhland (from Ziegler and Lüttge, 1966), (c) from Ziegler and Lüttge (1966) × 860; (d) with kind permission from Hill and Hill (1973b) × 560. C, collecting cell; CE, cuticular envelope; Ch, excretory channel; E, epidermis, M, mesophyll; P, excretory pores; Pr, cell wall protuberances; N, nucleus; V, vacuole.

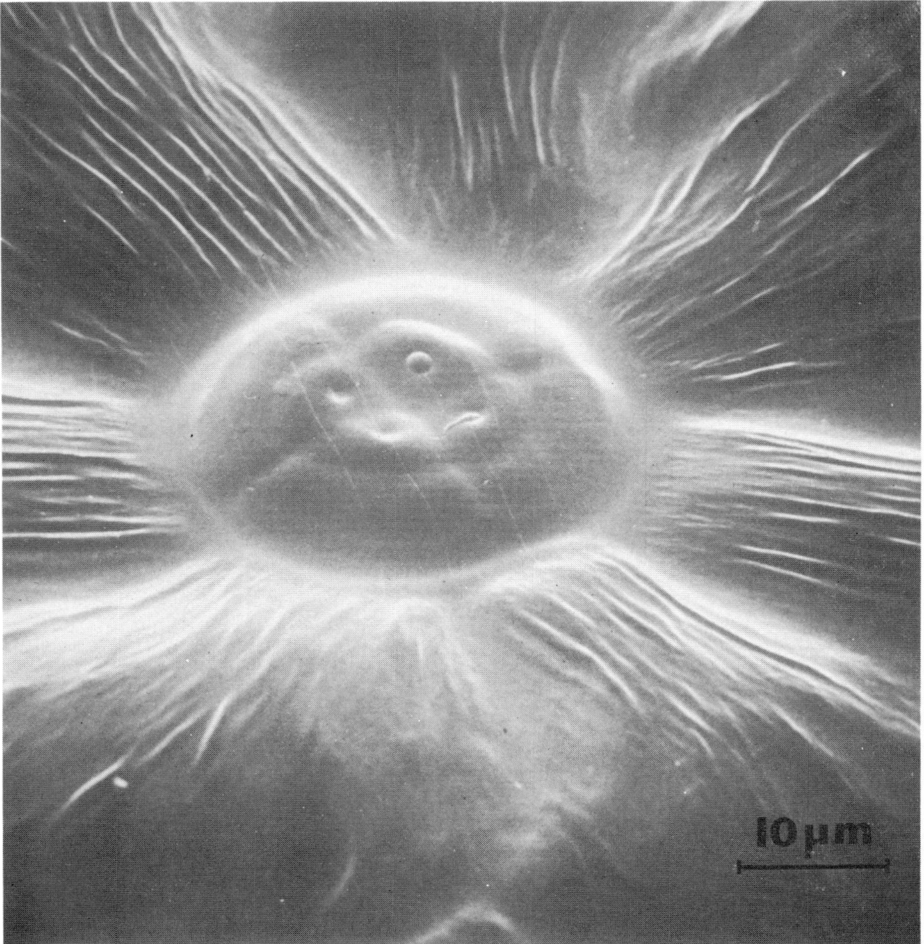


Fig. 11.7. Scanning electron micrograph of a polystyrene replica of a *Limonium* leaf surface with a salt gland. The four cuticular pores are clearly discernable. Original micrograph by courtesy of Bruria Hill, Cambridge (see also Hill and Hill, 1973b).

apparently very active cell organelles. In particular the numerous mitochondria must be mentioned (Figs 11.8 and 11.12). Increased numbers of mitochondria in gland cells lead to increased respiratory activity as shown, for instance, in the nectary glands of *Abutilon* (Ziegler, 1956). Respiratory activity may be required to provide energy for the excretion process (see section 11.3). It is interesting in this context that salt gland cells do not contain plastids that develop into photosynthetically active chloroplasts (see also

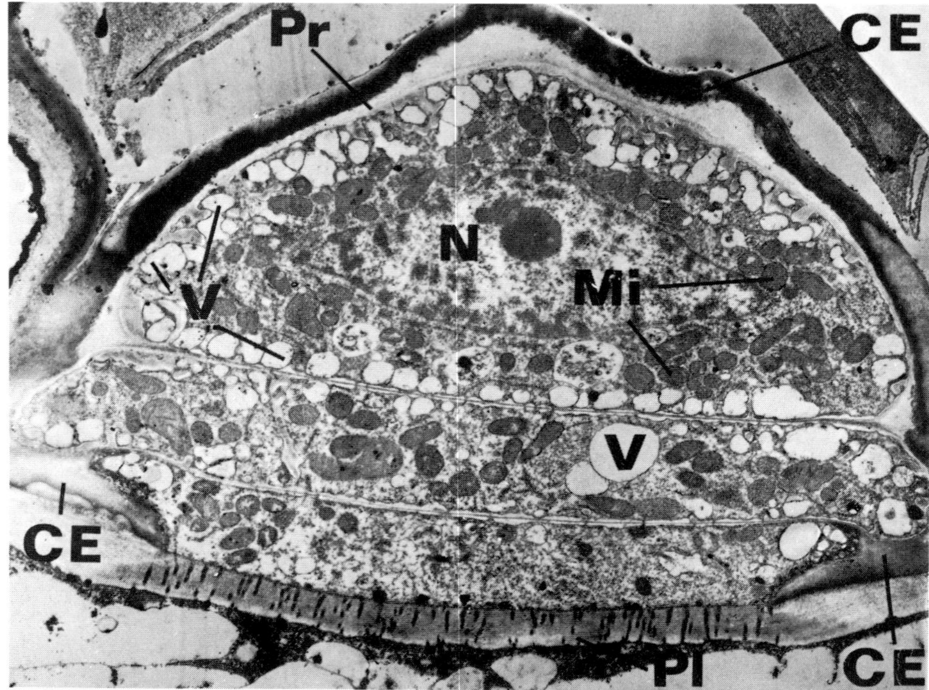

Fig. 11.8. Salt gland of *Tamarix aphylla*. The typical cytological features of salt glands are clearly seen: dense cytoplasm, numerous mitochondria, relatively large nuclei, cell wall protuberances and cuticular envelope. In addition this electron micrograph shows very well the numerous plasmodesmata between the gland cells and between the lower gland cell and the mesophyll, and numerous small vacuoles around the periphery of the excretory cells (with kind permission from Thomson and Liu, 1967). CE, cuticular envelope; Mi, mitochondria; N, nucleus, Pl, plasmodesmata; Pr, cell wall protuberances; V, vacuoles. Magnification × 5400.

sections 11.3 and 11.4). In this respect, in contrast to stomatal guard cells, salt gland cells resemble normal epidermal cells. Notwithstanding the lack of chloroplasts in gland cells, photosynthetic energy transfer reactions, at least under certain experimental conditions, can drive salt excretion (see Section 11.3, Table 11.6 and Fig. 11.14).

These general aspects of gland cytology can be clearly seen in Figs 11.6 and 11.8. There are, however, some particular cytological features of glands which appear to be of great physiological importance and therefore need a separate discussion relating them to gland function.

*Co-operation between various tissues and cell types of the leaf in a symplasmic continuum*

A number of cytological features of salt glands appear to have the specific function of establishing a symplasmic continuum between the gland cells and surrounding mesophyll cells.

First of all, cell walls around the glands or at strategic positions in the vicinity of glands are regularly modified by incrustations with hydrophobic material, i.e. cutins and/or suberins (Figs 11.6, 11.8 and 11.15). These cell wall modifications have been called the cuticular envelope of the glands. Cuticular envelopes of glands are analogous to the Casparian strips in the endodermis of roots and of some stems (see Ch. 8). It is clear from the drawings (Figs 11.6a, b and 11.15) that cuticular envelopes prevent continuous apoplasmic contact between the leaf mesophyll via the gland cells to the excreted salt solution.

However, as can be seen in Figs 11.6a, 11.8 and 11.15, the cuticular envelope does not totally isolate the gland cells from their neighbours. There are interruptions or gaps in the cuticular envelope. The cell walls at these gaps are penetrated by a strikingly large number of plasmodesmata facilitating symplasmic contact with adjacent mesophyll cells (Fig. 11.8). However, the gaps are arranged in such a way that apoplasmic coupling remains impossible (Fig. 11.15). Thus, contact between gland cells and surrounding parenchyma cells via symplasmic pathways is enforced by the cuticular envelope. With the latter statement, of course, we make the assumption that plasmodesmata are indeed pathways for transport between the various cells of our system, i.e. pathways for symplasmic transport and for co-operation between leaf mesophyll cells and gland cells.

Electron microscopists have so far not been able to reach an agreement in the controversy as to whether plasmodesmata are plugged by membranes establishing a high resistance to transport of solutes, or whether they are penetrated by ER membranes facilitating symplasmic contact between adjacent cells (see Fig. 8.2).

Cytologists have tried to tackle the problem by localisation techniques, such as precipitation of $Cl^-$ ions with silver ions, and detecting the AgCl precipitations as electron-dense material in the electron microscope. Indeed, such precipitations are found within plasmodesmata, suggesting that $Cl^-$ can at least move into these structures (Fig. 11.9; Ziegler and Lüttge, 1967; Hill and Hill, 1973b; Van Steveninck et al., 1973).

Physiologists have used bioelectrical phenomena to demonstrate symplasmic coupling between cells of a parenchyma. The transfer of exogenously applied or endogenously triggered electrical signals from cell to cell has been studied (see Lüttge, 1973; Brinckmann and Lüttge, 1974; Hill and Hill, 1975;

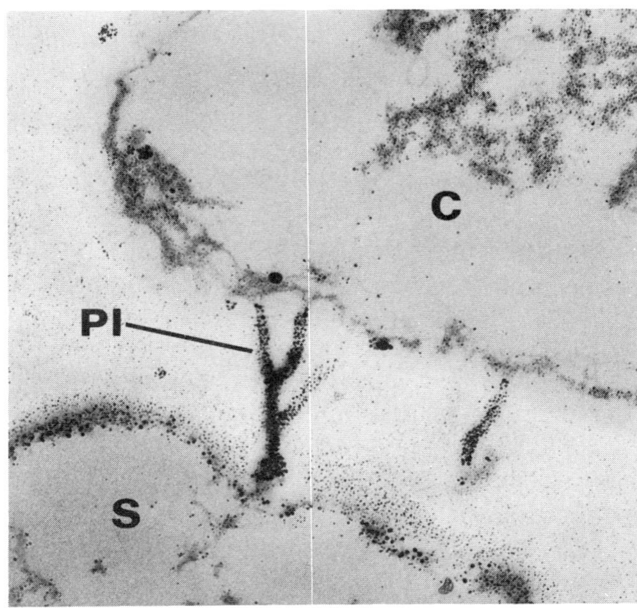

Fig. 11.9. Localisation of AgCl in plasmodesmata between a sieve tube and a companion cell of a *Limonium* leaf excreting NaCl from a 550 Mol m$^{-3}$ solution. Fixation for electronmicroscopy in the presence of Ag$^+$. (From Ziegler and Lüttge, 1967.) C, companion cell; Pl, plasmodesmos; S, sieve tube. Magnification × 15 250.

Spanswick, 1975). Salt hairs of chenopods have been a useful system in such investigations (Osmond et al., 1969; Lüttge and Pallaghy, 1969; Pallaghy and Lüttge, 1970b). Green cells of the leaf lamina show photosynthesis-dependent oscillations of the membrane potential when the light is switched on or off (Fig. 11.10a). The photosynthetically inactive stalk and bladder cells do not display this phenomenon if isolated epidermal strips are investigated (Fig. 11.10c), but they do if intact leaf strips are used (Fig. 11.10b). This means that in the situation of Fig. 11.10b the endogenously triggered signal is transported symplasmatically from the green mesophyll to the bladder cell. Apoplasmic contact is ruled out by the cuticular envelope.

Hill (1967a) has used leaf discs of *Limonium* and, by scanning the surface with voltage pulses of constant amplitude, has shown that the region of the glands provides the only pathway of high conductance across a leaf (Fig. 11.11). This confirms the conclusion drawn above from the consideration of the arrangement of cell wall incrustations. It shows that the cuticular envelope is a good insulator and clearly demonstrates that it enforces a symplasmic pathway via the glands.

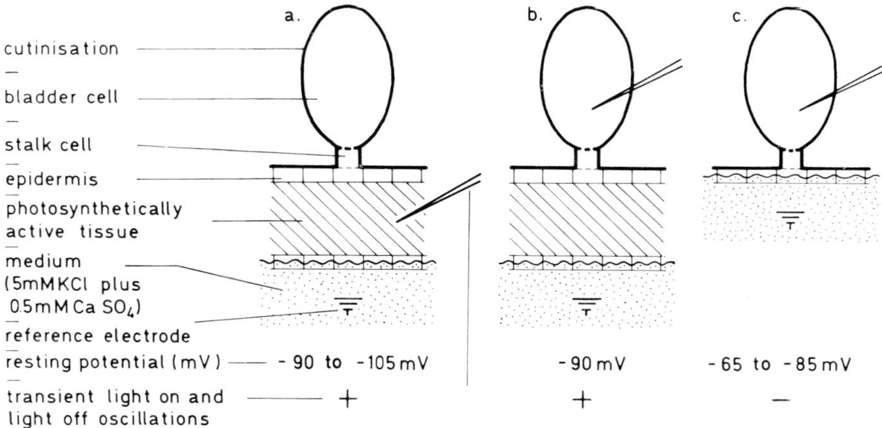



Fig. 11.10. Electric coupling between vacuoles of salt hairs and mesophyll cells of *Chenopodium album*. An electrical signal (i.e. transient light-on and light-off oscillations of membrane potential) generated in the mesophyll cells (a) is picked up in the bladder cells attached to leaf mesophyll (b) but not in bladder cells of isolated epidermis (c). (From Lüttge, 1973.)

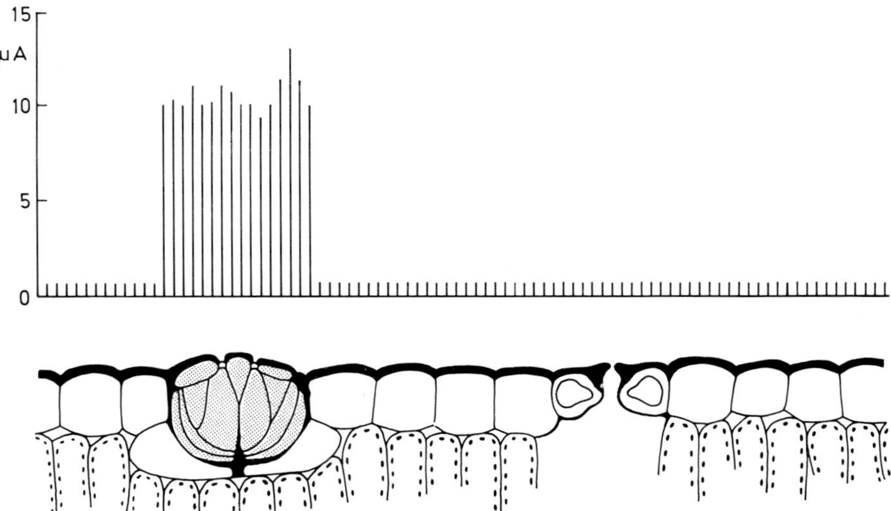



Fig. 11.11. Probing the surface of a *Limonium* leaf with constant voltage pulses. Current pulses of 10–13 $\mu$A are measured only above the glands. The ionic conductivity above other parts of the leaf surface (i.e. stomata, normal epidermal cells) is less than a tenth, i.e. the cuticle has negligible ionic conductivity. The schematical cross section of the leaf surface is drawn after the dimensions given in Ziegler and Lüttge (1966, 1967), cutinisation is indicated by thick black lines. Electrical properties from Hill (1967a).

Electric coupling in systems such as that shown in Fig. 11.10 is very fast. In one specific case, coupling has been observed over a distance of 0.9 mm without discernable lag (Brinckmann and Lüttge, 1974). These electrical measurements showing transport of charge do not actually prove that solutes are transported through the plasmodesmata. However, there are other reasons, arising particularly from investigations of transport along the length of *Vallisneria* leaves (Arisz, 1954, 1960, 1969; Arisz and Wiersema, 1966; see Lüttge, 1973; Spanswick, 1975), which demonstrate that symplasmic transport of low molecular solutes indeed does occur, and that its velocity is in the order of 10–60 mm h$^{-1}$.

Radioisotopes, e.g. $^{22}$Na$^+$ and $^{36}$Cl$^-$, can be used to study half-times of exchange of various compartments in plant cells with an external solution. Using this approach, Hill (1970a) found that the half-time for loading the cytoplasm of *Limonium* leaf cells with labelled ions was identical to the half-time of transit across a leaf and into the glandular exudate:

|  | $t_{\frac{1}{2}}$ for cytoplasmic exchange (min) | $t_{\frac{1}{2}}$ for transit (min) |
|---|---|---|
| $^{22}$Na$^+$ | $29 \pm 4$ | $26 \pm 5$ |
| $^{36}$Cl$^-$ | $10 \pm 2$ | $13 \pm 4$ |

This provides clear evidence that transit occurs in a symplasmic continuum of which the gland cytoplasm is an integral part. Hence, the conclusion appears justified that the numerous plasmodesmata regularly found between salt gland cells and adjacent mesophyll cells and also between adjacent gland cells, facilitate co-operation via the symplasmic continuum and that the cell wall cutinisations enforce use of this pathway.

*Proliferations of the cell wall and enlargement of the plasmalemma surface*
Plasmalemma surfaces of plant cells, where massive short distance transport of low molecular weight solutes occurs, are very often enlarged considerably by cell wall protuberances which are coated by the plasmalemma. Gunning and Pate (1969) and Pate and Gunning (1972) have given these cells the name transfer cells, which aptly describes the function of these cells in fulfilling various tasks of transport. Cell wall protuberances are regularly found in salt gland cells (see Figs 11.6, 11.8 and 11.13), which thus can be considered as transfer cells.

It is evident that the enlarged plasmalemma surface facilitates the transfer of solutes. In some nectary glands formation of cell wall proliferations coincides with physiological activity of secretion. Cell wall protuberances are formed a short time before the onset of nectar secretion; they constitute a plasmalemma-coated labyrinth during active nectar secretion and they are re-absorbed in the post-secretory period (Schnepf, 1964).

However, the anatomical correlation of transfer cells with strategic positions in plant tissues and organs where massive short distance transport occurs and the cytological correlation of formation of cell wall protuberances with secretory activity leave open the question of how the enlarged plasmalemma surface serves increased transport capacity. There are two possible and rather straightforward explanations: firstly, the enlarged plasmalemma surface may provide more space for more carriers facilitating transport or for more pumps actively eliminating solutes. Secondly, it may just provide an increased area across which diffusion can occur. The latter would also lead to increased transport, since according to Fick's law (Ch. 1) diffusion of solutes is not only proportional to the concentration gradient and the membrane permeability, but also to the effective area across which particles move. It is not easy to make a distinction between the two alternatives. However, this problem already leads to more physiological questions and to the structure–function models discussed below.

Hill and Hill (1973a, b) have observed that in *Limonium* the plasmalemma invaginations often appear inflated (Fig. 11.13b). These authors propose that the invaginations function as osmotic coupling spaces for the salt pumped out of the gland cells and the water molecules moving passively. Thus the exudate is formed in the coupling spaces before moving by mass flow in the apoplast and via the cuticular pores out of the gland.

Partitioning of the basal cell of the salt gland of *Spartina* by fingerlike infoldings of the plasmalemma (Fig. 11.12) is interpreted by Levering and Thomson (1972) in a similar way. These infoldings reach deeply into the cell. They constitute channels which are closed at their terminal ends within the cell and which open into the apoplast at the cell surface. Numerous mitochondria are in close contact with the channels. The channels are inflated under conditions of NaCl-loading. Levering and Thomson suggest that the channels provide the structural basis of a standing-gradient flow mechanism of excretion where salt transport into the channels at equilibrium maintains a standing osmotic gradient with a mass flow of water and solute in the outward direction (Diamond and Bossert, 1967; Pate and Gunning, 1972).

Before we leave this section it should be noted that the enormous ion flux rates reported for salt glands (see Section 11.1) might come down to rather normal values, if the surface of the cell wall protuberances rather than the cell surface were taken into account.

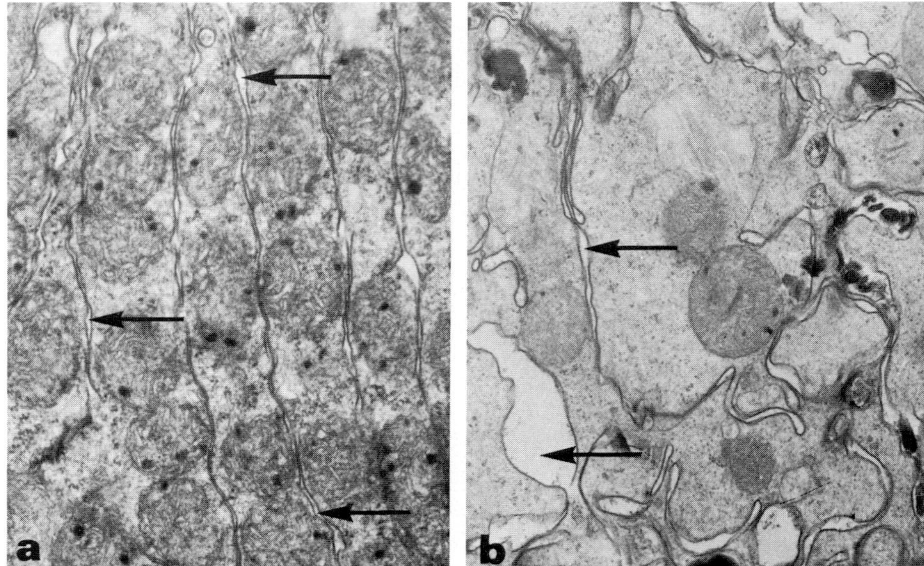

Fig. 11.12. Plasmalemma invaginations partitioning the cytoplasm of the basal cell of *Spartina* salt glands. Arrows: extracellular channels opening into the apoplast. (a) numerous mitochondria in contact with the plasmalemma infoldings; no salinity was applied and the infoldings are closely aligned; (b) inflated extracellular channels under conditions of salt loading: 3% NaCl for 6 h. Magnification a: × 12000; b: × 14625. (With kind permission from Levering and Thomson, 1972.)

*Vesicles*

The last cytological property we have to consider here is vesiculation of the cytoplasm of salt gland cells. We have mentioned above that typical gland cells do not have large central vacuoles (Section 11.2). Instead there are apparently many small vacuoles which often accumulate near the secretory surface of the gland cells (Fig. 11.8). When *Tamarix* plants are supplied with $Rb^+$ ions, $Rb^+$ is excreted by the salt glands. Since Rb is electron opaque, accumulations of Rb can be seen in the electron microscope, and Thomson et al. (1969, Fig. 11.13a) have shown that the vesicles of gland cells excreting $Rb^+$ contain large amounts of this metal. This led to the assumption that these vesicles empty their contents out of the gland cells by fusing with the plasmalemma, i.e. that excretion would occur by a mechanism of vesicular extrusion. This proposal was also supported by kinetic investigations with algal coenocytes (MacRobbie 1969, 1970, 1971a) and with higher plant tissues (Pallaghy and Lüttge, 1970a; Lüttge and Pallaghy, 1972) suggesting transport of ions in vesicles (see Lüttge, 1973; Baker and Hall, 1973).

Hill and Hill (1973b) discard this possibility for salt excretion by *Limonium* glands. Their electronmicrographs suggest that the vesicles at the gland surface are mostly just inflated invaginations of the plasmalemma supported by the cell wall protuberances (Fig. 11.13b). The vacuoles seen apparently within the cells may often be cross-sections of these invaginations. (For a possible function of the invaginations see above.) Furthermore, Hill and Hill calculate that plasmalemma membrane turnover in the gland cells would have to be 30 times an hour to account for a vesicular extrusion of the observed amounts of salt excreted, which is a rather high rate. Finally, vesicular extrusion in *Limonium* seems to be ruled out by the high specificity of the excretion process (see below).

## 11.3 Active transport and sources of metabolic energy in salt excretion

The question as to whether a transport process is passive or active is usually tackled by biophysical approaches. Investigations are carried out to study whether the transport process obeys thermodynamical laws describing the physical driving forces or whether it works against these forces (see Ch. 1). In a more biochemical approach experiments are designed to investigate linkage of the transport processes to particular energy transfer reactions of metabolism. Both approaches have been applied to the problem of salt excretion.

### 11.3.1 Salt excretion against gradients of electrochemical potential

Possible physical driving forces in salt excretion are concentration gradients of the salts and ions involved and electrical gradients since the moving particles are charged electrically. Hence, the passive or active nature of salt excretion can be determined by investigating whether or not the ion fluxes involved obey the Nernst criterion (Eqs. 1.6 and 1.7), the Goldman equation (Eq. 1.19), or the Ussing–Teorell criterion (Eq. 1.16), or whether they deviate from these laws describing passive ion distribution.

The Nernst and the Goldman equations only apply to systems which are in flux equilibrium, i.e. systems where no net fluxes occur ($\phi_j^{oi} = 0$ in Eq. 1.12). In most studies with salt glands this is a situation which is difficult to obtain and the Ussing–Teorell criterion must be used in addition. Table 11.3 gives an example for $Cl^-$ excretion into the bladder vacuoles of *A. spongiosa* salt hairs. It can be seen that the Nernst potential (Eq. 1.7) for the distribution of

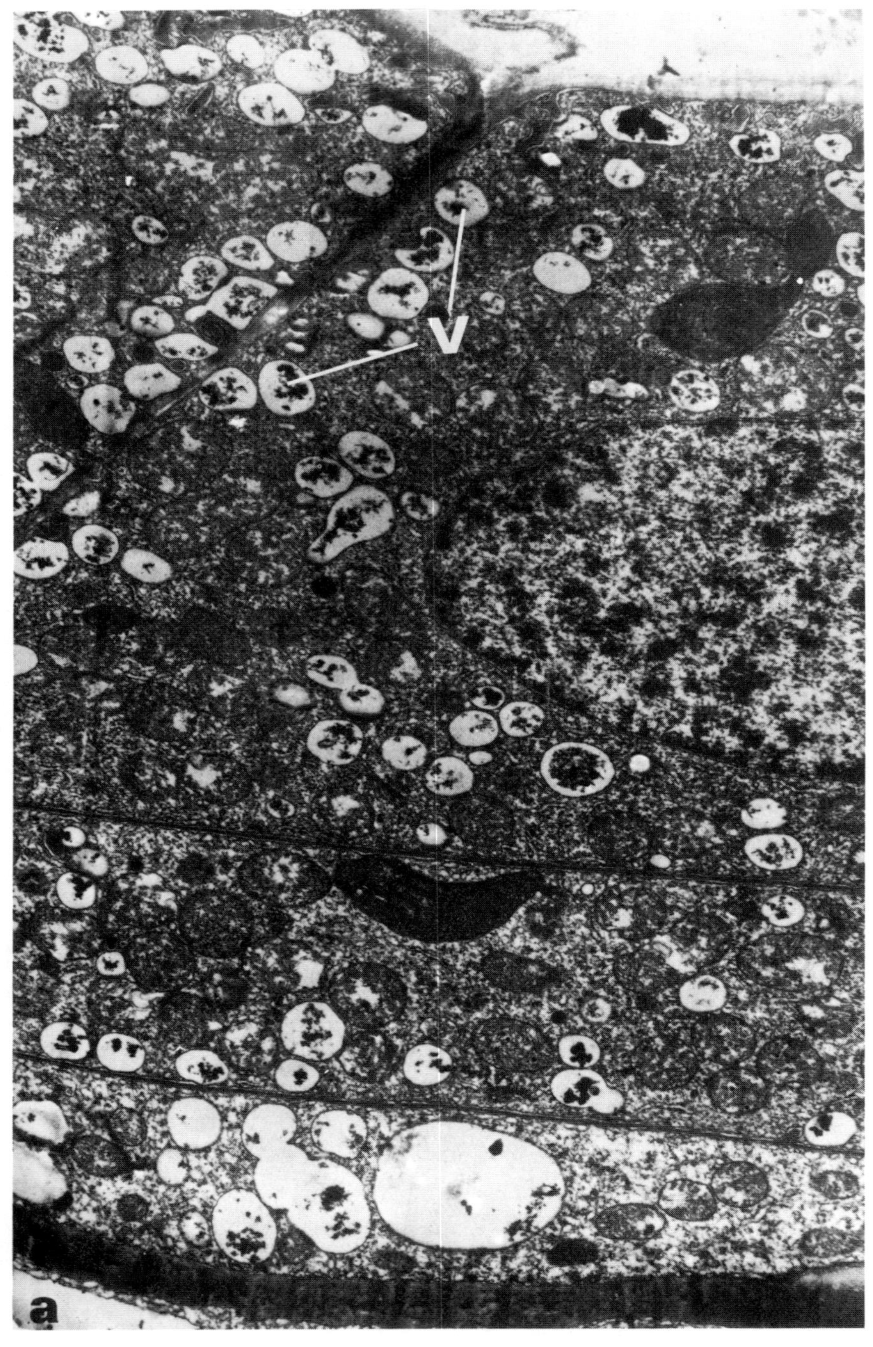

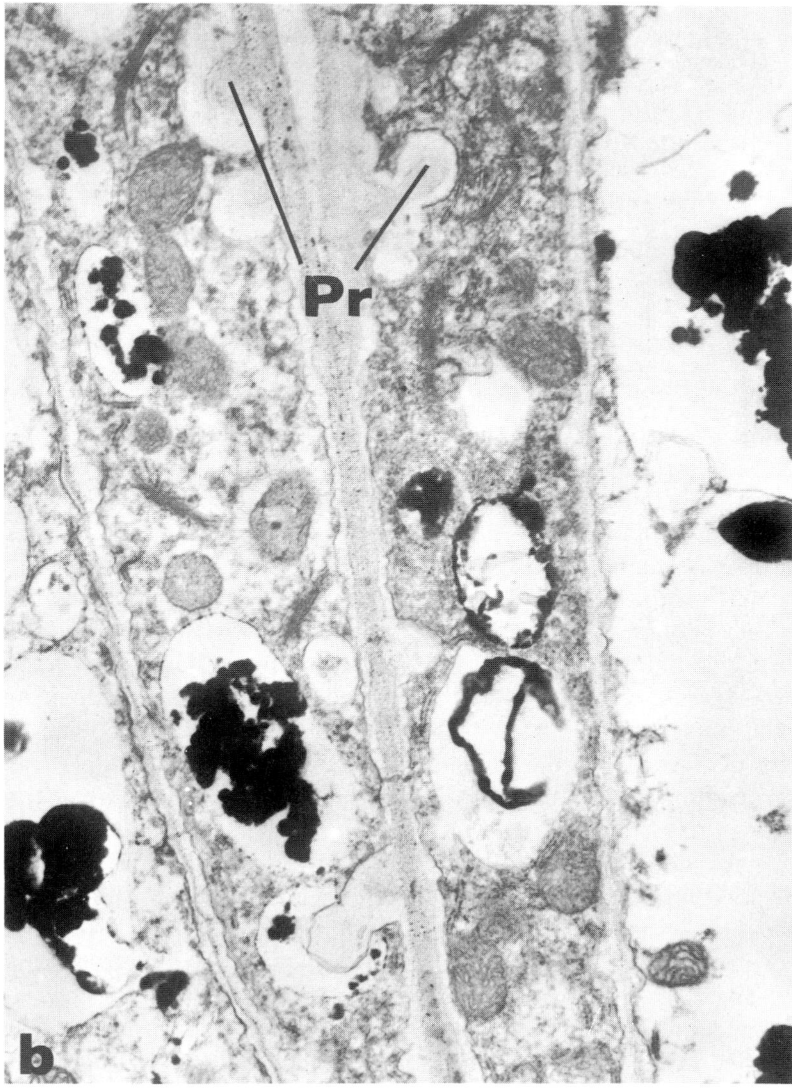



Fig. 11.13. The role of small vacuoles in ion excretion? (a) Cytoplasm of a *Tamarix aphylla* salt gland excreting Rb$^+$. Small vacuoles (V) contain electron-dense material (with kind permission from Thomson, Berry and Liu, 1969). Magnification × 11 500. (b) Inflated plasmalemma invaginations of a salt-excreting *Limonium vulgare* gland cell, supported by cell wall protuberances (Pr) (with kind permission from Hill and Hill, 1973b). Magnification × 12 000.

Table 11.3.

$Cl^-$ concentrations in bladders of *A. spongiosa* leaf slices ($Cl_i^-$) kept floating for 42 h on a solution of 5 Mol m$^{-3}$ KCl + 0.1 Mol m$^{-3}$ CaSO$_4$ in the light and in the dark respectively, Nernst potential ($E_{N_{Cl}-}$) for the distribution of $Cl^-$ between the bladders and the medium, and potential difference ($E_M$) measured with glass microelectrodes. (Data from Osmond et al., 1969.)

| | $Cl_i^-$ [$\mu$equiv. gFW$^{-1}$] | $E_{N_{Cl}-}$ [mV] | $E_M$ [mV] |
|---|---|---|---|
| Initial | 26.2 | +43 | $-\ 89 \pm 23$ |
| Light, 42 h | 33.8 | +49 | $-105 \pm\ \ 3$ |
| Dark, 42 h | 18.3 | +33 | $-\ 83 \pm 13$ |

$Cl^-$ between the external solution and the bladders $E_{N_{Cl}-}$ is positive, whereas the measured electropotential difference ($E_M$) is highly negative. According to the Nernst criterion this would imply active transport of $Cl^-$ at least at one stage between uptake from the medium and excretion into the bladder vacuoles. However, as can also be seen in Table 11.3 in the light there was a net $Cl^-$ influx into the bladder vacuoles during the experiment and even after 42 h so that perhaps Eq. 1.7 does not apply, i.e. the system is not in flux equilibrium. Considering Eq. 1.16 in addition shows immediately that with a net influx the Nernst criterion would even underestimate the gradient against which $Cl^-$ must be transported in the light and that active transport is clearly involved. Conversely, in the dark the net efflux from the bladders observed suggests that the Nernst criterion overestimates the gradient against which $Cl^-$ appears to be transported.

Similar investigations have been made with other salt gland systems. The most detailed thermodynamical and electrochemical investigation is available for *Limonium*. Distribution of $Cl^-$ and of the cations K$^+$ and Na$^+$ between the medium and the excreted salt solution does not obey the Nernst criterion (Eq. 1.7). However, if chloride in the medium is replaced by other ions such as $SO_4^{2-}$, benzenesulphonate, borate or gluconate, no exudation is observed and no secretory potential is developed irrespective of large amounts of Na$^+$ present. This suggests that the system is highly specific for active transport of $Cl^-$. Only the other halides Br$^-$ and I$^-$ can replace $Cl^-$. Na$^+$ can be replaced by other alkali cations (Hill and Hill, 1973b). Removal of alkali cations from the medium does not inhibit gland activity, whereas in $Cl^-$-free media gland activity totally breaks down. Ratios of gland activities in various media observed by Hill (1967b) were: 100 Mol m$^{-3}$ NaCl : 100 Mol m$^{-3}$ sodium-benzenesulphonate : 100 Mol m$^{-3}$ choline chloride (1.00 : 0.015 : 1.89).

In addition to the criteria discussed above Hill and Hill (1973a; Hill, 1967b) used the short circuit technique (see Ch. 1) to investigate the thermodynamics of $Cl^-$ excretion in *Limonium*. Short circuiting the exudate on the gland surface against the medium eliminates the physical driving forces (Eqs 1.7, 1.16 and 1.19). Any current flowing under these conditions must be due to active transport of particles (ions) carrying electrical charge. It is also possible to clamp the electrical potential between the exudate and the medium at a certain positive or negative value and to investigate the fluxes of ions under such conditions of constant voltage clamp, i.e. under conditions of an artificial (i.e. experimental) physical driving force. The application of these techniques to salt excretion is somewhat sophisticated theoretically because the excretory surface is not simply the outer surface of the gland. In fact, the spatially complex borderline between the gland cytoplasm and the apoplasmic space outside the cuticular envelope must be considered. Notwithstanding these difficulties, on the basis of such investigations it becomes quite clear that in *Limonium* $Cl^-$ is transported actively out of the gland cytoplasm with $Na^+$ following passively along the gradient established by the electrogenic $Cl^-$ pump and water following osmotically.

Such a dependence on $Cl^-$ transport may not apply in general. For instance Berry's analyses of *Tamarix aphylla* exudates (Table 11.4) suggest a more

Table 11.4.

Relative cation contents (% of total cation equivalents) of salt excreted by salt glands of *Tamarix aphylla*. (From Berry, 1970.)

| Additions to basal culture medium* [100 mequiv. cation] | Relative cation content (%) | | | |
|---|---|---|---|---|
| | $Na^+$ | $K^+$ | $Mg^{2+}$ | $Ca^{2+}$ |
| NaCl | 65 | 15 | 7 | 13 |
| $Na_2SO_4$ | 55 | 25 | 8 | 12 |
| $NaNO_3$ | 28 | 25 | 23 | 24 |
| KCl | 2 | 87 | 8 | 3 |
| $K_2SO_4$ | 3 | 75 | 15 | 7 |
| $KNO_3$ | 3 | 73 | 18 | 6 |

* Basal culture medium: $Na^+$, 0.5; $K^+$, 2.9; $Mg^{2+}$, 1.9; $Ca^{2+}$, 4.7; $NO_3^-$, 7.1; $Cl^-$, 0.5; $SO_4^{2-}$, 1.9; $H_2PO_4^-$, 0.5 equiv. $m^{-3}$.

complex dependence of excretion on the anion in the medium, although his experiments were designed in a different way and are not directly comparable to the *Limonium* investigations described above. In *T. aphylla* the amount of salt excreted per unit of plant fresh weight is much reduced in 100 Mol $m^{-3}$

KCl as compared to 100 Mol m$^{-3}$ NaCl, so that in contrast to *Limonium*, Na$^+$ is not fully replaceable by K$^+$ (Table 11.5).

Table 11.5.
Effect of alkali cation (K$^+$ or Na$^+$) on salt excretion by *Tamarix aphylla* glands. (Data from Waisel, 1972.)

| Irrigation solution | Ions excreted (mequiv. 100 gFW$^{-1}$ 3 days$^{-1}$) | | | | |
|---|---|---|---|---|---|
| | Ca$^{2+}$ | K$^+$ | Na$^+$ | total cations | Cl$^-$ * |
| Tap water | 2.19 | 0.45 | 11.42 | 14.06 | 1.55 |
| 100 Mol m$^{-3}$ NaCl in tap water | 2.30 | 0.44 | 20.07 | 22.81 | 3.56 |
| 100 Mol m$^{-3}$ KCl in tap water | 2.09 | 0.59 | 6.74 | 9.47 | 5.39 |

* Cl$^-$ does not balance the charge of Ca$^{2+}$, K$^+$ and Na$^+$, suggesting that other anions must be involved.

### 11.3.2 Metabolic energy sources

Respiration and photosynthesis are the major sources of metabolic energy generally considered to drive active ion fluxes in plant cells (see Ch. 2). In the past decade considerable effort was devoted to elucidating the associations of active ion transport processes in plants with particular partial reactions of energy transfer in photosynthesis and respiration, respectively. Inhibitor and light-regime experiments of increasing sophistication were designed to allow only one or at least only an extremely limited number of possibilities of metabolic energy supply.

Table 11.6 lists a number of examples selected from studies of salt excretion or of ion transport in leaf cells of plants carrying salt glands or salt hairs. As in investigations of metabolically driven ion fluxes in plant tissues, in general, it appears that the energy supply is not very specific. ATP seems to be able to drive ion transport irrespective of the nature of the ATP-providing partial reaction of energy metabolism (e.g. oxidative phosphorylation, non-cyclic photophosphorylation, cyclic photophosphorylation or to some extent even glycolysis). In other words, ion transport mechanisms seem to escape being pinned down by inhibitor experiments to a particular energy-providing reaction. Instead they use whatever energy is available and they cease when energy metabolism is minimized, e.g. in the dark with uncouplers or inhibitors of mitochondrial electron flow or in green tissues in the light under anaerobiosis plus uncoupler plus inhibitor of photosynthetic electron flow (see Table 11.6).

Only in an exceptional case so far has it been suggested that an ion excretion may be driven by a particular source of energy. $Cl^-$ excretion into the bladder vacuoles of *A. spongiosa* salt hairs is highly stimulated by photosynthesis (Fig. 11.14). The cytoplasm of the salt hair cells is itself photosynthetically

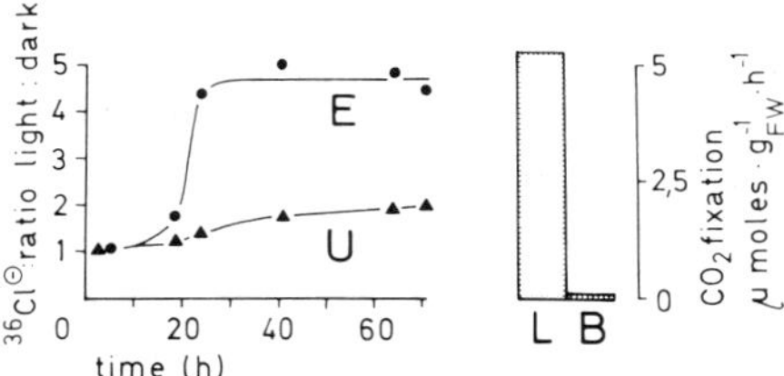

Fig. 11.14. Stimulation of chloride ($^{36}Cl^-$) elimination (E) to the bladder vacuoles of *A.spongiosa* by light. U, uptake of $^{36}Cl^-$ by leaf blades; L, photosynthesis of lamina; B, photosynthesis of bladders (salt hairs). Data are from experiments with 3-mm wide leaf slices floating on a solution of 0.5 Mol m$^{-3}$ CaSO$_4$, 5 Mol m$^{-3}$ K$^{36}$Cl, 0.25 Mol m$^{-3}$ KH$^{14}$CO$_3$. (Reproduced from Osmond et al., 1969; and Lüttge, 1974.)

inactive (see Section 11.2 and Fig. 11.14). Hence metabolic reactions in the photosynthetically active leaf mesophyll tissue must be important for light-dependent $Cl^-$ excretion. The light-dependent fraction of $Cl^-$ uptake by leaf slices of *A. spongiosa* is DCMU-sensitive, i.e. dependent on non-cyclic photosynthetic electron flow. However, it is much less affected by the uncoupler FCCP inhibiting ATP production by non-cyclic photophosphorylation than the ATP-consuming $CO_2$ assimilation (Table 11.6). Thus it looks as if this $Cl^-$ transport and consequently probably also $Cl^-$ excretion can be powered specifically by non-cyclic photosynthetic electron flow or reducing equivalents generated by it (Lüttge et al., 1970). This situation resembles very much that of some algal coenocytes described earlier (Ch. 5). Nevertheless, this does not prove that $Cl^-$ excretion in *A. spongiosa* is energetically linked exclusively to non-cyclic electron flow. At best this linkage appears as a possibility under the particular experimental conditions described but a different energy supply might be effective in the light in the absence of the uncoupler.

## 11.4 Structure–function models

In Section 11.2 we have developed a fairly detailed picture of physiological anatomy and cytology of salt glands. In section 11.3 we have seen that active transport is required at least at one stage in the excretory system. We can now

Table 11.6.

Effects of metabolic inhibitors on salt-excreting systems. Inhibitor concentrations giving 50 % inhibition are listed in all cases where this was possible either directly from the data given in the references or by redrawing of such data. In other cases inhibitor concentrations are given together with degrees of inhibition in % of controls. The supposed action of the inhibitors on metabolism of higher plant leaves is given below. For abbreviations see p. xi. Uncouplers of oxidative phosphorylation, non-cyclic and cyclic photophosphorylation: CCCP, FCCP, DNP, $AsO_3^{3-}$, $AsO_4^{3-}$. Inhibitors of respiratory electron-flow: $NaN_3$, KCN, anaerobiosis. Inhibitors of non-cyclic photosynthetic electron-flow: DCMU, $CO_2$-free $N_2$. Inhibitors of glycolysis: NaF, IA.

| Plant species | Experimental details and *process studied | Light stimulation | Experiment in light or dark | Inhibitors ($Mol\ m^{-3}$) | Inhibition (% of control) |
|---|---|---|---|---|---|
| *Aegialitis annulata*[1] | cut leaves; 100 Mol $m^{-3}$ NaCl applied via the petiole *$Cl^-$ excretion by glands | + | L | $5 \times 10^{-2}$ CCCP | 9–17 |
| *Limonium latifolium*[2] | leaf discs 30 mm diameter; 470 Mol $m^{-3}$ NaCl applied *$Cl^-$ excretion by glands | ± | ? | 7–8 KCN | 50 |
| | | | | 2–4 IA | 50 |
| | | | | 2.5–7.5 $NaN_3$ | 50 |
| | | | | 7.5 NaF | 50 |
| | | | | 1–5 $AsO_3^{3-}$ | 50 |
| | | | | 0.5 DNP | 50 |
| *Limonium* sp.[3] | leaf discs; 100 Mol $m^{-3}$ NaCl applied *short circuit current | | D | 1 KCN | 0 |
| | | | D | $10^{-2}$–$10^{-1}$ DNP | inhibitory |
| | | | D | $10^{-2}$ CCCP | 43 |
| | | | L | $10^{-2}$ DCMU | 100 |
| | | | D | anaerobiosis | 0 |
| | | | L | anaerobiosis | 100 |
| | | | L | $CO_2$-free $N_2$ + $10^{-2}$ DCMU | 100 |
| | | | L | $CO_2$-free $N_2$ + $10^{-2}$ DCMU + $10^{-2}$ CCCP | 0 |

| | | | | |
|---|---|---|---|---|
| *Atriplex spongiosa*[4] | 3 mm wide leaf slices;<br>5 Mol m$^{-3}$ KCl applied | | | |
| | *Cl$^-$ uptake by mesophyll and salt hairs | + | L | 2 × 10$^{-3}$ CCCP | 71 |
| | | | L | 2 × 10$^{-3}$ FCCP | 67 |
| | | | D | 2 × 10$^{-3}$ CCCP | 17 |
| | | | D | 2 × 10$^{-3}$ FCCP | 30 |
| | *light-dependent fraction of Cl$^-$ uptake : | | | | |
| | mesophyll 5 Mol m$^{-3}$ KCl | + | L | 10$^{-3}$ DCMU | 46 |
| | mesophyll 100 Mol m$^{-3}$ KCl | + | L | 10$^{-3}$ DCMU | 31 |
| | salt hairs 5 Mol m$^{-3}$ KCl | + | L | 10$^{-3}$ DCMU | 48 |
| | salt hairs 100 Mol m$^{-3}$ KCl | + | L | 10$^{-3}$ DCMU | 0 |
| *Nepenthes* sp.[5] | pitcher tissue discs 10 mm diameter;<br>NaCl solutions applied | | | | |
| | *Cl$^-$ excretion by glands | | | | |
| | 1 Mol m$^{-3}$ NaCl | | | 5 KCN | 50 |
| | 0.37 Mol m$^{-3}$ NaCl | | | 0.2–0.3 AsO$_4^{3-}$ | 50 |
| | *Cl$^-$ uptake and excretion | | | | |
| | 0.37 Mol m$^{-3}$ NaCl | | | 0.5 KCN | 50 |
| | *Cl$^-$ uptake | | | | |
| | 0.37 Mol m$^{-3}$ NaCl | | | 3 KCN | 50 |

References : 1) Atkinson et al., 1967; 2) Arisz et al., 1955; 3) Hill and Hill, 1973c; 4) Lüttge and Osmond, 1970; 5) Lüttge, 1964b, 1966a.

condense the salient structural features into simplified models of co-operating cell types in excretory systems and we can try to understand how transport works in such systems.

### 11.4.1 Models of co-operation in excretory systems

We will distinguish between leaves with salt glands and leaves with salt hairs, respectively. In addition we will draw a model of the plant root, which naturally is the plant organ where co-operation of various tissues in ion transport has been investigated most extensively.

Looking at the schemes shown in Fig. 11.15, we can perhaps begin with stressing the analogies of the 3 systems. Firstly, in all three cases the living cells take up ions from an outer space. For the roots this outer space is the soil (or an external medium when roots are in water culture) plus the apoplasmic space of the root cortex up to the Casparian strip in the root endodermis. For the salt glands and salt hairs the outer space is the transpiration stream plus the apoplasmic space in the leaf mesophyll up to the cuticular envelope. Secondly, in all 3 cases, more or less distant from the site of ion uptake from the outer apoplasmic space into the symplasm of the system, there is a release of ions from the symplasm into another apoplasmic space, i.e. into the xylem vessels in the root, to the gland surface in the salt glands and into the bladder vacuoles in the salt hairs. Thirdly, Casparian strips or cuticular envelopes block a continuous apoplasmic pathway from the site of uptake to the site of release so that symplasmic coupling is the only way via which co-operation of the various tissues and cell types becomes possible. Fourthly, ion transport into the vacuoles of leaf and root cells is a process diversive to release; however it is also clear that efflux of ions from the vacuoles may add to loading the symplasmic transport system with ions.

A straightforward dissimilarity between the two excretory systems and the root is the role of long-distance transport in the xylem. In the roots the transpiration stream in the xylem removes the eliminated ions, but in leaves it constitutes the supply of ions. More important than this fact itself is its consequence. During uptake, roots directly select ions from the soil or from a culture solution; by contrast, leaves (and all other aerial parts of plants) only see ions which have already passed the root system and which were subject to selection and control by the living cytoplasm of the root cells during passage through the root (see Chs 8 and 9).

This has led to the question whether leaves have intrinsic mechanisms whereby they can control their ionic relations or whether control of ionic relations of the whole plant is exclusively at the level of the root. Indeed, in a

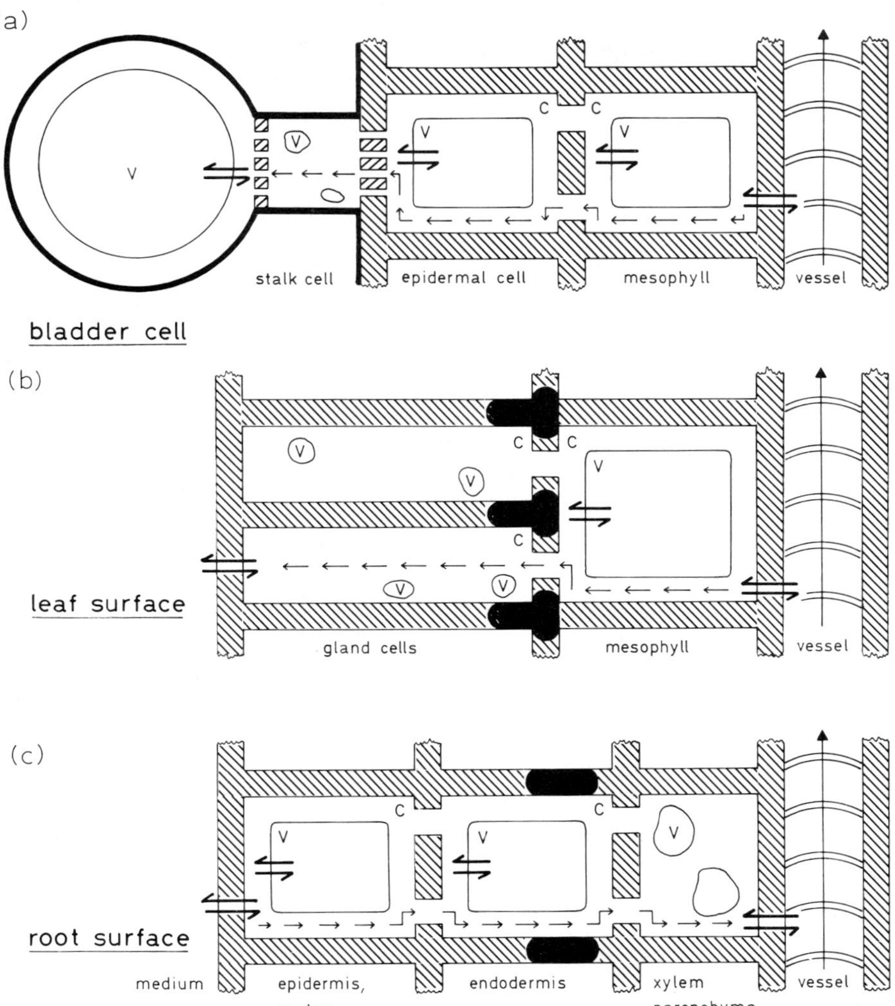



Fig. 11.15. Simplified structural schemes of a leaf with salt hairs, of a leaf with salt glands, and of a root. The schemes show (i) short distance transport at membranes, i.e. firstly, uptake into the symplast from the ambient apoplast; secondly, release into bladder vacuoles, to the leaf surface or into the xylem; and thirdly, fluxes into and out of the vacuoles; (ii) symplasmic transport between co-operating cells of leaf and root tissues respectively; (iii) long distance transport in the xylem. (After Lüttge, 1971, 1973, 1974; reproduced, with permission, from 'Structure and Function of Plant Glands' by U. Lüttge, Annu. Rev. Plant Physiol., 22. Copyright © 1971 by Annual Reviews Inc. All rights reserved.)

368 U. Lüttge

number of investigations with young barley plants, the physiological status of the leaves affected the supply of ions from the roots to the leaves by feedback to the roots; however, it had little direct consequence for ion uptake by the leaf cells themselves (Cram and Pitman, 1972; Pitman, 1972; Pitman et al., 1974a, b). Thus, one answer to the problem of co-operation between roots and leaves in regulation of ionic relations is that the leaves are dependent on the roots and do not possess intrinsic regulatory mechanisms (see Ch. 9).

However, salt glands and salt hairs provide an additional answer. Active salt-excretion systems are, in fact, mechanisms by which leaves can directly regulate their ionic balance (see Section 11.1). Again this is particularly well documented in *Limonium*. When leaves of *Limonium* plants grown on a dilute culture solution are loaded with salt from 100 Mol m$^{-3}$ NaCl or KCl solutions, ion concentration in the chloroplasts increases for about 3 h and then remains constant (Larkum and Hill, 1970). At the same time the salt glands develop their maximal excretory activity which is low initially and rises during about 3 h of salt treatment (Fig. 11.16). This coincidence elucidates the function of the salt glands in control of the ionic balance of metabolically active systems in the leaf cytoplasm. Hence, intrinsic mechanisms of ionic regulation, though probably absent in normal leaves, can develop as an adaptation to salinity stress in leaves of halophytes.

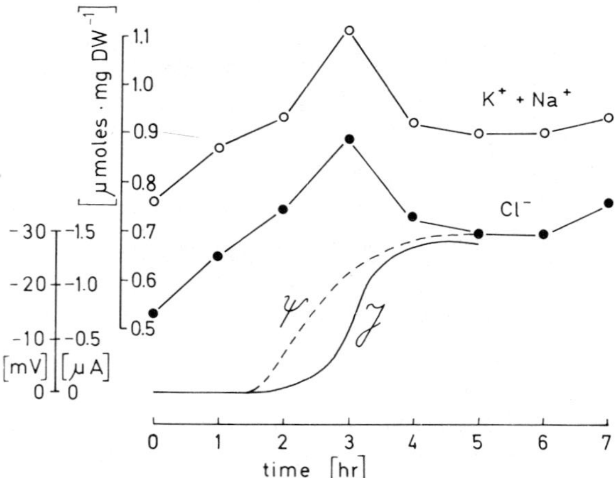


Fig. 11.16. Rise of excretory gland activity and ion content in the chloroplasts of *Limonium* leaves treated with 100 Mol m$^{-3}$ NaCl. Excretory gland activity is expressed as potential ($\psi$[mV]) and as short circuit current (J[$\mu$A]) respectively across a leaf disc (redrawn from Hill, 1970b). Ion content of the chloroplasts ($\mu$Mol mgDW$^{-1}$) was obtained by analysis of non-aqueously isolated chloroplasts (Larkum and Hill, 1970).

### 11.4.2 Sites of active ion pumps as driving forces of excretion

The models of Fig. 11.15 can also be applied to the localisation of ion pumps which drive the excretion and which are required thermodynamically (Section 11.3). It is thermodynamically sufficient, but not necessarily required, that pumping occurs only at one stage in the system. The site of the ion pump is most likely at a membrane, but it is very difficult to deduce at which membrane in the system. One possibility is the plasmalemma where ions are taken up into the living cells of the system (in all three cases of Fig. 11.15). Alternative or additional possibilities are the membranes where the ions are released from the symplasm, i.e. the plasmalemma of the root xylem parenchyma cells, the plasmalemma at the surface of the gland cells, or the tonoplast of the salt hair bladders.

This question cannot be decided thermodynamically. Electrochemical potential gradients have only been obtained for the entire system, i.e. between the excreted salt solution and the medium and not between the various cell types of the system. Some experiments of Field and co-workers (Bostrom and Field, 1973; Billard and Field, 1974) appear to suggest that the salt glands of the mangrove *Aegiceras* are electrically isolated and that probably the pump resides somewhere in the gland. However, in other examples there is mostly good electric coupling (section 11.2) and thermodynamic considerations do not indicate at which of the membranes bordering the symplasmic continuum the active step might be localised.

It is dangerous to conclude too much on the basis of the cytological observations. True, the dense cytoplasm, large nuclei, numerous mitochondria and other cell organelles of the gland cells suggest that active pumps may reside in the gland plasma membranes. However, in *A. spongiosa* photosynthesis drives excretion into the bladder vacuoles, but the photosynthetically active chloroplasts are in the mesophyll far distant from the stalk cell of the salt hair with its gland-like cytoplasm (Section 11.3, Fig. 11.14). Similarly, in *Limonium* under anaerobic conditions light can drive salt excretion, but the glands do not have chloroplasts. Of course, it is possible that photosynthetic energy is carried to the glands by symplasmic transport of metabolites and utilized by a pump in the gland plasmalemma. Conversely, however, the pump may reside in the mesophyll cells concentrating the ions in the symplast of the leaf from where the ions may then eventually leak.

Although there is clearly active uptake of ions into leaf tissue of *A. spongiosa* (Lüttge et al., 1970), the possibility that the important step in active $Cl^-$ excretion into the bladder vacuoles resides in the mesophyll cells is perhaps ruled out because $Cl^-$ accumulation in the bladders, but not $Cl^-$ uptake by the mesophyll cells, is highly stimulated by light (Fig. 11.14). In salt excretion by *Limonium* glands the kinetic data given above also strongly suggest that the pumping occurs from the symplast, i.e. out of the symplast and not into the symplast of the system. Firstly, the transit half-time of $Cl^-$ from the leaf apoplast into the excreted fluid correlates with the half-time of filling of the cytoplasm (Section 11.2). Secondly, the ion concentration in cytoplasmic constituents such as chloroplasts stops rising when the pump reaches maximum activity (Section 11.4, Fig. 11.16). If pumping occured only into the symplast, one would expect an increase in symplasmic ion concentration as pumping activity becomes most effective.

Cytological techniques of localisation of ions, i.e. micro-autoradiography, electron microscopy of electron-dense precipitations and electron probe micro-analysis so far have suggested that there is either no uphill gradient towards the glands in the symplasm of excreting systems (Lüttge, 1966b; Ziegler and Lüttge, 1967) or that there may be even a downhill gradient (Shimony et al., 1973). This is consistent with the view which assumes that pumping occurs from the symplast of the system, i.e. that the site of active transport in excretion is at the plasmalemma of the gland cells. However, these techniques are not yet perfect enough to give unequivocal answers.

## 11.5 *New concepts: salt excretion as an example of biochemically mediated ecological adaptation*

### 11.5.1 *Inducibility of ion excretion in salt glands*

It has been shown that light-dependent $Cl^-$ accumulation in the bladders of *A. spongiosa* salt hairs has a distinct lag phase (Fig. 11.14). Despite the spatial problems of diffusion in the 3-mm wide leaf slices used in these experiments, the rather long lag phase might be interpreted as suggesting an induction phenomenon.

When leaves of *Limonium* plants grown on a $Cl^-$-free dilute nutrient solution are supplied with $Cl^-$, salt excretion activity rises in a sigmoidal fashion attaining maximum activity within about 3 h (Figs 11.16 and 11.17). Hill and co-workers (Hill, 1970b; Hill and Hill, 1973c) have investigated this phenomenon in detail. The lag phase after incubation in $Cl^-$-containing solution is sensitive to inhibitors of transcription and translation of the genetic code. If inhibitors such as actinomycin or fluorouracil, puromycin and 5-fluorophenylalanine are added before salt loading, no excretory activity is induced. When the inhibitors are added at the end of the lag phase they have no effect

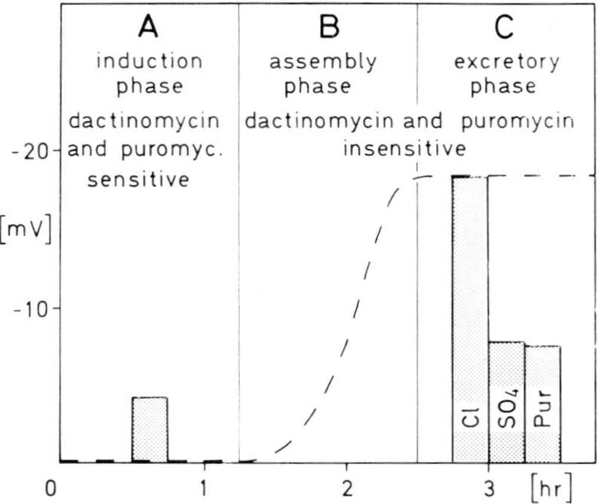



Fig. 11.17. Various phases during induction of excretory gland activity by treatment of *Limonium* leaves with 100 Mol m$^{-3}$ NaCl. Broken line : gland activity ($\psi$[mV] : c.f. Fig. 11.16) (from Hill and Hill, 1973b). Bars represent relative Cl$^-$ ATPase activities compiled from Hill and Hill (1973c). These ATPase data were normalised for the ATPase activity of microsomes of induced leaves tested in the presence of Cl$^-$ ( = large bar in phase C : 0.05–0.5 $\mu$Mol mg protein$^{-1}$ h$^{-1}$ depending on the leaf sample and preparation). Bar in phase A : ATPase activity of non-induced leaves; bar-SO$_4^{2-}$ in phase C : ATPase activity of induced leaves tested in the presence of SO$_4^{2-}$ instead of Cl$^-$ ; bar-Pur in phase C : ATPase activity of leaves treated with NaCl in the presence of puromycin.

on the rate of salt pumping which is established eventually. So it appears that the lag phase (A in Fig. 11.17) is an induction phase. The phase of sigmoidal rise (B) is an 'assembly phase', where the induced principle is incorporated into the system, so that salt elimination becomes possible in a phase (C) of constant excretory activity. Conversely, a reduction of the salt available in the medium results in a decrease of the rate of pumping. Thus it is clear that in *Limonium* salt excretion is a genetically controlled mechanism which readily adapts to salinity as an environmental factor.

### 11.5.2 *The biochemical basis of ecological adaptation*

In its natural ecological habitat the euryhaline (or facultatively halophilic) plant *Limonium* may depend to a considerable extent on the ability to adapt readily to frequently changing degrees of salinity in its medium. In the coastal areas of the moderate zone where *Limonium* grows such changes may be brought about by, for example, frequent rain on the one hand and tidal fluctuations on the other. The genetic induction phenomenon described above suggests that there must be a biochemical (enzymatical) basis for this ecological adaptation.

Kylin and Gee (1970) have demonstrated the presence of an alkali-cation-dependent ATPase in leaf extracts of the salt-excreting mangrove *Avicennia nitida*. In *Limonium* the enzymatical basis for adaptation to changing degrees of salinity is most likely an inducible $Cl^-$-dependent ATPase. A microsomal membrane fraction has been isolated from *Limonium* leaves which contains a $Cl^-$-stimulated ATPase whose activity responds to the history of the leaves (i.e. non-induced or induced in the absence and presence of puromycin, respectively) in exactly the same way as salt excretion (Fig. 11.17).

### 11.5.3 Future research

This latter finding makes us look back to the first section where we considered the phenomenon of salt excretion as an ecological adaptation. At the same time, it allows us to look forward to future research which may elucidate the phenomenon of ecological adaptation by an understanding of its mechanisms at the biochemical and molecular level.

Much work needs still to be done to further characterize the $Cl^-$-ATPase enzyme protein of *Limonium* and perhaps to extend the approach to other systems which are promising in this respect (e.g. *A. spongiosa*). Induction appears to be associated with increased polyribosome activity in the cytoplasm of *Limonium* glands (Shachar-Hill and Hill, 1970; Fig. 11.18). If the ATPase itself can be clearly shown to be associated with particular membranes in the gland system, this would also solve the question of localisation of the pump (see Section 11.4).

Although the inducible $Cl^-$-pumping ATPase of *Limonium* is one of the most clear cut examples of an ecological adaptation at the biochemical level, it is not unique. Reference can also be made, for example, to osmoregulation of some algae (Ben-Amotz and Avron, 1973; Kauss, 1973; Schobert et al., 1972). More complicated adaptations to an economical use of water are the $C_4$- and crassulacean acid metabolism (CAM)-pathways of photosynthesis. In CAM, $CO_2$ is fixed by phosphoenolpyruvate carboxylase during the night when the stomata are open (see Ch. 12). The malate synthesized is stored in the vacuoles, released again during the day (when the stomata are closed) and decarboxylated, with the resulting $CO_2$ refixed in the Calvin cycle (Ranson and Thomas, 1960). It is very interesting in the context of a discussion of salt excretion, that the facultative halophyte *Mesembryanthemum crystallinum* can be induced by salinity to switch from normal $C_3$-photosynthesis to CAM (Winter and Von Willert, 1972; Winter et al., 1974). This plant has large

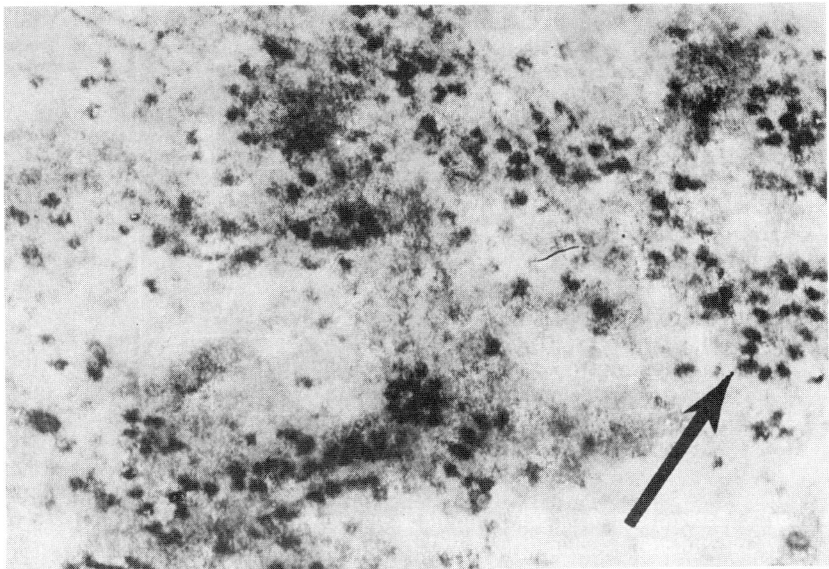

Fig. 11.18. Polyribosome activity of a salt-pumping *Limonium* gland cell. Many polyribosomes with RNA threads can be seen (arrow). Magnification × 134 000. (Reproduced with kind permission from Shachar-Hill and Hill, 1970.)

epidermal bladder cells which, under conditions of NaCl salinity, serve as a compartment for salt and water storage (unpublished data).

CAM is not only induced by a high salt concentration in the root medium of *M. crystallinum* but generally by factors which impose conditions of water stress (Winter, 1973, 1974). All of these factors are known to affect levels of hormones in leaves (in particular ABA and cytokinins). Phytohormones have been shown to regulate ion movements in response to water stress of the leaves. Examples are $K^+$ fluxes in stomatal regulation (Ch. 12) or ion export from the roots regulated by hormonal feedback from the leaves (Ch. 9). Little is known about the role of hormones in adaptation of halophytes to salinity, and nothing is known about phytohormones as messengers in adaptive salt excretion systems. However, this is an exciting possibility to be tested in future investigations.

## 11.6 Literature review

As an outstanding example of effective ion transport salt excretion by salt glands is now reviewed more or less briefly in almost any book on transport in plants (e.g. Epstein, 1972; Lüttge, 1973). Chapter 8 in the book of Waisel (1972) provides a more comprehensive collection of material and data, selected with particular emphasis on ecological implications. *Limonium* salt glands are the system for which the most detailed investigation is available and the monographical review of Hill and Hill (1973b) is very informative. More general reviews of biophysics and biochemistry of salt excretion are available by Lüttge (1972) and by Hill and Hill (1975).

## Acknowledgements

Thanks are due to Frau D. Schäfer for making the drawings and to Frau G. Zirke for taking care of half-tone photographs. I thank all colleagues who kindly made available original figures from their work and who gave suggestions for improvement of the text. My own experiments on salt glands were supported by research grants from the Deutsche Forschungsgemeinschaft, Bad Godesberg.

## References

W.H. ARISZ, Nature, 174 (1954) 223.

W.H. ARISZ, Protoplasma, 52 (1960) 309.

W.H. ARISZ, Acta Bot. Neerl., 18 (1969) 14.

W.H. ARISZ, I.J. CAMPHUIS, H. HEIKENS and A.J. VAN TOOREN, Acta Bot. Neerl., 4 (1955) 322.

W.H. ARISZ and E.P. WIERSEMA, Proc. K. Ned. Acad. Wet. C., 69 (1966) 223.

M.R. ATKINSON, G.P. FINDLAY, A.B. HOPE, M.G. PITMAN, H.D.W. SADDLER and K.R. WEST, Aust. J. Biol. Sci., 20 (1967) 589.

D.A. BAKER and J.L. HALL, New Phytol., 72 (1973) 1281.

A. BEN-AMOTZ and M. AVRON, Plant Physiol., 51 (1973) 875.

W.L. BERRY, Am. J. Bot., 57 (1970) 1226.

B. BILLARD and C.D. FIELD, Planta, 115 (1974) 285.

T.E. BOSTROM and C.D. FIELD, in W.P. Anderson (Ed.) Ion Transport in Plants (1973) Academic Press, London, p. 385.

E. BRINCKMANN and U. LÜTTGE, Planta, 119 (1974) 47.

S. CARDALE and C.D. FIELD, Planta, 99 (1971) 183.

W.J. CRAM and M.G. PITMAN, Aust. J. Biol. Sci., 25 (1972) 1125.

J.M. DIAMOND and W.H. BOSSERT, J. Gen. Physiol., 50 (1967) 2061.

F. EHRENDORFER, in D. von Denffer, W. Schumacher, K. Mägdefrau and F. Ehrendorfer (Eds.) Lehrbuch der Botanik für Hochschulen begründet von E. Strasburger, E. Noll, H. Schenck und A.F.W. Schimper (1971) 30th Ed., Gustav Fischer Verlag, Stuttgart, p. 584.

A. ENGLER and L. DIELS, Syllabus der Pflanzenfamilien (1936) 11th Ed., Verlag von Gebrüder Borntraeger, Berlin.

E. EPSTEIN, Mineral Nutrition of Plants: Principles and Perspectives (1972) John Wiley and Sons, New York.

A. FREY-WYSSLING, Die Stoffausscheidung der höheren Pflanzen (1935) Springer Verlag, Berlin.

A. FREY-WYSSLING and E. HÄUSERMANN, Ber. Schweiz. Bot. Ges., 70 (1960) 150.

B.E.S. GUNNING and J.S. PATE, Protoplasma, 68 (1969) 107.

A.E. HILL, Biochim. Biophys. Acta, 135 (1967a) 454.

A.E. HILL, Biochim. Biophys. Acta, 135 (1967b) 461.

A.E. HILL, Biochim. Biophys. Acta, 196 (1970a) 66.

A.E. HILL, Biochim. Biophys. Acta, 196 (1970b) 73.

A.E. HILL and B.S. HILL, J. Membrane Biol., 12 (1973a) 129.

A.E. HILL and B.S. HILL, Int. Rev. Cytol., 35 (1973b) 299.

A.E. HILL and B.S. HILL, in U. Lüttge and M.G. Pitman (Eds.) Transport: Cells and Tissues, Encyclopedia of Plant Physiology, New Series (1975) Vol. 2, Springer Verlag, Berlin, in press.

B.S. HILL and A.E. HILL, J. Membrane Biol., 12 (1973c) 145.

H. KAUSS, Plant Physiol., 52 (1973) 613.

G.J. KENAGY, Science, 178 (1972) 1094.

G.J. KENAGY, Oecologia, 12 (1973) 383.

A. KYLIN and R. GEE, Plant Physiol., 45 (1970) 169.

A.W.D. LARKUM and A.E. HILL, Biochim. Biophys. Acta, 203 (1970) 133.

C.A. LEVERING and W.W. THOMSON, Planta, 97 (1971) 183.

C.A. LEVERING and W.W. THOMSON, in C.J. Arceneaux (Ed.) 30th Annu. Proc. Electron Microscopy Soc. Am. (1972) Los Angeles.

U. LÜTTGE, Protoplasma, 59 (1964a) 157.

U. LÜTTGE, Ber. Dtsch. Bot. Ges., 77 (1964b) 181, suppl.

U. LÜTTGE, Planta, 66 (1965) 331.

U. LÜTTGE, Planta, 68 (1966a) 44.

U. LÜTTGE, Planta, 68 (1966b) 269.

U. LÜTTGE, Naturwissenschaften, 53 (1966c) 96.

U. LÜTTGE, Annu. Rev. Plant Physiol., 22 (1971) 23.

U. LÜTTGE, Stofftransport der Pflanzen (1973) Springer Verlag, Berlin.

U. LÜTTGE, in U. Zimmermann and J. Dainty (Eds.) Membrane Transport in Plants and Plant Organelles (1974) Springer Verlag, Berlin.

U. LÜTTGE, in U. Lüttge and M.G. Pitman (Eds.) Transport: Cells and Tissues, Encyclopedia of Plant Physiology, New Series (1975) Vol. 2, Springer Verlag, Berlin, in press.

U. LÜTTGE and G. KRAPF, Cytobiologie, 1 (1969) 121.

U. LÜTTGE and C.B. OSMOND, Aust. J. Biol. Sci., 23 (1970) 17.

U. LÜTTGE and C.K. PALLAGHY, Z. Pflanzenphysiol., 61 (1969) 58.

U. LÜTTGE and C.K. PALLAGHY, Z. Pflanzenphysiol., 67 (1972) 359.

U. LÜTTGE, C.K. PALLAGHY and C.B. OSMOND, J. Membrane Biol., 2 (1970) 17.

U. LÜTTGE, C.K. PALLAGHY and K. VON WILLERT, J. Membrane Biol., 4 (1971) 395.

E.A.C. MACROBBIE, J. Exp. Bot., 20 (1969) 236.

E.A.C. MACROBBIE, J. Exp. Bot., 21 (1970) 335.

E.A.C. MACROBBIE, J. Exp. Bot., 22 (1971a) 487.

E.A.C. MACROBBIE, Biol. Rev., 46 (1971b) 429.

S. MORRISSEY, Nature, 176 (1955) 1220.

O. NEMČEK, K. SIGLER and A. KLEINZELLER, Biochim. Biophys. Acta, 126 (1966) 73.

W. NULTSCH, Allgemeine Botanik (1971) 4th Ed., G. Thieme Verlag, Stuttgart.

C.B. OSMOND, U. LÜTTGE, K.R. WEST, C.K. PALLAGHY and B. SHACHAR-HILL, Aust. J. Biol. Sci., 22 (1969) 797.

C.K. PALLAGHY and U. LÜTTGE, Z. Pflanzenphysiol., 62 (1970a) 51.

C.K. PALLAGHY and U. LÜTTGE, Z. Pflanzenphysiol., 62 (1970b) 417.

J.S. PATE and B.E.S. GUNNING, Annu. Rev. Plant Physiol., 23 (1972) 173.

M.G. PITMAN, Aust. J. Biol. Sci., 25 (1972) 905.

M.G. PITMAN, U. LÜTTGE, A. LÄUCHLI and E. BALL, Z. Pflanzenphysiol., 72 (1974a) 75.

M.G. PITMAN, U. LÜTTGE, A. LÄUCHLI and E. BALL, Aust. J. Plant Physiol., 1 (1974b) 377.

G. POLLAK and Y. WAISEL, Ann. Bot., 34 (1970) 879.

S.C. RANSON and M. THOMAS, Annu. Rev. Plant Physiol., 11 (1960) 81.

E. SCHNEPF, Protoplasma, 58 (1964) 137.

B. SCHOBERT, E. UNTNER and H. KAUSS, Z. Pflanzenphysiol., 67 (1972) 385.

B. SHACHAR-HILL and A.E. HILL, Biochim. Biophys. Acta, 211 (1970) 313.

C. SHIMONY, A. FAHN and L. REINHOLD, New Phytol., 72 (1973) 27.

R.M. SPANSWICK, in U. Lüttge and M.G. Pitman (Eds.) Transport : Cells and Tissues, Encyclopedia of Plant Physiology, New Series (1975) Vol. 2, Springer Verlag, Berlin, in press.

R.F.M. VAN STEVENINCK, A.R.F. CHENOWETH and M.E. VAN STEVENINCK, in W.P. Anderson (Ed.) Ion Transport in Plants (1973) Academic Press, London, p. 25.

W.W. THOMSON and L.L. LIU, Planta, 73 (1967) 201.

W.W. THOMSON, W.L. BERRY and L.L. LIU, Proc. Natl. Acad. Sci. U.S., 63 (1969) 310.

J. TROUGHTON and L.A. DONALDSON, in Probing Plant Structure (1972) A.H. and A.W. Reed, Wellington.

Y. WAISEL, Biology of Halophytes (1972) Academic Press, New York.

K. WINTER, Planta, 114 (1973) 75.

K. WINTER, Plant Sci. Lett., 3 (1974) 279.

K. WINTER and D.J. VON WILLERT, Z. Pflanzenphysiol., 67 (1972) 166.

K. WINTER, U. LÜTTGE and E. BALL, Biochim. Biophys. Acta, 343 (1974) 465.

H. ZIEGLER, Planta, 47 (1956) 447.

H. ZIEGLER and U. LÜTTGE, Planta, 70 (1966) 193.

H. ZIEGLER and U. LÜTTGE, Planta, 74 (1967) 1.

*Ion transport in plant cells and tissues*
*edited by D.A. Baker and J.L. Hall*
© *North-Holland Publishing Company, 1975*

CHAPTER 12

# Stomata

D.A. Thomas

## Contents



## 12.1 Introduction

Stomata are the main sites at which plants equilibrate with their aerial environment. Adjustments of their pore size play a major role in determining the latent heat flux component of the heat energy balance of plants and hence the rate of water flow through the plant. Within the limitations set by the energy balance, the extent of stomatal opening determines the flux of $CO_2$ reaching the photosynthetic sites. By their major control of these fluxes stomata protect the plant from dessication and control the rate of carbon fixation. Hence they play a dominating role in plant survival and growth rate. Early realizations that this might be so stimulated interest in how stomatal aperture is controlled. Today an understanding of how stomata work in response to environmental change is a major problem in modelling predictions of plant production and performance.

Basically the concept that osmotic swelling by the accumulation of solutes in the guard cells, leading to guard cell swelling and stomatal opening originally proposed by Lloyd (1908) and probably best supported by the experiments of Heath (1938) is still valid. A survey of how our concepts of the stomatal mechanism have developed is a good example of a philosopher's description of Science in general – 'as a constantly changing series of approximations'. Our current knowledge of the role of ions in the workings of stomata is due to a union of (i) a histochemical method for staining $K^+$ and the observation that $K^+$ is accumulated in the guard cells of open stomata (Macallum, 1905); (ii) an extension of Lloyd's starch-to-sugar hypothesis (1908) for the provision of osmotic solutes in the guard cell; (iii) the use of floating epidermal strips as an experimental tool. The re-adoption of the $K^+$-staining method and confirmation of Macallum's original observations have led to the use of modern techniques such as radioactive tracers, electron beam and microprobe analysis to quantify the ion fluxes associated with guard cell volume changes and stomatal movement.

With the finding that a considerable proportion of the charge balance for the $K^+$ accumulated in guard cells is due to organic acids, mainly malate, Lloyd's original starch-to-sugar hypothesis has been extended to give starch-to-organic acids. Epidermal strips floated on simple solutions of dilute potassium salts can approximate the behaviour of stomata on leaves in their response to such factors as light and dark, changing environmental $CO_2$ concentrations and hormones.

This chapter covers in the main the role of ions in the regulation of the stomatal mechanism. More general reviews on stomatal physiology are provided by Stålfelt (1956), Heath (1959), Ketellapper (1963), Zelitch (1963) and Meidner and Mansfield (1968).

## 12.2 Structure of the stomatal apparatus

The stomatal complex might be defined as cells that have differentiated from protoderm cells to form two guard cells which are characterized by unevenly thickened walls. The guard cells form the stomatal pore by changes in shape and make up the basic stomatal unit. In many plants, the guard cells are associated with two or more cells, termed subsidiary or accessory cells, that are morphologically different from the surrounding epidermal cells and in such plants, the accessory cells plus the guard cells form the stomatal unit (Esau, 1965). The mature guard cell appears to contain the same components as those found in many other functional plant cells as they include a nucleus, chloroplasts, mitochondria and spherosomes. Less than a dozen electron microscope studies on guard cells have been made and the interpretation of the observations is controversial. Many interesting features such as invaginations of the plasma membrane are thought to be artifacts of the fixation and embedding procedures. Fig. 12.1 shows an electron micrograph of a guard cell pair from tobacco.

Plasmodesmata have been observed between guard cell pairs and the associated epidermal cells. They were found in large numbers ($15$–$20$ $\mu m^{-2}$) and almost exclusively in pit fields of end walls from functioning guard cells of tobacco and *Vicia faba*. Pallas and Mollenhauer (1972) consider that the small volume of guard cells together with their full complement of organelles gives them an advantage over the larger neighbouring cells in turgor generation. The large surface-to-volume ratio of the guard cells could result in a larger flux-to-volume ratio than in surrounding cells and allows rapid changes in turgor brought about by a water influx coupled to an active transport of ions together with the production of osmotically active compounds and energy by the chloroplasts and mitochondria.

## 12.3 Role of ions and ion fluxes in stomatal movements

### 12.3.1 Cations

#### Potassium
The accumulation of $K^+$ in the guard cells of open stomata observed by Macallum (1905) using his own histochemical method, was confirmed and extended by Imamura (1943), Yamashita (1952) and Fujino (1967). Although it is difficult to quantify accurately histochemical methods, the large differences found between the level of $K^+$ accumulated in the guard cells of closed, partially closed and open stomata led Fujino (1967) to postulate that a light-activated $K^+$ accumulation in the guard cells could play a major role in the

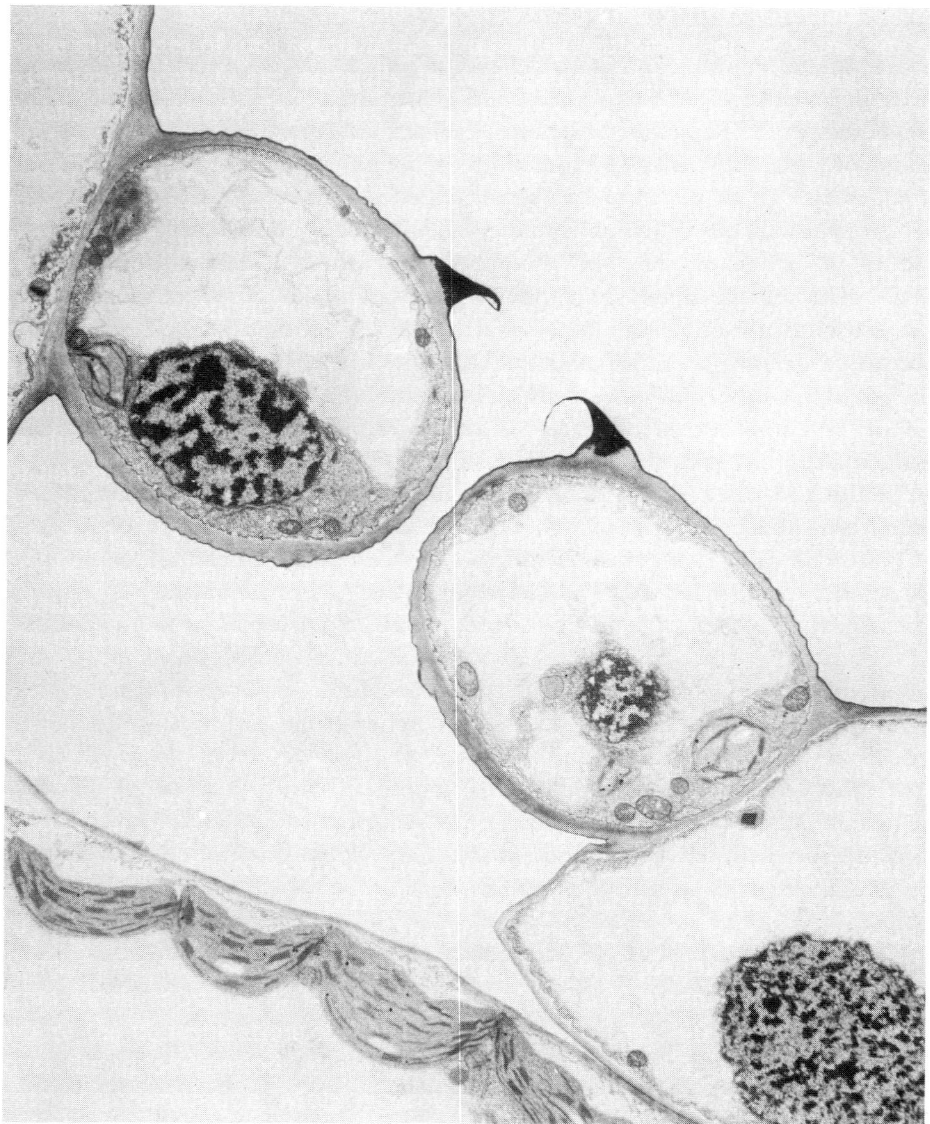

Fig. 12.1. Guard cells from *Nicotiana tabacum*, showing typical distribution of cellular constituents. Magnification × 4030. (From Pallas and Mollenhauer, 1972.)

mechanism of opening and, conversely, the efflux of $K^+$ was connected with closure. Following this work, the emphasis has been to quantify the $K^+$ fluxes and to try and define the transport mechanisms driving these fluxes which could be connected, together with an associated movement of water, to increases and decreases in the guard cell volume and stomatal opening and closing.

Using abaxial epidermal strips from *V. faba* leaves in which the stomata showed normal type responses to light, dark and $CO_2$-free air, Fischer and Hsiao (1968) showed that there was a quantifiable relation between the amount of $^{86}Rb^+$ (used as a tracer for $K^+$) accumulated and the degree of opening. The accumulation of $K^+$ together with an associated anion could account for the decrease in osmotic potential and thus the influx of water to increase guard cell volume. Environmental conditions which lead to reductions in aperture e.g. dark and normal $CO_2$-containing air, also reduced the amount of $^{86}Rb^+$ accumulated. The concentration of $K^+$ in the guard cells at maximum opening was 300 Mol m$^{-3}$. The rate of uptake was estimated at 90 nMol m$^{-2}$ s$^{-1}$. There was no specificity shown for the anion associated with $K^+$; and Fischer and Hsiao suggested that organic anions might balance the accumulated $K^+$.

The light opening of stomata on *V. faba* and tobacco epidermal strips is highly specific for low concentrations of $K^+$ ($<10$ Mol m$^{-3}$) in the bathing solution (Humble and Hsiao, 1969; Willmer and Mansfield, 1969; Thomas, 1970a). For *V. faba* Li$^+$, Cs$^+$ and Na$^+$ did not support similar light openings until the concentration in the bathing solution reached 100 Mol m$^{-3}$ (Fig. 12.2). Concentrations greater than 10 Mol m$^{-3}$ of $K^+$, Rb$^+$, Li$^+$, Cs$^+$ and

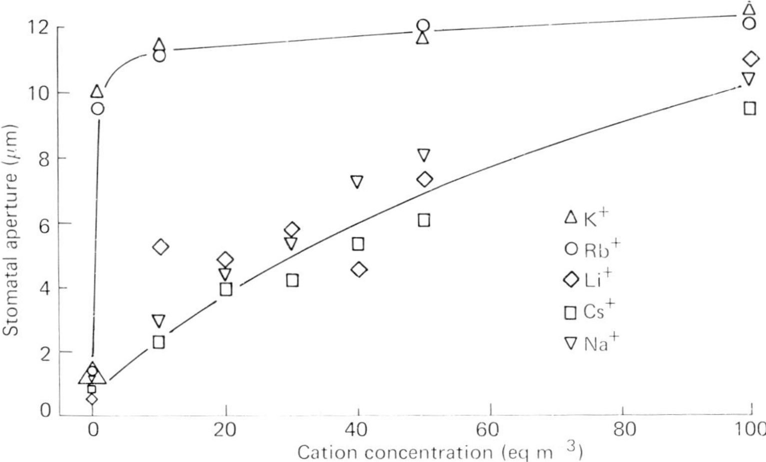


Fig. 12.2. Stomatal opening in response to the cations $K^+$, Rb$^+$, Cs$^+$, Ki$^+$ and Na$^+$, with Cl$^-$ as the associated anion, in light. (From Humble and Hsiao, 1969.)

$Na^+$ all lead to increases in aperture in the dark. These results emphasize the importance of low concentrations ($\leqslant 10$ Mol m$^{-3}$) of $K^+$ in the bathing solution for the light opening and dark closing responses normally found in vitro. Willmer and Mansfield (1969) found that stomata on *Commelina communis* strips require much higher concentrations of $K^+$, ($> 120$ Mol m$^{-3}$) in the bathing solution to enable them to open in the light.

Using epidermal strips from *V. faba*, Humble and Hsiao (1970) showed that stomatal opening in the light and closing in the dark was paralleled by an influx and efflux of $K^+$ respectively. Staining for $K^+$ showed that $K^+$ taken up during opening was substantially localized in the guard cells. Electron probe microanalysis of stomata from tobacco leaf discs by Sawhney and Zelitch (1969) showed that $K^+$ increased in the guard cells after exposure of the discs to light and conversely decreased in the dark. The concentration in closed stomata (width approx. 1 $\mu$m) was 210 Mol m$^{-3}$ which increased to 500 Mol m$^{-3}$ in fully opened stomata (width 8 $\mu$m). Together with an accompanying anion, the increase in solute concentration of the guard cell could be 580 Mol m$^{-3}$. Humble and Raschke (1971) also analysed the amount of $K^+$ accumulated in guard cells of *V. faba* during the light-opening process. For a stomatal opening from 2 to 12 $\mu$m, an average of $4.0 \times 10^{-12}$ Mol $K^+$ were accumulated in the cell. The calculated rate of accumulation was $10 \times 10^{-15}$ Mol $K^+$ min$^{-1}$. $Na^+$, $Mg^{2+}$ and $Ca^{2+}$ were not accumulated to any measurable extent during opening and it was found that the amount of $Cl^-$ taken up was very much less than that of $K^+$ (Fig. 12.3). It is hard to exactly localize the region of $K^+$ accumulation employing the electron microprobe technique, but as observations with the light microscope showed that in open guard cells the protoplasm was reduced to a thin layer around the periphery, it was then suggested that most of the cell volume was a vacuole. The only components of intracellular space not considered to contribute to the turgor regulation of the guard cell were the chloroplasts and the nucleus. Table 12.1 gives the estimates of the increase in $K^+$, $Na^+$ and $Cl^-$ content, volume and osmotic potential for an increase of stomatal aperture from 2 to 12 $\mu$m.

The linear relationship between the amount of $K^+$ accumulated and the increase in stomatal aperture found in tobacco (Sawhney and Zelitch, 1969) has also been found in *Phaseolus vulgaris* using the electron probe technique (Turner, 1973) and in *V. faba* using radioactive methods (Fischer, 1972) and by flame photometry (Allaway and Hsiao, 1973). The estimates of Fischer give 26 $\mu$Mol $K^+$ accumulated m$^{-2}$ $\mu$m$^{-1}$ of aperture, equivalent to a change in $K^+$ concentration of 40 Mol m$^{-3}$ in the guard cells for each micrometer change in stomatal aperture. The initial and steady state net fluxes and the

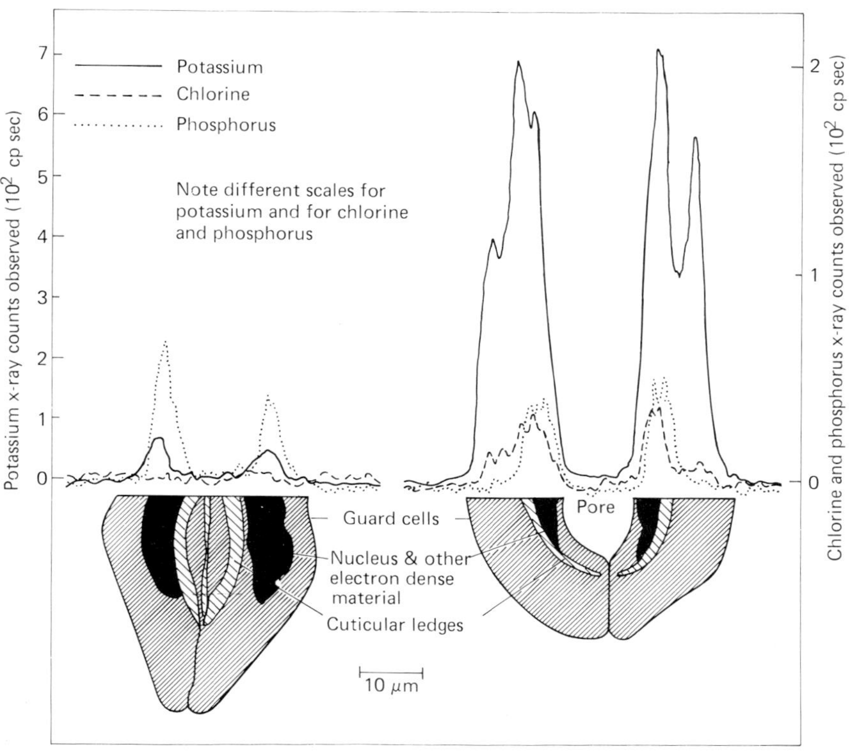


Fig. 12.3. Profiles of relative amounts of $K^+$, $Cl^-$ and P across an open and a closed stoma measured by electron probe microanalysis. (From Humble and Raschke, 1971).

Table 12.1.

The amounts of various elements, measured with the electron probe, in open and closed guard cells, the guard cell volumes, and the changes in stomatal aperture and osmotic potential (Humble and Raschke, 1971).

| | Amounts per stoma ($10^{-14}$ Mol) | | | Stomatal aperture ($\mu$m) | Guard cell volume per stoma ($10^{-15}$ m$^{-3}$) | Guard cell osmotic potential from incipient plasmolysis (bars) |
|---|---|---|---|---|---|---|
| | $K^+$ | $Na^+$ | $Cl^-$ | | | |
| Open stoma | 424 | 0 | 22 | 12 | 4.8 | −35 |
| Closed stoma | 20 | 0 | 0 | 2 | 2.6 | −19 |
| Difference between open and closed stoma | 404 | 0 | 22 | 10 | 2.2 | −16 |

Table 12.2.
Estimates of the $K^+$ fluxes associated with guard cells (Fischer, 1972).

| Time and conditions | Type of $K^+$ flux | $K^+$ flux rate (nMol m$^{-2}$ s$^{-1}$) |
|---|---|---|
| A) Time zero | net flux (tracer) | +165 |
| B) After 300 min in | influx | + 20 |
|    light and $CO_2$-free air | efflux | − 37 |
|    on 10 Mol m$^{-3}$ KCl + 0.1 Mol m$^{-3}$ CaCl$_2$ | net flux | − 18 |
| C) After 300 min in dark on water | net flux | +122 |
| A) Time zero | net flux (tracer) | +110 |
| B) After 200 min | influx | + 27 |
|    in light and CO-free air | efflux | − 26 |
|    on 10 Mol m$^{-3}$ KCl + 0.1 Mol m$^{-3}$ CaCl$_2$ | net flux | + 16 |
| C) After 200 min in dark on water | net flux (tracer) | + 85 |

steady state influx and efflux are given in Table 12.2. Compared to the high initial net influx rate, the steady state influx and efflux rates are low. The reduction in influx in the steady state condition was not due to a time effect since the net influx rate still remained high in guard cells on epidermal strips stored in the dark on distilled water before the uptake determinations were made. The lowered influx rate under the steady state conditions might be due to a feed-back inhibition of influx caused by the accumulation of $K^+$ or due to the associated increase in turgor as suggested by Gutknecht (1968). These fluxes of experimental necessity have been determined for guard cells that do not have the back pressure that could be expected from a full complement of associated epidermal cells (Heath, 1938) as epidermal strips were prepared so that the number of viable epidermal cells would be the least possible. Without the back pressure of the epidermal cells on the guard cells, the fluxes, particularly the steady state fluxes, could be changed appreciably (Penny and Bowling, 1974). A flux that has not been measured is the amount and rate of $K^+$ effluxed from the guard cells during stomatal closing, which must be known before a model for the mechanism of stomatal closing can be constructed (see Section 12.6.3).

The linear relationship between the estimated guard cell $K^+$ content and stomatal aperture can be obtained in dark, low light and light conditions (Allaway and Hsiao, 1973) suggesting that this relation can be independent of environmental light conditions and independent of how the stomata are opened.

The increase in guard cell $K^+$ concentration of 50 Mol m$^{-3}$ for each micrometer increase in aperture is similar to that made by others for guard cells on *V. faba* epidermal strips (Humble and Raschke, 1971; Fischer, 1972). Electron microprobe measurements show that the relation between the increase in $K^+$ content and aperture is also linear in the stomata of abaxial epidermal strips taken directly from *P. vulgaris* suggesting that the linear relation found in vitro also holds in vivo. No change was found in the $K^+$ content of the subsidiary cells as the stomatal aperture increased (Turner, 1973). Until recently the in vivo supply of $K^+$ to the guard cells of dicotyledons such as *C. communis, V. faba, P. vulgaris* and tobacco has been uncertain, as no definite movement of $K^+$ between the guard cells and subsidiary or epidermal cells has been observed (Willmer and Pallas, 1973) and stomata on epidermal strips taken from leaves (except *Z. mays*) all require $K^+$ in the bathing solution for full light opening to occur. Using $K^+$-sensitive microelectrodes, Penny and Bowling (1974) have determined the steady state concentrations of $K^+$ in the guard, inner lateral subsidiary, outer lateral subsidiary, terminal subsidiary and epidermal cells around the stomatal complex on *C. communis* leaves when the stomatal pore is open and closed. The determinations showed that when the stomata are open there is a stepwise decrease in vacuolar $K^+$ content from the guard cells outwards and when they are closed the gradient is reversed (Fig. 12.4). It therefore seems that in dicotyledons there may be a shuttle of $K^+$ between the guard and subsidiary cells during opening and closing as has been found in *Z. mays* (Raschke and Humble, 1971) where the subsidiary cells act as a reservoir of both $K^+$ and $Cl^-$ for the guard cells. It still remains to be determined whether the transfer to ions to and from the guard cells is apoplastic or symplastic.

Using both the electron probe microanalyser and staining, Raschke and Fellows (1971) found that $K^+$ and $Cl^-$ moved into the guard cells from the subsidiary cells within 1–2 min following exposure to light and a reversal in the direction of movement could be detected in the same time period after darkening. The estimated rate of $K^+$ uptake into the guard cells upon illumination was $10^{-14}$ Mol min$^{-1}$ and the final concentration reached in the guard cells of open stomates was 400 Mol m$^{-3}$. This is less than the concentration of $K^+$ found in the guard cells of open *V. faba* stomates (Humble and Raschke, 1971) but is still sufficient to account for the estimated changes in cell turgor because, unlike the guard cells of *V. faba* in which $Cl^-$ accumulation only accounts for 5% of the positive charge of $K^+$, in those of *Z. mays* it accounts for at least 40%. With an estimate of $2 \times 10^{-15}$ m$^3$ of lumen of a pair of guard cells, the rate of $K^+$ and $Cl^-$ uptake and the concentrations reached were sufficient to account for the rate and extent of opening.

In studies employing *V. faba* epidermal strips in the light, Fischer and Hsiao (1968) found little change of stomatal opening in the range of $K^+$ concen-

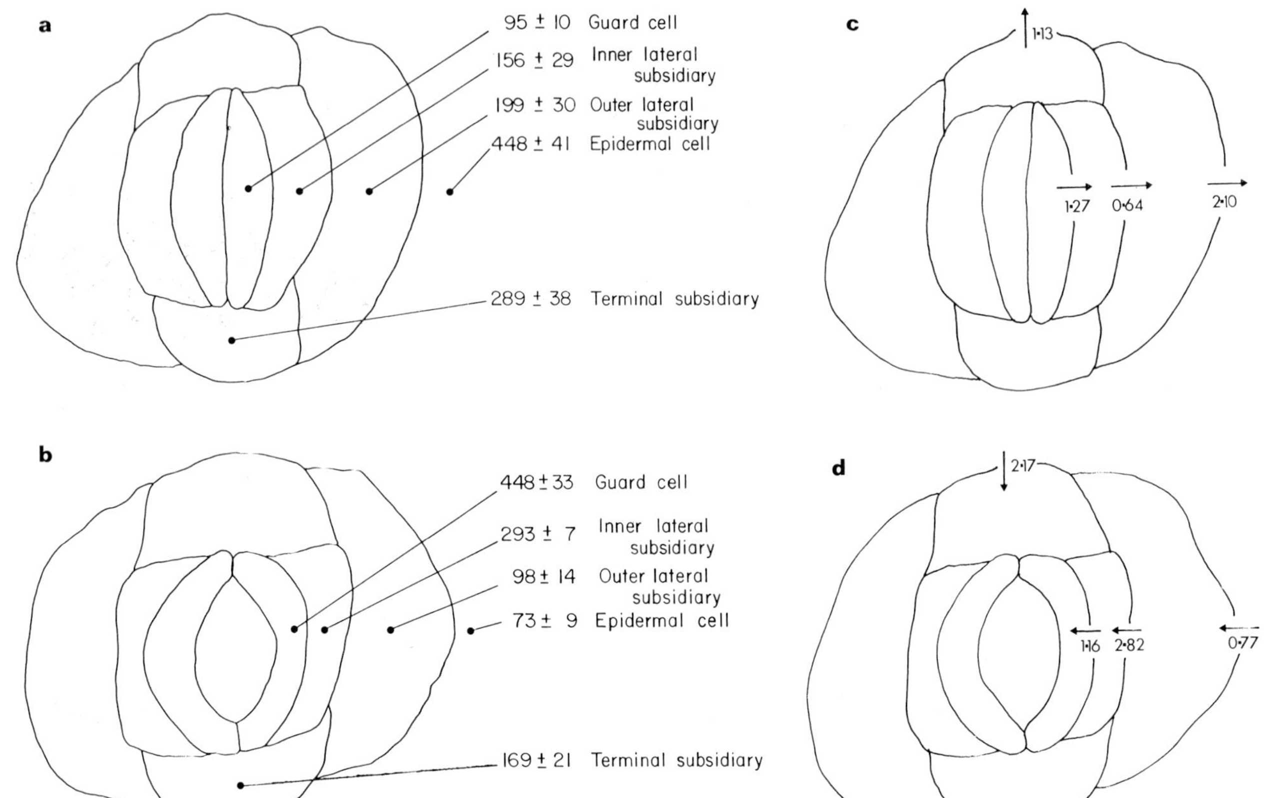


Fig. 12.4. Vacuolar $K^+$ concentrations (Mol m$^{-3}$) in various cells of the leaf epidermis of *Commelina communis* measured when the stomata were (a) closed, and (b) open ($\pm$ standard error of the mean) and the calculated driving forces on $K^+$ (kJ Mol$^{-1}$) required to maintain the differences in cellular $K^+$ concentration in (c) closed, and (d) open stomata. Arrows indicate the direction of active transport. (From Penny and **Bowling**, 1974.)

tration 10–100 Mol m$^{-3}$, reductions in aperture only occurring when K$^+$ was increased beyond 100 Mol m$^{-3}$. Using tobacco leaf epidermal strips, where as many as possible of the epidermal cells were kept intact, in a solution flow porometer in which rapid changes in stomatal aperture could be measured, Thomas (1970a) found that increases in the concentration of K$^+$ in the bathing solution greater than 10 Mol m$^{-3}$ caused decreases in aperture. However on returning to a concentration of 10 Mol m$^{-3}$, the aperture increased. Similar results have been found by Brag (1972) using whole plants of *Triticum aestivum* and *Pisum sativum*. If with increasing external concentrations of K$^+$ the accumulation of K$^+$ in the epidermal cells started to increase while that in the guard cell remained at a steady level, the turgor pressure of the epidermal cells could increase relative to that of the guard cells. This increase in pressure around the guard cells causes decreases in stomatal aperture (Stålfelt, 1967; Raschke, 1970). Nevertheless, increasing concentrations of K$^+$ greater than 100 Mol m$^{-3}$ have been found to decrease the stomatal aperture (Fischer and Hsiao, 1968) suggesting that the increases in K$^+$ in the environment of the guard cells result in reductions of the aperture. This decrease is not explained solely in terms of osmotic water potential balance between the bathing solution and that in the guard cells because the concentration of K$^+$ in open guard cells of *V. faba* has been estimated to reach 400 Mol m$^{-3}$ (Table 12.1). It is well known that water uptake can be influenced to different degrees by different types and concentrations of ions, at concentrations too low to significantly affect the osmotic driving force (Slatyer, 1967). K$^+$ may not only regulate stomatal aperture depending on the extent to which it is accumulated in the guard cells, but it may also have a role in regulating the permeability influx–efflux characteristics of guard cell membranes. For example, it has been shown that when low concentrations of K$^+$ are added to stomata that have opened in the dark while bathed in a 10 Mol m$^{-3}$ NaCl solution, they return to the normal dark closed state (Thomas, 1970a). Furthermore, the transpiration of plants deficient in K$^+$ has been found to be higher and their stomatal apertures larger (Peaslee and Moss, 1968). Under conditions of transpirational stress it could be expected that the concentration of ions in the plant would increase. Changes in ion concentration, particularly K$^+$, may be the means by which water deficits are sensed and stomatal adjustments made, both in short- and long-term responses to water stress. In short-term adjustments stomatal aperture may increase and decrease with increases and decreases in K$^+$ concentration, e.g. the well known phenomenon of stomatal closure at mid-day. When the effects of water stress are prolonged and ion concentrations in the plant remain elevated for longer periods, this elevated concentration may stimulate other adjustment mechanisms controlling stomata, e.g. hormonal (see Section 12.4).

*Sodium*

In studies with detached epidermal strips of *C. communis*, incubating solutions of between 100 to 400 Mol m$^{-3}$ NaCl resulted in greater stomatal opening than KCl solutions of the same concentrations. At the lowest concentration, 100 Mol m$^{-3}$, the stimulation of opening by NaCl was more than 40% greater than that by KCl (Willmer and Mansfield, 1969) and Na$^+$ could have a definite physiological role in the opening of *C. communis* stomata. Reductions in aperture were only found after the concentrations of the NaCl and KCl incubating media were increased to more than 400 Mol m$^{-3}$. The estimated content of K$^+$ reached in the guard cells of stomata showing maximum opening is between 400 and 600 Mol m$^{-3}$ (Sawhney and Zelitch, 1969; Humble and Raschke, 1971; Fischer and Hsiao, 1968). If this was the maximum concentration that could be reached in the guard cells, it might be expected on osmotic grounds that concentrations greater than 500 Mol m$^{-3}$ in the external solution could lead to reduction in aperture.

In *V. faba* epidermal strips exposed to light it is only at concentrations of 100 Mol m$^{-3}$ that NaCl supports approximately the same stomatal opening as a 0.1 Mol m$^{-3}$ KCl incubating solution. In the dark, floating epidermal strips on solutions of 100 Mol m$^{-3}$ K$^+$, Rb$^+$, Li$^+$, Cs$^+$ and Na$^+$ with Cl$^-$ as the associated anion, there is an extensive stomatal opening which does not occur if the concentrations of the solutions are kept at or below 10 Mol m$^{-3}$ (Humble and Hsiao, 1969). Extensive dark opening of stomata other than those of CAM plants is not considered normal. High ($> 100$ Mol m$^{-3}$) concentrations of ions in the solution external to the guard cells might lead to a swamping of a specific high affinity mechanism for the accumulation of K$^+$ from low environmental concentrations. At high external concentrations non-specific and possibly passive diffusional mechanisms might be favoured to balance the negative charges associated with the formation of organic acids in the guard cells (see Section 12.4).

The response of stomata on tobacco epidermal strips to 10 Mol m$^{-3}$ KCl or 10 Mol m$^{-3}$ NaCl in the light and dark over the same interval of time showed marked differences. Opening in the light in the K$^+$ solution after a slow initial phase of opening was very rapid, as was the closing response when the stomata were darkened. In the K$^+$ solution the darkened stomata remained shut. The light opening of stomata bathed in 10 Mol m$^{-3}$ NaCl was much slower and the aperture reached over the time period much less. The opening in the light proceeds in small steps and remains constant over relatively long periods and does not show the rapid rise once stomatal opening has started which is a feature of the opening when the bathing medium contains K$^+$. On darkening, stomata bathed in NaCl show a small initial decrease in aperture which remains constant for some time but then is followed by a slow increase in aperture, whereas in KCl once the aperture is reduced in the dark, there is no subsequent increase (Thomas, 1970a). The effect of $K^+$ addition to stomata that have opened in the dark while bathed in 10 Mol $m^{-3}$ NaCl is to reduce dark opening and further opening in the dark is stopped, suggesting that the normal dark closing of tobacco stomata is only maintained in the presence of $K^+$. In the presence of $K^+$ either the influx of $Na^+$ into the guard cells is substantially reduced or the efflux increased or both occur simultaneously (Thomas, 1970a).

The reduction by $K^+$ of the $Na^+$-supported opening cannot be explained on the basis of a net exchange in which an internal $Na^+$ is exchanged for an external $K^+$ because such a one-for-one exchange cannot explain the observed decreased aperture unless the difference in the response of guard cells to $Na^+$ and $K^+$ is more than just purely osmotic in nature. The presence of $K^+$ could change the membrane permeability to $Na^+$ by altering either or both the electrical mobility of $Na^+$ in the membranes and the $Na^+$ partition coefficient.

The opening that occurs in the dark when strips are bathed in a solution containing $Na^+$ alone can be reduced on exposure to light. The normal sequence of light opening and dark closing can be reversed in the presence of $Na^+$ alone and the normal sequence can be re-established on the addition of $K^+$ to the bathing solution (Thomas, 1970a). Thus, in the guard cells of tobacco, $Na^+$ and $K^+$ give reverse responses and there may be metabolically dependent light-activated extrusion processes for $Na^+$ and accumulation processes for $K^+$. Support for metabolic dependence comes from the finding that in the presence of $K^+$ the addition of ATP can initiate opening in the dark, whereas in the presence of $Na^+$ alone ATP will neither initiate, support nor maintain opening in the light or the dark (Thomas, 1971).

*$Na^+$ in relation to stomatal regulation in CAM plants*
Plants with a crassulacean acid metabolism show metabolic and physiological adaptations to change in their environment. One such adaptation is that their stomata can open in the dark and close in the light. Net $CO_2$ fixation takes place by the carboxylation of PEP to form malate via PEP carboxylase. The dark opening of stomata found in the presence of $Na^+$ might be another of the adaptive mechanisms selected for adjustment to saline and arid environments (see Ch. 10). Fig. 12.5 shows the response of stomata on *Kalanchoe marmorata* (known to open its stomata in the dark and close them in the light) epidermal strips to $Na^+$ (Nishida, 1963). The stomata open in the dark and close in the light, similar to those on tobacco epidermal strips in the presence of $Na^+$ alone (Thomas, 1970a).

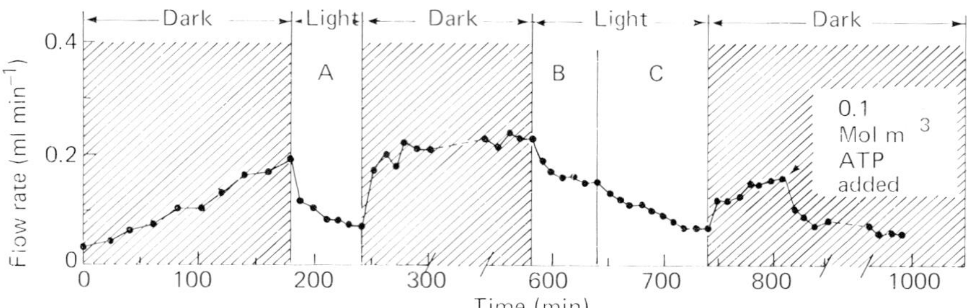

Fig. 12.5. Effect of light, dark, and ATP on the opening of *Kalanchoe marmorata* stomata bathed in 10 Mol m$^{-3}$ NaCl. Light fluxes at A and B were 226 W m$^{-2}$ and at C 334 W m$^{-2}$. (From Thomas, 1970a.)

Increasing the incident light flux increases the extent of closure. The addition of ATP in the dark, as in the case of the Na$^+$-supported opening of tobacco stomates, causes stomatal closure. The plants of *K. marmorata* used were grown on a saline river gravel and no sensitivity of the stomata to K$^+$ in the light or dark was found. In *K. marmorata* adaptations might occur which could gear the dark opening and light closing to Na$^+$ influxes and effluxes. A dark Na$^+$ influx might be caused by a process of H$^+$/Na$^+$ exchange diffusion in which H$^+$ was effluxed from the guard cells in a process of pH regulation and Na$^+$ influxed to balance the negative charges of the acids formed during CAM. In the light the increased supply of energy substrates through photosynthetic reactions e.g. ATP, stimulates a Na$^+$-efflux mechanism which leads to closure (Thomas, 1970a). The action of light would also result in the removal of the organic acids by their incorporation into polysaccharides. Very similar results have been found in *Mesembryanthemum crystalinum* when subjected to saline conditions by Winter and von Willert (1972), suggesting that not only are metabolic changes possible in adaptation to environmental salinity but also that the physiology of the stomata can adapt to give a reversed cycle of opening and closing. Willmer and Pallas (1973) could only detect K$^+$ accumulation in the guard cells of non Na$^+$-adapted *Kalanchoe blossfeldiana* and *K. pinnate*. It would be interesting to study in CAM plants the balance between the K$^+$ and Na$^+$ concentrations in the environment which lead to changes in metabolism and stomatal behaviour.

*Calcium and magnesium*
Though there seems to be little or no uptake of the two major physiologically active divalent cations, Ca$^{2+}$ and Mg$^{2+}$, by guard cells (Humble and Raschke, 1971) their presence in bathing solutions can effect opening. Reductions in

aperture have been associated with the presence of both $Ca^{2+}$ and $Mg^{2+}$. Using epidermal strips of *C. communis*, Fujino (1967) and Willmer and Mansfield (1969) found that $Ca^{2+}$ at a concentration of 1 Mol m$^{-3}$ caused almost complete closure. With tobacco epidermal strips, the addition of 1.0, 0.5 and 0.1 Mol m$^{-3}$ $Ca^{2+}$ to 10 Mol m$^{-3}$ KCl bathing solutions gave 74, 58 and 40% reductions in aperture, respectively (Thomas, 1970a). In contrast, in *V. faba* epidermal strips floated on 10 Mol m$^{-3}$ KCl, $Ca^{2+}$ at a concentration of 1.0 Mol m$^{-3}$ (Willmer and Mansfield, 1969) and 0.5 Mol m$^{-3}$ in the initial work reported by Fischer (1968) did not change the aperture. Subsequent work by Fischer (1972) revealed that $Ca^{2+}$ did have an inhibitory effect on stomata of *V. faba*. The inhibitory effect of $Ca^{2+}$ was most marked at low $K^{+}$ concentrations (0.01 Mol m$^{-3}$) and tended to decrease at high external $K^{+}$ concentrations (100 Mol m$^{-3}$) suggesting that $Ca^{2+}$ might compete with and exclude $K^{+}$ from uptake sites or from a region such as a Donnan free space (see Ch. 1) near the site of $K^{+}$ uptake. Using epidermal strips of *V. faba*, Pallaghy (1970) found that at concentrations of 10 Mol m$^{-3}$ KCl or NaCl, the addition of $Ca^{2+}$ reduced the light-stimulated opening in NaCl, but at higher concentrations (50 Mol m$^{-3}$) opening was increased in NaCl with respect to that obtained in the KCl solutions, which does not suggest that $Ca^{2+}$ aids consistently in a discrimination against $Na^{+}$ compared to $K^{+}$. It seems that the range of responses shown by stomata to the presence of $Ca^{2+}$ may be similar to the range of effects shown by $Ca^{2+}$ on $K^{+}$ uptake in different plant species or the adaptations in response to $Ca^{2+}$ possible within a single plant species due to differences in the growing conditions, age, etc. (see Ch. 9).

$Ca^{2+}$ is not very mobile in plants and it seems that to establish its importance in stomatal movement, some idea should be obtained of its range of concentrations in vivo in the guard cell environment. The effect of $Mg^{2+}$ on the stomatal mechanism has not been investigated as extensively as that of $Ca^{2+}$, though it would be interesting to compare its effect with that of $Ca^{2+}$ within a range of species. With tobacco epidermal strips it was found that the presence of 0.5 Mol m$^{-3}$ $Mg^{2+}$ reduced aperture by 37% (Thomas, 1970a). In *C. communis* epidermal strips 1.0 Mol m$^{-3}$ $Mg^{2+}$ reduced the aperture by 50% (Willmer and Mansfield, 1969). In the same species it was found that $Mg^{2+}$ caused no reduction in aperture until the external concentration reached 10 Mol m$^{-3}$ (Fujino, 1967).

### 12.3.2 Anions

One of the features found with the $K^{+}$-stimulated opening of stomata on epidermal strips of *V. faba* and tobacco was that although a high specificity was found for opening at low concentrations of $K^+$, little specificity was shown for the anion associated with $K^+$, opening being equally supported by equivalent concentrations of $K^+$ in solutions of KCl, $K_2SO_4$, KBr and $KNO_3$. Even when $K^+$ was associated with anions that are considered to be non-absorbed (e.g. iminodiacetate, 4,4-dimethyl-4,7-diazadecane-1,10-disulphonate and benzenesulphonate) the opening was the same as that on an equivalent concentration of $K^+$ supplied as KCl (Raschke and Humble, 1973). In the electron microprobe studies of *V. faba* guard cells it was found that only 5% of the positive charge of $K^+$ was balanced by inorganic anions (Humble and Raschke, 1971). Findings such as these have led many workers to suggest that electroneutrality for the accumulated $K^+$ is provided mainly by organic anions resulting from starch catabolism. In *Z. mays* approximately 40% of the charge balance has been found to be provided by $Cl^-$ (Raschke and Fellows, 1971).

Increases in the organic acid content of the guard cells of *V. faba* have been found with opening (Allaway, 1973; Fig. 12.6). The relationship between $K^+$ accumulation and increase in aperture is linear as has been found by other workers (see Section 12.3.1). This is also true for the appearance of malate though it does not parallel the accumulation of $K^+$. The discrepancy in the balance between $K^+$ and malate becomes greater with increasing aperture. It was estimated that for an accumulation in the guard cell of 390 equiv. $m^{-3}$ $K^+$, malate was present at 230 equiv. $m^{-3}$ with respect to $K^+$. The remaining 160 equiv. $m^{-3}$ of $K^+$ was not accounted for in this balance.

Measurements of the malate content of epidermal tissue from *Commelina cyanea* and *V. faba* by Pearson (1973) gave the minimum and maximum guard cell malate concentrations as 0.8 and 2.0 pMol guard cell$^{-1}$. These concentrations were sufficient to balance the charge over the range of $K^+$ concentration found by other workers in the guard cells of closed and open stomata. Concentrations of organic compounds in the epidermis showed little association with those in the mesophyll and thus the epidermis functioned as an isolated unit. This stresses the independence of the guard cells from the metabolism of the mesophyll and the fluxes associated with the guard cells must take place between them and the surrounding epidermal cells, or more probably their free-space (or apoplast) solution in the guard and epidermal cell walls.

As most of the epidermal chloroplasts and mitochondria are found in the guard cells, these must be metabolically the most active cells of the epidermis and the source of malate and other organic acids needed to balance $K^+$ uptake by the guard cells. If $CO_2$ fixation in the guard cells is not rapid enough to synthesise anions at the rate required to balance the $K^+$ uptake, catabolism

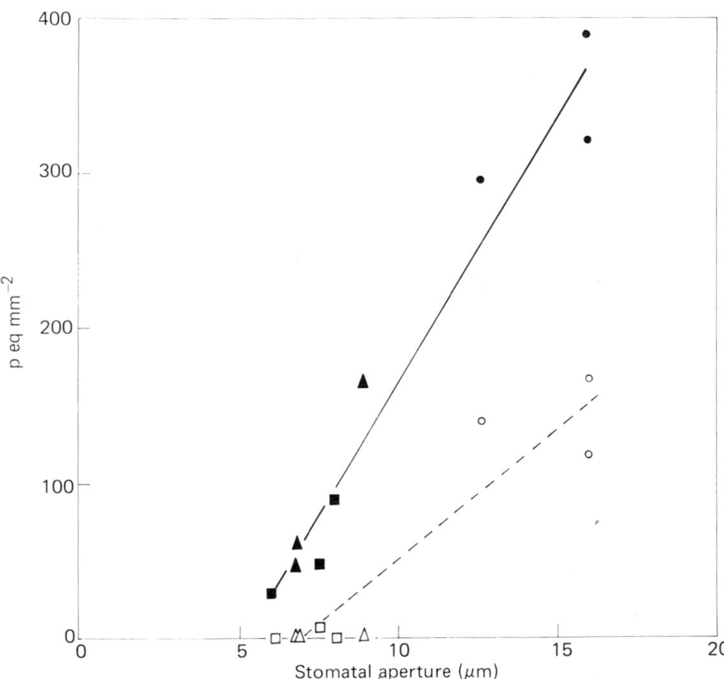

Fig. 12.6. Potassium (filled symbols and solid line) and malate levels (open symbols and broken line) in epidermal strips of *Vicia faba*, taken from leaves treated with light (circles), darkness (squares), or from leaves at the start of experiments (triangles). The fitted regression lines are somewhat conjectural, since points for intermediate stomatal apertures were not obtained. (From Allaway, 1973.)

of starch and polysaccharide reserves is the most likely source. However, if the $C_4$ pathway was the major synthetic way for $CO_2$ fixation in the guard cells, the synthesis of malate via PEP carboxylase could make an important if not a major contribution to the anion balance.

For a full understanding of how charge neutrality is attained with $K^+$ accumulation in the guard cells it seems that a full anionic balance must be made of guard cells in the open and closed state. Malate may not be the only balancing anion and a range of organic anions could be envisaged, depending on the metabolic fluxes taking place in the guard cell at any one time. The extent to which organic or inorganic anions balance $K^+$ may be adaptive in nature and depend on environmental conditions, the metabolic state prevailing in the guard cells or on species differences. In their studies of *V. faba* stomata, Pallaghy and Fischer (1974) found that the guard cells accumulated approximately 3 $K^+$ for every $Cl^-$ during opening, a much higher $Cl^-$-to-$K^+$ balance than that found in the same species by Humble and Raschke (1971).

There are still many aspects that have to be considered with respect to the metabolism of organic anions. What happens to them when $K^+$ is effluxed from the guard cells during closure? Are they re-incorporated into the guard cell polysaccharides? If so, is the rate of polysaccharide synthesis fast enough to account for closure, as this latter process is very rapid? Or are they effluxed together with $K^+$? If they are effluxed, are they stored in the guard cell walls or in the surrounding subsidiary cells? From such a storage are they readily accessable to the guard cells, or are they translocated to and metabolized by other cells? (see Section 12.5). Probably the hardest question to answer is whether it is the start of $K^+$ accumulation that triggers the synthesis of organic anions or if the reverse occurs.

*Organic anions and $H^+/K^+$ exchange*
Associated with the accumulation of $K^+$ and the appearance of malate and other organic acid ions in the guard cells is an efflux of $H^+$ which is proportional to the extent of opening (Raschke and Humble, 1973). It has been shown that, though the exchange capacity of epidermal strip material for $H^+$ is large, $H^+$ ions are exchanged one-for-one for $K^+$ during opening and the malic acid formed in the guard cells during stomatal opening was the most likely source of the $H^+$ effluxed.

The exchange of $K^+$ for $H^+$ must be part of a mechanism to regulate the internal pH of the guard cells. On osmotic grounds the accumulation of organic acids e.g. malic, would be approximately as effective as the accumulation of $K^+$ and malate in reducing the guard cell water potential but the accumulation of probably greater than 500 Mol m$^{-3}$ $H^+$ would reduce the internal pH of the guard cell solution to seemingly impossible low levels. Increases in guard cell pH from 5 to 6 have been associated with stomatal opening (Fujino, 1967) which might be explained by the accumulation of $K^+$ and malate.

### 12.3.3 Ion accumulation and guard cell osmotic relations

All recent work has shown that stomatal opening is assisted by an accumulation of $K^+$ and anions in the guard cells. Can this accumulation reduce the guard cell water solute potential sufficiently to give the influx of water required to bring about the increase in the guard cell volume required to cause opening? Using conventional methods to determine the incipient plasmolysis, Humble and Raschke (1971) found that the osmotic potential was $-19$ bars for those of closed and $-35$ bars for those of open stomatal guard cells of *V. faba*. For this 16-bar decrease in guard cell osmotic potential, there was an $84\%$ increase in guard cell volume and a 10-$\mu$m increase in aperture. The associated increase in guard cell $K^+$ was $404 \times 10^{-14}$ Mol (Table 12.1). The relationship between guard cell osmotic potential and stomatal aperture has been found to be linear with a 1.24 bar $\mu m^{-1}$ change in aperture for stomata on *V. faba* epidermal strips (Allaway and Hsiao, 1973). It therefore seems that the $K^+$ accumulation together with that of an associated anion can adequately account for the decrease in guard cell osmotic potential and the increase in volume found on stomatal opening. The results suggest that $K^+$ must be free in the guard cell solution to exert its full osmotic potential. Penny and Bowling (1974) have calculated that the influx of water into the guard cells during stomatal opening is well within the electro-osmotic efficiency of $K^+$.

## 12.4 Compounds regulating stomatal aperture

### 12.4.1 Carbon dioxide

The presence of $CO_2$ reduces both $K^+$ accumulation and the extent to which stomata open. Fischer and Hsiao (1968) found that stomata on *V. faba* epidermal strips opened to a greater extent in the dark in $CO_2$-free air than in the light with normal air when floated on KCl solutions, suggesting that $CO_2$ at the low concentrations found in normal air might have as great an influence on stomatal opening as light. Table 12.3 gives the response in aperture, $K^+$ and $Cl^-$ content of stomata on *V. faba* epidermal strips floated on 10 Mol $m^{-3}$ KCl under light and dark conditions surrounded by normal or $CO_2$-free air (Pallaghy and Fischer, 1974). As in the other studies (see Section 12.3) there is a correspondence between the extent of opening and the $K^+$ content. Light and the absence of $CO_2$ give similar increases in aperture and $K^+$ content. The only difference found was that the light response saturates at a rate 3 times or greater than the response to $CO_2$-free air.

Increasing concentrations of $HCO_3^-$ added to a 10 Mol $m^{-3}$ KCl bathing solution in which stomata on tobacco epidermal strips were opened in the light, caused a reduction in aperture (Fig. 12.7). When the strip was flushed and returned to a $HCO_3^-$-free 10 Mol $m^{-3}$ KCl solution there was a rapid increase in aperture. This is similar to the response of stomata on leaves to changing $CO_2$ concentrations in the atmosphere (Raschke, 1966). Collectively the evidence shows that the effect of $CO_2/HCO_3^-$ on opening is rapid, quantitative and reversible and that it is effective by reducing the concentration of $K^+$ in the guard cells. This might occur by (a) increasing the $K^+$ efflux,

Table 12.3.

Responses of stomatal aperture and of $K^+$ and $Cl^-$ content of guard cells to light to $CO_2$-free air. Epidermal strips of *Vicia faba* floated on 10 Mol m$^{-3}$ $^{42}$K $^{36}$Cl for times shown. Treatments: 1 = dark + normal air; 2 = dark + $CO_2$-free air; 3 = light + normal air; 4 = light + $CO_2$-free air. (Pallaghy and Fischer, 1974.)

| Experiment (replicates) | Duration of treatment (min) | Treatment no. | Aperture ($\mu$m) | Ion content ($\mu$Mol m$^{-2}$) | | |
|---|---|---|---|---|---|---|
| | | | | $K^+$ | $Cl^-$ | $Cl^-/K^+$ |
| 37 | 400 | 1 | 6.1 | 69 | 7.0 | 0.10 |
| (4) | | 2 | 8.2 | 143 | 9.3 | 0.07 |
| | | 3 | 8.2 | 157 | 15.7 | 0.10 |
| | | 4 | 10.3 | 223 | 23.0 | 0.10 |
| 64 | 270 | 1 | 7.8 | 51 | 5.6 | 0.11 |
| (2) | | 2 | 8.5 | 99 | 11.5 | 0.12 |
| | | 3 | 8.0 | 76 | 11.8 | 0.16 |
| | | 4 | 11.0 | 173 | 39.7 | 0.23 |
| 65 | 115 | 1 | 7.3 | 43 | 3.1 | 0.07 |
| (4) | | 2 | 7.6 | 67 | 12.0 | 0.18 |
| | | 3 | 8.4 | 72 | 12.0 | 0.18 |
| | | 4 | 10.0 | 126 | 30.3 | 0.24 |
| 71 | 90–135 | 1 | 5.7 | 45 | 15.4 | 0.34 |
| (3) | | 2 | 6.4 | 74 | 22.8 | 0.31 |
| | | 3 | 7.3 | 65 | 21.5 | 0.33 |
| | | 4 | 10.2 | 159 | 76.7 | 0.48 |

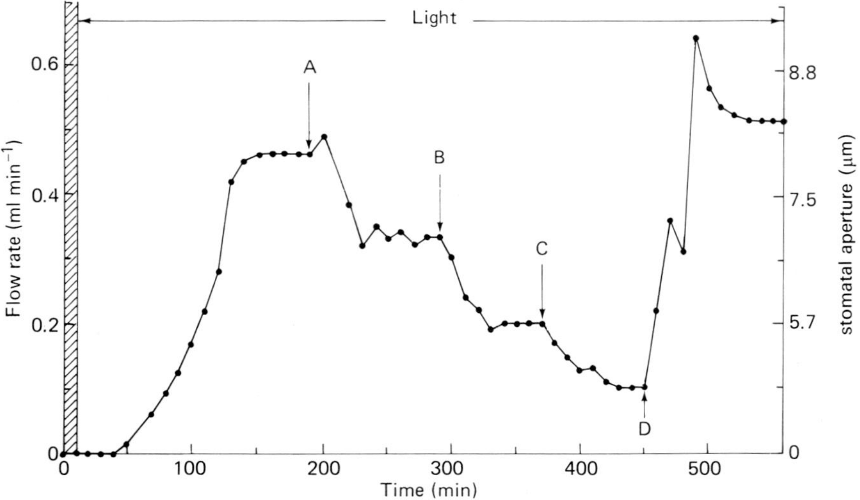


Fig. 12.7. Effect of adding increasing concentrations of $KHCO_3$ (A, $0.47 \times 10^{-2}$ Mol m$^{-3}$; B, $0.94 \times 10^{-2}$ Mol m$^{-3}$; C, $1.4 \times 10^{-2}$ Mol m$^{-3}$) and its subsequent removal (D) on stomatal aperture. Basal bathing solution 10 Mol m$^{-3}$ KCl. (From Thomas, 1972.)

(b) decreasing the $K^+$ influx, or by a combination of (a) and (b) which could occur either by effects on uptake mechanisms or by alterations in the guard cell permeability or (c) diverting a limited supply of ATP and/or other energy substrates in the guard cells to carbon fixation and away from ion-accumulation processes and metabolic requirements that might be needed for ion accumulation. As the fixation of $CO_2$ has a higher energy and ATP-requirement (114–115 Kcal Mol$^{-1}$ $CO_2$ reduced) than that estimated by Raven (1967) for processes of ion uptake (15–16 Kcal Mol$^{-1}$ of ions accumulated) it might be considered that while carbon is being fixed in the guard cells the supply of energy substrates, such as ATP, to a $K^+$-accumulation process could be reduced. This then leads to an efflux of $K^+$ until a new equilibrium between influx and efflux is reached depending on the amount of energy available to the accumulation process. Added energy substrates (e.g. ATP, PEP) are effective in increasing stomatal aperture in the presence of $CO_2/HCO_3^-$ both in the light and dark and it seems fairly certain that increasing the energy substrate available to guard cells (also by increasing the flux of light energy) can reverse the stomatal closing caused by $CO_2/HCO_3^-$). Increases in the energy available could increase both the net influx of $K^+$ and the amount of carbon fixation and by these means help reverse the effects of $CO_2/HCO_3^-$. However, this does not explain how $CO_2/HCO_3^-$ can also reduce the opening supported by $Na^+$ (Thomas, 1972). The closing of $Na^+$-supported opening requires energy (see p. 389) and if $CO_2/HCO_3^-$ drew energy away from a $Na^+$-efflux mechanism it might be expected that more $Na^+$ would accumulate in the guard cells and cause an increase in aperture rather than the observed decrease. The effect of $CO_2/HCO_3^-$ might be due to increases in guard cell membrane permeability. Increasing concentration of $CO_2/HCO_3^-$ might also reduce the concentration of malate in the guard cells as a result of a combination to form pyruvate, PGA and then sugars. In this way some of the anion charge balance provided by malate (see p. 392) for $K^+$ and $Na^+$ could be lost which might lead to an efflux of $K^+$ and $Na^+$, decreases in guard cell volume and reductions in aperture. As stomatal aperture and the accumulation of $K^+$ in the guard cells is so sensitive to the concentration of $CO_2/HCO_3^-$ in the environment, it is often very hard to judge if other compounds that regulate do so in their own right or by in some way regulating the $CO_2/HCO_3^-$ concentration.

### 12.4.2 Abscisic acid (ABA)

This naturally occurring hormone, the synthesis of which is increased markedly under stress conditions e.g. water and salinity (Wright and Hiron, 1969;

Mizrahi et al., 1970) has been found to reduce stomatal aperture and the accumulation of $K^+$ in guard cells (Mittleheuser and Van Stevenink, 1969; Jones and Mansfield, 1970; Horton and Moran, 1972). A list of species in which ABA has been found to be effective in reducing stomatal aperture and transpiration has been tabulated by Livne and Vaadia (1972). No case has been reported of the ineffectiveness of applied ABA in reducing transpiration. The effect on stomata of ABA applied at a concentration of 0.1 Mol m$^{-3}$ to tobacco leaves through the petiole is rapid, causing almost complete closure within less than an hour. This effect is not due to an increase in the intra-cellular space $CO_2$ concentration as it is not reversed by $CO_2$-free air (Jones and Mansfield, 1970). Similar results have been reported by Loveys et al. (1972).

A genetic inability to synthesise ABA in a tomato mutant has been correlated with a lack of stomatal control under water-stress conditions. Supplying ABA to the leaves as a spray or in the root medium to the mutant resulted in ability to reduce stomatal aperture with continued treatment. If treatment was not continued, stomatal control was again lost in a few days (Tal, 1966; Imber and Tal, 1970) indicating that ABA might be the hormone synthesized and released into the general circulation of plants to enable them to adjust to water stress.

Of interest here is how ABA could regulate the accumulation of ions, particularly $K^+$ in the guard cells. Horton and Moran (1972) showed that the presence of ABA inhibited the influx of $K^+$ into the guard cells and similar effects have been found in other plant tissues (Reed and Bonner, 1974), leading to the suggestion that ABA is acting specifically on ion uptake mechanisms and that the effect of ABA is not due to a general change in membrane perm-eability since the uptake of organic molecules was affected to a much lesser extent.

ABA has been reported to reduce the activity of $\alpha$-amylase (Hemberg, 1967). Thus ABA, by its effect on $\alpha$-amylase activity, could inhibit starch hydrolysis which might prevent the synthesis of the organic anions such as malate, required to balance the charge of the $K^+$ accumulated during opening.

The action of ABA on stomata can be very rapid, closing and opening being initiated within 5 min of its addition and withdrawal at concentrations of 10$^{-4}$ Mol m$^{-3}$ ABA, specific for the *cis-trans* isomer. The effect of ABA was not due to an inhibition of photosynthesis, hence it was considered that ABA acted directly on the stomata (Cummins et al., 1971). The fast action of ABA on the stomata suggested its action was unlikely to be on enzyme activity or on a membrane function. In a bioassay of the effect of ABA concentration on the transpiration rates of excised barley leaves, Cooper and Digby (1972) found that the effect of ABA on transpiration is strikingly similar to that found by Thomas (1970b) for the effect of ouabain on stomatal apertures. This leads to an idea that ABA may be exercising its effect at the level of ion transport mechanisms such as ATPases and proton transfer processes.

It seems that the effect of ABA on stomatal movement is at the level of the guard cells, but the actual site within the guard cell is a matter of conjecture. A knowledge of how such a natural regulation of stomatal aperture affects guard cells would be of great value in understanding the major guard cell functions effective in regulating stomatal movement.

### 12.4.3 Phytotoxins

*Fusicoccin*

Produced by *Fusicoccum anygdali* Del., fusicoccin gives wilt symptoms in several plant species, induces stomatal opening and increases transpiration (Graniti, 1964; Turner, 1972; Heichel and Turner, 1972). Fusicoccin induces stomatal opening in both the light and dark. In a study using *P. vulgaris* epidermal strips, the presence of $10^{-2}$ Mol m$^{-3}$ fusicoccin greatly increased the rate of K$^+$ accumulation in the light and caused K$^+$ to accumulate in the guard cells in the dark, a phenomenon normally only found in the light when epidermal strips are floated on 10 Mol m$^{-3}$ KCl. Using leaves on whole plants the same result was found. Opening the stomata with fusicoccin in either the light or dark also increased the K$^+$ concentration of the guard cells, the K$^+$ concentration in the subsidiary cells remaining unchanged. Addition of ABA could not reverse the opening caused by fusicoccin, though it quickly reduced the light-stimulated opening in the controls (Tucker and Mansfield, 1971; Turner, 1973). Turner (1973) proposes that the mechanism by which fusicoccin stimulates K$^+$ uptake may be similar to that of ionophores which facilitate the transfer of alkali ions across plant and animal membranes.

*Helminthosporium maydis pathotoxin*

Stressing the importance of genetic variation in K$^+$-accumulation mechanisms present in guard cells, it has been found in *Helminthosporium maydis*-susceptible Texas male-sterile cytoplasm maize that *H. maydis* toxin inhibited the light-stimulated K$^+$ uptake into the guard cells. No such inhibition was found in a resistant variety which contained normal cytoplasm (Arntzen et al., 1973). The action of the pathotoxin and those of ABA and a reported rapid partial depolarization in the membrane potential of corn root cells by the presence of the toxin are similar and its effects could be directly on the mechanisms by which K$^+$ is accumulated in the guard cells.

*12.4.4  Metabolic inhibitors*

*DCMU*

This is an inhibitor of Photosystem II open-chain or non-cyclic photophos-phorylation in chloroplasts and an inhibitor of photosynthetic $CO_2$ fixation. In *C. communis* epidermal strips the partial stomatal closure in the light caused by the presence of DMCU can be reversed in $CO_2$-free conditions, suggesting that the effect of DCMU is through an inhibition of $CO_2$ fixation which increases the $HCO_3^-/CO_2$ concentration in the environment of the guard cells (Willmer and Mansfield, 1970). For *V. faba* epidermal strips in a $N_2$ atmosphere in the light, $10^{-3}$ Mol m$^{-3}$ DCMU reduced stomatal aperture by about 20% and $K^+$ uptake by 50% compared to controls kept in the light and supplied with $CO_2$-free air (Humble and Hsiao, 1970). Using the same type of material, Pallaghy and Fischer (1974) report higher inhibitions of both opening and guard cell $K^+$ uptake, though they point out that there is a wide range in the reported effects of DCMU on stomata.

From the range of reported effects of DCMU it might be concluded that non-cyclic photophosphorylation could supply some of the energy for $K^+$ accumulation in the guard cells, but its contribution does not seem to be major and could depend on the environmental conditions under which the plants were raised, age of tissue, etc.

*DNP*

For *C. communis* epidermal strips floated on incubating media containing high ion concentrations, 132 Mol m$^{-3}$ $K^+$, the presence of 0.1 Mol m$^{-3}$ DNP reduced both the rate and extent of opening. After 3 h the aperture was similar to those on strips incubated in darkness in the absence of DNP. DNP failed to suppress dark ion-stimulated opening (Willmer and Mansfield, 1970) suggesting that dark, ion-stimulated opening is not dependent on metabolism and only occurs in the presence of high concentrations of ions in the bathing medium, as found by Humble and Hsiao (1969). Using a bathing solution of 10 Mol m$^{-3}$ KCl for *P. vulgaris* epidermal strips, Turner (1973) showed that 0.1 Mol m$^{-3}$ DNP totally inhibited light and the fusicoccin dark-stimulated opening. In *V. faba* epidermal strips, 0.1 Mol m$^{-3}$ DNP also completely inhibited light, ion-stimulated opening in $CO_2$-free air.

*FCCP and CCCP*

As potent inhibitors of both oxidative and photosynthetic phosphorylations, FCCP and CCCP completely inhibit light, ion-stimulated opening in $CO_2$-free air of stomata on *V. faba* epidermal strips. FCCP at a concentration of $4 \times 10^{-4}$ Mol m$^{-3}$ completely inhibits K$^+$ accumulation in the guard cells (Pallaghy and Fischer, 1974). Collectively these results suggest that phosphorylation reactions are important in the light-stimulated K$^+$ accumulation process of guard cells.

## 12.5 Energy coupling and ion transport mechanisms

### 12.5.1 Energy estimates for K$^+$ accumulation

Estimates of the osmotic work ($W$) required for the accumulation of an equivalent of K$^+$ in the guard cells during stomatal opening can be calculated from the relation

$$W = RT \ln (K_i^+/K_o^+) \tag{12.1}$$

where $R$ = gas constant, $T$ = absolute temperature, $K_i^+$ = concentration of K$^+$ inside the cell, $K_o^+$ = concentration of K$^+$ outside the cell. Experimental values of $K_o^+$ from 0.1 to 10 Mol m$^{-3}$ and resulting $K_i^+$ values of between 300 and 500 Mol m$^{-3}$ for the guard cells in a number of species gives a value of $W$ of 2000 to 5200 cal. Hence an estimated accumulation of about 2 pMol K$^+$ guard cell$^{-1}$ of an open stomate, requires approximately $6 \times 10^{-9}$ cal guard cell$^{-1}$ of energy expended during the process of opening. Humble and Raschke (1971) calculated that approximately $3 \times 10^{-9}$ cal guard cell$^{-1}$ are expended during opening. Stomata can open in about 30 min, requiring an energy expenditure of about $10^{-10}$ cal min$^{-1}$ guard cell$^{-1}$. Estimating the surface area of a guard cell to be $2 \times 10^{-8}$ m$^2$ and 0.1% efficiency of light energy conversion in the guard cell for K$^+$ accumulation, this amount of energy could be supplied by an incident flux density of approximately 10 cal m$^{-2}$ min$^{-1}$ which seems reasonable. Photosynthetically generated ATP could thus account for the rate of K$^+$ accumulation in the guard cells (Pallas and Dilley, 1972). These estimates only consider the energy expended in K$^+$ accumulation, the energy requirements of other processes connected with guard cell swelling and stomatal opening being unknown.

### 12.5.2 Energy sources

#### Light
In all plants so far studied, except for CAM types, light is the environmental stimulus associated with stomatal opening. It is now almost certain that the

opening effect of light is independent of a photosynthetic reduction of the $CO_2$ concentration in the environment of the guard cells, since $K^+$ accumulation in the light and efflux in the dark can occur in $CO_2$-free air, showing that guard cell $K^+$ fluxes and stomatal movement are independent of changes in the $CO_2/HCO_3^-$ concentration. Light specifically stimulates high $K^+$ accumulation from low external concentrations in the guard cells of *V. faba*, maize and tobacco and it appears there is a specific light-driven $K^+$-accumulation mechanism present in the guard cells of most plants. Light-driven mechanisms for ion uptake, particularly $K^+$ have been found in other plant cells (see Chs 4 and 5).

*Respiration and oxygen*
There are marked differences in reports of the effect of $O_2$ on stomatal movement. Both opening and closing can be inhibited by anaerobic conditions (Walker and Zelitch, 1963; Louguet, 1968) although Heath and Orchard (1956) found that anaerobic conditions did not affect stomated closing. Stomatal movements in maize are less affected by anaerobiosis than those of *Pelargonium* where both opening and closing in darkness or in the presence of $CO_2$ is almost completely inhibited in the absence of $O_2$ (Louguet, 1968). In *V. faba* leaf discs floated on water opening was markedly inhibited in a $N_2$ atmosphere, although the opening on epidermal strips floating on a 10 Mol m$^{-3}$ KCl solution in a $N_2$ atmosphere was only slightly reduced (Humble and Hsiao, 1970). Stomata on tobacco epidermal strips opened in the light in an $O_2$-containing 10 Mol m$^{-3}$ KCl solution, the equilibrium aperture attained was reduced by 34% in $O_2$-free conditions. The rate of opening in the light and closing in the dark is also reduced under $O_2$-free conditions (Thomas, unpublished results). In *V. faba* epidermal strips, both guard cell $K^+$ uptake and the light-opening response of the stomata are substantial and not affected by the absence of $O_2$ (Pallaghy and Fischer, 1974).

Taken overall, it seems that respiration can supply some of the energy required for stomatal opening and this respiration might be light stimulated. The energy for stomatal opening and closing may be supplied from a pool which receives supplies from many sources, the importance of each contributary source varying with different plant species and environmental conditions.

### 12.5.3 Energy coupling

Willmer and Mansfield (1970) have suggested that energy for active transport into the guard cells may be obtained from a direct linkage to mitochondrial electron transport together with an additional linkage to electron transport of Photosystem I. Using epidermal strips from *C. communis*, Fujino (1967) found that if ATP was added to a bathing medium containing $K^+$ opening was stimulated, suggesting that ATP was directly involved in providing the energy for ion-stimulated opening. Staining showed higher ATPase activity associated with the guard cells than that of the surrounding epidermal cells, suggesting that ATPases might gear the energy released by the hydrolysis of ATP to ion transport into the guard cells. Using ouabain, a specific inhibitor of membrane-bound ATPase transport systems in animal cells which does not affect the general energy metabolism of tissues, at concentrations greater than $10^{-2}$ Mol m$^{-3}$ the light-stimulated opening on tobacco epidermal strips could be reduced by 80%. The addition of ouabain causes rapid reductions and its removal a rapid recovery in aperture. This is consistent with the way in which ouabain is thought to inhibit cation transport i.e. by blocking attachment to a transport site such as a membrane-bound ATPase (see Ch. 2). When $K^+$ influx is blocked e.g. by ouabain, the efflux reduces the guard cell content of $K^+$ and results in a reduction of aperture or, in the presence of ouabain, the rate of efflux is increased (Thomas, 1970b). Similar results with ouabain have been found for light- or fusicoccin-stimulated opening on *P. vulgaris* epidermal strips (Turner, 1973). In contrast, Pallaghy and Fischer (1974) report slight but not significant reductions in stomatal aperture on *V. faba* epidermal strips using 0.5 Mol m$^{-3}$ ouabain. This is consistent with the range of response found for ion uptake processes by ouabain in both plant and animal cells (Thomas, 1970b; MacRobbie, 1971). When added to bathing solutions containing $K^+$, ATP can stimulate opening both in the light and dark on *C. communis* (Fujino, 1967) and tobacco (Thomas, 1971) epidermal strips. In the dark the extent of opening was dependent on the concentration of ATP, while the pattern of opening was similar to that in the light (Thomas, 1971). This suggests that increasing concentrations of ATP in the dark can simulate the effect of increases in the light flux density on opening. Other energy substrates (PEP in the dark and light, ADP in the light) can also stimulate $K^+$-supported opening. PEP might also help in the opening process by increasing the malate content of the guard cells (see Section 12.3). UTP did not support the $K^+$-stimulated opening in the light or dark. Replacing UTP by ATP resulted in opening, suggesting that a stimulation of opening was specific for ATP. These results and the ouabain inhibition referred to above suggest that $K^+$ uptake into guard cells might, to a certain extent, be coupled to the free energy released on the hydrolysis of ATP via a transport ATPase. In the presence of ATP, $Na^+$ can neither initiate nor maintain opening on tobacco epidermal strips. The addition of ATP to a NaCl solution in which stomata have already opened, brings about rapid closure and, while ATP is present in the solution, no opening is supported by $Na^+$. Similar results were observed when stomata having opened in the dark, while bathed in a 10 Mol m$^{-3}$ NaCl solution, are exposed to light (Thomas, 1971).

## 12.6 *Ion transport mechanisms*

It seems that during the process of stomatal opening there is more than one mechanism by which ions can enter the guard cells. Very few inhibitors completely prevent opening and if they do so it is at concentrations which could cause irreversible damage to cell metabolism and membrane structure.

On epidermal strips with closed stomata, the opening that occurs in a 10 Mol m$^{-3}$ NaCl solution is very slow but if partially opened stomata are exposed to this same solution or to one containing only $Na^+$, very rapid opening occurs under light or dark environmental conditions and these quick $Na^+$-induced openings are only reversible by the addition of ATP to the bathing solution (Thomas, 1971). Thus, once opening has reached a certain level, selectivity for $K^+$ is lost and $Na^+$ can also support rapid opening. Fast opening after an initial lag phase is a feature of the $K^+$-specific light opening (Thomas, 1970a). This might be due to an increase in permeability of the guard cell membrane as a result of the associated stretching of the membranes or due to the increased concentration of ions in the cytoplasm. With an increase in permeability, ions may move rapidly into the guard cells and diffusion may become important in controlling the influx and efflux of ions. The driving force for the influx of cations under these conditions might be the synthesis of organic anions such as malate (Allaway, 1973). An active efflux from the subsidiary cells may also contribute to the guard cell influx (Penny and Bowling, 1974). The general picture that evolves is that of an initial phase of opening associated with a light-activated $K^+$-specific mechanism, during which the contribution of a diffusion flux increases. When the stomata are almost fully open both the light-activated and diffusional influxes become important in determining the extent to which $K^+$ is accumulated.

The relationship between aperture and external $K^+$ within the range 0–10 Mol m$^{-3}$ shows the same type of saturation curve as that obtained for ion accumulation within cells with respect to increasing external concentration (e.g. Thomas, 1970a) and has been almost universally taken as indicative of a specific carrier mechanism. This mechanism is at least a process of facilitated diffusion, if not of active transport (see Ch. 1). The concentration giving half-maximal opening was estimated as 0.32 Mol m$^{-3}$ for tobacco (Thomas, 1970a) and 0.10 Mol m$^{-3}$ for *V. faba* stomata (Fischer, 1972), indicating that a high-affinity mechanism as defined by Epstein (1966) is operative in the accumulation of $K^+$ by the guard cells. The electrochemical determinations of Penny and Bowling (1974) indicate that the movement of $K^+$ to and from the guard cells of *C. communis* involve active transport mechanisms.

### 12.6.1 The $K^+$ uptake and accumulation process

Light energy is absorbed in various wave bands and this energy is used to increase the amount of energy substrate available in the guard cell by processes of photophosphorylation and photoreduction e.g. ATP, NADH, etc., which could help stimulate mitochondrial activity and the production of organic acids, particularly malate. $CO_2$ fixation is enhanced and contributes to an increase in organic acids and reduces the breaking effect of $CO_2$ on $K^+$ accumulation. However, the energy absorbed may also be utilized to bring about changes in membrane and cytoplasmic organization which aid ion transport to the vacuole. With the increase in ATP a $(H^+/K^+)$ transport ATPase situated in the plasmalemma which transfers $H^+$ to the outer surface and $K^+$ to the inner, is fuelled, these vectorial changes being brought about by the free energy of ATP hydrolysis. This is then repeated as the basic cycle for $K^+$ accumulation. Energy might be needed to form and reform sites specific for $H^+$ on the inside and $K^+$ on the outside. However, the influx of $K^+$ and efflux of $H^+$ need not be coupled (see Ch. 1). The primary process might be the electrogenic extrusion of $H^+$ by an $H^+$-translocating reversible ATPase setting up a negative potential inside the guard cell which generates a gradient for the movement of $K^+$ to the inside through a $K^+$-specific path. The extent to which the $K^+$ and $H^+$ fluxes are coupled may depend on species differences and adaptions to the environment in which plants grow; for example, an $H^+$-efflux mechanism might be connected to the regulation of pH in the cell which could account for some unexplained stomatal openings in the dark. When organic acid does not provide a major proportion of the charge balance for $K^+$ in the guard cells there may be an electrogenic mechanism associated with $K^+$ accumulation which could explain differences in behaviour of maize and *V. faba* (see Section 12.3.1).

Following uptake, $K^+$ and its associated anion (e.g. malate) would be transported through the cytoplasm. With ion accumulation and the swelling of the guard cell, permeability may increase and a diffusional influx of ions may become important during rapid stomatal opening which may not be specific for $K^+$ (Thomas, 1970a). Entry of $K^+$ by this means could account for $K^+$ accumulation which is not counterbalanced by malate production.

The site of action of compounds such as ABA and fusicoccin might be at the sites of ion uptake and transfer through the plasmalemma. For example, ABA has a selective action on protein and enzyme synthesis (Addicott and Lyon, 1969) which might affect carrier proteins and/or the activation and reactivation of a transport ATPase. The effects of fusicoccin have been considered to be associated with a stimulation of $H^+$ extrusion (Iado et al., 1974). This could lead to a stimulation of $K^+$ accumulation by the $H^+/K^+$ exchange mechanisms discussed above. However they might also effect the inter-relationships between ion and water transfer through membranes.

## 12.6.2 $Na^+$-associated stomatal movements

The characteristics of opening supported by $Na^+$ alone are different from those of $K^+$ (see Section 12.3). With $Na^+$ as the only cation present, stomata can open in the dark. In the presence of ATP no $Na^+$-supported opening occurs and the addition of ATP to stomata opened by the presence of $Na^+$ alone brings about rapid reductions in aperture. Light can bring about decreases in aperture of stomata opened by the presence of $Na^+$ in the dark. With the addition of $K^+$, the $Na^+$-induced dark opening can be reduced. It does not seem that the influx of $K^+$ and the efflux of $Na^+$ are necessarily coupled, as ATP can cause reductions of the $Na^+$-supported opening in the absence of $K^+$. However, $Na^+/H^+$ exchanges might occur on the hydrolysis of ATP. The influx of $Na^+$ might be associated with a mechanism for the regulation of the cytoplasmic pH of the guard cells. With the respiratory synthesis of organic acids in the dark or light, the dissociation of these acids would decrease the pH of the cytoplasm which could stimulate an $H^+$ efflux mechanism leaving the cytoplasm negatively charged, which would in turn give the potential for the diffusion of $Na^+$ into the guard cells. Again, with swelling of the guard cells diffusional entry of $Na^+$ into the cell could increase, giving the observed increased rate of opening. When the level of ATP is adequate, a $Na^+$-efflux mechanism is fuelled and either the formation of organic acids slowed or resynthesis to polysaccharides takes place. $K^+$ also has a role in controlling the $Na^+$-supported opening which might be by a regulation of respiration and organic acid synthesis in the dark. $Na^+$-supported opening in the dark and a light-activated efflux of $Na^+$ might be important in the dark opening – light closing in CAM plants (see Section 12.3).

## 12.6.3 Guard cell ion efflux and stomatal closing

The characteristics and mechanism of the ion effluxes from the guard cells associated with closing have not been adequately studied. The process of

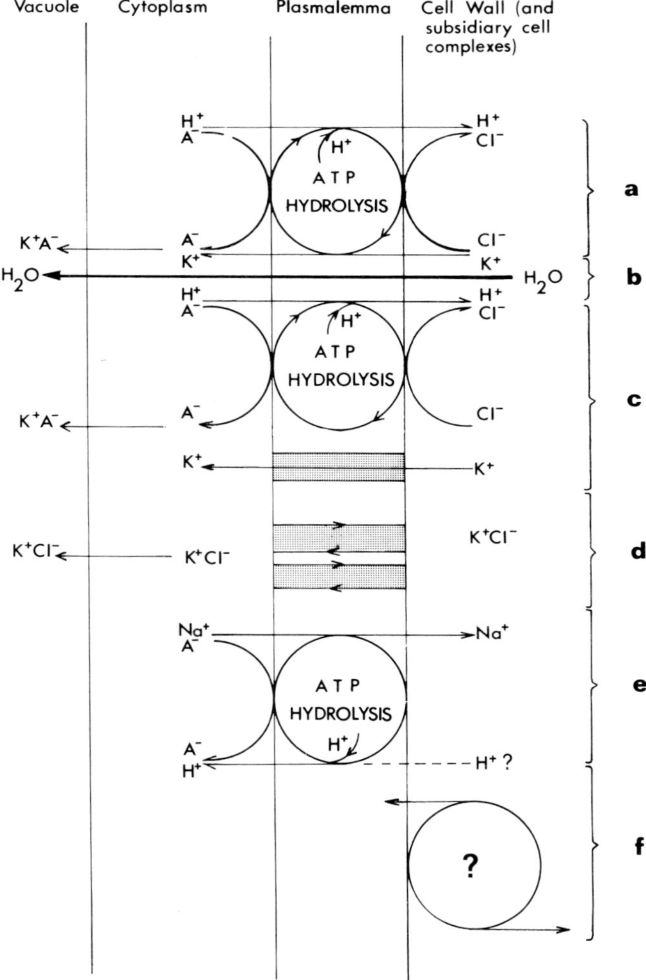



Fig. 12.8. A tentative model for the mechanisms controlling the ion fluxes associated with guard cells. ATP is provided by phosphorylation in mitochondria and chloroplasts within the guard cells. Anions ($A^-$) are provided by organic acid synthesis in these organelles. The postulated fluxes are: (a) direct ($H^+/K^+$) transfer ATPase; (b) electro-osmotic $H_2O$ transfer; (c) $H^+$ transport ATPase plus $K^+$ exchange through a $K^+$ 'specific pore': (d) diffusional paths of ion entry; (e) $Na^+$-efflux mechanism; (f) unknown contribution of subsidiary cell ion influx and efflux processes to guard cell fluxes.

closing is just as important to our understanding of stomatal control as that of opening. In certain stress situations e.g. a sudden water stress, the rate of closure could mean the difference between plant survival and death. In stomata surrounded by intact epidermal cells, the initial phase of closing can be as rapid as that of the fastest phase of opening, after which the rate of closing slows down.

From the closing effects caused by darkening and metabolic inhibitors, the sequence of closing might be modelled as follows. The equilibrium aperture for opening is reached when the influx and efflux of ions in the guard cells is balanced. When the supply of energy to the influx mechanisms is cut, there is a net efflux of ions and associated water which reduces the guard cell volume and aperture. The initial fast phase of closing may correspond to a fast diffusional efflux of ions. This might be expected from (i) the high concentration gradient, 400 Mol m$^{-3}$ for K$^+$, that can initially exist between the guard cell and its surroundings; (ii) the high pressure existing on the guard cells due to turgor and the reaction of the surrounding epidermal cells; (iii) the possibility that the permeability of the guard cell is high when it is fully swollen.

During the fast closing phase there could be an efflux of both K$^+$ and anions so that the charge balance in the guard cell is maintained. Possibly the anions effluxed during this stage would be mainly inorganic e.g. Cl$^-$, though some organic might also be effluxed depending on the relative proportions of inorganic/organic anions in the guard cell and also on the adaptional and species differences e.g. *V. faba* and maize. The slowing of the closing rate could correspond not only to reductions in the efflux rate due to decreases in the diffusion potential, the pressure on the guard cells and their permeability but also may correspond to the time required to incorporate the organic anions associated with K$^+$ into polysaccharides. For charge to be balanced at this stage, an influx of H$^+$ for the K$^+$ effluxed might become important which may need some metabolic assistance from respiratory energy to bring about charge separations and ion transfers or even for the production of H$^+$. This might explain the inhibition of closure by anoxia reported by some workers e.g. Louguet (1968). Differences in species may again be important. In maize, for example, much more of the accumulated K$^+$ is balanced by Cl$^-$ (Raschke and Fellows, 1971) and anoxia is less effective in preventing closure.

As rates of K$^+$ efflux from open guard cells are low, Fischer (1972) considered that equilibrium opening was not brought about when the influx equalled the efflux but by a slowing down of the influx due to the accumulation of ions. However, his measured rates of efflux were from guard cells on epidermal strips in which most of the surrounding epidermal cells were destroyed and did not provide the back pressure that would normally be expected from these cells. Stomata opened in the light on epidermal strips with a low percentage of intact epidermal cells, showed greatly reduced rates and extents of closing on darkening (Humble and Hsiao, 1970). The work of Penny and Bowling (1974) suggests that active accumulation by the subsidiary cells might contribute to the efflux from the guard cells. This raises the interesting question as to whether environmental changes e.g. light, dark, concentration of $CO_2$ etc. regulate the direction of ion, and perhaps energy substrate, fluxes between the guard and subsidiary cells.

## 12.7 *Concepts, conclusions and future research*

Two broad concepts come to mind:

(i) Guard cells are excitable cells and excitability is coupled to ion fluxes. In this respect, they are similar to nerve and muscle cells and many of the methods used and concepts developed in neuromuscular physiology might be applicable to the study of guard cells as part of a comparative study of the evolutionary development of excitable cells.

(ii) Guard cells in many ways behave like younger plant cells in that they remain plastic, respond to hormones, accumulate and excrete ions, etc. It seems that a comparative study of these cells would also be very worthwhile. The response of stomata exposed to similar conditions can be variable, a feature we might expect to find in excitable cells that have been selected to respond to a large range of changing environmental conditions and in plants kept in an equilibrium state. In such cells we might expect a range of adaptive and inbuilt regulatory systems. At the enzyme and membrane level there could be rapid adaptive changes in response to environmental fluctuations.

Inbuilt natural rhythms in stomatal activity are also present which could change the response of stomata to a given stimulus at any given time. These inbuilt rhythms may have a daily or annual periodicity. For example, under constant environmental conditions, daily endogenous rhythms have been found in stomatal aperture (Pallas, 1969). The stomata of plants grown in constant environmental chambers open more quickly and reach wider apertures in plants grown in the spring-to-autumn period than those grown in the winter (Seidman and Riggan, 1968). The sensitivity of stomata to ouabain has been reported to be greater in plants grown in the mid-spring to mid-autumn period (Thomas, 1970a).

In the field of adaption, it would be interesting to study the workings of the 'circadian clock' that has been found in stomatal activity, in the light of the ideas put forward by Njus et al. (1974) of a membrane model for this clock.

Possibly the most useful research in the near future would be to try to define the range of adaptations that are possible in the stomatal mechanism in response to environmental changes. It seems that this would involve studying the epidermal, subsidiary and guard cell metabolic interrelations, hormone levels and their effect together with the ion fluxes that occur during adaptation to different environments e.g. light, ion concentration and composition (e.g. conditions of salinity), changes in water potential, etc.

Studies of the possible changes could be made at the cytological, physiological and biochemical level. Electron microscope studies on the changes that occur in the guard cell during opening and how these could change in various environments, are badly needed. At the physiological level, the changes in response to various ions and ion concentrations could be studied. At the biochemical level, changes in enzyme activities and characteristics, particularly of the ATPases, would be very interesting. These studies would have great practical importance as they could give us a clearer understanding of how plants adjust to environmental fluctuations and provide us with metabolically reasonable and ecologically acceptable methods to regulate stomatal aperture to some extent.


## Acknowledgements

I deeply thank Mlle. Wilma Ord to whom I owe a great debt for her constant help, support and encouragement. I thank Dr. Phillippe Louguet for helpful discussions and the loan of his bibliography after my manuscript and bibliography were lost.


## References


F.T. ADDICOTT and J.L. LYON, Annu. Rev. Plant Physiol., 20 (1969) 139.

W.G. ALLAWAY and T.C. HSIAO, Aust. J. Biol. Sci., 26 (1973) 309.

W.G. ALLAWAY, Planta, 110 (1973) 63.

C.J. ARNTZEN, M.F. HAUGH and S. BOBICK, Plant Physiol., 52 (1973) 569.

H. BRAG, Physiol. Plant., 26 (1972) 250.

R.S. COCKRELL, E.J. HARRIS and B.C. PRESSMAN, Biochemistry 5 (1966) 3919.

M.J. COOPER and J. DIGBY, Planta, 105 (1972) 43.

W.R. CUMMINS, H. KENDE and K. RASCHKE, Planta, 99 (1971) 347.

E. EPSTEIN, Nature, 212 (1966) 1324.

K. ESAU, Plant Anatomy (1965) 2nd Ed., John Wiley and Sons, New York.

R.A. FISCHER, Science, 160 (1968) 784.

R.A. FISCHER and T.C. HSIAO, Plant Physiol., 43 (1968) 1953.

R.A. FISCHER, Aust. J. Biol. Sci., 25 (1972) 1107.

M. FUJINO, Sci. Bull. Fac. Educat. Nagasaki Univ., 18 (1967) 1.

A. GRANITI, in Z. Kiraly and G. Ubrizsy (Eds.) Host–Parasite Relations in Plant Pathology (1964) Research Institute for Plant Protection, Budapest, p. 211.

J. GUTKNECHT, Science, 160 (1968) 68.

O.V.S. HEATH, New Phytol., 37 (1938) 385.

O.V.S. HEATH, in F.C. Steward (Ed.) Plant Physiology (1959) Vol. II, Academic Press, New York, p. 193.

O.V.S. HEATH and B. ORCHARD, J. Exp. Bot., 7 (1956) 313.

G.H. HEICHEL and N.C. TURNER, Physiol. Plant Pathol., 2 (1972) 375.

T. HEMBERG, Acta Chem. Scand., 21 (1967) 1665.

R.F. HORTON and I. MORAN, Z. Pflanzenphysiol., 66 (1972) 13.

T.C. HSIAO, W.G. ALLAWAY and L.T. EVANS, Plant Physiol., 51 (1973) 82.

G.D. HUMBLE and T.C. HSIAO, Plant Physiol., 44 (1969) 230.

G.D. HUMBLE and T.C. HSIAO, Plant Physiol., 46 (1970) 483.

G.D. HUMBLE and K. RASCHKE, Plant Physiol., 48 (1971) 447.

S. IMAMURA, Jap. J. Bot., 12 (1943) 251.

R.J. JONES and T.A. MANSFIELD, J. Exp. Bot., 21 (1970) 714.

H.J. KETELLAPPER, Annu. Rev. Plant Physiol., 14 (1963) 249.

P. IADO, F. RASI-CABLOGNO and R. COLOMBO, Physiol. Plant., 31 (1974) 14.

F.E. LLOYD, in The Physiology of Stomata (1908) Vol. 82, Publ. Carnegie Inst., Washington, p. 1.

B.R. LOVEYS, P.E. KRIDERMANN and E. TÖROKFALVY, Plant Sci. Lett., 1 (1973) 335.

A. LIVNE and Y. VAADIA, in T.T. Kozlowski (Ed.) Water Deficits and Plant Growth (1972) Vol. III, Academic Press, London, p. 255.

P. LOUGUET, Physiol. Veg., 6 (1968) 83.

A.B. MACALLUM, J. Physiol., 32 (1905) 95.

E.A.C. MACROBBIE, Annu. Rev. Plant Physiol., 22 (1971) 75.

T.A. MANSFIELD and R.J. JONES, J. Exp. Bot., 21 (1971) 714.

H. MEIDNER and T.A. MANSFIELD, Physiology of Stomata (1968) MacGraw–Hill, London.

C.J. MITTLEHAUSER and R.F.M. VAN STEVENINCK, Nature, 221 (1969) 281.

Y. MIZRAHI, A. BLUMENFIELD and A.E. RICHMOND, Plant Physiol., 46 (1970) 169.

K. NISHIDA, Physiol. Plant., 61 (1963) 281.

D. NJUS, F.M. SULZMAN and J.W. HASTINGS, Nature, 248 (1974) 116.

C.K. PALLAGHY, Z. Pflanzenphysiol., 62 (1970) 58.

C.K. PALLAGHY, Planta, 101 (1971) 287.

C.K. PALLAGHY and R.A. FISHER, Z. Pflanzenphysiol., 71 (1974) 332.

J.E. PALLAS and R.A. DILLEY, Plant Physiol., 49 (1972) 649.

J.E. PALLAS, Physiol. Plant., 22 (1969) 546.

J.E. PALLAS and H.H. MOLLENHAUER, Am. J. Bot., 59 (1972) 504.

C.J. PEARSON, Aust. J. Biol. Sci., 26 (1973) 1035.

D.E. PEASLEE and D.N. MOSS, Crop Sci., 8 (1968) 427.

M.G. PENNY and D.J.F. BOWLING, Planta, 119 (1974) 17.

K. RASCHKE, Planta, 68 (1966) 111.

K. RASCHKE, Science, 167 (1970) 189.

K. RASCHKE and M.P. FELLOWS, Planta, 101 (1971) 296.

K. RASCHKE and G.D. HUMBLE, Planta, 115 (1973) 47.

412       *D.A. Thomas*

J.A. RAVEN, J. Gen. Physiol., 50 (1967) 1607.
J.A. RAVEN, New Phytol., 68 (1969) 1089.
N.R. REED and B.A. BONNER, Planta, 116 (1974) 173.
B.L. SAWHNEY and I. ZELITCH, Plant Physiol., 44 (1969) 1350.
G. SEIDMAN and W.B. RIGGAN, Nature, 217 (1968) 684.
R.O. SLATYER, in W. Ruhland (Ed.) Plant-Water Relationships (1967) Vol. III, Academic Press, London.
M.G. STÅLFELT, Encyclopedia of Plant Physiology (1956) Vol. III, Springer Verlag, Berlin, p. 351.
M.G. STÅLFELT, Physiol. Plant., 20 (1967) 57.
M. TAL, Plant Physiol., 41 (1966) 1387.
M. TAL and D. IMBER, Plant Physiol., 47 (1971) 849.
D.A. THOMAS, Aust. J. Biol. Sci., 23 (1970a) 961.
D.A. THOMAS, Aust. J. Biol. Sci., 23 (1970b) 981.
D.A. THOMAS, Aust. J. Biol. Sci., 24 (1971) 689.
D.A. THOMAS, Aust. J. Biol. Sci., 25 (1972) 877.
D.J. TUCKER and T.A. MANSFIELD, Planta, 98 (1971) 157.
N.C. TURNER, Nature, 235 (1972) 341.
N.C. TURNER, Am. J. Bot., 60 (1973) 717.
D.A. WALKER and I. ZELITCH, Plant Physiol., 38 (1963) 390.
C.M. WILLMER and T.A. MANSFIELD, Z. Pflanzenphysiol., 61 (1969) 398.
C.M. WILLMER and T.A. MANSFIELD, New Phytol., 69 (1970) 639.
C.M. WILLMER and J.E. PALLAS, Can. J. Bot., 51 (1973) 374.
K. WINTER and D.J. VON WILLERT, Z. Pflanzenphysiol., 67 (1972) 166.
S.T. WRIGHT and R.W. HIRON, Nature, 224 (1969) 719.
T. YAMASHITA, Sieboldia, 1 (1952) 51.
I. ZELITCH, in I. Zelitch (Ed.) Stomata and Water Relations in Plants (1963) Conneticut Agric. Exp. Stn., Newhaven, Conn., U.S.A.

# Author index

**DATE DUE**